Dr. [illegible]

[illegible]

27. Dezember 1983

Fortschritte der Chemie organischer Naturstoffe

# Progress in the Chemistry of Organic Natural Products

# 44

Founded by L. Zechmeister
Edited by W. Herz, H. Grisebach, G. W. Kirby

Authors:
J. G. Buchanan, P. Crews, B. Epe,
F. J. Evans, F. J. Hanke, L. V. Manes,
A. Mondon, S. Naylor, S. E. Taylor

Springer-Verlag
Wien New York 1983

Dr. W. Herz, Professor of Chemistry, Department of Chemistry,
The Florida State University, Tallahassee, Florida, U.S.A.

Prof. Dr. H. Grisebach, Biologisches Institut II, Lehrstuhl für Biochemie der Pflanzen,
Albert-Ludwigs-Universität, Freiburg i. Br., Federal Republic of Germany

G. W. Kirby, Sc. D., Regius Professor of Chemistry, Chemistry Department,
The University, Glasgow, Scotland

With 72 partly coloured Figures



Library of Congress Catalog Card Number AC 39-1015

Printed in Austria

ISSN 0071-7886

ISBN 3-211-81754-9 Springer-Verlag Wien-New York
ISBN 0-387-81754-9 Springer-Verlag New York-Wien

# Contents

**List of Contributors** . . . . . . . . . . . . . . . . . . . . . IX

**Pro-Inflammatory, Tumour-Promoting and Anti-Tumour Diterpenes of the Plant Families Euphorbiaceae and Thymelaeaceae.** By F. J. EVANS and S. E. TAYLOR . . . . . . . . . . 1

I. Introduction . . . . . . . . . . 2

II. Classification of Structural Types . . . . . . . . . . 3

III. Botanical Considerations . . . . . . . . . . 4

1. The Family Euphorbiaceae . . . . . . . . . . 4
2. The Family Thymelaeaceae . . . . . . . . . . 5

IV. Biosynthetic Relationships . . . . . . . . . . 6

V. Isolation of Diterpenes . . . . . . . . . . 9

VI. The Macrocyclic Diterpenes . . . . . . . . . . 13

1. Casbane Type . . . . . . . . . . 13
2. Jatrophane Type . . . . . . . . . . 13
3. Lathyrane Type . . . . . . . . . . 16
4. Jatropholane and Crotofolane Types . . . . . . . . . . 25
5. Rhamnofolane . . . . . . . . . . 26

VII. Tigliane Diterpenes . . . . . . . . . . 27

1. Phorbol and Its Esters . . . . . . . . . . 27
2. Distribution of Phorbol Esters in Plants . . . . . . . . . . 28
3. Identification of Phorbol Esters . . . . . . . . . . 30
4. 4-Deoxyphorbol Esters . . . . . . . . . . 34
5. Other 4-Deoxyphorbol Derivatives . . . . . . . . . . 36
6. Identification of 4-Deoxyphorbol Esters . . . . . . . . . . 37
7. 12-Deoxyphorbol Esters . . . . . . . . . . 47
8. Identification of 12-Deoxyphorbol Esters . . . . . . . . . . 48
9. 12-Deoxy-16-Hydroxyphorbol Esters . . . . . . . . . . 52
10. Other 12-Deoxyphorbol Derivatives . . . . . . . . . . 54
11. 16-Hydroxyphorbol Esters . . . . . . . . . . 56

VIII. Ingenane Derivatives ... 58
1. Ingenol ... 58
2. Distribution of Ingenol Esters in Plants ... 66
3. 5-Deoxyingenol ... 67
4. 16-Hydroxyingenol ... 68
5. 13-Hydroxyingenol ... 69
6. 13,19-Dihydroxyingenol ... 71
7. 20-Deoxyingenol ... 72
IX. Daphnane Derivatives ... 73
1. Daphnetoxin Type ... 73
2. 12-Hydroxydaphnetoxin Type ... 76
3. Resiniferonol Type ... 79
4. 1-Alkyldaphnane Type ... 84
X. Closing Remarks ... 87
Acknowledgements ... 89
References ... 90

**Bitter Principles of Cneoraceae.** By A. MONDON and B. EPE ... 101
I. Introduction ... 102
II. Nomenclature and Classification ... 104
III. Constitution and Configuration of Cneorins **A**, **B**, **C** and **D** ... 106
1. Structural Studies by Ozonolysis ... 106
2. Establishment of Relative and Absolute Configuration ... 108
3. Carbon Skeleton of Cneorins **A**–**D**: Numbering of C-Atoms ... 112
4. Configuration of Cneorin-$\mathbf{C}_{I}$ ... 112
5. Configuration of Cneorin **B**, $\mathbf{B}_{I}$ and $\mathbf{B}_{III}$ ... 114
a) Elucidation of the Configuration at Carbon Atoms 7 and 9 ... 114
b) Elucidation of the Configuration at Carbon Atoms 5, 10, 13 and 17 ... 116
c) Hydrogenolysis of Cneorins $\mathbf{B}_{III}$ and $\mathbf{C}_{III}$ ... 118
d) The Steric Series **B** and **C** with (17 *R*)- and (17 *S*)-Configuration ... 118
6. Intramolecular Cyclisation to Cneorins $\mathbf{B}_{II}$ and $\mathbf{C}_{II}$ ... 119
7. Constitution and Configuration of Cneorins **A** and **D** and Pyrolysis of Cneorin-**D** ... 122
IV. The Series of Stereoisomeric $\Delta^{8(30)}$-Olefins ... 124
V. Stereoisomeric Alcohols with an 8-OH Group and (5*S*, 10*R*)-Configuration ... 126
VI. Cneorins and Tricoccins with a C-7 Carbonyl Group ... 128
1. Cneorin-**F** and Tricoccin-$\mathbf{S}_{14}$ ... 128
2. The Tricoccins $\mathbf{R}_{9}$, $\mathbf{R}_{12}$ and Their Epi-compounds ... 131
VII. C-7 Hemiacetals and Methylacetals with (5*S*, 10*R*, 17*R*)-Configuration ... 133
1. Cneorin **K** and $\mathbf{K}_{1}$ ... 133
2. Tricoccin $\mathbf{R}_{1}$ and $\mathbf{R}_{10}$ ... 134

VIII. C-9 Hemiacetals with (5*S*, 8*S*, 10*R*)-Configuration ........ 136
IX. Cyclic Peroxides ........ 141
X. Bitter Principles with a γ-Lactol Ring (C-15 Hemiacetals) ........ 144
1. Precursors of Known Cneoroids ........ 144
2. C-15 Cycloacetals ........ 147
3. γ-Lactols with the Partial Structure of Ring A of Obacunone ........ 150
XI. Tetranortriterpenoids from Cneoraceae ........ 155
1. 3,4-*seco*-Meliacans ........ 155
2. 7,8-*seco*-Meliacans ........ 162
3. 3,4-16,17-*seco*-Meliacans (Limonin Group) ........ 166
4. 3,4-7,8-16,17-*seco*-Meliacans ........ 168
XII. Protolimonoids ........ 168
1. Tirucallan-(20*S*)-Triterpenoids ........ 168
2. Apotirucallan-(20*S*)-Triterpenoids ........ 172
XIII. Comments on Biosynthesis of Cneorins and Tricoccins ........ 173
XIV. Tables of Natural Bitter Principles from Cneoraceae ........ 178
1. Cneorins from *Neochamaelea pulverulenta* (Vent.) Erdtm. of Known Constitution and Configuration ........ 178
2. Cneorins of Unknown Structure ........ 179
3. Tricoccins from *Cneorum tricoccon* L. of Known Constitution and Configuration ........ 179
4. Tricoccins of Unknown Structure ........ 181
References ........ 181

**Chemical and Biological Aspects of Marine Monoterpenes.** By S. NAYLOR, F. J. HANKE, L. V. MANES, and P. CREWS ........ 189

I. Introduction ........ 190
II. Structural Variation ........ 193
III. The Role of Halogens in Biogenesis ........ 195
1. Introduction ........ 195
2. Acyclic Structures ........ 196
3. Monocyclic Structures ........ 199
IV. Relationships Between Taxonomy and Occurrence of Structural Types ........ 200
1. Introduction ........ 200
2. The Plocamiaceae Family ........ 201
3. The Rhizophyllidaceae Family ........ 206
4. The Ceramiaceae Family ........ 207
5. Degraded and Mixed Biogenetic Monoterpenes ........ 209
6. Conclusion ........ 209
V. Metabolite Transfer and Biological Activity ........ 209

VI. Spectroscopic and Chemical Properties .... 212
1. Introduction .... 212
2. Halogen Content and Regiochemistry .... 215
3. Stereochemistry .... 219
4. Artifacts .... 221
5. Conclusions .... 222
VII. Physical and Spectroscopic Tables .... 223
1. Table 13A—H: Summary of Structures and Carbon-13 NMR Chemical Shifts 223
2. Table 14: Physical Properties .... 233
Acknowledgement .... 236
References .... 236

**The C-Nucleoside Antibiotics.** By J. G. BUCHANAN .... 243
I. Introduction .... 243
II. General Aspects of *C*-Nucleosides .... 246
III. Showdomycin .... 250
IV. The Formycins .... 264
V. Pyrazofurin (Pyrazomycin) .... 270
VI. Oxazinomycin (Minimycin) .... 276
VII. The Ezomycins .... 279
VIII. Biosynthesis of *C*-Nucleoside Antibiotics .... 284
References .... 290

**Author Index** .... 301

**Subject Index** .... 312

## List of Contributors

BUCHANAN, Prof. J. G., Department of Chemistry, Heriot-Watt University, Riccarton, Edinburgh EH14 4AS, U.K.

CREWS, Prof. Dr. P., Department of Chemistry and Center for Coastal Marine Studies, Thimann Laboratories, University of California, Santa Cruz, CA 95064, U.S.A.

EPE, Dr. B., Institut für Toxikologie, Julius-Maximilians-Universität, Versbacherstrasse 9, D-8700 Würzburg, Bundesrepublik Deutschland.

EVANS, Dr. F. J., Department of Pharmacognosy, The School of Pharmacy, University of London, 29/39, Brunswick Square, London WC1N 1AX, U.K.

HANKE, F. J., Department of Chemistry, Thimann Laboratories, University of California, Santa Cruz, CA 95064, U.S.A.

MANES, L. V., Department of Chemistry, Thimann Laboratories, University of California, Santa Cruz, CA 95064, U.S.A.

MONDON, Prof. Dr. A., Institut für Organische Chemie, Neue Universität, Olshausenstrasse 40-60, D-2300 Kiel, Bundesrepublik Deutschland.

NAYLOR, S., B.Sc., M.Sc., 26 Aller Park Road, Aller Park, Newton Abbot, Devon TQ12 4NG, U.K.

TAYLOR, Dr. S. E., Department of Pharmacognosy, The School of Pharmacy, University of London, 29/39, Brunswick Square, London WC1N 1AX, U.K.

# Pro-Inflammatory, Tumour-Promoting and Anti-Tumour Diterpenes of the Plant Families Euphorbiaceae and Thymelaeaceae

By F. J. Evans and S. E. Taylor, Department of Pharmacognosy, The School of Pharmacy, University of London, London, U.K.

With 69 Figures

**Contents**

I. Introduction .......... 2

II. Classification of Structural Types .......... 3

III. Botanical Considerations .......... 4

1. The Family Euphorbiaceae .......... 4
2. The Family Thymelaeaceae .......... 5

IV. Biosynthetic Relationships .......... 6

V. Isolation of Diterpenes .......... 9

VI. The Macrocyclic Diterpenes .......... 13

1. Casbane Type .......... 13
2. Jatrophane Type .......... 13
3. Lathyrane Type .......... 16
4. Jatropholane and Crotofolane Types .......... 25
5. Rhamnofolane .......... 26

VII. Tigliane Diterpenes .......... 27

1. Phorbol and Its Esters .......... 27
2. Distribution of Phorbol Esters in Plants .......... 28
3. Identification of Phorbol Esters .......... 30
4. 4-Deoxyphorbol Esters .......... 34
5. Other 4-Deoxyphorbol Derivatives .......... 36
6. Identification of 4-Deoxyphorbol Esters .......... 37
7. 12-Deoxyphorbol Esters .......... 47

8. Identification of 12-Deoxyphorbol Esters .............................. 48
9. 12-Deoxy-16-Hydroxyphorbol Esters .............................. 52
10. Other 12-Deoxyphorbol Derivatives .............................. 54
11. 16-Hydroxyphorbol Esters .............................. 56
VIII. Ingenane Derivatives .............................. 58
1. Ingenol .............................. 58
2. Distribution of Ingenol Esters in Plants .............................. 66
3. 5-Deoxyingenol .............................. 67
4. 16-Hydroxyingenol .............................. 68
5. 13-Hydroxyingenol .............................. 69
6. 13,19-Dihydroxyingenol .............................. 71
7. 20-Deoxyingenol .............................. 72
IX. Daphnane Derivatives .............................. 73
1. Daphnetoxin Type .............................. 73
2. 12-Hydroxydaphnetoxin Type .............................. 76
3. Resiniferonol Type .............................. 79
4. 1-Alkyldaphnane Type .............................. 84
X. Closing Remarks .............................. 87
Acknowledgements .............................. 89
References .............................. 90

This review is dedicated to the late J. W. FAIRBAIRN (Emeritus Professor of Pharmacognosy in the University of London).

## I. Introduction

Plants and plant products have been used as drugs for thousands of years and in recent history have provided a definite stimulus for the development of natural product chemistry. However, the contribution that plant chemicals have made to the understanding of biochemistry, physiology and pharmacology is often overlooked. Secondary plant metabolites are useful tools for the investigation of biological systems and even of diseased states. The isolation and final structure elucidation of the tumour-promoting phorbol-12,13-diesters from the seed oil of *Croton tiglium* by ERICH HECKER (*1*), nearly 15 years ago, initiated intensive research into the pro-inflammatory and tumour-promoting diterpenes of the plant families Euphorbiaceae and Thymelaeaceae. These pure compounds have been instrumental in furthering our understanding of the diseases of cancer and inflammation.

In most developed countries the incidence of cancer is high and could be responsible for every fifth death (*2*). The existence of chemical carcinogens in the environment was verified as long ago as 1914 (*3*), but more recently it has been realised that for the majority of human cancers solitary carcinogenesis engineered by a single chemical carcinogen may be an exception. It is more likely that the majority of cancers are the result of exposure to more than one carcinogenic risk factor, a process known as syncarcinogenesis (*4*), or possibly due to exposure to small quantities of a carcinogen and a tumour-promoter, known as co-carcinogenesis (*5*). Although a number of substances are known to be tumour-promoters, the diterpene esters of plant origin have been most widely studied to date (*6*) and the health hazards presented by such compounds in the environment have been fully realised.

Tumour-promoting agents do not themselves elicit tumours but they do promote tumour-growth following exposure to a subcarcinogenic dose of a solitary or chemical carcinogen. In the classical Berenblum (*7*) model of co-carcinogenesis, a subthreshold dose of a solitary carcinogen is applied to a group of subjects (normally mice) and a similar group receive repeat doses of a tumour-promoter. Neither of these groups develop tumours. A third group receive a single subthreshold dose of a carcinogen followed by repeated doses of a tumour-promoter. Tumours develop in this group. The tumour-promoting diterpenes are biologically active in submicrogram doses and in a situation of repeated or chronic exposure are second order carcinogenic risk factors. Their acute effect is the induction of inflammation of skin (*8*) and both their chronic and acute effects render them hazardous substances for normal isolation and handling in the laboratory.

Of recent interest is the observation (*9, 10*) that certain of these diterpenes or their biosynthetic precursors may possess anti-tumour activity. Investigation of such compounds will further elucidate the mechanism of action of the phorbol derivatives and also holds out a hope for the development of a new class of anti-tumour drugs.

## II. Classification of Structural Types

The toxic diterpenes of the plant families Euphorbiaceae and Thymelaeaceae are based upon the hydrocarbons, tigliane, daphnane and ingenane (Fig. 1) (*11*).

The first compound of this series isolated was phorbol (*1*), which belongs to the tigliane group. The tiglianes consist of a five-membered ring A, normally *trans* linked to a seven membered ring B. Ring C is six-membered and the cyclopropane ring D is linked to it in the *cis*-configuration. Tiglianes occur in plants as esters of polyhydroxylated diterpenes.

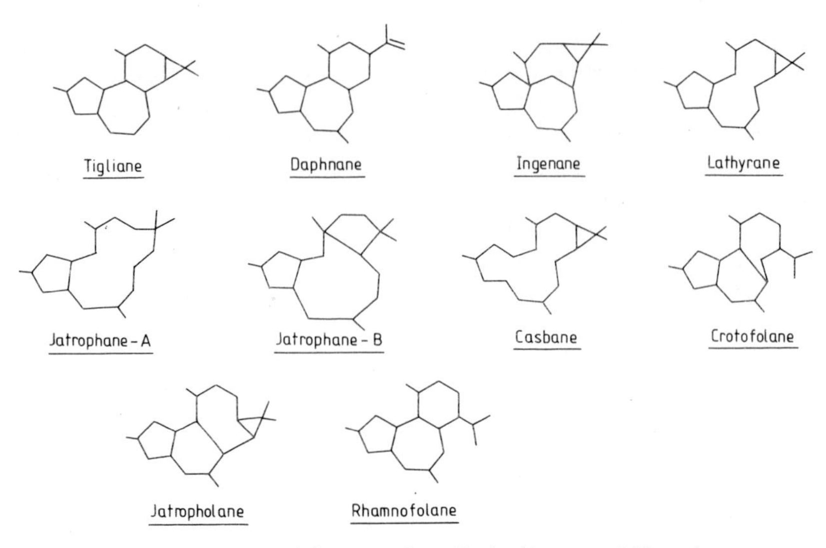


Fig. 1. Structural types of diterpenes from Euphorbiaceae and Thymelaeaceae

Daphnane diterpenes are tricyclic compounds in which ring D of the tigliane nucleus has opened to form an isopropenyl side chain. Daphnanes frequently occur in plants as orthoesters but may also be found in the O-acyl form (*12*). The ingenanes are a novel group of tetracyclic diterpenes related to the tiglianes. They differ in that ring C is seven membered, exhibiting a quaternary centre at C-10 of the structure (*1*).

A number of diterpenes bearing a close structural relationship to the toxic compounds are found in plants of the family Euphorbiaceae. These substances also exist in the plant as oxygenated derivatives, esterified at one or more positions. Many of these diterpenes are macrocyclic (Fig. 1). These various types are based upon the hydrocarbons casbane (*13*), jatrophane (*16, 17, 18*), lathyrane (*14, 15*), jatropholane (*20*), crotofolane (*19*) and rhamnifolane (*21*).

## III. Botanical Considerations

### III.1. The Family Euphorbiaceae

The Euphorbiaceae, or spurge family, is one of the largest and most unwieldy families of the angiosperms, and because of the range of morphological variation it may be polyphyletic in origin (*22*). The largest genera are *Euphorbia* consisting of over 1600 species and *Croton* of about 700 species. Thirteen other genera contain over 100 species each (*23*); these include *Phyllanthus* (480 species), *Acalypha* (430 species), *Glochidon* (280 species), *Macaranga* (240 species), *Manihot* (160 species), *Jatropha* (150 species) and *Tragia* (140 species). Several of these genera contain species which produce toxic diterpenes, but it is also of interest to note that several genera, including *Tragia,* are true stinging nettles in that they exhibit stinging trichomes similar to those of the common nettle, *Urtica dioica,* of the family Urticaceae. The chemical composition of the toxin from the stinging nettles is still unknown at the present time (*24*).

The family Euphorbiaceae is widely distributed throughout both hemispheres and ranges in morphological form from large desert succulents to trees and even small herbaceous types. Many species are weeds in that they invade cultivated land as secondary growth and provide a health hazard both to humans and grazing livestock. Many attempts have been made to classify this family, but the authoritative work is the classification put forward by PAX and HOFFMAN in 1931 (*25*). These workers divided the whole family into four sub-families on morphological grounds:

Phyllanthoideae
Crotonoideae
Poranteroideae
Ricinocarpoideae

The tumour-promoting and pro-inflammatory diterpenes have been obtained from the sub-family Crotonoideae. This sub-family was divided into two tribes known as the Crotoneae and the Euphorbieae (*25*). The tribe Euphorbieae is characterised by an inflorescence in the form of a cyathium (*26*) and has recently been sub-divided into three sub-tribes, Pimelodendrinae, Hippomanae and Euphorbiinae (*27*). The sub-tribe Euphorbiinae contains the diterpene yielding genera *Euphorbia* (*28*) and *Elaeophorbia* (*25*) together with *Pedilanthus* and *Synadenium;* the sub-tribe Hippomanae contains diterpene producing genera such as *Stillingia, Sapium* and *Excoecaria* together with *Sebastiania* and *Homolanthus* (*27*).

From the tribe Crotoneae, which contains the genus *Croton,* the first phorbol-12,13-diesters were isolated (*1*).

### III.2. The Family Thymelaeaceae

The family Thymelaeaceae, or the Daphnane family, consists of about 55 genera containing some 500 species. These plants are mainly erect shrubs with a tropical and a temperate distribution (*29*). The family has been divided into four sub-families:

Gonystyloideae
Aquilarioideae
Gilgiodaphoideae
Thymelaeoideae

The sub-family Thymelaeoideae contains the bulk of the genera from this family including *Daphnane, Gnidia, Pimelea* and *Lasiosiphon.* These genera contain toxic and anti-tumour daphnane derivatives.

## IV. Biosynthetic Relationships

It was recently demonstrated (*30*) that the macrocyclic diterpene, casbene (Fig. 2), was produced from geranyl-geranyl in cell free cultures of *Ricinus communis* of the family Euphorbiaceae.

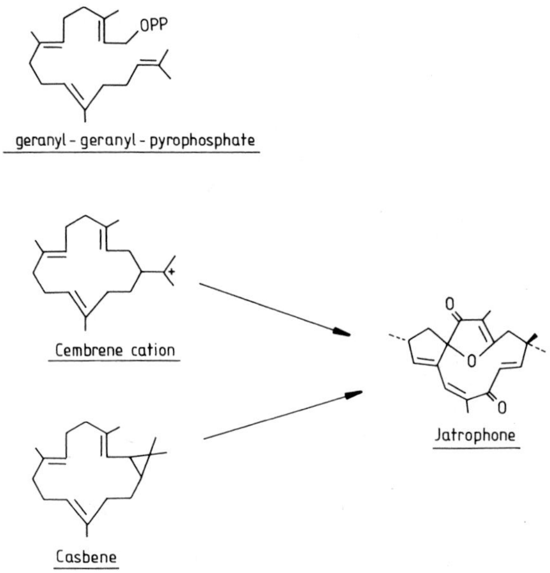

OPP
geranyl-geranyl-pyrophosphate
Cembrene cation
Casbene
Jatrophone

Fig. 2. Biosynthesis of jatrophane diterpenes

Casbene has been considered to be the precursor of the polycyclic and polyfunctional diterpenes of the tigliane, daphnane and ingenane types. Other macrocyclic diterpenes may be intermediates in this biosynthetic process (*31*). It is possible, according to Adolf and Hecker (*31*), that the macrocyclic jatrophane type may be formed in the plant from a cembrene cation, or alternatively by ring opening of the cyclopropane ring of a casbene precursor.

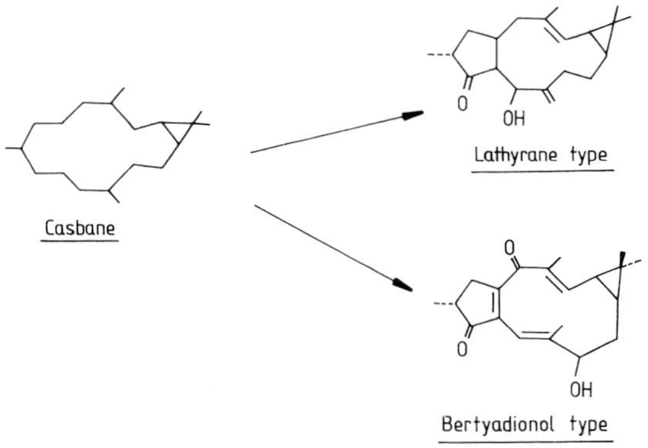

Fig. 3. Biogenic formation of macrocylic-type diterpenes form casbane

The hydrocarbon nucleus of casbene and its saturated analogue, casbane, may be considered as the biogenetic precursor of a number of macrocyclic diterpenes such as the lathyranes (Fig. 3). Some lathyranes may then be oxidatively produced from the jolkinols (*32*). The toxicologically important tigliane group could be derived from the lathyrane skeleton by a further cyclisation step (Fig. 4).

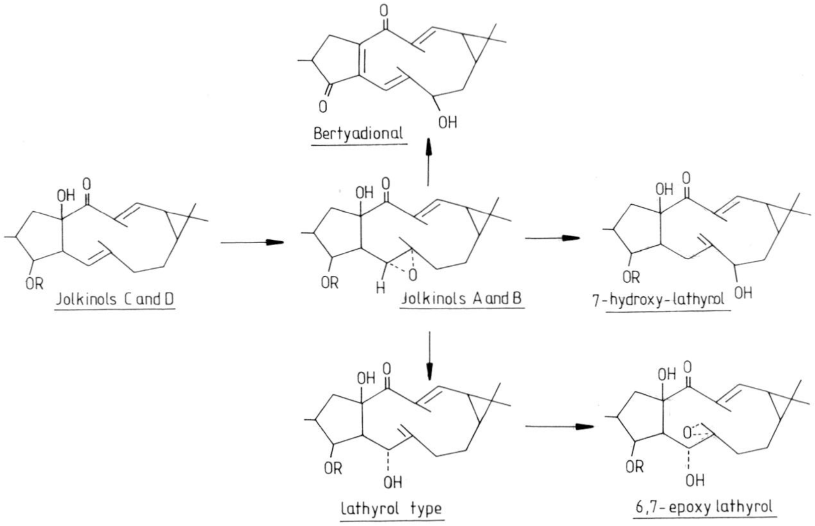

Fig. 4. Biosynthesis of lathyrane diterpenes

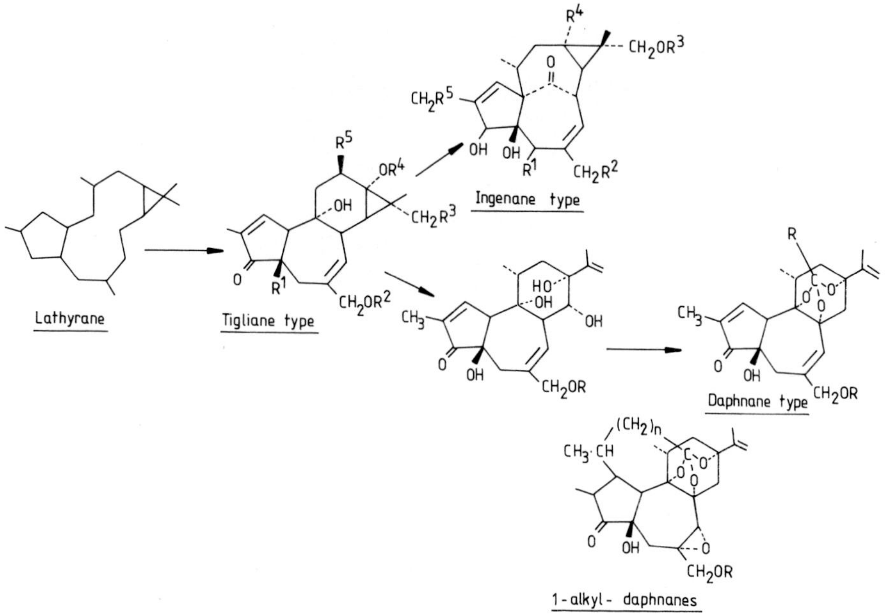

Lathyrane

Tigliane type

Ingenane type

Daphnane type

1-alkyl-daphnanes

Fig. 5. Biosynthesis of ortho-ester diterpenes

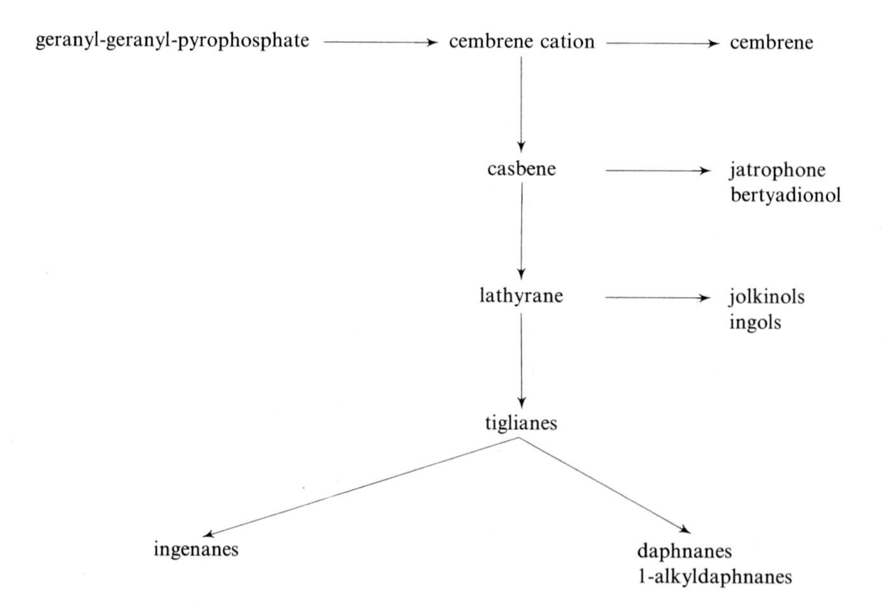


Fig. 6. Hypothetical biogenic interconversions of diterpene esters in Euphorbiaceae

*References, pp. 90—99*

A close relationship exists between the tigliane and the daphnane group of diterpenes. It has been suggested (*31*) that the daphnanes are formed by opening of the cyclopropane ring of a 12-deoxyphorbol derivative, followed by reduction of the resulting 13,14-ketol ester to the 13β-isopropenyl moiety. The 9α-, 13α- and 14α-hydroxy groups are oriented spatially in a manner which allows easy formation of an orthoester by reaction with a carboxylic acid (Fig. 5). This reaction may be carried out chemically in the presence of hydrogen ions (*12, 33*). Subsequent intramolecular Michael type addition of C-$9^1$ of the orthoester side chain to the 1,2-double bond of a ring A of the daphnane nucleus would then generate the 1-alkyldaphnanes.

The ingenanes are structurally related to the tiglianes and it has been postulated that they arise biogenetically from them. The various hypothetical interconversions of these diterpenes are summarised in Fig. 6.

## V. Isolation of Diterpenes

Many of the esterified forms of these diterpenes are unstable compounds and break down during isolation procedures. Hydrolysis or transesterification reactions are known during chromatographic separation (*1, 34*) and configurational changes may occur if crystallisation is attempted (*35*). Furthermore, the compounds are sensitive to heat, light, oxygen, acid and alkaline conditions (*11, 34*).

Extractions of plant materials are carried out at room temperature by maceration and evaporation of extracts is achieved at reduced pressure below 45° C. As the diterpene esters are neutral lipid components of plants purification generally requires involved separation methods. This task is often aggravated by the fact that the plant produces complex mixtures of esters of the same diterpene nucleus. To date, separation procedures have been based upon a combination of chromatographic and partition techniques. The number and position of hydroxy functions and the type and molecular weight of acyl substituents facilitate separation by these methods.

Plant material may be extracted with acetone or methanol and after removal of solvent the residue is dissolved in a methanol-water mixture for partition against hexane or petroleum spirit. The hexane fraction contains hydrocarbons, steroids, triterpenoids and some macrocyclic diterpenes of the less polar type (*13, 16, 18, 19, 20, 21*) (Fig. 7). Further extraction of the methanol-water phase with ether provides a solution rich in diterpene derivatives (*1, 11, 34*). When plant extracts have been prepared from herbaceous starting material a further partition of the ether-phase against 1% sodium carbonate solution may remove a considerable portion of the green colouring matter.

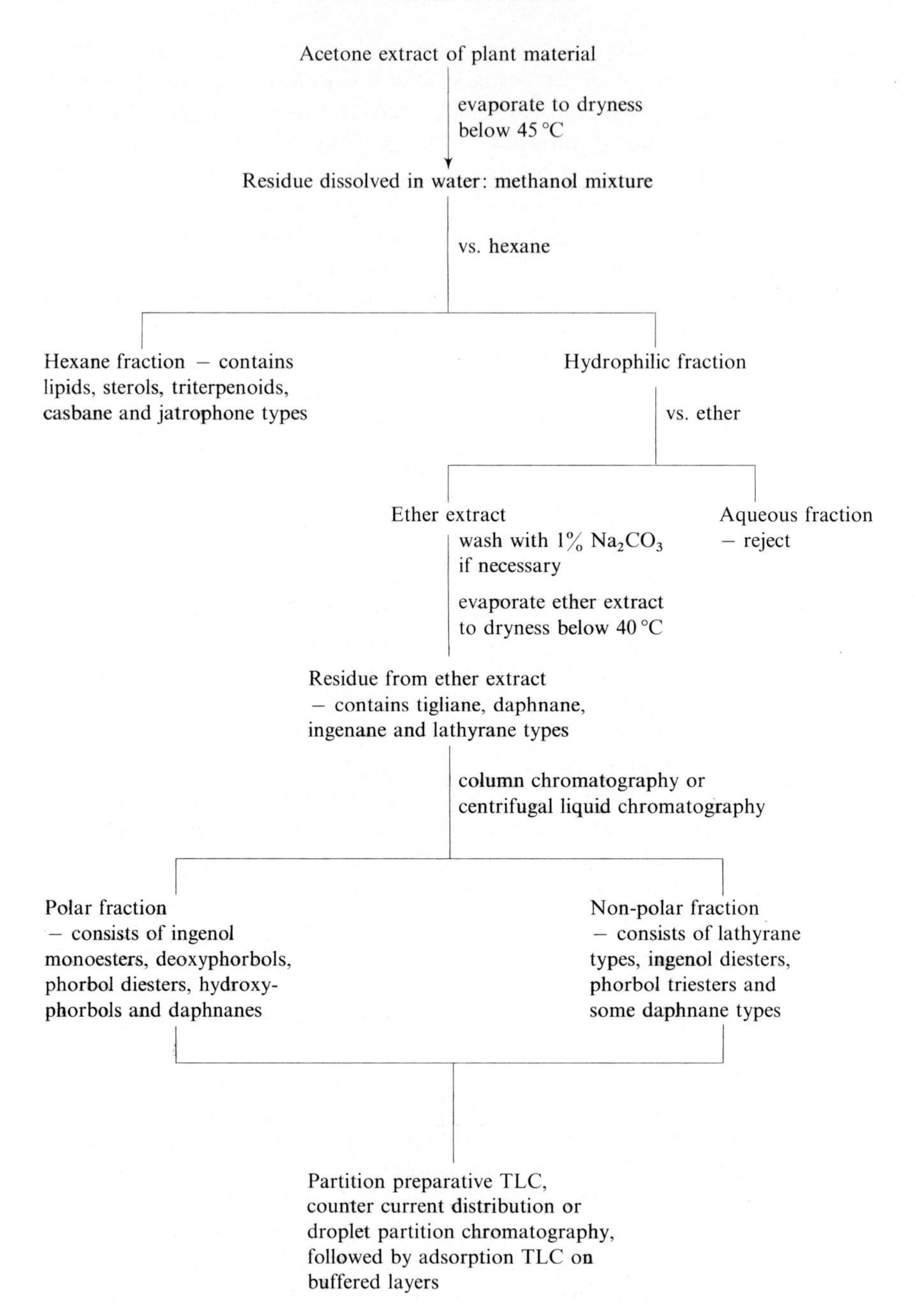


Fig. 7. General isolation procedure for diterpene esters

*References, pp. 90—99*

Table 1. *Thin Layer Chromatography of Diterpene Acetates*

| | Phorbol triacetate | 12-Deoxyphorbol diacetate | 4-Deoxyphorbol triacetate | 4α-Phorbol triacetate | 4α-Phorbol tetraacetate | Ingenol triacetate |
|---|---|---|---|---|---|---|
| *Silica gel G* | | | $hR_f$ *Values* | | | |
| Chloroform:ether (95:5) | 16 | 15 | 21 | 7 | 26 | 44 |
| Ether:ethyl acetate:cyclohexane (1:1:1) | 48 | 51 | 52 | 26 | 49 | 65 |
| Cyclohexane:isopropanol (2:1) | 64 | 68 | 55 | 44 | 57 | 72 |
| *Silica gel H* | | | | | | |
| Chloroform:ethyl acetate (2:3) | 49 | 50 | 53 | 39 | 55 | 59 |
| *Alumina E* | | | | | | |
| Toluene:ethyl acetate (9:2) | 5 | 6 | 12 | 1 | 11 | 32 |
| Chloroform:acetone:benzene (95:5:50) | 31 | 33 | 51 | 4 | 54 | 69 |
| Chloroform | 53 | 53 | 72 | 10 | 76 | 80 |
| Cyclohexane:ether:benzene (1:2:1) | 6 | 10 | 11 | 1 | 7 | 37 |
| Cyclohexane:benzene:ethyl acetate:ether (20:15:40:30) | 30 | 38 | 52 | 3 | 45 | 70 |
| Chloroform:ether:benzene (1:3:3) | 22 | 29 | 36 | 2 | 43 | 74 |
| Ethyl acetate:benzene (1:3) | 35 | 39 | 39 | 5 | 51 | 86 |

The ether soluble resin can be separated into fractions by means of column chromatography. Both silica gel and florosil (*37*) have been used in conventional gradient elution procedures. A dry column technique has also been advocated (*38*) but has not received universal acceptance. Repeated column separation utilising a number of different adsorbents has been used successfully for the purification of the major component of a complex mixture (*39*). More recently the Japanese technique of centrifugal-liquid-chromatography (CLC) has been applied to the initial fractionation of diterpene esters (*40*). This method has the advantage that separations requiring several days by conventional column methods can be achieved within a few hours. However, no improvement in resolution of fractions was obtained. This technique may be used with gradient elution, and because the built-in U.V. detector controls a collection apparatus, it provides an attractive alternative to column chromatography.

Further separation of a mixture of closely related diterpenes obtained from a column, requires the application of partition methods. Liquid-liquid distributions of the O'Keefe or Craig type have been used successfully on a large scale (*34*). More recently, an attempt to separate phorbol derivatives by means of droplet-counter current chromatography (*41*) was published. However our own experience with this method suggested that a more conventional technique using preparative partition TLC provides greater resolution (*42*). Partition TLC involves coating a support material, either kieselguhr or silica gel, with propylene glycol or diethylene glycol. This is achieved by developing the prepared plates their full length in a 20% glycol solution in acetone. After evaporation of the acetone at room temperature, the plates can be used in the normal preparative manner. Glycol can be separated from the purified diterpenes by adding an excess of sodium chloride solution and partitioning the liberated esters into chloroform or methylene chloride (*42*). The final purification of separated diterpene derivatives is best achieved by using adsorption TLC methods on buffered silica gel plates (*43*) (Table 1).

Other techniques used for the separation of these diterpenes include high-pressure-liquid chromatography (*44, 45*), and a gas-liquid chromatographic method which provides both qualitative and quantitative data on a micro scale (*46*).

The diterpenes in the pure state are clear colourless resins. Traces of solvents may be removed by exposure to a high vacuum for several days; the resins can then be stored in an atmosphere of nitrogen gas at —20° C until required for biological evaluation. The purity of the isolated resins may be monitored by means of analytical TLC (*47*), GLC (*46*), mass-spectrometry (*42*) and a quantitative biological test (*48*).

*References, pp. 90—99*

## VI. The Macrocyclic Diterpenes

Although compounds of this group were the first diterpenes to be isolated from the Euphorbiaceae, little interest has been shown in them until recently, compared with other classes of compounds such as the phorbols. The first macrocyclic diterpene to be isolated in pure form was "euphorbia-steroid" by DUBLYANSKAYA in 1937 (*49*). These substances do not exhibit the pronounced toxicological actions of other diterpenes (*14*) and remained only of chemical interest until 1970. As part of an intensive anticancer screening programme, KUPCHAN's group (*16*) demonstrated that macrocyclic diterpenes of the jatrophane type exhibited marked anti-leukemic activity against mouse P-388 lymphocytic leukemia *in vivo*. More recently, several classes of macrocyclic diterpenes from the Euphorbiaceae have been shown to possess either cytotoxic action *in vitro* (*50*) or anti-tumour action *in vivo* (*10, 18*). Accordingly, these compounds are potential chemotherapeutic agents.

### VI.1. Casbane Type

Casbene was isolated from an enzyme preparation of *Ricinus communis* seedlings (*30*). Due to the low yields, the structure and stereochemistry of casbene could not be assigned with assurance. Recently the total synthesis of casbene (*51*) by elaboration of 1 *R*, 3*S*(+)-*cis*-chrystanthemic acid to the bis-allylic bromide, followed by closure of the 14-membered ring with tetracarbonyl nickel, established that casbene possesses a 4,8,12-all-*E*-triene system and a *cis*-disubstituted cyclopropane ring. Casbene has been shown to possess anti-fungal properties and is a phytoalexin (*51*).

A novel casbene type of diterpenoid was isolated from the twigs and leaves of *Croton nitens* (*13*). This compound, known as crotonitenone, is shown in Fig. 8. The structure of crotonitenone was deduced by spectral, chemical and X-ray methods.

### VI.2. Jatrophane Type

Jatrophane macrocyclic diterpenes have been isolated from a number of species of Euphorbiaceae (Table 2). These compounds are characterised by the absence of a cyclopropane ring. They also exhibit pronounced anti-tumour activities. Jatrophone, the first member of this group, was isolated (*16*) from the roots of *Jatropha gossypiifolia*. X-ray analysis demonstrated that jatrophone contained a *cis* 5,6-double bond and a *trans* configuration for the remaining two double bonds in the macrocyclic structure (*52*) (Fig. 9). A second anti-tumour compound was recently isolated from *Jatropha macrorhiza*. This compound, known as jatrophatrione, consists of two five membered rings and a nine membered ring. The absolute configuration of jatrophatrione was determined by X-ray methods (*18*).

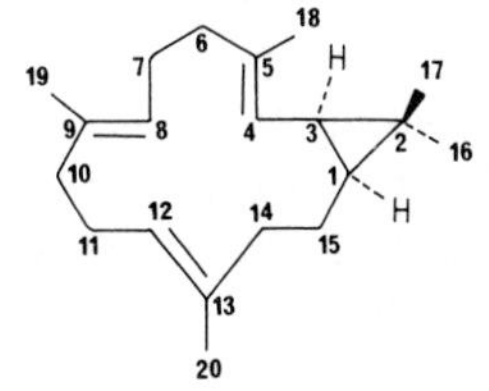

Casbene

(numbered according to (51))

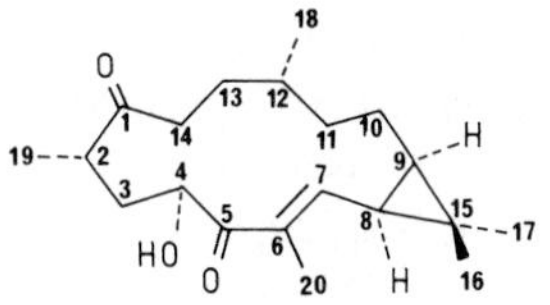

Crotonitenone

(numbered according to (13))

Fig. 8

Table 2. *Biological Sources of Jatrophane Macrolides*

| Name | Source | Literature |
|---|---|---|
| Jatrophone | *Jatropha gossypiifolia* | (*16*) |
| Jatrophatrione | *Jatropha macrorhiza* | (*56*) |
| Kansuinine A | *Euphorbia kansui* | (*57, 14*) |
| Kansuinine B | *Euphorbia kansui* | (*17*) |
| Euphornin | *Euphorbia maddeni* | (*53*) |

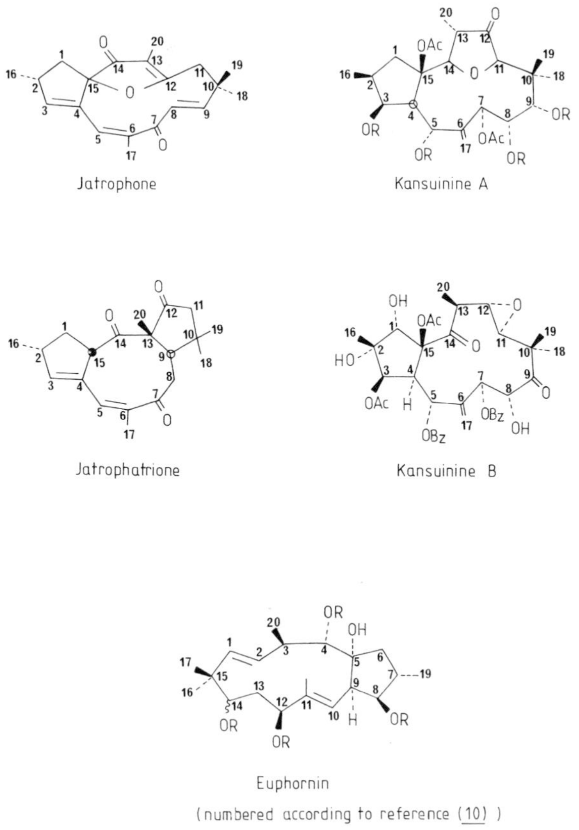

Jatrophone

Kansuinine A

Jatrophatrione

Kansuinine B

Euphornin

(numbered according to reference (10))

Fig. 9. Jatrophane diterpenes

The kansuinines (Fig. 9) represent two highly oxygenated diterpenes of the jatrophane type which were isolated from *Euphorbia kansui*. The structure of kansuinine B was obtained by X-ray methods (*53*) whilst that of kansuinine A was elucidated from spectroscopic and chemical data (*17, 54*). It has been claimed that both of these compounds demonstrate analgesic and anti-writhing properties in mice.

The most recent member of the jatrophane group of diterpenes, euphornin (*10*) (Fig. 9), was isolated from *Euphorbia maddeni*. This species is a small Himalayan herbaceous herb. Euphornin demonstrated anti-tumour activity in animals and is undergoing clinical trials.

The physical and spectroscopic data of jatrophane, the first member of this group to be isolated, and euphornin are summarised in Fig. 10.

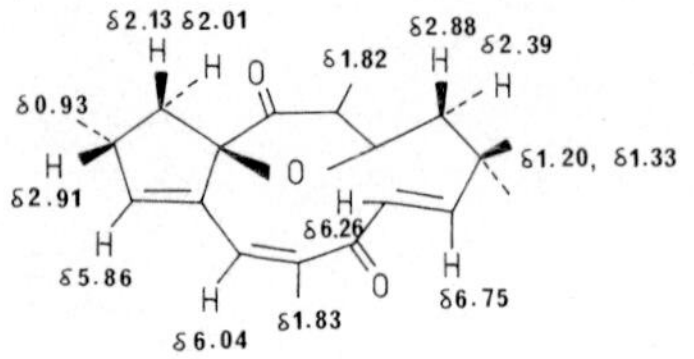

Jatrophone M. P. 152–153°C reference (52) $C_{20}H_{24}O_3$

U.V. $\lambda^{max}_{EtOH}$, n.m. ($\varepsilon$); 285 (10200)

I.R. (KBr) $\nu_{max}$ $\mu$; 3.35, 3.43, 3.46, 5.96, 6.05, 6.20, 7.10, 7.35, 8.07, 8.17, 10.10

M.S. m/z 312 ($M^{+\cdot}$)

$^1$H-N.M.R. (pyridine-$d_5$, 100 MHz) assigned according to (52)

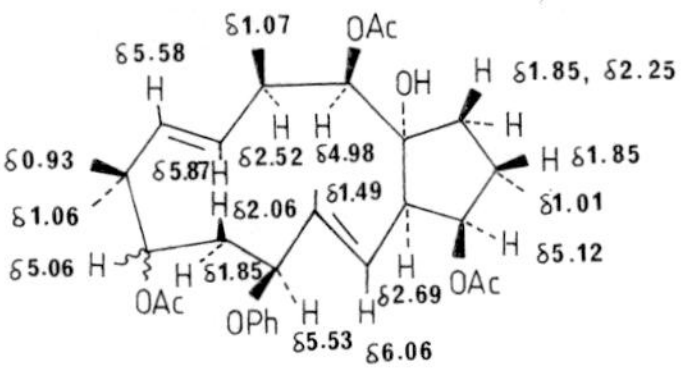

Euphornin M.P. 180°C reference (10)

U.V. $\lambda^{max}_{MeOH}$ n.m. (log $\varepsilon$) 228 (4.16)

I.R. $\nu_{max}$ $cm^{-1}$ 3500, 2980, 2880, 1740, 1705, 1600, 1500, 1370, 1280, 1245, 1138, 1040, 965, 720

M.S. (C.I.-butane) m/z 567 ($M^+$ + 1 − $H_2O$)

$^1$H-N.M.R. ($C_6D_6$, 80°, 270 MHz) assigned according to (10)

Fig. 10

## VI.3. Lathyrane Type

The lathyrane diterpenes are numerically the largest group of macrocyclic compounds isolated from the Euphorbiaceae. Three substances known as $L_1$, $L_2$ and $L_3$ were isolated from the seed oil of *Euphorbia lathyris* (*14*). Compound $L_1$ had a melting point identical with that of "euphorbia-steroid" previously obtained from the same plant (*49*). This diterpene was later characterised as an ester of 6,17-epoxylathyrol (Fig. 11) (*55*). An identical ester was later obtained from *Macaranga tanarius* (Euphorbiaceae) (*56*). Compound $L_2$ was identified as a tetra-ester of 7-hydroxylathyrol (*57*) and $L_3$ as a tri-ester of lathyrol (*58*).

Lathyrol consists of a cyclopropane ring, *cis* fused to C-9 and C-11 of the macrocyclic undecane ring, with H-9 and H-11 $\alpha$. The cyclopentane and the cycloundecane ring are linked *trans* with H-4 α. Thus lathyrol is said to possess the 4α-lathyrane skeleton. The botanical sources of the lathyrol groups of esters are summarised in Table 3. The positions of the ester functions in these compounds were not assigned (*57, 58*).

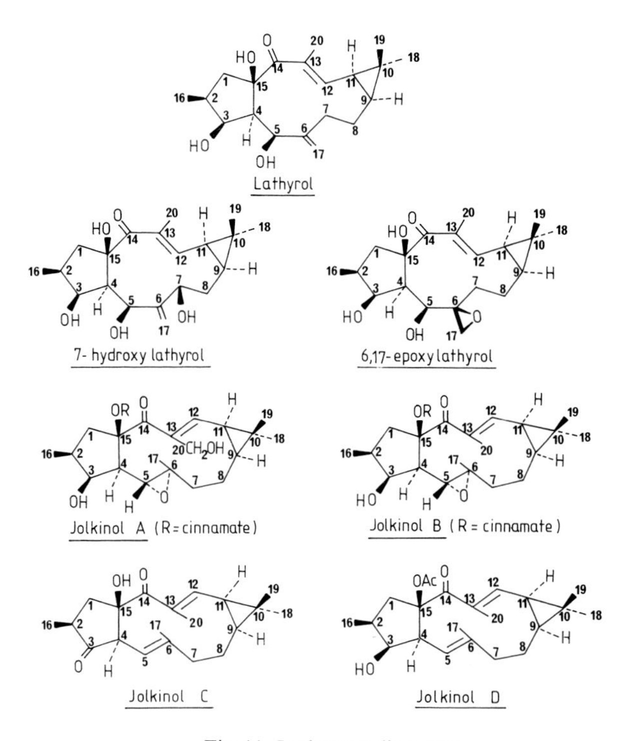


Fig. 11. Lathyrane diterpenes

*Table 3*

| Compound | Botanical Source | Literature |
|---|---|---|
| 6,17-Epoxylathyrol- | *Euphorbia lathyris* | (*14, 58*) |
| 3-Phenylacetate-5,10-diacetate | *Macaranga tanarius* | (*59*) |
| 7-Hydroxylathyrol-dibenzoate-diacetate | *Euphorbia lathyris* | (*60*) |
| Lathyrol-dibenzoate-diacetate | *Euphorbia lathyris* | (*61*) |
| Lathyrol-diacetate-nicotinate | *Euphorbia lathyris* | (*61*) |
| Bertyadional | *Bertya cuppressoidea* | (*15*) |
| 15-OH-4-hydro-bertyadional | *Bertya cuppressoidea* | (*63*) |
| 1,2-Dehydro-4,15-dihydro-bertyadional | *Bertya cuppressoidea* | (*63*) |
| Jolkinols A, B, C and D | *Euphorbia jolkini* | (*32*) |

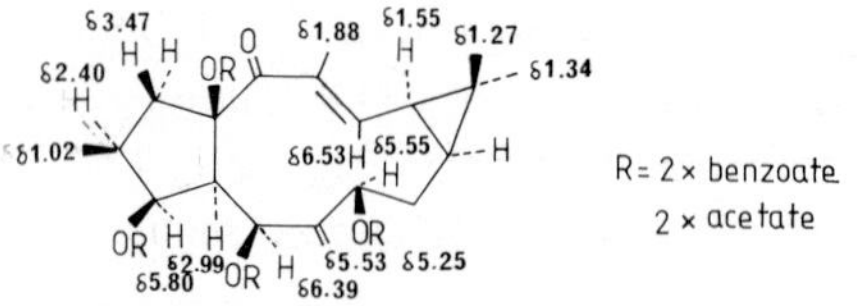

R = 2 × benzoate
2 × acetate

7-hydroxy-lathyrol-diacetate-dibenzoate

Empirical formula $C_{38}H_{42}O_9$
M.P 205–207°C
U.V. (MeOH) $\lambda_{max}$ ($\varepsilon$) nm 229 (29800), 273 (14200)
I.R. (KBr) $\nu_{max}$ cm$^{-1}$, 1735, 1713, 1650, 1623, 907, 712
$^1$H-NMR ($CDCl_3$) assigned according to reference (57)

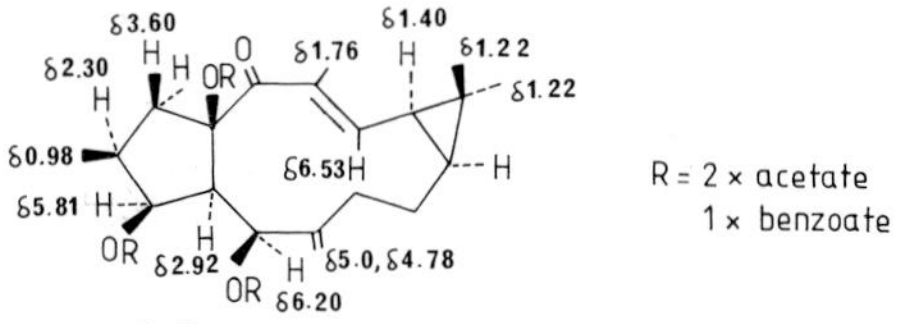

R = 2 × acetate
1 × benzoate

Lathyrol-benzoate-diacetate

Empirical formula $C_{31}H_{38}O_7$
M.P. 156–158°C
U.V. (MeOH) $\lambda_{max}$ ($\varepsilon$) nm 229 (16400) 275 (15300)
I.R (KBr $\nu_{max}$ cm$^{-1}$, 1730, 1705, 1640, 1613, 897, 705
$^1$H-NMR ($CDCl_3$) assigned according to reference (58)

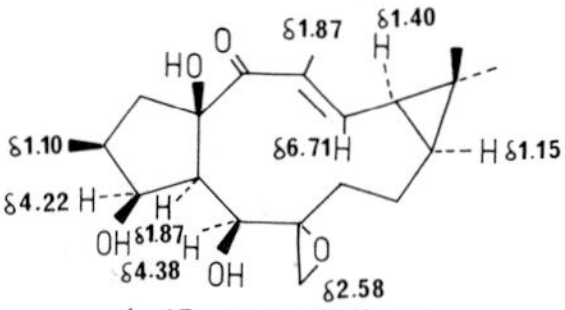

6, 17-epoxy lathyrol

Empirical formula $C_{20}H_{30}O_5$
M.P. 204–207°C
U.V. $\lambda_{max}$ 273 nm ($\varepsilon$ = 15000)
$^1$H-NMR assigned according to reference (14)

Fig. 12

The physical and spectral data of the lathyrol derivatives are summarised in Fig. 12.

An unusual lathyrane diterpene, bertyadionol, was isolated from *Bertya cuppressoidea* (Euphorbiaceae) in 1970 (*15*). This compound was characterised by the presence of a double bond in the 4,15-position, presumably as the result of dehydration of a 15-hydroxy-3-ketone. The stereochemistry of bertyadionol was deduced later (*66*). NOE measurements indicated a *trans* configuration for the 5,6-double bond. The possible precursor, 15-hydroxy-4-hydro-bertyadionol and another analogue, 1,2-dehydro-4,15-dihydro-bertyadionol, were also isolated from *Bertya* species (*60*). These compounds had the typical *trans* ring junction of the five and eleven membered rings of lathyrol (Fig. 13).

Four other closely related lathyrane diterpenes were later obtained from extracts of *Euphorbia jolkini*, and are known as jolkinols A, B, C and D (Fig. 11). Jolkinols A and B were characterised by the presence of an α-orientated 5,6-epoxide. The structure of jolkinol B was elucidated by chemical conversion to the known lathyrol. In the case of the jolkinols C and D, NOE measurements confirmed the *trans* configuration of the 5,6-double bond (Fig. 11). The physical and spectroscopic data of the jolkinol diterpenes are summarised in Fig. 14.

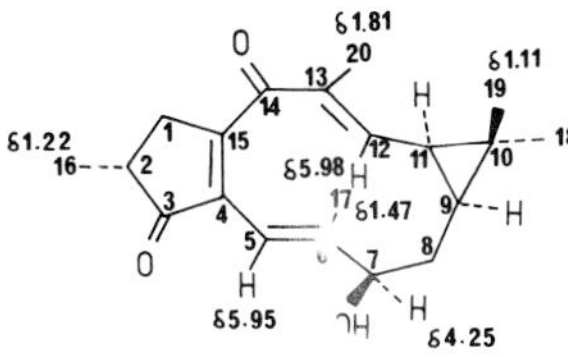

Bertyadionol

Empirical formula $C_{20}H_{26}O_3$

M. P. 159-160° C

U.V. $\lambda_{max}$ (ε) nm; 220 (10800), 263 (9400), 332 (3400)

I.R. $\nu_{max}$ cm$^{-1}$; 3610, 1710, 1665

$^1$H-NMR ($CDCl_3$) assigned according to reference (15)

Numbering system according to (31)

Fig. 13

Jolkinol A

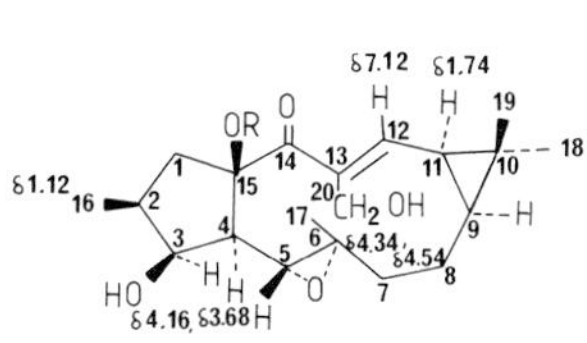

Empirical formula (R = cinnamate) $C_{29}H_{36}O_6$

I.R. ($CHCl_3$) $\nu_{max}$ cm$^{-1}$, 3520, 1720, 1640, 1590

M.S m/z 480 ($M^+$)

$^1$H-NMR ($CDCl_3$, 100 MHz) according to reference (32)

Fig. 14

Jolkinol B

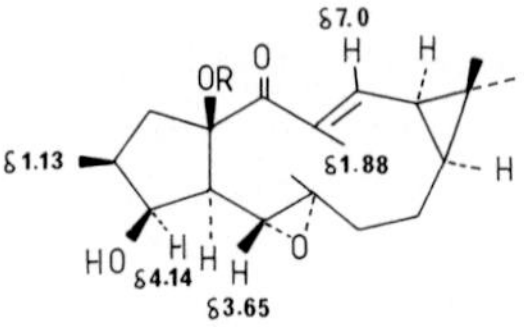

Empirical formula (R = cinnamate) $C_{29}H_{36}O_5$

I.R. ($CHCl_3$) $\nu_{max}$ $cm^{-1}$, 3520, 1720, 1630, 1580,

M.S. m/z 464 ($M^+$)

$^1H$-NMR ($CDCl_3$, 100 MHz) according to reference (32)

Jolkinol C

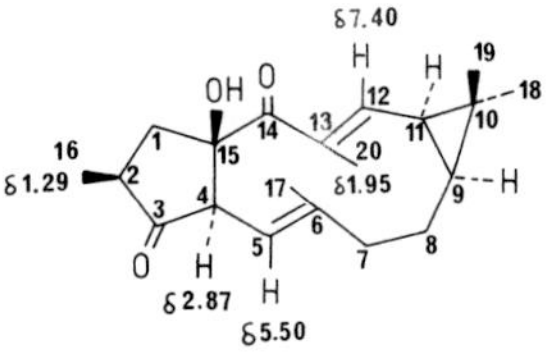

Empirical formula $C_{20}H_{28}O_3$

I.R. ($CHCl_3$) $\nu_{max}$ $cm^{-1}$, 3460, 1740, 1630, 1610

M.S. m/z 316 ($M^+$)

$^1H$-NMR ($CDCl_3$ 100 MHz) according to reference (32)

Jolkinol D

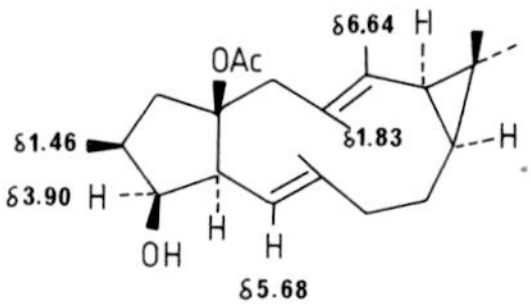

Empirical formula $C_{22}H_{32}O_4$

I.R. ($CHCl_3$) $\nu_{max}$ $cm^{-1}$ 3580, 1740, 1640, 1610

M.S m/z 360 ($M^+$)

$^1H$-NMR ($CDCl_3$ 100 MHz) according to reference (32) see figure

Fig. 14 (continued)

The genus *Euphorbia* has also yielded a large variety of esters of a further lathyrane diterpene, ingol (*61*). Ingol was originally isolated from the latex of *Euphorbia ingens* (*65, 66*) in the form of a triacetate-nicotinate tetra-ester.

However at that time no stereochemical assignments were made. More recently the crystal structure and stereochemistry of ingol-tetraacetate, obtained by acetylation of ingol, was elucidated by X-ray methods *(61)*. In contrast to lathyrol, ingol (Fig. 15) has a 4β-lathyrane nucleus, *i. e.* while the ingol nucleus is composed of the same three ring system as lathyrol the epoxide bridging C-4 and C-15 confers *cis* configuration on the cyclopentane and cycloundecane ring junction. The cyclopropane ring junction is *cis* as in the 4α-lathyranes and the 5,6-double bond is *trans*. The absolute configuration was assigned by analogy to the other lathyrane diterpenes (Fig. 15).

Ingol - 3,7,8,12 - tetraacetate

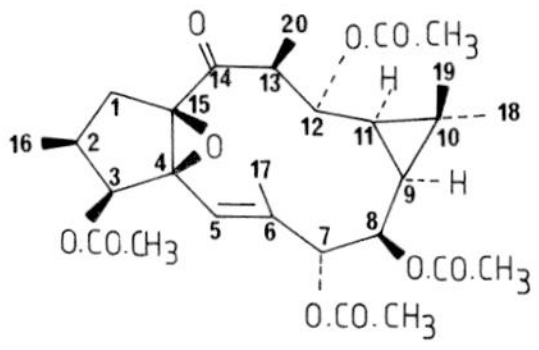

Empirical formula $C_{28}H_{38}O_{10}$

M.S. $M^+$ m/z 534 (14%)

$^1$H-NMR see figure 16

U.V $\lambda_{max}^{MeOH}$ nm, (log ε) 212 (3.41), 226 (3.15)

I.R. $\nu_{max}$ $cm^{-1}$ 2910, 1735, 1700, 1650, 1365, 1230

Fig. 15. Ingol-3,7,8,12-tetraacetate

*Table 4*

| Compound | Botanical Source | Literature |
|---|---|---|
| Ingol, 3,7,12-triacetate-8-nicotinate | *Euphorbia ingens* | *(66)* |
| Ingol, 3,8,12-triacetate-7-phenylacetate | Euphorbium | *(67)* |
| Ingol, 3,8,12-triacetate-7-*p*-methoxyphenylacetate | Euphorbium | *(67)* |
| Ingol, 3,12-diacetate-8-tigliate | *E. lactea* | *(68)* |
| Ingol, 3,7,8-triacetate-12-tigliate | *E. kamerunica* | *(51)* |
| Ingol, 3,7-diacetate-12-tigliate | *E. kamerunica* | *(51)* |
| Ingol, 3,7,8,12-tetraacetate | *E. kamerunica* | *(51)* |
| Ingol, 3,7,12-triacetate-8-benzoate | *E. kamerunica* | *(69)* |
| Ingol, 3,7,12-triacetate-8-angelate | *E. kamerunica* | *(69)* |
| Ingol, 3,7,12-triacetate-8-tigliate | *E. kamerunica* | *(69)* |
| Ingol, 3,12-diacetate-7-tigliate | *E. kamerunica* | *(70)* |

Two tetra-esters of ingol were reported (*64*) from euphorbium, a pharmaceutical resin used to manufacture ointments and most probably derived from *Euphorbia resinifera* latex which has been allowed to dry in the sun. A tri-ester of ingol was isolated from the latex of *Euphorbia lactea* (*65*) and a series of tetra- and tri-esters were recently obtained from the fresh latex of *Euphorbia kamerunica* (*50, 66, 67*). The biological sources of the various esters of ingol are summarised in Table 4.

Ester functions are present on the ingol nucleus at positions 3, 7, 8 and 12. The esterifying groups consist of aliphatic acids such as acetate, angelate or tiglate as well as aromatic moieties such as benzoate and nicotinate. From the mass spectra alone, it is not possible to assign the positions of acylation in the ingol tetra-ester series (*50*). Electron-impact (e.i.) mass spectra of ingol tetraesters normally exhibit reasonable $M^{+\cdot}$ ions of 1%–40% relative abundance depending upon the temperature of the source. The tri- and di-esters of ingol exhibit less intense $M^{+\cdot}$ ions, but their intensity is sufficient for accurate mass measurements to be made. The main feature of the fragmentation by e.i.-m.s. of ingol esters is the sequential loss of the acyl groups in the carboxyl form, which permits determination of the molecular weight and the number of acyl substituents (Fig. 16).

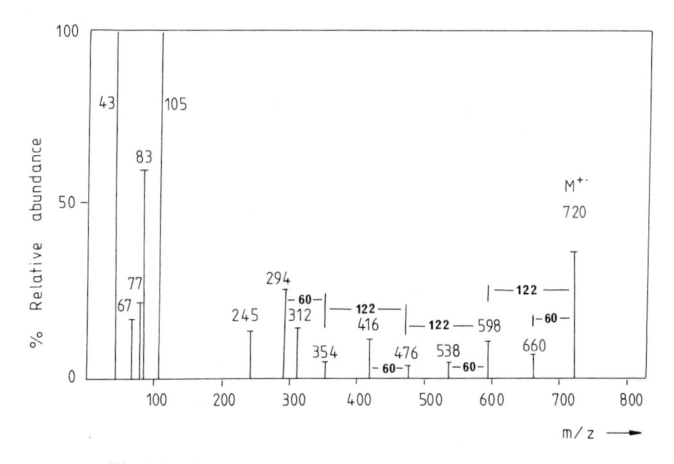


Fig. 16. e.i.-m.s. of 12-O-acetyl-ingol-3,7,8-tribenzoate

The chemical-ionisation (c.i.) mass spectra of ingol esters are essentially similar to the e.i.-m.s. However a quasi-molecular ion ($M^{+\cdot}+1$) is the main feature of these spectra, with a relative abundance of 50%–95%. Depending upon the nature of the reagent gas ($NH_3$, methane or isobutane), further quasi-molecular ions may also be exhibited. The fragmentation pattern produced by c.i.-m.s. involves elimination of acyl substituents as RCOOH· from one or more quasi-molecular ions. For structure elucidation of ingol esters it may therefore be necessary to utilise mass-spectrometry in both the e. i. and the c. i. mode.

In the $^1$H-NMR spectrum of ingol-3,7,8,12-tetraacetate (Fig. 17), the protons geminal to the acyl functions on carbons 3, 7, 8 and 12 are found downfield. The assignments for these protons were confirmed by decoupling experiments, using an F. T. 250 MHz instrument. Absence of one or more of the acyl groups at these positions results in an upfield shift of the appropriate geminal proton on the order of 1 ppm and distinguishes acylated from free secondary hydroxy groups in the ingol nucleus.

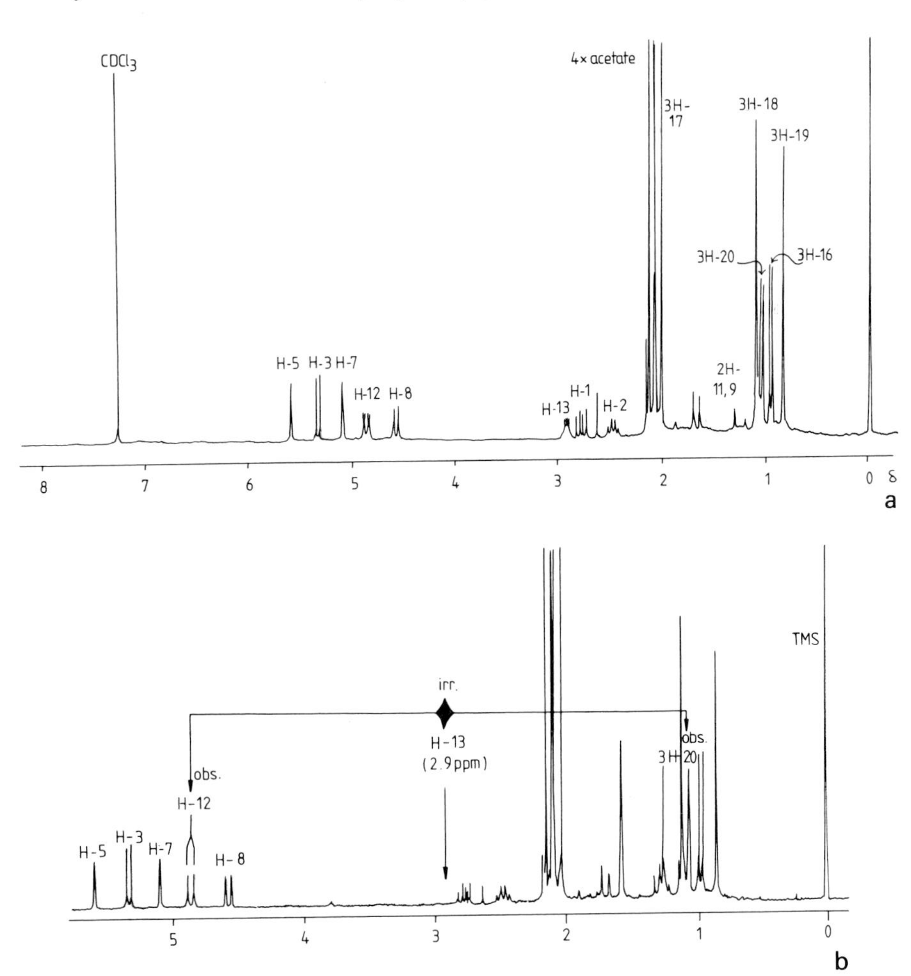


Fig. 17. 250 MHz $^1$H-NMR spectrum of ingol-3,7,8,12-tetraacetate

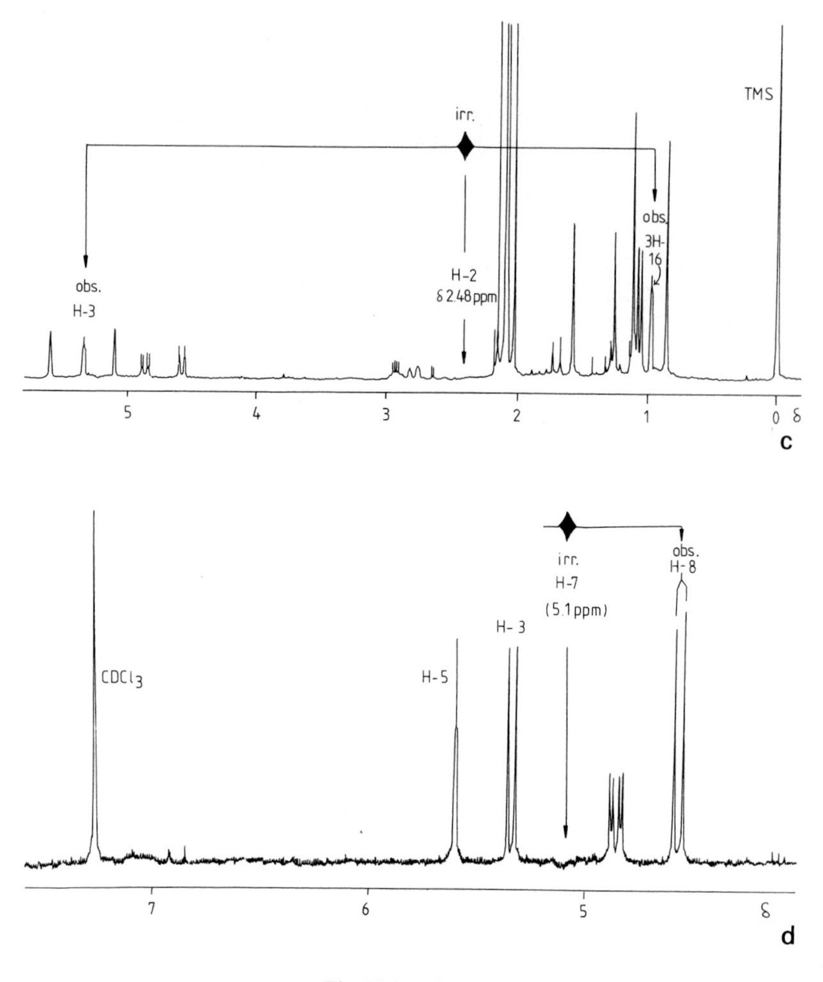


Fig. 17 (continued)

The assignment of various acyl substituents to the 3, 7, 8 or 12 positions of ingol tetraesters has been achieved by means of stepwise hydrolysis. Mild alkaline hydrolysis using 0.1 M KOH in methanol of an ingol-tetraester produces a mixture of ingol tri- and diesters. More vigorous hydrolysis with 0.5 M KOH in methanol produces an ingol monoester. This derivative is usually acylated at C-12 of the nucleus. The parent polyol, ingol, is also produced by this reaction, but in less than 20% yield. The various reaction products may be separated by preparative-TLC (*50*) and assignments of acyl substituents made on the basis of their $^{1}$H-NMR spectra (see Fig. 17).

Further derivatives of ingol might be expected to be found in species of the Euphorbiaceae in the near future, and in fact it has been claimed (*31*) that 18-hydroxyingol was isolated from *Euphorbia marginata* and *Euphorbia unispina* although no chemical evidence was provided for this.

In general, biological activities have not been reported for the lathyrane group of diterpenes. However, a lathyrol derivative was shown to exhibit slight cytotoxic activity at 132 μg/ml (*68*) *in vitro* against 7288 Morris rat hepatoma cells. More recently, several ingol esters of the tetra- and triester type were isolated from column fractions of *Euphorbia kamerunica* which demonstrated pronounced cytotoxic action against mouse TLX/5 lymphoma cells (*50*). The activity of the pure esters has not been confirmed using *in vivo* methods. However, an interesting observation was recently reported by ROTHSCHILD and others (*69*) who demonstrated that extracts of the pupae of the spurge hawk moth (*Hyles Euphorbiae* L.) exhibited pronounced antitumour activity *in vivo* and cytotoxic activity *in vitro*. This species of moth produces larvae which feed on *Euphorbia* species and chemical analysis of the pupae demonstrated the presence of ingol esters. It is possible that this species of moth sequesters diterpenes for defence purposes.

### VI.4. Jatropholane and Crotofolane Types

The jatropholane and crotofolane diterpenes are closely related diterpenes obtained from *Jatropha* and *Croton* species. They differ in the fact that in the crotofolanes the cyclopropane ring of the jatropholane nucleus has opened out to an isopropylidene side chain (Fig. 1).

The jatropholanes A and B were isolated from the roots of *Jatropha gossypiifolia* (*20*). The structure of jatropholane B was elucidated by X-ray analysis of the corresponding acetate. Jatropholane A was recognised as the C-2 epimer of jatropholane B because treatment with mild base of either compound afforded an equimolar mixture of both (Fig. 18).

| INGOL ESTERS | Chemical shifts (δ ppm) of acyl protons at: | | | |
|---|---|---|---|---|
| | C-3 | C-7 | C-8 | C-12 |
| 3,7,8-Triacetyl-ingol-12-tiglate | 5.25 | 5.23 | 4.60 | 4.88 |
| 12-Tigloyl-ingol-7,8-diacetate | 4.29 | 5.37 | 4.60 | 4.88 |
| 3,7-Diacetyl-ingol-12-tiglate | 5.27 | 5.10 | 3.54 | 4.88 |
| 7-O-Tigloyl-ingol-3,12-diacetate | 5.26 | 5.10 | 3.55 | 4.90 |
| 8-Tigloyl-ingol-12-acetate | 4.37 | 4.30 | 4.58 | 4.87 |
| 8-Acetyl-ingol-12-tiglate | 4.29 | 4.30 | 4.60 | 4.88 |
| Ingol-12-acetate | 4.36 | 4.34 | 2.87 | 4.84 |

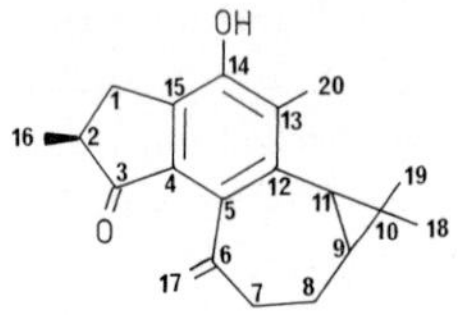

Jatropholone-A

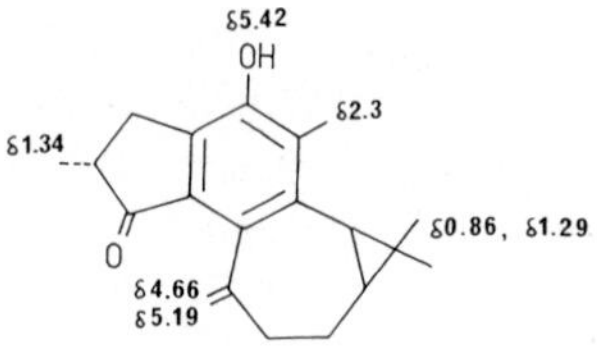

Jatropholone - B $C_{20}H_{24}O_2$

U.V. $\lambda^{max}_{EtOH}$ nm 225, 238, 275, 335

I.R. $\lambda^{CCl_4}_{max}$ $cm^{-1}$, 3603, 1715

$^1$H-NMR according to reference (20)

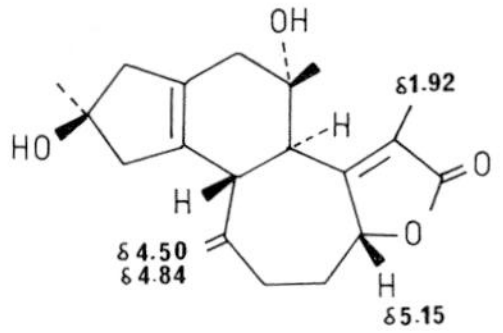

Crotofolin $C_{20}H_{24}O_5$ M.P. 277–279° C

M.S. $M^+$ m/z 344

U.V $\lambda^{max}_{MeOH}$ n.m. (ε), 227 (23,800)

I.R. $\nu_{max}$ $cm^{-1}$ Nujol., 3465, 1740, 1705, 1650, 900

$^1$H-NMR (DMSO-$d_6$) as reference (19)

Fig. 18

The crotofolane nucleus is represented by crotofolin (*19*) from *Croton corylifolius*. The structure and relative configuration of crotofolin was obtained through a single crystal X-ray analysis. The cyclohexane ring is *trans* fused to the seven membered ring (Fig. 18). The spectroscopic and physical data of the jatropholanes and crotofolin are summarised in Fig. 18.

## VI.5. Rhamnofolane

A diterpene with the rhamnofolane nucleus has been reported as occurring in *Croton rhamnifolious* (*21*). This compound as its synthetic acetate was identified as 13,15-seco-4,12-dideoxy-4α-phorbol-20-acetate (Fig. 19) by comparison of its spectral data with those of phorbol triacetate (Fig. 23).

*References, pp. 90—99*

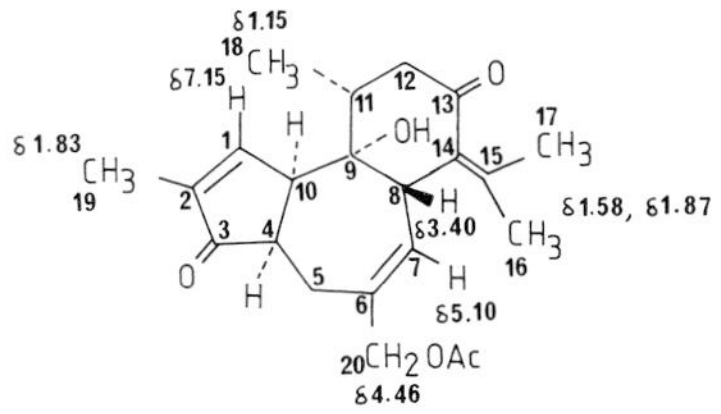

13, 15-seco-4, 12-dideoxy-4α-phorbol-20-acetate

M.P. 171° C $C_{22}H_{28}O_5$

UV $\lambda_{max}^{EtOH}$ n.m. (ε), 205 (14180), 235 (12210)

I.R. $\nu_{max}$ Nujol $cm^{-1}$, 3410, 1736, 1689, 1661, 1626

$^1$H-NMR according to reference (21)

Fig. 19

## VII. Tigliane Diterpenes

The tiglianes consist of a group of tetracyclic compounds which occur in plants as acylated polyhydroxy diterpenes (*11*). They are the toxic tumour-promoting and pro-inflammatory agents of many species of Euphorbiaceae (*2*). Over recent years these substances have been increasingly utilised as pharmacological and biochemical tools for investigating the mechanisms of tumour-promotion in both *in vivo* and *in vitro* (*70*) systems. In most biological systems 12-O-tetradecanoyl-phorbol-13-acetate has been shown to be the most potent tigliane ester and is certainly the one most widely used in cancer reseach (*2, 11, 70*). Accordingly its name has been abbreviated to A′, PMA or TPA in biological research communications. It would, however, lead to less confusion in the literature if TPA were generally accepted as the abbreviation for this tumour-promoting agent. Furthermore, use of the term "tigliane esters", as proposed by ADOLF and HECKER (*31*) to cover esters of hydroxylated, deoxy and epoxy derivatives of phorbol, would be a distinct advantage in literature searching for biological activities of this diverse group of toxic compounds.

The first tigliane diterpenes to be isolated from plants were the 12,13-diesters of phorbol. These compounds were the toxic components of the seed oil of *Croton tiglium* (*1, 34*).

### VII.1. Phorbol and Its Esters

Phorbol (Fig. 20) is a tetracyclic diterpene. It consists of a five membered ring A, a seven membered ring B, a six membered ring C and a cyclopropane system D. An α,β-unsaturated keto group is located at C-3 of ring A, a glycol at C-12 and C-13, a primary hydroxyl at C-20 and tertiary hydroxyls at C-9 and C-4. Rings A and B are *trans* fused and ring D is *cis* fused to the six membered ring C.

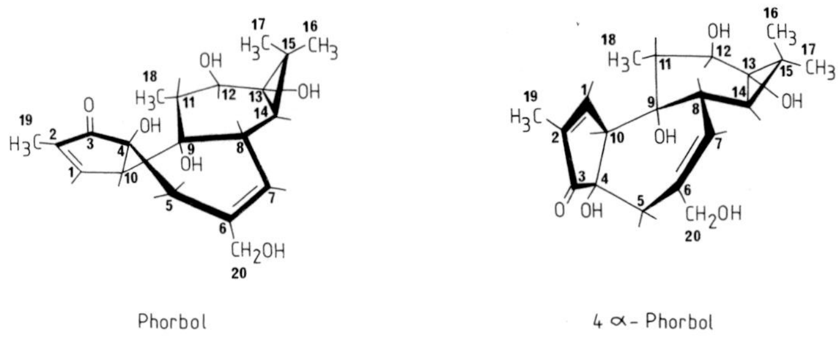

Phorbol

4α-Phorbol

Fig. 20

Phorbol was isolated as a hydrolysis product of *Croton* oil as early as 1935 (*71*). Its correct empirical formula was obtained in 1966 (*72*). Several structures were proposed for this novel compound (*73, 74, 75, 76*) before its structure was finally elucidated by HECKER's group (*77*) by means of X-ray crystallography of neophorbol-13,20-diacetate-$3_p$-bromobenzoate. The structure was later confirmed by X-ray analysis of phorbol bromofuroate as the chloroform solvate (*78*) and by X-ray studies of phorbol as the ethanol solvate (*79*). The structure and stereochemistry of phorbol were also confirmed by chemical methods at about the same time (*80*). Chemistry and reactivity of phorbol have been fully reviewed in the 1971 volume of this series (*34*). More recently the geometry of phorbol has been totally optimised and the charge distribution calculated (*81*).

The isomer 4α-phorbol (Fig. 20) was also isolated from *Croton* oil (*1, 82*). This compound is characterised by *cis* fusion of rings A and B, the 4-hydroxyl group being α-orientated. Dreiding models of this isomer (Fig. 20) show that the shape is quite different from that of phorbol and this may account for the lack of biological activity shown by esters of 4α-phorbol (*1, 34*).

## VII.2. Distribution of Phorbol Esters in Plants

Phorbol occurs naturally in the form of 12,13-diesters and 12,13,20-triesters of the polyhydroxylated tigliane nucleus. The triesters are known as "cryptic irritants" because they do not exhibit pro-inflammatory activity on mammalian skin unless the C-20 acyl group is removed by hydrolysis (*83*). Because of their pronounced toxicological properties, the 12,13-diesters of phorbol have been extensively investigated biochemically (*70*) but their distribution in higher plants is limited when compared to other tigliane derivatives. Phorbol esters have been identified in extracts of *Croton tiglium* (*34*), *Croton sparciflorus* (*84*), *Sapium japonicum* (*39*), *Sapium indicum* (*94*), *Euphorbia tirucalli* (*85*), *Euphorbia frankiana* (*86*) and *Euphorbia coerulescens* (*36*). Phorbol esters are therefore confined to three genera of the family Euphorbiaceae at the present time. The naturally occurring esters of phorbol are summarised in Fig. 21.

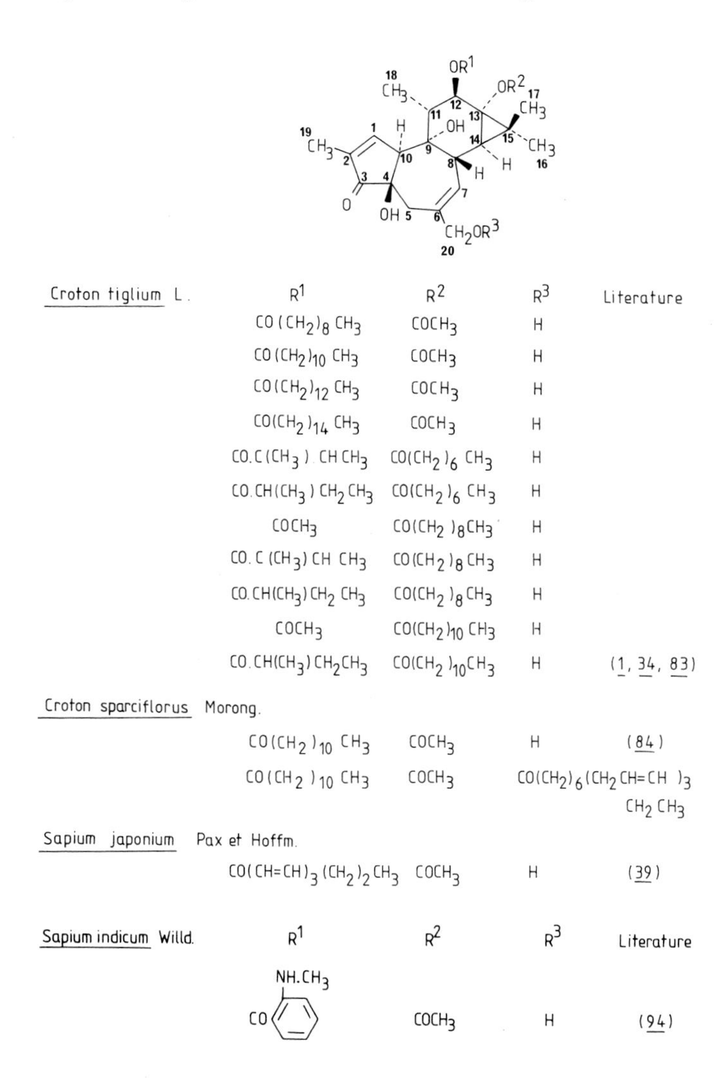


| | $R^1$ | $R^2$ | $R^3$ | Literature |
|---|---|---|---|---|
| Croton tiglium L. | | | | |
| | $CO(CH_2)_8CH_3$ | $COCH_3$ | H | |
| | $CO(CH_2)_{10}CH_3$ | $COCH_3$ | H | |
| | $CO(CH_2)_{12}CH_3$ | $COCH_3$ | H | |
| | $CO(CH_2)_{14}CH_3$ | $COCH_3$ | H | |
| | $CO.C(CH_3).CHCH_3$ | $CO(CH_2)_6CH_3$ | H | |
| | $CO.CH(CH_3)CH_2CH_3$ | $CO(CH_2)_6CH_3$ | H | |
| | $COCH_3$ | $CO(CH_2)_8CH_3$ | H | |
| | $CO.C(CH_3)CHCH_3$ | $CO(CH_2)_8CH_3$ | H | |
| | $CO.CH(CH_3)CH_2CH_3$ | $CO(CH_2)_8CH_3$ | H | |
| | $COCH_3$ | $CO(CH_2)_{10}CH_3$ | H | |
| | $CO.CH(CH_3)CH_2CH_3$ | $CO(CH_2)_{10}CH_3$ | H | (1, 34, 83) |
| Croton sparciflorus Morong. | | | | |
| | $CO(CH_2)_{10}CH_3$ | $COCH_3$ | H | (84) |
| | $CO(CH_2)_{10}CH_3$ | $COCH_3$ | $CO(CH_2)_6(CH_2CH{=}CH)_3CH_2CH_3$ | |
| Sapium japonium Pax et Hoffm. | | | | |
| | $CO(CH{=}CH)_3(CH_2)_2CH_3$ | $COCH_3$ | H | (39) |
| Sapium indicum Willd. | $R^1$ | $R^2$ | $R^3$ | Literature |
| | $CO$-$C_6H_4$-$NH.CH_3$ | $COCH_3$ | H | (94) |

Fig. 21. Distribution of phorbol esters in plants

*Euphorbia frankiana* Berger

| | | |
|---|---|---|
| $CO.CH.(CH_3)CH_2CH_3$ | $CO.C(CH_3).CH.CH_3$ | H |
| $CO.CH(CH_3)_2$ | $COCH_3$ | $CO.C(CH_3)CH.CH_3$ (36, 86) |

*Euphorbia coerulescens* Haw.

| | | |
|---|---|---|
| $CO\,CH(CH_3)CH_2CH_3$ | $COC(CH_3)\,CH\,CH_3$ | H |
| $CO\,CH(CH_3)_2$ | $COCH_3$ | $CO\,C(CH_3)\,CH\,CH_3$ |

*Euphorbia tirucalli* L.

| | | |
|---|---|---|
| $CO.(CH{=}CH)_n(CH_2)_mCH_3$<br>m = 2, n = 2, 3, 4, 5<br>m = 4, n = 1, 2, 3, 4 | $COCH_3$ | H |
| $COCH_3$ | $CO(CH{=}CH)_n(CH_2)_mCH_3$<br>m = 2, n = 2, 3, 4, 5<br>m = 4, n = 1, 2, 3, 4 | H (85) |

Fig. 21 (continued)

## VII.3. Identification of Phorbol Esters

Naturally occurring esters of phorbol are unstable compounds and may be isolated from plants using neutral separation methods (see Section V). Identification of the various acylated derivatives of phorbol may be achieved by a combination of spectroscopic and chemical methods.

The 12,13-diesters of phorbol can exist as a series of esters in which a high molecular weight acyl derivative is attached to C-12 of the nucleus and a low molecular weight acyl derivative to C-13 (the A-series) or this may be

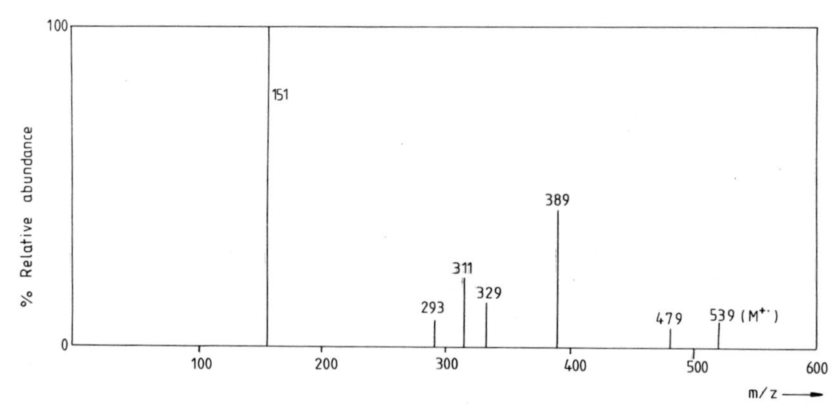


Fig. 22. e.i.-m.s. (upper region) of a phorbol diester, 12-O-[2-methylaminobenzoyl]-phorbol-13-acetate

reversed as in the B-series of diesters (Fig. 21). The mass spectra of these isomers have distinct fragmentation patterns (*34*). In the A-series, the high molecular weight acyl group is eliminated from the molecular ion in the form of an acyloxy radical, RCOO·, whilst in the B-series this acyl group is eliminated as RCOOH. In both series the molecular ion in the e. i.-m. s. has a relative abundance of 1% or less, but in the c. i.-m. s. a larger quasi molecular ion ($M^{+\cdot}+1$) is characteristically exhibited (Fig. 22).

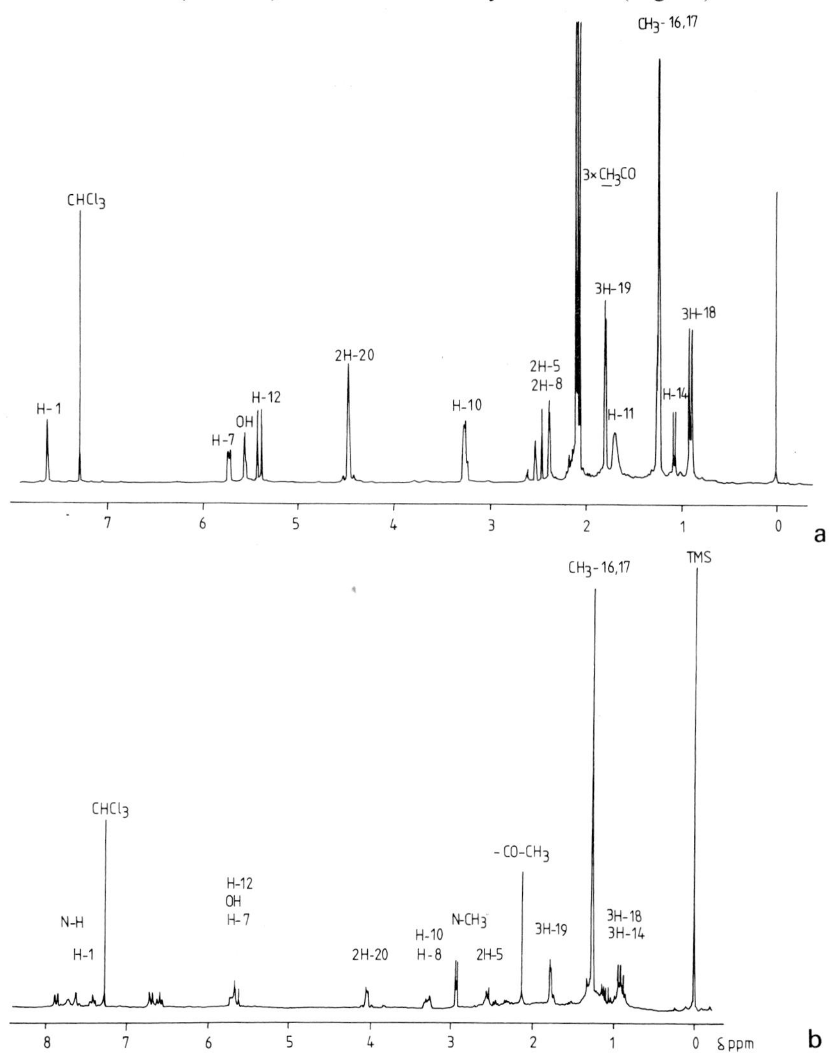


Fig. 23. a) 250 MHz $^1$H-NMR spectrum of phorbol triacetate ($CDCl_3$). b) 250 MHz $^1$H-NMR spectrum of a phorbol diester, 12-O-[2-methylaminobenzoyl]-phorbol-13-acetate ($CDCl_3$)

The fragmentation of phorbol esters by both e. i.- and c. i.-m. s. involves the sequential elimination of both acyl groups to produce a fragment ion at m/z 328 (20 – 40% relative abundance) accompanied by an ion at m/z 310 (m/z 328 – 18). Accordingly these spectra are of assistance in identifying the molecular weight of acyl substituents and may provide evidence as to the position of acylation in the phorbol nucleus.

Comparisons of the $^1$H-NMR spectra of phorbol esters have been useful in the identification of naturally occurring esters, because the absolute configuration of phorbol was determined at an early stage by X-ray analysis (*79*). Phorbol triacetate is a fully acylated structure within the tigliane series and all the distinguishing chemical shifts for the protons of phorbol esters are exhibited in its $^1$H-NMR spectrum (Fig. 23). Comparison of the spectrum of phorbol triacetate with that of a phorbol diester (Fig. 23) demonstrates that when the C-20 acyl group of a triester is removed to give a C-20 primary hydroxy group, there is an upfield shift in the allylic 2 proton signal of H-20. Phorbol triesters may be distinguished from the corresponding diesters on this basis. Triesters of phorbol can be chemically converted to 12,13-diesters by the action of perchloric acid in methanol. Hydrolysis of tertiary acyl groups on C-13 of phorbol, on the other hand, requires the action of 0.5 M KOH in methanol. The $^1$H-NMR spectra of the C-12 monoesters so formed show a shift in the signal for H-12 from 5.63 ppm in the diester spectrum to 5.21 ppm in the spectrum of the monoester. This observation, also confirmed upon synthesis (*87*), provides a means for distinguishing the A and B isomer series. The total hydrolysis of phorbol esters to the parent alcohol may be achieved by the action of alcoholic barium hydroxide over a period of 10 – 12 hours (*34*). It is possible that small amounts of 4α-phorbol may be produced as a result of this reaction. The physical and spectral properties of 4α-phorbol are summarised in Fig. 24.

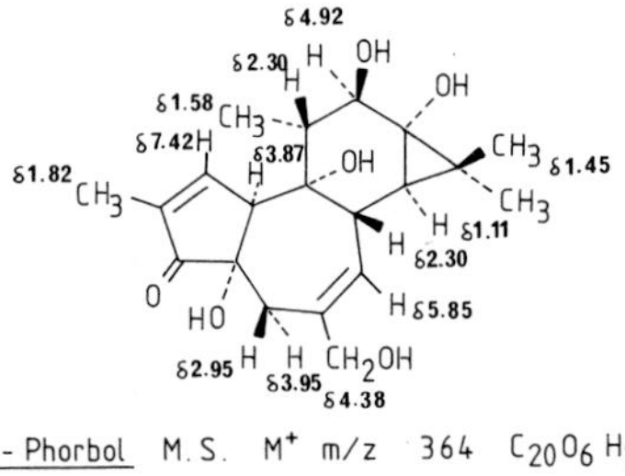

4α - Phorbol M. S. $M^+$ m/z 364 $C_{20}O_6H_{28}$

U.V. $\lambda_{max}^{MeOH}$ (ε) nm. 241 (5300), 336 (71)

I.R. $\nu_{max}^{KBr}$ $cm^{-1}$. 3415, 1700, 1628

$^1$H - NMR (pyridine-$d_5$) according to reference (82)

Fig. 24

The $^{13}$C-NMR spectra of phorbol and phorbol triacetate were first described by NEEMAN and SIMMONS (*125*). Signals were assigned by analogy to model compounds, with those for C-4 and C-9 at 80.0 ppm and 75.3 ppm being only tentative. However, $^{13}$C-NMR data for 4-deoxyphorbols (see Section VII.6.) show that the signal formerly at 75 ppm is shifted upfield, whilst the signal at 80 ppm remains unaltered in the spectra of 4-deoxyphorbol analogues. On this basis, the assignments of NEEMAN and SIMMONS for C-4 and C-9 can be reversed. The $^{13}$C-NMR spectrum of phorbol triacetate is shown in Fig. 25.

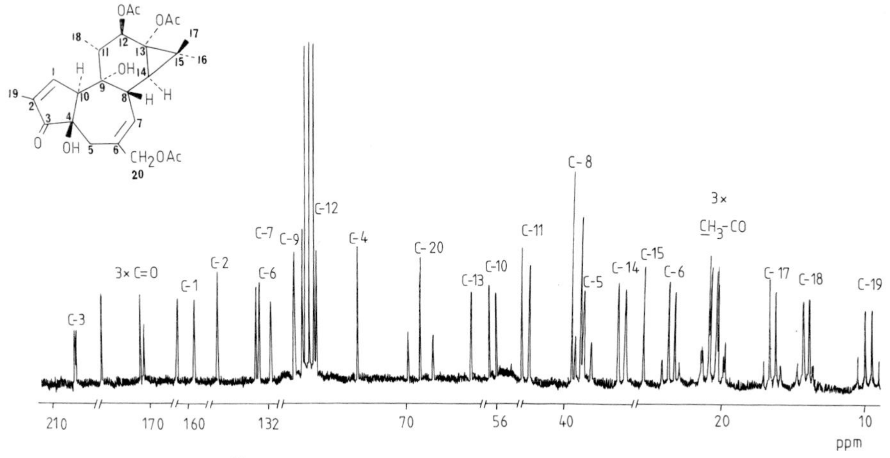


Fig. 25. 100.6 MHz $^{13}$C-NMR spectrum of phorbol triacetate off resonance decoupled ($CDCl_3$)

Both the U.V. and I.R. spectra of the various phorbol esters are similar. In the U.V. spectrum characteristic inflections are exhibited at 230 and 330 n.m. due to the α,β-unsaturated carbonyl group at C-3. The I.R. spectrum exhibits bands in the hydroxyl region at 3500 $cm^{-1}$, carbonyl bands due to the ester groups at 1720 – 1740 $cm^{-1}$, and ketone bands at 1695 – 1715 $cm^{-1}$. The circular dichroism spectra of phorbol and 4α-phorbol esters (Fig. 26) are particularly diagnostic for these isomers and this technique provides a useful adjunct to mass spectrometry and thin-layer chromatography for identification of small quantities of naturally occurring esters (*88*).

The C.D. spectrum of phorbol triacetate exhibits a pronounced positive extremum between 225 – 250 nm and less intense negative extremum between 330 – 340 nm. These Cotton effects are due respectively to $\pi-\pi^*$ and $\eta-\pi^*$ transitions of the α,β-unsaturated carbonyl group of ring A. 4α-phorbol triacetate, on the other hand, exhibits a much less intense Cotton effect between 225 – 250 nm.

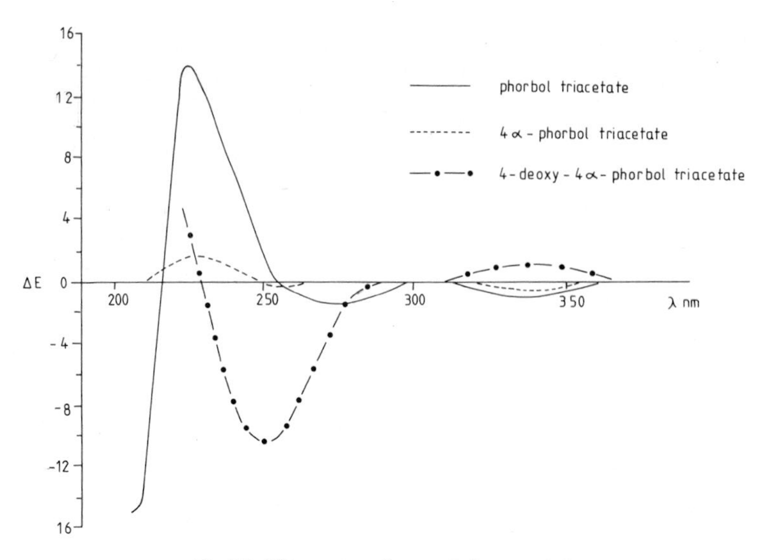


Fig. 26. CD spectra of some tigliane acetates

## VII.4. 4-Deoxyphorbol Esters

A group of highly irritant tigliane diterpenes isolated from the latex of *Euphorbia tirucalli* (*89, 90*) was shown to consist of esters of 4-deoxyphorbol which were analogues of the phorbol diesters of *Croton tiglium* (Fig. 27). They occurred as isomers at the 12,13-positions, one series having a long chain acyl residue at C-12 and an acetate at C-13, whilst in a second series the relative positions of these acyl groups were reversed. The higher molecular weight acyl group ranged from $C_8$ to $C_{14}$ (*89, 90, 91*). A corresponding series of 4-deoxyphorbol triesters was later obtained from extracts of *Euphorbia biglandulosa* (*92*). These compounds contained a similar high molecular weight acyl derivative at C-12 of 4-deoxyphorbol, an acetate at C-20 and either acetate, propionate or α-methylbutyrate at the C-13 position. Recently a further ester of 4-deoxyphorbol was isolated from the Indian poisonous plant *Sapium indicum* (*93*). This compound, sapintoxin A, was distinguished by the presence of a 2-methylaminobenzoate moiety at C-12 of 4-deoxyphorbol and an acetate function at C-13. Consequently sapintoxin A exhibits a pronounced bright blue fluorescence under U.V. light. Sapintoxin A was isolated from the unripe fruits of *Sapium indicum* together with 12-O-[*n*-deca-2,4,6-trienoyl]-4-deoxyphorbol-13-acetate (*95*), an aliphatic ester previously obtained from *Euphorbia tirucalli* (*90*) (Fig. 27).

| | $R^1$ | $R^2$ | $R^3$ | Literature |
|---|---|---|---|---|
| Euphorbia tirucalli | $COCH_3$ | $CO(CH{=}CH)_n(CH_2)_2CH_3$ n = 2, 3, 4, 5 | H | |
| | $COCH_3$ | $CO(CH{=}CH)_n(CH_2)_2CH_3$ n = 1, 2, 3, 4 | H | |
| | $CO(CH{=}CH)_n(CH_2)_2CH_3$ n = 2, 3, 4, 5 | $COCH_3$ | H | |
| | $CO(CH{=}CH)_n(CH_2)_4CH_3$ n = 1, 2, 3, 4 | $COCH_3$ | H | (89, 90, 91) |
| Sapium indicum | $CO(CH{=}CH)_3(CH_2)_2CH_3$ | $COCH_3$ | H | (95) |
| | CO-(2-$NHCH_3$-phenyl) | $COCH_3$ | H | (93) |
| Euphorbia biglandulosa | $CO(CH{=}CH)_2(CH_2)_2CH_3$ | $COCH_3$ | $COCH_3$ | |
| | $CO(CH{=}CH)_2(CH_2)_2CH_3$ | $COCH_2CH_3$ | $COCH_3$ | |
| | $CO(CH{=}CH)_2(CH_2)_2CH_3$ | $COCH(CH_3)CH_3$ | $COCH_3$ | (92) |

Fig. 27. Esters of 4-deoxyphorbol

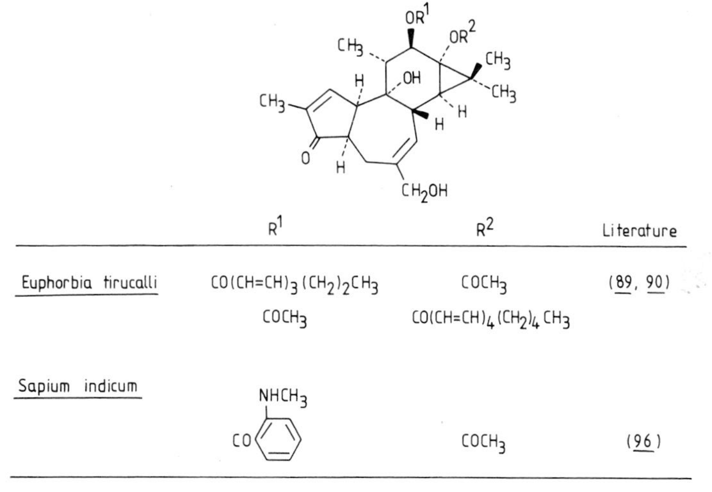

| | $R^1$ | $R^2$ | Literature |
|---|---|---|---|
| Euphorbia tirucalli | $CO(CH{=}CH)_3(CH_2)_2CH_3$ | $COCH_3$ | (89, 90) |
| | $COCH_3$ | $CO(CH{=}CH)_4(CH_2)_4CH_3$ | |
| Sapium indicum | CO-(2-$NHCH_3$-phenyl) | $COCH_3$ | (96) |

Fig. 28. Esters of 4α-deoxyphorbol

4α-Deoxyphorbol is the AB *cis* analogue of 4-deoxyphorbol. Esters of 4α-deoxyphorbol have also been obtained from natural sources but it is uncertain at the present time whether these derivatives are produced from the unstable AB *trans*-isomers during the extraction and subsequent work up of the extract. Two aliphatic esters of 4α-deoxyphorbol were originally isolated from *Euphorbia tirucalli* (*89, 90*) and an aromatic diester was later isolated from an extract of *Sapium indicum* (*96*) (Fig. 28).

### VII.5. Other 4-Deoxyphorbol Derivatives

Several other closely related derivatives of 4-deoxyphorbol have been isolated from plants of the family Euphorbiaceae. The first of these were two esters of 4-deoxy-16-hydroxyphorbol, obtained from *Croton flavens* (*97, 98*) (Fig. 29). This tigliane diterpene occurred as both a 12,13-diester and a 12,13,20-triester, with the hydroxy group at C-16 being free in both cases (*97, 98*). Two esters of 4-deoxy-5-hydroxyphorbol were recently isolated from the fruits of *Sapium indicum*, one of which had an aliphatic acyl group on C-12 whilst the second compound contained a 2-methylaminobenzoate at this position (*95, 99*). These compounds were isolated together with two similar esters of the tigliane derivative 4,20-dideoxy-5-hydroxyphorbol (*95, 99*) (Fig. 29).

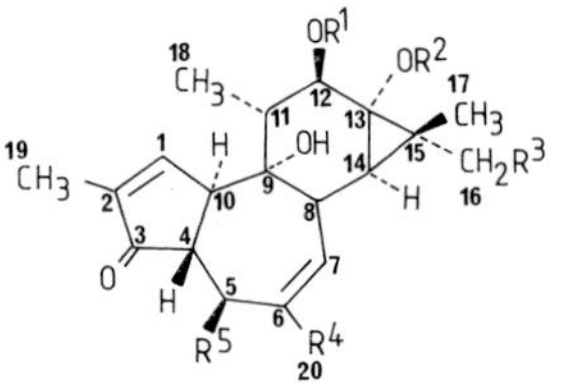

| | $R^1$ | $R^2$ | $R^3$ | $R^4$ | $R^5$ | Literature |
|---|---|---|---|---|---|---|
| Croton flavens | $CO(CH_2)_{14}CH_3$ | $COCH_3$ | OH | $CH_2OH$ | H | (97) |
| | $CO(CH_2)_{14}CH_3$ | $COCH_3$ | OH | $CH_2OCO(CH_2)_8CH_3$ | H | (98) |
| Sapium indicum | $CO(CH{=}CH)_3(CH_2)_2CH_3$ | $COCH_3$ | H | $CH_2OH$ | OH | (95) |
| | 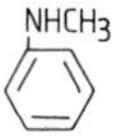 $COC_6H_4NHCH_3$ | $COCH_3$ | H | $CH_2OH$ | OH | (99) |
| | $CO(CH{=}CH)_3(CH_2)_2CH_3$ | $COCH_3$ | H | $CH_3$ | OH | (95) |
| | 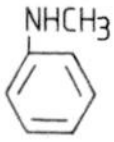 $COC_6H_4NHCH_3$ | $COCH_3$ | H | $CH_3$ | OH | (99) |
| | 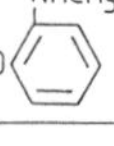 $COC_6H_4NHCH_3$ | $COCH_3$ | H | CHO | H | (100) |

Fig. 29. Other derivatives of 4-deoxyphorbol

A further unusual diterpene from *Sapium indicum*, a C-20 aldehyde, was termed 4-deoxyphorbaldehyde (*100*). 4-Deoxyphorbaldehyde occurred in the fruits of this plant together with its *cis*-isomer, 4α-deoxyphorbaldehyde; both substances occurred as 12,13-diesters (*100*). The structures of both 4-deoxyphorbaldehyde and 4α-deoxyphorbaldehyde were confirmed by sodium borohydride reduction to 4-deoxyphorbol and 4α-deoxyphorbol respectively (*100*).

## VII.6. Identification of 4-Deoxyphorbol Esters

The naturally occurring 4-deoxyphorbol esters can be identified by a combination of spectroscopic techniques and selective hydrolysis reactions. They are the most unstable of the tigliane type of diterpenes and both

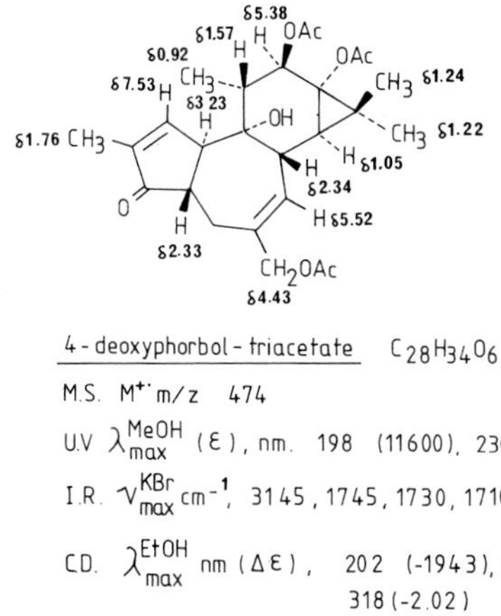

4 - deoxyphorbol - triacetate $C_{28}H_{34}O_6$

M.S. $M^{+\cdot}$ m/z 474

U.V $\lambda_{max}^{MeOH}$ (ε), nm. 198 (11600), 230 (6050), 310 (140)

I.R. $\nu_{max}^{KBr}$ cm$^{-1}$, 3145, 1745, 1730, 1710, 1635

C.D. $\lambda_{max}^{EtOH}$ nm (Δε), 202 (-19.43), 241 (+3.10), 318 (-2.02)

$^1$H-NMR ($CDCl_3$) according to reference (89)

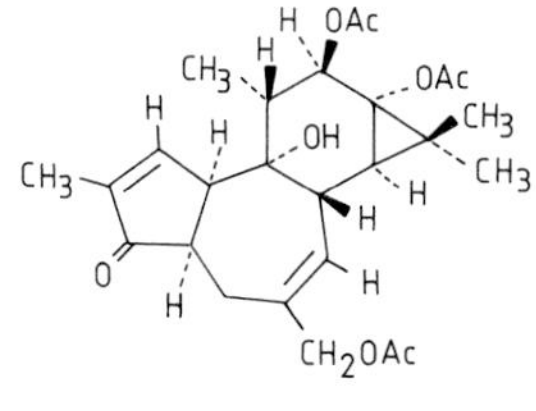

4α- deoxyphorbol-triacetate $C_{28}H_{34}O_6$

M.S. $M^{+\cdot}$ m/z 474

Fig. 30

alkaline and acid catalysed hydrolysis (*89*) induces epimerization of the 4β-oriented proton to form the 4α-isomers. This conversion together with consequent removal of the C-13 tertiary acyl group to form the C-12 monoester may be achieved by using 0.1 M KOH in methanol (*93*). Removal of the C-12 ester function of 4-deoxyphorbol derivatives requires the use of 1% sodium methoxide over a period of 16 hours. The AB *cis* parent polyol 4α-deoxyphorbol is produced from this reaction in low yield, but can be isolated from the reaction mixture in the form of its more stable triacetate after acetylation with acetic anhydride and pyridine (*89*). Consequently 4-deoxyphorbol itself has not been obtained by hydrolysis of its naturally occurring esters; 4-deoxyphorbol triacetate has, however, been obtained by partial synthesis from 3,12,13,20-tetra-acetyl-3-deoxo-4-deoxy-3(ξ)-hydroxyphorbol (*89*). Spectral and physical data of both 4-deoxyphorbol and 4α-deoxyphorbol triacetates are summarised in Fig. 30.

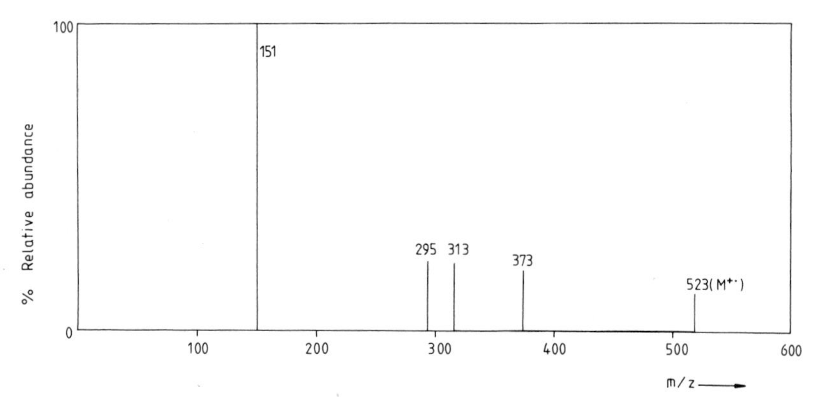


Fig. 31. e.i.-m.s. (upper region) of a 4-deoxyphorbol diester, 12-O-[2-methylaminobenzoyl]-4-deoxyphorbol-13-acetate

In the e.i.-m.s., esters of 4-deoxyphorbol exhibit either an $M^{+\cdot}$-18 fragment ion (*89*) as the ion of highest molecular weight or they show a small $M^{+\cdot}$ ion with 1–4% relative abundance. The occurrence of a molecular ion in spectra of these compounds depends upon the molecular weight of the C-12 acyl function and the inlet temperature of the mass spectrometer. In the c.i.-m.s., however, strong $M^{+\cdot}$ ions together with $M^{+\cdot}+1$ quasi-molecular ions are commonly found (*93*). Fragmentation of 4-deoxyphorbol esters by means of e.i.-m.s. and c.i.-m.s. involves the sequential loss of acyl derivatives from the $M^{+\cdot}$ ion or the $(M^{+\cdot}+1)$ ion respectively, to generate typical fragment ions at m/z 312 or m/z 313, together with the dehydroxylated fragment ion at m/z 294 or m/z 295 (Fig. 31). In a similar manner the mass spectra of 4-deoxyphorbol analogues, such as esters of 4,20-dideoxyphorbol and 4-deoxy-5-hydroxyphorbol, are useful both for identification of the number and molecular weight of acyl substituents and for identification of the tigliane nucleus involved (Fig. 32).

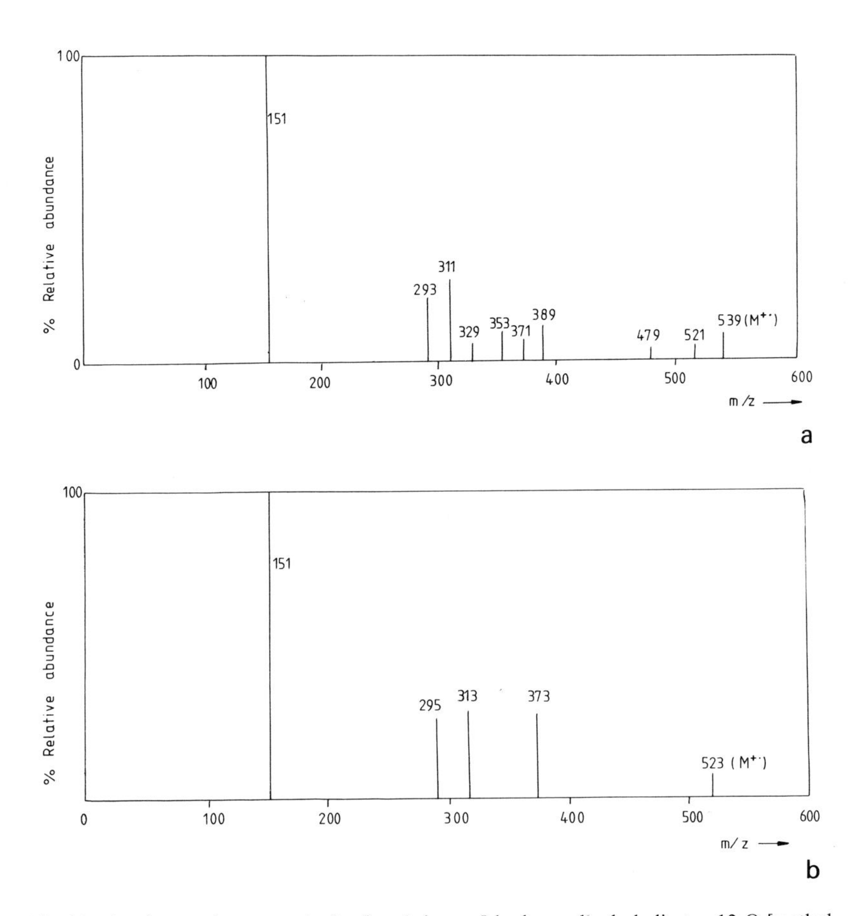


Fig. 32. a) e.i.-m.s. (upper region) of a 4-deoxy-5-hydroxyphorbol diester 12-O-[methylaminobenzoyl]-4-deoxy-5-hydroxyphorbol-13-acetate. b) e.i.-m.s. (upper region) of a 4,20-dideoxy-5-hydroxyphorbol diester, 12-O-[2-methylaminobenzoyl]-4,20-dideoxy-5-hydroxyphorbol-13-acetate

The $^1$H-NMR spectra of esters of 4-deoxyphorbol and 4α-deoxyphorbol are of particular use for structure elucidation of these natural products. The essential and characteristic chemical shifts of the protons from these two isomers are represented in the $^1$H-NMR spectra of 4-

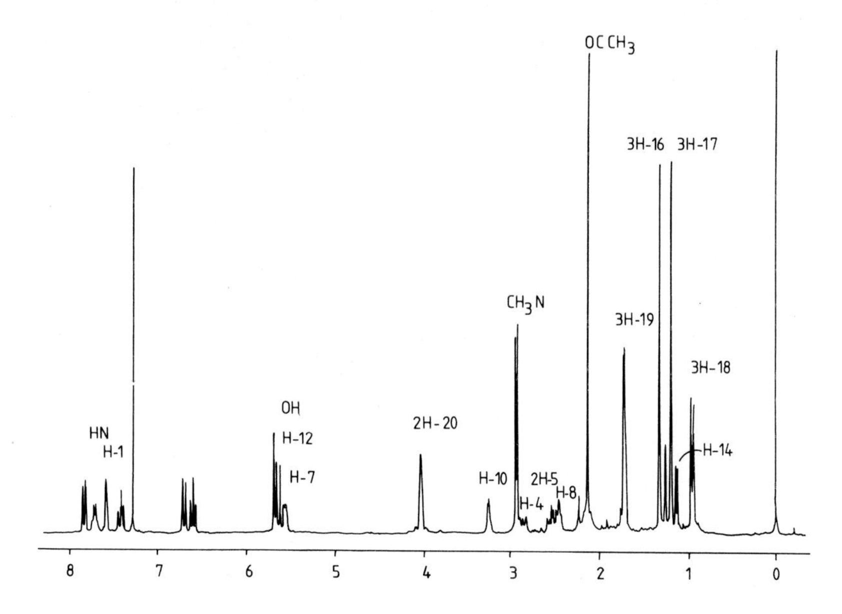


Fig. 33. 250 MHz $^{1}$H-NMR of 12-O-[2-methylaminobenzoyl]-4-deoxyphorbol-13-acetate ($CDCl_3$)

deoxyphorbol and 4α-deoxyphorbol triacetate (Fig. 30). 12,13-Diesters of 4-deoxyphorbol exhibit the characteristic chemical shifts of phorbol diesters (Fig. 33). These spectra are distinct from those of the triesters in that the signals for the allylic protons at C-20 are exhibited upfield at about 4.00 ppm rather than at about 4.44 ppm as is the case for the triesters. The $^{1}$H-NMR spectra of phorbol and 4-deoxyphorbol esters may be distinguished by the occurrence of a one proton multiplet at about 2.85 ppm in the spectra of the latter compounds due to H-4. Irradiation of this multiplet at 2.85 ppm induces the H-10 multiplet at 3.28 ppm to become a broad singlet. There are significant differences in the chemical shifts of certain protons in the $^{1}$H-NMR spectra of 4-deoxyphorbol and 4α-deoxyphorbol esters. These diagnostic differences permit identification of the two isomeric series of esters. The most striking chemical shift differences are those of H-1 found at 7.57 ppm in the $^{1}$H-NMR spectra of the AB-*trans* isomers and at 7.09 ppm in the spectra of the *cis*-analogues, for H-7 shifted from 5.59 to 5.15 ppm and for H-8 shifted from 2.45 to 2.03 ppm. There is also a downfield shift for the H-10 signal, from 3.28 ppm in the *trans* isomers to 3.54 ppm in the *cis* isomers. The characteristic chemical shifts are summarised in Table 5.

*References, pp. 90—99*

Table 5. *Characteristic Changes in the $^{1}$H-NMR Spectra of 12,13-Diesters of 4-Deoxy- and 4α-Deoxyphorbol*

| Protons on carbon number | Chemical shift δ ppm | |
|---|---|---|
| | AB trans | AB cis |
| 1H- 1 | 7.57 s. | 7.09 s. |
| 1H - 7 | 5.59 d. | 5.15 d. |
| 2H - 20 | 4.04 s. | 3.97 ABq. |
| 1H - 10 | 3.28 m. | 3.54 m. |
| 1H- 4 | 2.85 m. | 3.00 m. |
| 2H- 5 | 2.54 m. | 3.46.d.d. and 2.51 d.d. |
| 1H - 8 | 2.45 m. | 2.03 m. |
| 1H- 11 | 2.18 m. | 1.87 d.d. |
| 1H - 14 | 1.13 d. | 0.88 d. |

Other 4-deoxyphorbol analogues can be readily characterised by comparing their $^{1}$H-NMR spectra with that of 4-deoxyphorbol triacetate. The $^{1}$H-NMR spectra of a number of closely related 4-deoxyphorbol derivatives are shown in Fig. 34.

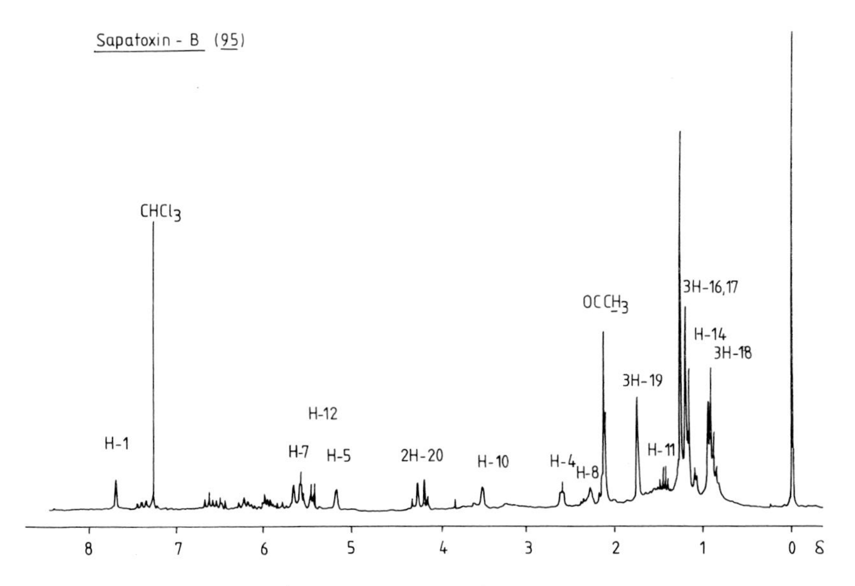


Fig. 34. 250 MHz $^{1}$H-NMR spectrum of 4-deoxyphorbol derivatives

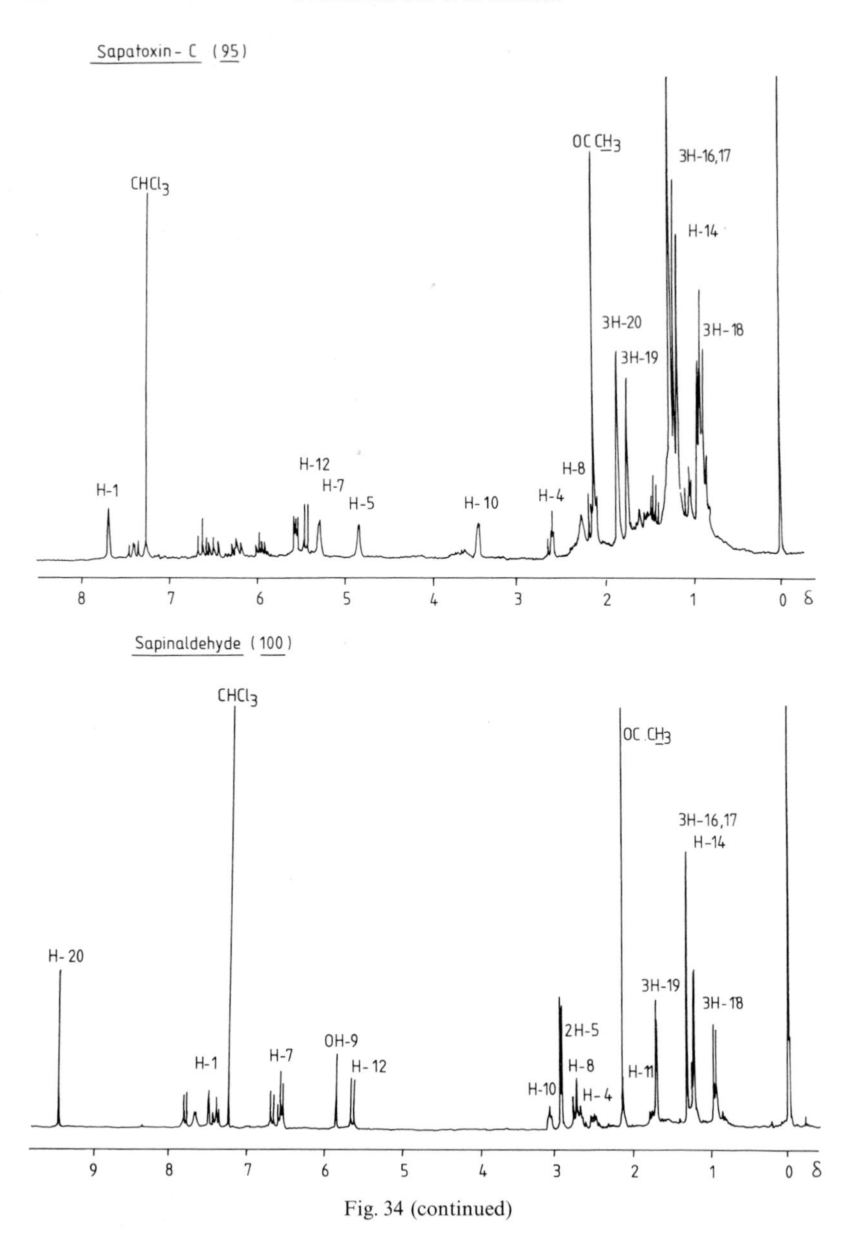


Fig. 34 (continued)

The $^{13}$C-NMR spectra of esters of 4-deoxyphorbol (Fig. 35) may be interpreted by comparing them with the spectrum of phorbol triacetate (*125*). The main difference in the $^{13}$C-NMR spectrum of a 4-deoxyphorbol ester compared with that of a phorbol ester is the upfield shift of the C-4 signal from a singlet at about 73 ppm for phorbol triacetate to a doublet about 43 ppm for a 4-deoxyphorbol derivative. There is also an upfield shift of the C-5 signal, from about 39 ppm for the 4-deoxyphorbol derivative to 30 ppm for phorbol triacetate, but other differences are not really significant. Differences in the spectra of a 4-deoxyphorbol ester and a 4α-deoxyphorbol analogue are due to changes in the molecular shapes. These differences are demonstrated in Fig. 35, which shows the $^{13}$C-NMR spectra of 12-O-[2-methylaminobenzoyl]-4-deoxyphorbol-13,20-diacetate and its

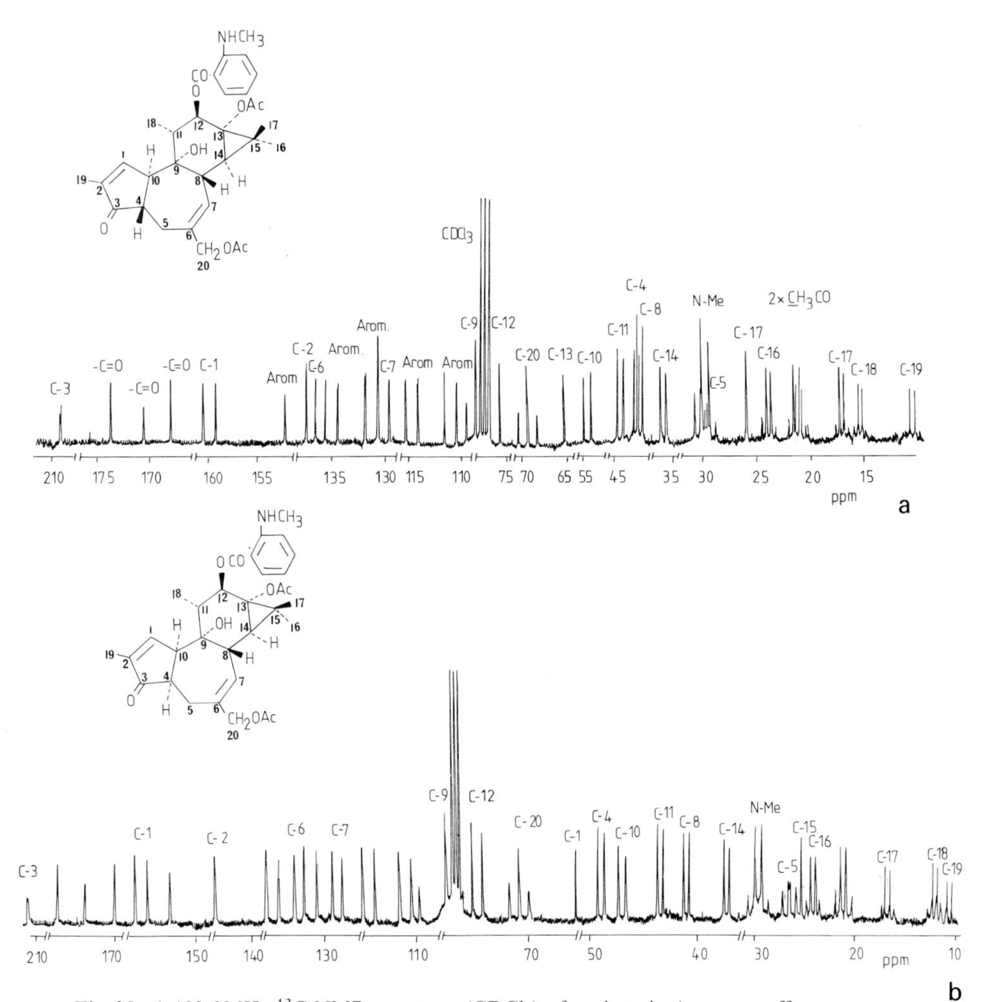


Fig. 35. a) 100.6 MHz $^{13}$C-NMR spectrum ($CDCl_3$) of sapintoxin-A-acetate off resonance decoupled ($CDCl_3$). b) 100.6 MHz $^{13}$C-NMR spectrum of 4α-sapinine-acetate off resonance decoupled ($CDCl_3$)

4α-analogue 12-O-[2-methylaminobenzoyl]-4α-deoxyphorbol-13,20-diacetate. The most notable differences in the spectrum of the 4α-compound compared with that of the 4β-isomer are the downfield shifts of the signals for C-2 (143 ppm from 137 ppm) and C-4 (47 ppm from 42 ppm) and the upfield shifts for the signals due to C-1 (155 ppm from 159 ppm), C-5 (26 ppm from 30 ppm), C-6 (133 ppm from 136 ppm), C-10 (49 ppm from 54 ppm) and C-18 (12 ppm from 15 ppm).

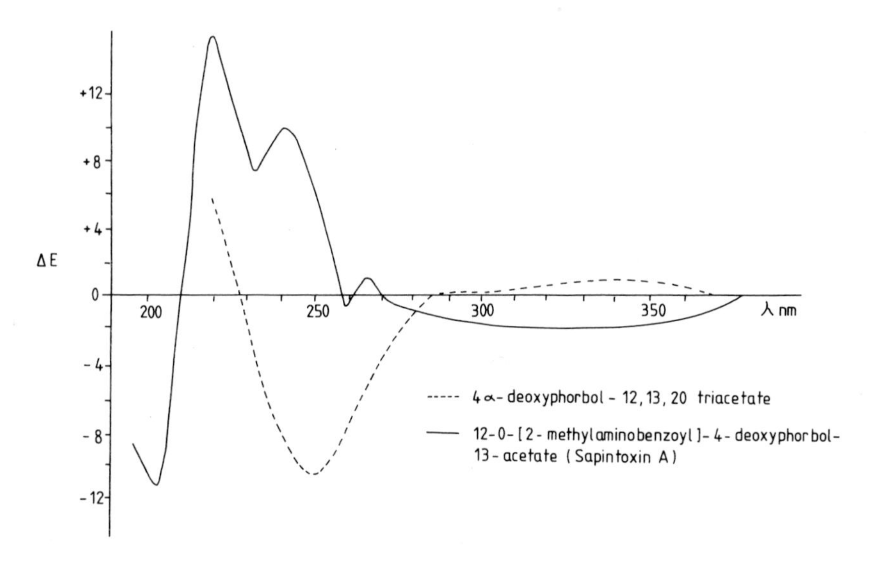


Fig. 36. Circular dichroism spectra of a 4-deoxyphorbol ester and a 4α-deoxyphorbol ester

In the C.D. spectra the AB *trans* and AB *cis* isomers of 4-deoxyphorbol are readily distinguished. Diesters of 4-deoxyphorbol like those of phorbol (Fig. 26) exhibit a pronounced positive Cotton effect between 230 and 240 nm and a negative Cotton effect at about 300 to 330 nm due to the α,β-unsaturated 3-oxo function. In the case of 4α-deoxyphorbol diesters a large negative extremum is shown at 250 nm. This difference in the sign of the Cotton effect in the region of 240 to 250 nm has been ascribed (*34*) to the difference in the fusion of the AB ring systems; and the effect is more marked in the 4α-deoxyphorbol esters than in esters of 4α-phorbol. The 4α-deoxyphorbol diesters also exhibit positive Cotton effects between 350 and 380 nm and at 320 nm, whereas the AB-*trans* derivatives show a negative extremum in this region of the spectrum (Fig. 36).

*References, pp. 90—99*

Esters of 12-deoxyphorbol

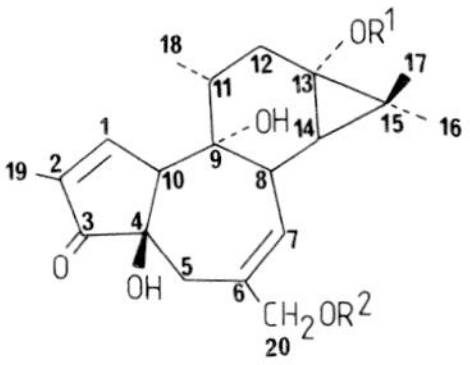

| | $R^1$ | $R^2$ | Reference |
|---|---|---|---|
| Euphorbia triangularis Desf. | | | |
| | $CO.CH(CH_3)CH_2CH_3$ | H | |
| | $CO.CH(CH_3)CH_3$ | H | |
| | $CO.C(CH_3){=}CH(CH_3)$ | H | |
| | $CO.CH(CH_3)CH_2CH_3$ | $COCH_3$ | |
| | $COCH(CH_3)CH_3$ | $COCH_3$ | |
| | $COCH(CH_3)CH_2CH_3$ | $COCH_3$ | (101, 102, 103) |
| Euphorbia resinifera Berg. | | | |
| | $CO.C(CH_3){=}C(CH_3)CH_3$ | $COCH_3$ | |
| | $CO.CH(CH_3)CH_3$ | $COCH_3$ | |
| | $COCH_2C_6H_5$ | $COCH_3$ | (64) |

| | $R^1$ | $R^2$ | Reference |
|---|---|---|---|
| Euphorbia fortissima Leach. | | | |
| | $CO.C(CH_3){=}CH(CH_3)$ | H | |
| | $COCH(CH_3)CH_3$ | $CO.CH_3$ | |
| | $CO.C(CH_3){=}CH(CH_3)$ | $CO.CH_3$ | |
| | $CO.CH(CH_3)CH_2CH_3$ | $CO.CH_3$ | |
| | $CO(CH_2)_{10}CH_3$ | H | |
| | $CO.C_{11}H_{21}$ | H | |
| | $CO.C_{11}H_{21}$ | $CO.CH_3$ | (37, 110) |

Fig. 37

| | $R^1$ | $R^2$ | Reference |
|---|---|---|---|
| *Euphorbia coerulescens* Haw. | | | |
| | $CO.\underset{CH_3}{CH}CH_2CH_3$ | H | |
| | $CO.\underset{CH_3}{C}=C{<}^{H}_{CH_3}$ | $CO.CH_3$ | |
| | $CO.\underset{CH_3}{CH}CH_2CH_3$ | $CO.CH_3$ | |
| | $CO.C_6H_{13}$ | H | |
| | $CO.(CH_2)_{10}CH_3$ | H | |
| | $CO.(CH_2)_{10}CH_3$ | $CO.CH_3$ | (36, 109, 110) |
| *Euphorbia polyacantha* Boiss. | | | |
| | $CO.C_7H_{13}$ | $CO.CH_3$ | |
| | $CO\ C_9H_{15}$ | $CO.CH_3$ | (109, 110) |
| *Euphorbia poissonii* Pax. | | | |
| | $CO.CH_2C_6H_5(OH)$ | H | |
| | $CO\ CH_2C_6H_5$ | H | |
| | $CO.\underset{CH_3}{CH}CH_3$ | H | |
| *Euphorbia poissonii* Pax (contd.) | | | |
| | $CO.\underset{CH_3}{CH}CH_2CH_3$ | H | |
| | $CO.\underset{CH_3}{C}=C{<}^{CH_3}_{H}$ | H | |
| | $CO.CH_2C_6H_5\ OH$ | $CO.CH_3$ | |
| | $CO.CH_2C_6H_5$ | $CO.CH_3$ | |
| | $CO.\underset{CH_3}{CH}CH_3$ | $CO.CH_3$ | |
| | $CO.\underset{CH_3}{C}=C{<}^{CH_3}_{H}$ | $CO.CH_3$ | |
| | $CO.\underset{CH_3}{CH}CH_2CH_3$ | $CO.CH_3$ | (33, 111, 112, 113, 114) |
| *Euphorbia helioscopia* L. | | | |
| | $CO.\underset{CH_3}{C}=C{<}^{H}_{CH_3}$ | H | |
| | $CO.CH_2C_6H_5$ | $CO.CH_3$ | |
| | $CO.\underset{CH_3}{C}=C{<}^{H}_{CH_3}$ | $CO\ CH_3$ | |
| | $CO.C_{11}H_{19}$ | $COCH_3$ | (115) |

Fig. 37 (continued)

*References, pp. 90—99*

| | $R^1$ | $R^2$ | Reference |
|---|---|---|---|
| Euphorbia unispina N.E. Br. | | | |
| | $CO.CH_2C_6H_5(OH)$ | $COCH_3$ | |
| | $CO.CH_2C_6H_5$ | $COCH_3$ | |
| | $CO.C(CH_3)=C(CH_3)H$ | $COCH_3$ | |
| | $CO.CH(CH_3)CH_2CH_3$ | $COCH_3$ | |
| | $CO.CH_2C_6H_5$ | H | |
| | $CO.C(CH_3)=C(CH_3)H$ | H | |
| | $CO.CH(CH_3)CH_2CH_3$ | H | (111) |
| Euphorbia balsimifera | | | |
| | $CO.C_{11}H_{19}$ | $CO.C_7H_{11}$ | (105) |
| Baliospermum montanum | | | |
| | $CO.(CH_2)_{14}CH_3$ | H | (116) |
| Pimelea prostrata | $COCH_3$ | H | (106, 107, 108, 118) |

Fig. 37 (continued)

## VII.7. 12-Deoxyphorbol Esters

12-Deoxyphorbol is the parent diterpene of a large number of pro-inflammatory esters from the genus *Euphorbia*. Esters of this tigliane derivative were initially isolated from *Euphorbia triangularis* (*101, 102, 103*) where they occurred as a series of C-13 monoesters (Fig. 37) and a corresponding series of 13,20-diesters. These compounds are more widely distributed in the genus *Euphorbia* than either phorbol or 4-deoxyphorbol esters (*104*). Routinely the C-20 acyl function of such 12-deoxyphorbol diesters has been identified as acetate, with the possible exception of an ester from the latex of *Euphorbia balsamifera* (*105*). The parent diterpene, 12-deoxyphorbol, is more unstable than phorbol but can be isolated as its stable diacetate after alkaline hydrolysis followed by acetylation (*101*). The C-13 acyl group of 12-deoxyphorbol may be aliphatic in nature varying in chain length from butyric to dodecanoic acid (Fig. 37). A small number of aromatic 12-deoxyphorbol esters have also been isolated from plants of the genus *Euphorbia*. The novel compound 12-deoxyphorbol-13-acetate, known as prostratin, was recently isolated from *Pimelea prostrata* of the family Thymelaeaceae (*106*). This is the only ester of 12-deoxyphorbol retaining a $\Delta^{6,7}$ bond to have been obtained from this plant family to date. Species from the genus *Euphorbia* of the family Euphorbiaceae which have been shown to contain esters of 12-deoxyphorbol include *Euphorbia resinifera* (*64*), *E. fortissima* (*37, 110*), *E. coerulescens* (*36, 109, 110*), *E. polyacantha* (*109, 110*), *E. poissonii* (*33, 111*), *E. helioscopia* (*115*), *E. unispina* (*111*) and *E. balsamifera* (*105*). An ester of 12-deoxyphorbol has been isolated from *Baliospermum montanum* also of the family Euphorbiaceae (*116*).

## VII.8. Identification of 12-Deoxyphorbol Esters

Esters of 12-deoxyphorbol obtained from natural sources can be characterised as 12-deoxyphorbol analogues after complete hydrolysis and subsequent acetylation of the parent alcohol. The spectral and physical data of 12-deoxyphorbol diacetate are summarised in Fig. 38.

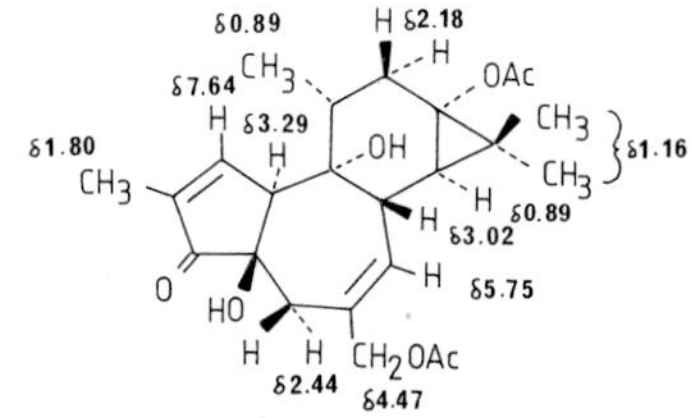

Empirical formula : $C_{24}H_{32}O_7$ , M.P. 138°C (101)

M.S. $M^{+\cdot}$, m/z 432

U.V. $\lambda_{max}^{MeOH}$, nm., (ε), 196 (12300), 235 (5200), 334 (65)

I.R. $\nu_{max}^{KBr}$ cm$^{-1}$, 3400, 1715, 1700, 1628

C.D. $\lambda_{max}^{Dioxane}$ nm., (Δε), 276 (-0.194), 334 (-0.583), 343 (-0.616)

$^1$H-NMR ($CDCl_3$ 60 MHz) assigned according to (109)

Fig. 38. Structure and spectral data of 12-deoxyphorbol-diacetate

In the mass spectra of 12-deoxyphorbol esters the ions above the fragment m/z 294 are useful for the identification of acyl substituents. Below the ion m/z 294 the spectra of all 12-deoxyphorbol esters are similar.

Both 12-deoxyphorbol monoesters and diesters can be characterised by their fragmentation patterns (*42*). In the spectra of the monoesters of 12-deoxyphorbol, the fragment ions at m/z 330 and m/z 312 are typical and are derived from the molecular ion by loss of the C-13 acyl group followed by elimination of a hydroxyl group as water (Fig. 39). The diesters of 12-deoxyphorbol exhibit fragment ions in their mass spectra due to the loss of both the C-13 and the C-20 acyl substituents. The ion at m/z 312 ($C_{20}H_{24}O_3$) is prominent and corresponds to the deacylated tigliane nucleus (Fig. 39) (*42*).

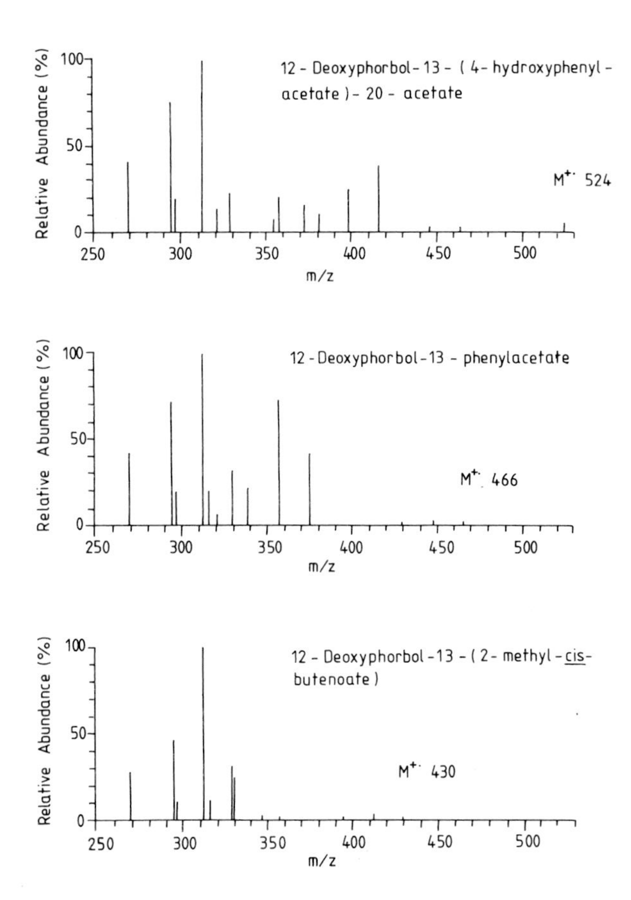


Fig. 39. Low resolution electron impact mass spectra (high mass regions) of 12-deoxyphorbol esters (after *120*)

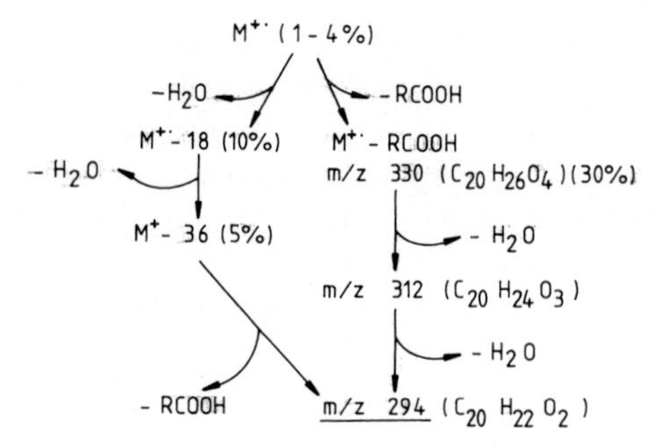


Fragmentation of 12-deoxyphorbol - monoesters by EI-MS (42)

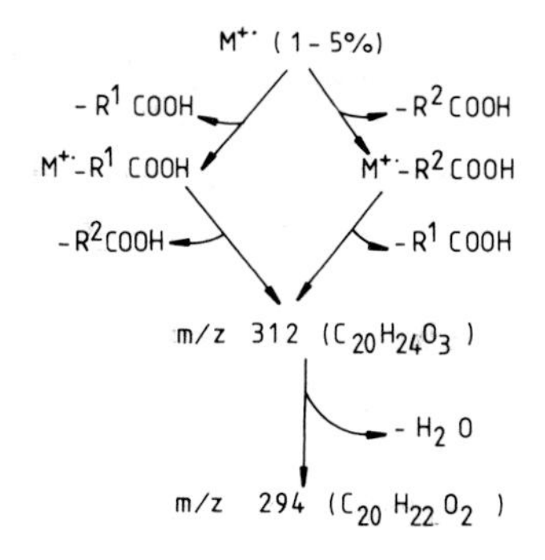


Fragmentation of 12-deoxyphorbol-diesters by EI-MS (42)

Fig. 39 (continued)

The $^{1}$H-NMR spectra of esters of 12-deoxyphorbol, like those of 4-deoxyphorbol, are reminiscent of the spectra of phorbol esters (Fig. 40). The major difference between 12-deoxyphorbol and phorbol ester $^{1}$H-NMR spectra is the fact that the two proton H-12 signal is exhibited in the former as a multiplet at about 2.0 ppm, whereas the 12-O-acyl phorbol esters display it at about 5.63 ppm. Furthermore, 13,20-diesters of 12-deoxyphorbol exhibit the H-20 signal at about 4.50 ppm compared with about 4.00 ppm in the spectra of C-13 monoesters (*114*).

The $^{13}$C-NMR spectrum of an ester of 12-deoxyphorbol is shown in Fig. 41. The most obvious change compared with the spectrum of phorbol triacetate (*125*) is the upfield shift and change in multiplicity of the C-12 signal from a doublet at 77 ppm for phorbol triacetate to a triplet at 32 ppm for the 12-deoxy compound. The signal for C-11 also appears further upfield, at 32 ppm for the 12-deoxyphorbol derivative compared with 43 ppm for the phorbol ester. Other differences in the two spectra are minor.

*References, pp. 90—99*

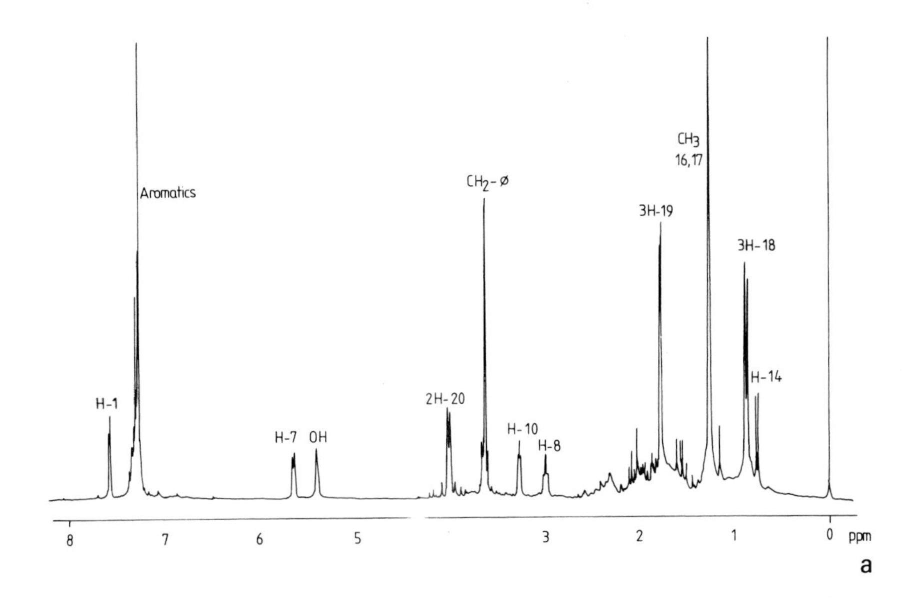


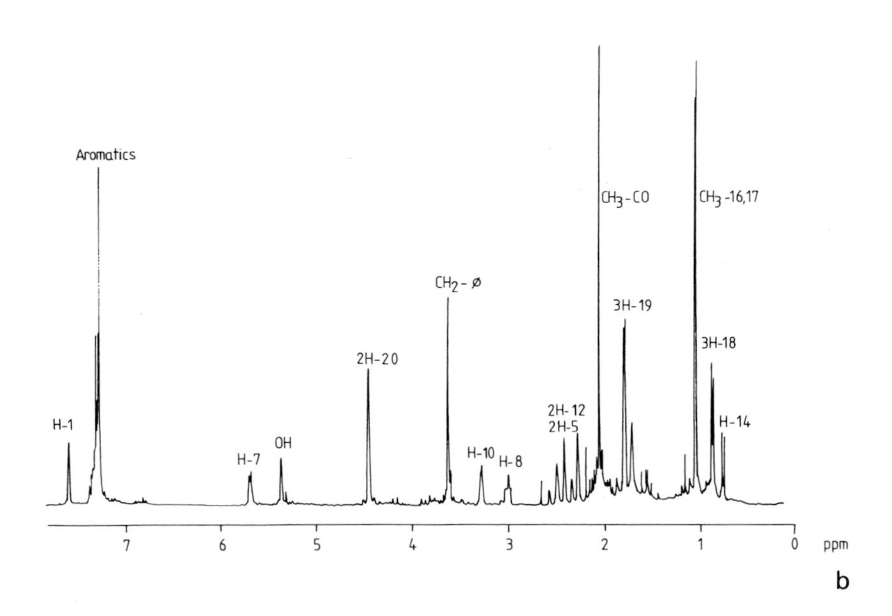


Fig. 40. a) 250 MHz $^{1}$H-NMR spectrum of 12-deoxyphorbol-13-phenylacetate. b) 250 MHz $^{1}$H-NMR spectrum of 12-deoxyphorbol-13-phenylacetate-20-acetate ($CDCl_3$)

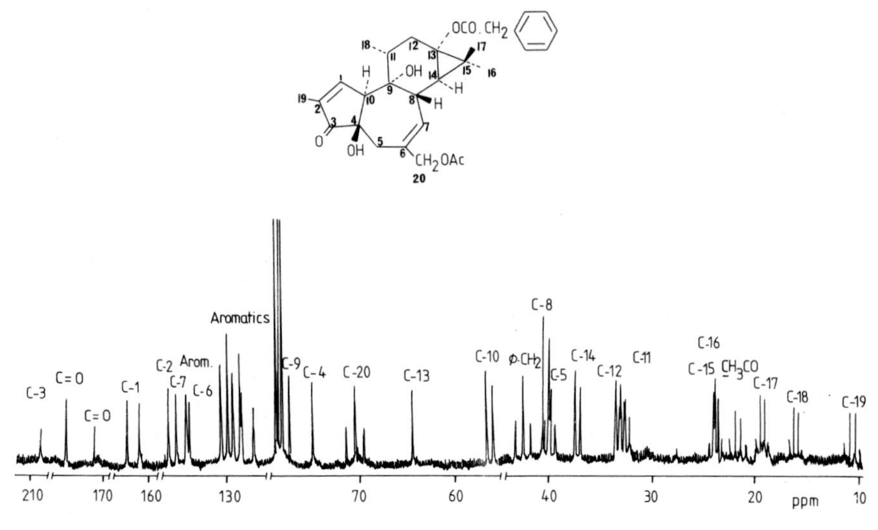


Fig. 41. 100.6 MHz $^{13}$C-NMR spectrum of 12-deoxyphorbol-13-phenylacetate-20-acetate off resonance decoupled ($CDCl_3$)

### VII.9. 12-Deoxy-16-Hydroxyphorbol Esters

Irritant diterpene esters based upon 12-deoxy-16-hydroxyphorbol were first isolated from the latex of *Euphorbia cooperii* (*102, 117, 118*). These compounds differ from other 12-deoxyphorbol diterpenes found in the genus *Euphorbia* in that an extra primary hydroxy group is present at C-16. The C-13 ester group of the toxins from *Euphorbia cooperii* was finally assigned as angelate (*118*) whereas the C-16 acyl function was isobutyrate, and the C-20 hydroxy group was either free or acetylated. At a later date (*119*) two further 12-deoxy-16-hydroxyphorbol esters were isolated from the latex of *E. poissonii*. In these esters the C-13 acyl group was identified as phenylacetate (Fig. 42).

The mass spectra of 12-deoxy-16-hydroxyphorbol esters are similar to those of phorbol esters in that they exhibit a sequential loss of acyl derivatives to form a fragment ion at m/z 328 which represents the deacylated tigliane nucleus (Fig. 43).

The $^{1}$H-NMR spectra of 12-deoxy-16-hydroxyphorbol esters are characterised by the presence of an AB quartet at about 4.15 ppm, which is due to the two allylic protons on C-16 (Fig. 44).

Esters of 12-deoxy-16-hydroxyphorbol are unstable on exposure to alkali and readily undergo intramolecular translocation leading to the formation of crotophorbolone (*118*).

*References, pp. 90—99*

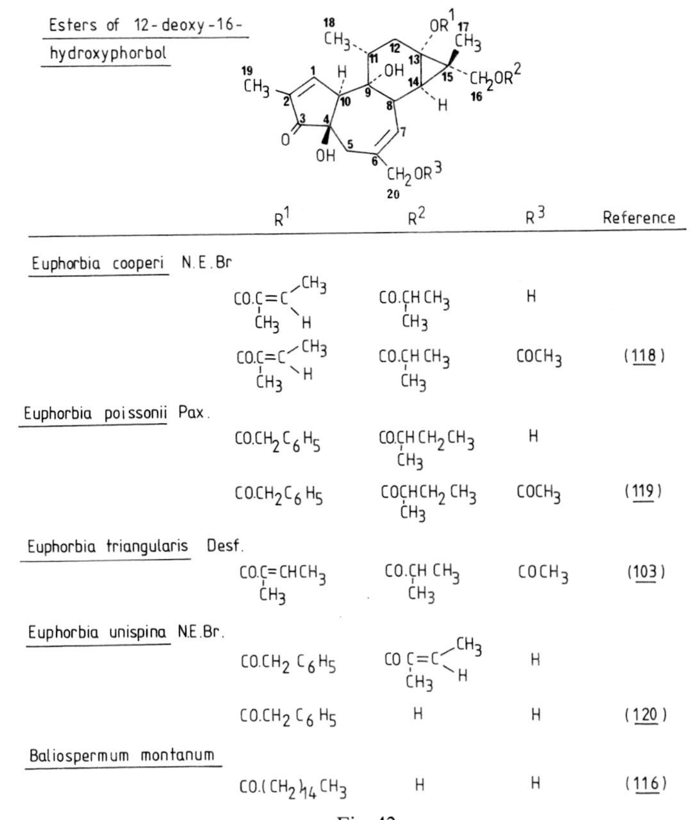


| | $R^1$ | $R^2$ | $R^3$ | Reference |
|---|---|---|---|---|
| Euphorbia cooperi N.E.Br | CO.C(CH3)=CH.CH3 | CO.CH(CH3)CH3 | H | |
| | CO.C(CH3)=CH.CH3 | CO.CH(CH3)CH3 | COCH3 | (118) |
| Euphorbia poissonii Pax. | CO.CH2C6H5 | CO.CH(CH3)CH2CH3 | H | |
| | CO.CH2C6H5 | CO.CH(CH3)CH2CH3 | COCH3 | (119) |
| Euphorbia triangularis Desf. | CO.C(CH3)=CHCH3 | CO.CH(CH3)CH3 | COCH3 | (103) |
| Euphorbia unispina N.E.Br. | CO.CH2C6H5 | CO.C(CH3)=CH.CH3 | H | |
| | CO.CH2C6H5 | H | H | (120) |
| Baliospermum montanum | CO.(CH2)14CH3 | H | H | (116) |

Fig. 42

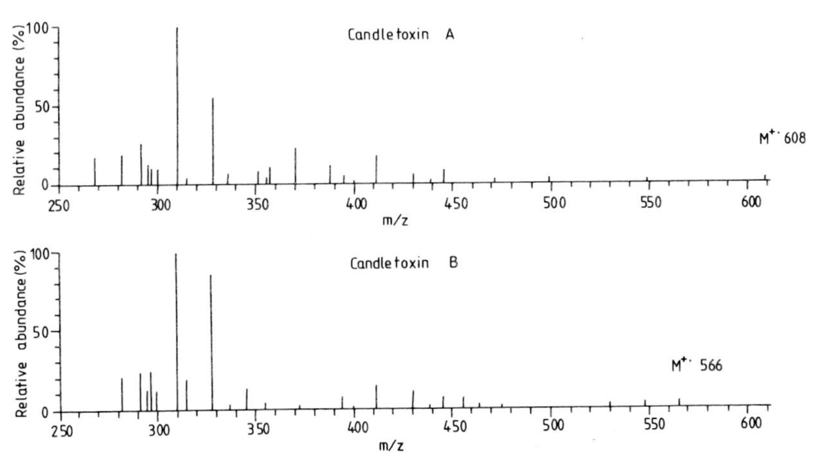


Fig. 43. Low resolution electron impact mass spectra (high mass regions) of candletoxins A and B (after *120*)

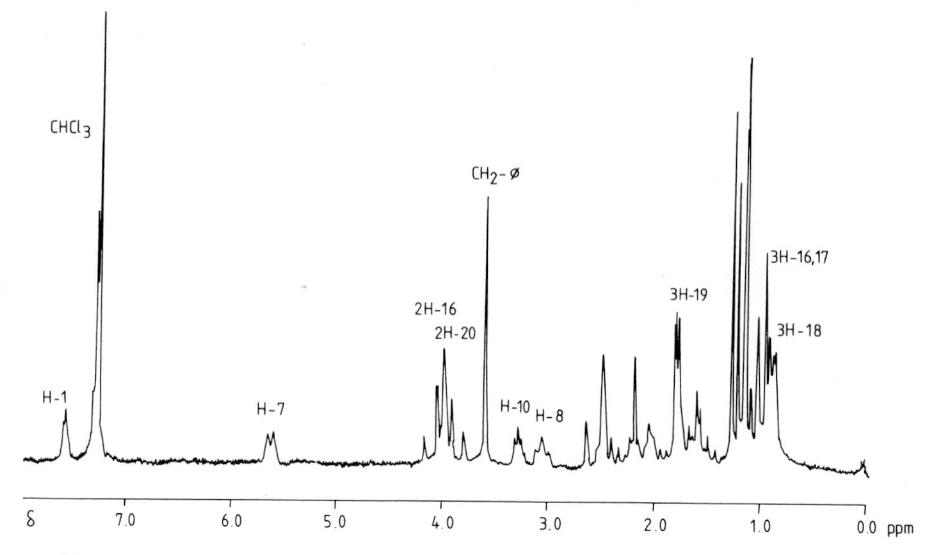


Fig. 44. 100 MHz PMR spectrum of 12-deoxy-16-hydroxyphorbol-13-phenylacetate-16-(2-methylbutanoate) (candletoxin B) in $(CDCl_3) + D_2O$

## VII.10. Other 12-Deoxyphorbol Derivatives

Two further derivatives of 12-deoxyphorbol have been isolated from natural sources. These compounds were esters of 12,20-dideoxyphorbol and 12-deoxy-5-hydroxyphorbol. Two C-13 monoesters of 12,20-dideoxyphorbol were isolated from *Euphorbia resinifera* (*12*), the C-13 ester groups being identified as isobutyrate and angelate. From extracts of *Baliospermum montanum* a C-13 monoester of 12-deoxy-5-hydroxyphorbol (*116*) was later isolated, and a similar ester was found in the roots of *Stillingia sylvativa* (*121*). Both of these latter species are from the plant family Euphorbiaceae (Fig. 45).

An unusual tigliane ester was isolated in 1975 from extracts of *Hippomane mancinella* (Euphorbiaceae) (*122*). This compound, mancinellin, was based upon the 12-deoxyphorbol skeleton, but had a β-orientated secondary hydroxyl group on C-5 and a 6,7-epoxy function in place of the $\Delta^{6,7}$ double bond typically found in the naturally occurring tigliane esters. It was the first of several 5-hydroxy-6,7-epoxy derivatives to be isolated, and contained an unsaturated aliphatic ester function at C-13 of the tigliane nucleus (Fig. 46). Recently a similar ester, baliospermin (*124*), was isolated from *Baliospermum montanum*, but in this case the C-13 ester function was a saturated aliphatic acid. A higher molecular weight analogue of baliosper-

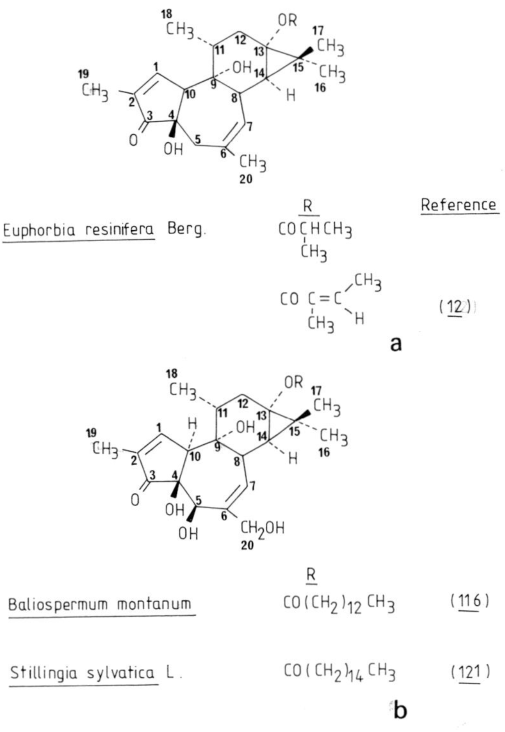


Fig. 45. a) Esters of 12,20-dideoxyphorbol. b) Esters of 12-deoxy-5-hydroxyphorbol

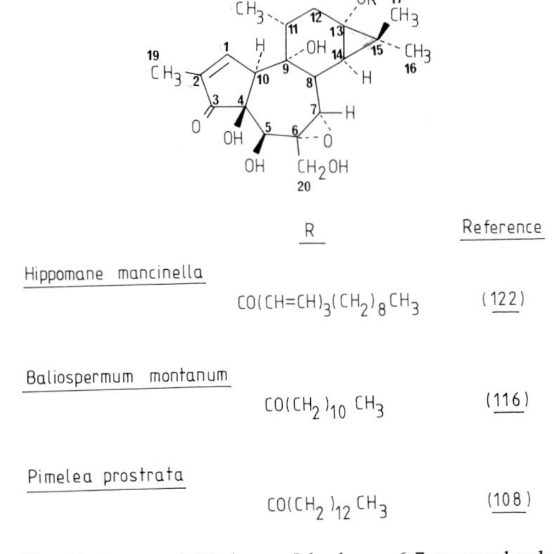


Fig. 46. Esters of 12-deoxy-5-hydroxy-6,7-epoxyphorbol

min, known as Pimelea factor P5, was isolated from extracts of *Pimelea prostrata* of the plant family Thymelaeaceae (*108*). The spectroscopic and physical data of mancinellin and baliospermin are summarised in Fig. 47.

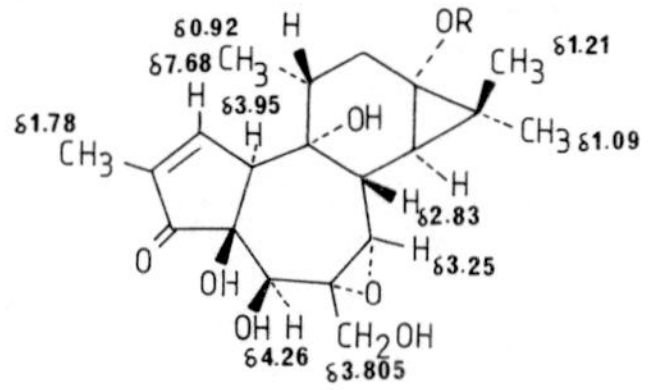

Mancinellin (R = $CO(CH{=}CH)_3(CH_2)_8CH_3$)

M.S. m/z 612 ($M^{+\cdot}$), base peak m/z 233

U.V $\lambda_{max}^{MeOH}$, n.m. ($\varepsilon$), 194 (12000), 253 (12000), 306.5 (27000)

I.R. $\nu_{max}^{CH_2Cl_2}$ $cm^{-1}$, 3520, 3380, 1685, 1610

$^1$H-NMR, assigned according to (122)

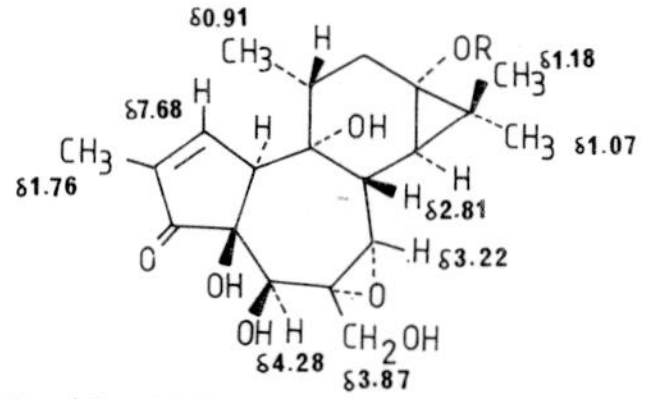

Baliospermin (R = $CO(CH_2)_{10}CH_3$)

U.V. $\lambda_{max}^{MeOH}$ n.m. ($\varepsilon$), 246 (6410)

I.R. $\nu_{max}^{NaCl}$, $cm^{-1}$, 3400, 1730, 1700, 1630, 750

$^1$H-NMR, assigned according to (116)

Fig. 47

## VII.11. 16-Hydroxyphorbol Esters

16-Hydroxyphorbol is a highly oxygenated tigliane derivative which was originally isolated in esterified form from *Aleurites fordii* (Euphorbiaceae) (*123, 124*). Two compounds were obtained from this plant, the first being a 12,13-diester of 16-hydroxyphorbol which contained a long chain acyl derivative at C-12 and an acetate at C-13, whilst the second compound was a monoester containing a C-13 acetate group only (Fig. 48). Recently, two similar 12,13-diesters of 16-hydroxyphorbol were isolated from *Croton flavens* (*97, 98*), together with two novel triesters in which a decanoate acyl group was present at C-20 of the tigliane nucleus.

The spectral and physical data of a 12,13-diester and a 13-monoester of 16-hydroxyphorbol are summarised in Fig. 49.

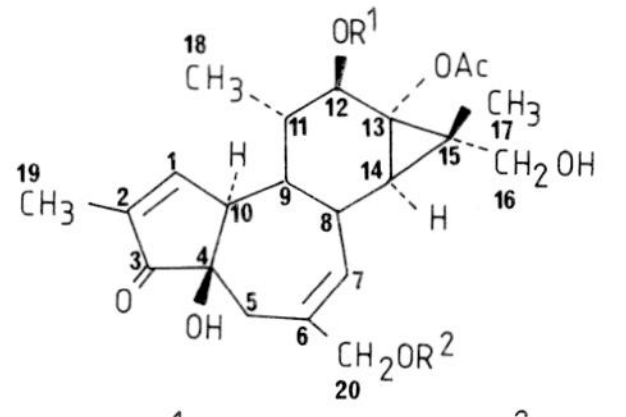

| | $R^1$ | $R^2$ | Reference |
|---|---|---|---|
| *Aleurites fordii* Hemsl. | | | |
| | $CO(CH_2)_{14}CH_3$ | H | |
| | H | H | (123, 124) |
| *Croton flavens* L. | | | |
| | $CO(CH_2)_{14}CH_3$ | H | |
| | $CO(CH_2)_{14}CH_3$ | $CO(CH_2)_8CH_3$ | |
| | $CO(CH_2)_{12}CH_3$ | H | |
| | $CO(CH_2)_{12}CH_3$ | $CO(CH_2)_8CH_3$ | (97, 98) |

Fig. 48. Esters of 16-hydroxyphorbol

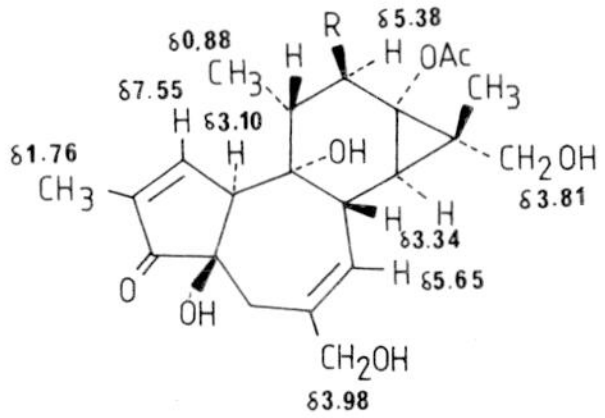

Croton factor $F_1$ from *Croton flavens* & *Aleurites fordii* m. p. 177–178° C

M.S. m/z 660 ($M^{+\cdot}$) R = Palmitate

U.V $\lambda_{max}^{MeOH}$, n m. (ε), 196 (12410), 231 (5000), 334 (130)

I.R. $\nu_{max}^{KBr}$ $cm^{-1}$, 3408, 3340, 1750, 1730, 1715, 1638

$^1H$-NMR ($CDCl_3$) assigned according to (98)

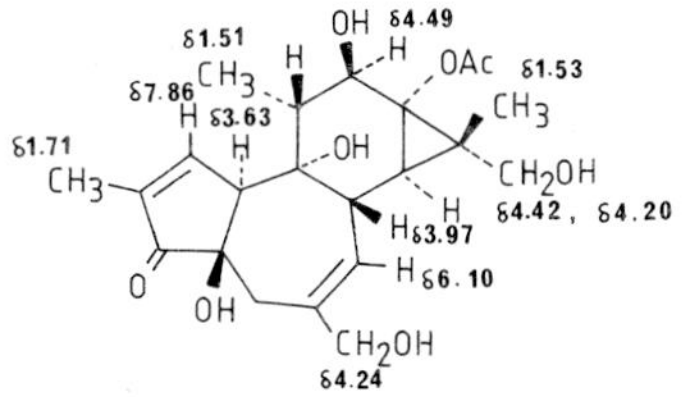

13-acetyl-16-hydroxyphorbol from *Aleurites fordii* m.p. 278–282 °C

M. S. m/z 422 ($M^{+\cdot}$, $C_{22}H_{30}O_8$)

U.V $\lambda_{max}^{MeOH}$, n.m. (log ε), 232 (3.70)

I.R. $\nu_{max}^{KBr}$ $cm^{-1}$, 3550, 3500, 3400, 3250, 1705, 1695, 1630

$^1H$-NMR ($C_5D_5N$) assigned according to (123)

Fig. 49

## VIII. Ingenane Derivatives

### VIII.1. Ingenol

In 1968, HECKER's group isolated an irritant principle from *Euphorbia lathyris* (*126*) which was identical with factor $I_1$ later isolated from *E. ingens* (*127*). Using X-ray crystallographic techniques, the triacetate of the parent alcohol was identified as ingenol triacetate (*128*). The parent alcohol, ingenol (Fig. 50), is a tetracyclic diterpene having many structural features in common with phorbol. It differs from phorbol in that ring C is seven-membered, C-8 being linked to C-10 by means of a keto bridge. Ingenol, like phorbol has a $\Delta^{1-2}$ bond in ring A, a $\Delta^{6-7}$ bond in ring B, a C-20 primary alcoholic group, a 4β-hydroxyl group fixing rings A and B in the *trans* configuration and a cyclopropane ring, D. The C-12 and C-13 OH groups are missing in ingenol, but secondary hydroxyl groups are present at C-3 and C-5 in rings A and B respectively.

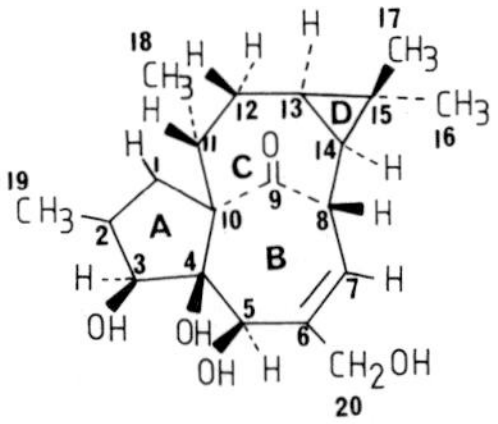

Fig. 50. Structure of ingenol

Ingenol triacetate has been isolated as a semi-synthetic product from several species of Euphorbiaceae following hydrolysis and acetylation of the irritant principles. Such species include *Euphorbia desmondi* (*129*), *E. deightonii* (*130*), *E. kamerunicà* (*130*), *E. seguieriana* (*131*), *Elaeophorbia grandifolia* (*130*) and *Elaeophorbia drupifera* (*130*). The structure and spectral characteristics of ingenol triacetate are shown in Fig. 51.

The $^1$H-NMR spectrum (Fig. 52) reveals the structural similarity of ingenol triacetate to phorbol triacetate (Fig. 23). The $\Delta^{1-2}$ and $\Delta^{6-7}$ bonds are common to both, and the olefinic protons are exhibited at 6.09 ppm and 6.25 ppm respectively. The main points of difference in the $^1$H-NMR spectrum of ingenol triacetate compared with that of phorbol triacetate are the absence in the former of the signal for H-10, and the presence at C-3 and C-5 of two secondary acetates, their geminal protons being exhibited at 4.96 ppm and 5.39 ppm respectively (*127, 136*). The environments of these

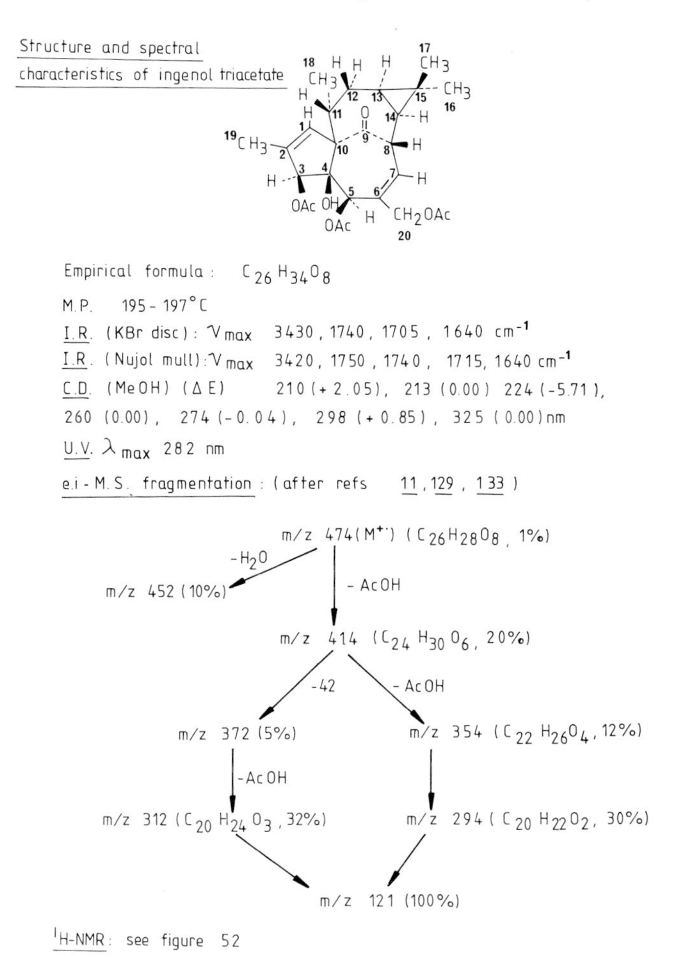


Fig. 51. Structure and spectral characteristics of ingenol triacetate

protons at C-3 and C-5 are very similar, and it is difficult to assign with certainty the acyl moiety of a monoester to either C-5 or C-3 by reference to the $^1$H-NMR spectrum alone. However, it is possible to differentiate tentatively between 3-O-acyl- and 5-O-acylingenol monoesters by comparing the chemical shift of the geminal proton with the shifts reported by Opferkuch and Hecker (*127*) for ingenol triacetate and 16-hydroxyingenol tetraacetate. Hirota *et al.* (*132*) assign the signals for H-3 and H-5 on the basis of their appearance. They point out that the signal for H-3 is a sharp singlet and the signal for H-5 a broad singlet (*9, 127, 128*). The C-20 protons geminal to the primary acetate are observed in the $^1$H-NMR

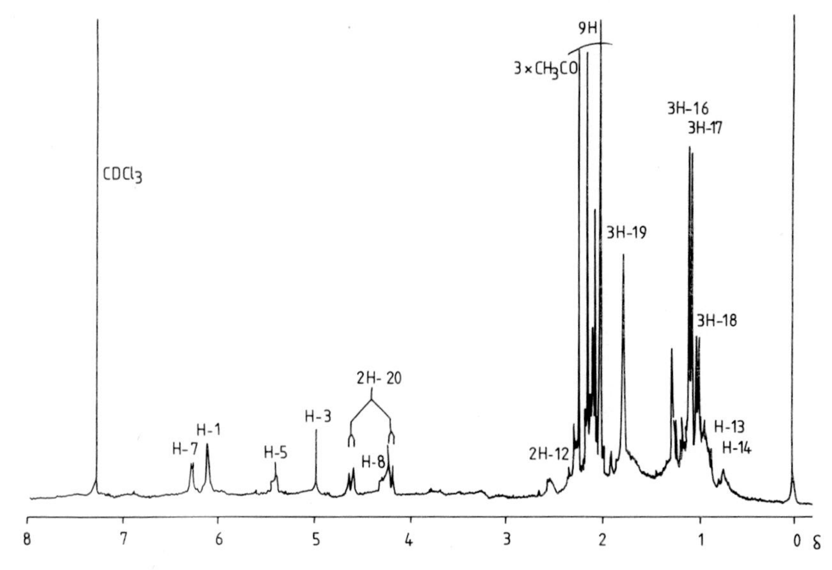


Fig. 52. 250 MHz $^1$H-NMR spectrum of ingenol-3,5,20-triacetate

spectrum of ingenol triacetate as an AB quartet at 4.58 ppm and 4.15 ppm in contrast to phorbol triacetate, where they occur at 4.43 ppm as a singlet.

The mass spectrum of ingenol triacetate is typical of the ingenane group of diterpenes (*11, 129, 133*). The molecular ion is weak (about 1%) and fragmentation by electron impact involves the sequential loss of the acyl group as acetic acid. The deacylated ingenane nucleus manifests itself at m/z 294 ($C_{20}H_{22}O_2$) as an intense fragment ion (~30% relative abundance). Another diagnostic fragment ion found at m/z 372 arises by the loss of 42 mass units from the molecular ion, possibly by loss of ketene (*134*). The base peak which occurs at m/z 121, and has the composition $C_8H_9O$, differentiates ingenol triacetate from the phorbol acetates and may represent the stable ion shown in Fig. 53; the major fragments of lower mass number at m/z 81, m/z 83 and m/z 97 are predominantly hydrocarbon.

In the infra-red spectrum of ingenol triacetate, there is some difference in the frequency of absorption maxima, depending on whether the compound is examined in KBr or in a mull, but characteristic absorbances are observed for the tertiary OH group at C-4 (3420 – 3430 $cm^{-1}$), ester functions (1740 – 1750 $cm^{-1}$), the β,γ-unsaturated ketone at C-9 (1705 – 1715 $cm^{-1}$) and the olefinic region (1640 $cm^{-1}$). The circular dichroism spectrum of ingenol triacetate shows a characteristic positive Cotton effect at about 300 nm due to the carbonyl bridge between rings B and C, which differentiates it from tigliane-type compounds. The CD spectrum is shown in Fig. 54.

*References, pp. 90—99*

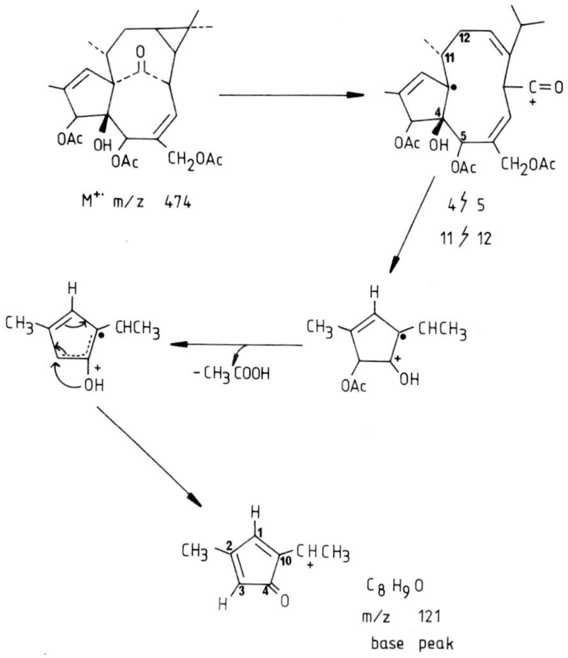

Fig. 53. Base peak fragment in ingenol triacetate (after *105*)

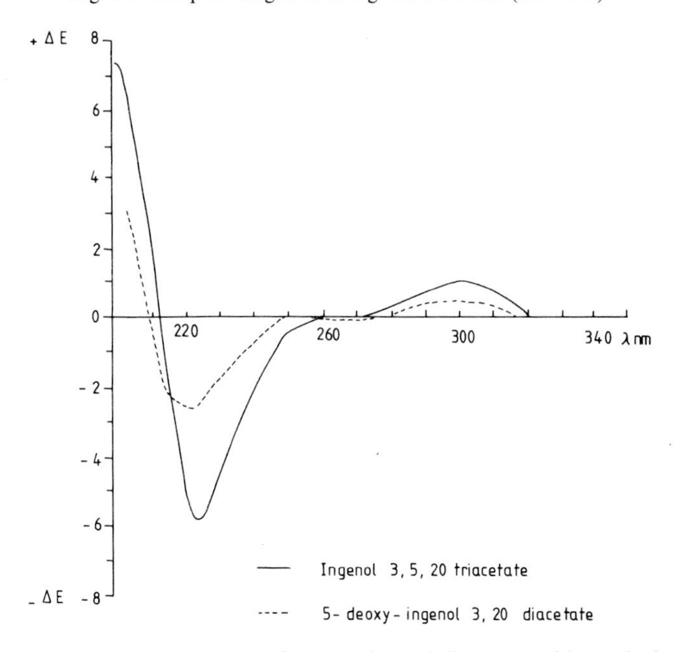


Fig. 54. Circular dichroism spectra of 5-deoxy-ingenol diacetate and ingenol triacetate

Table 6. *Diterpenes of Euphorbia species*

| Genus *Euphorbia* L. section | Subsection and series | Species[1] | Geographical origin | Diterpenes present |
|---|---|---|---|---|
| I. *Anisophyllum* (Haw.) Roeper | 3. *Hypericifoliae* Boiss. | *E. hirta* L. | America | None |
| III. *Poinsettia* (Graham) Boiss | | *E. geniculata* Orteg. | India | None |
| | | *E. pulcherrima* Willd. | America | None |
| VII. *Euphorbium* (Bentham) | 2. *Tirucalli* Boiss. | *E. tirucalli* L. | W. Africa | 4-Deoxy-4α-phorbol |
| | | *E. stenoclada* Baillon | Madagascar | None |
| | | *E. lateriflora* Schum. and Thonner | W. Africa | Unknown |
| | | *E. nubica* N. E. Br. | E. Africa | None |
| | 5. *Diacanthium* Boiss. | | | |
| | (i) *Splendentes* Berger | *E. milii* Desm. var. *milii* | Madagascar | Ingenol |
| | (ii) *Grandifoliae* Berger | *E. nerifolia* L. | | Ingenol |
| | | *E. nivulia* Buch.-Ham. | India | Ingenol and unknown ingenane |
| | (v) *Trigonae* Berger | *E. antiquorum* L. | | Unknown ingenane |
| | | *E. lactea* Haw. | India and Africa | Ingenol |
| | | *E. royleana* Boiss. | | Ingenol |
| | (vi) *Polygonae* Berger | *E. canariensis* L. | Canary Is. | Unknown ingenane |
| | | *E. candelabrum* Trem. | | Ingenol |
| | | *E. coerulescens* Haw. | | 12-Deoxy-phorbol |
| | | *E. cooperi* N. R. Br. | | 12-Deoxy-16-hydroxy phorbol |
| | | *E. deightonii* Croizat | | Ingenol |
| | | *E. desmondii* Keay | | Ingenol |
| | | *E. erythraea* Hemsley | | Ingenol |
| | | *E. fortissima* Leach | | 12-Deoxy-phorbol |
| | | *E. frankiana* Berger | Africa | Phorbol: 12-Deoxy-phorbol |

References, pp. 90—99

| | | | | |
|---|---|---|---|---|
| | | *E. kamerunica* Pax | | Ingenol |
| | | *E. ledienii* Berger | | 12-Deoxy-16-hydroxy phorbol: 12-Deoxy phorbol |
| | | *E. polyacantha* Boiss. | | Ingenol: 12-Deoxy phorbol |
| | | *E. resinifera* Berger | | Ingenol: 12-Deoxy phorbol: Resiniferatoxin |
| | | *E. triangularis* Desf. | | 12-Deoxy phorbol |
| | 6. *Anthacantha* Lem. | *E. pentagona* Haw. | Africa | Ingenol |
| IX. *Tithymalus* (Boiss.) ..... | 2. *Decussatae* Boiss. | *E. lathyris* L. | Europe | Ingenol |
| | 8. *Pachycladae* Boiss. | *E. balsamifera* Aiton | Canary Is. | 12-Deoxy phorbol |
| | 10. *Galarrhaei* Boiss. | *E. coralloides* L. | | Ingenol |
| | | *E. helioscopia* | | 12-Deoxy phorbol and unknowns |
| | | *E. hiberna* L. | | Unknown |
| | | *E. palustris* L. | Europe | Unknown |
| | | *E. pilosa* L. | | Unknown |
| | | *E. polychroma* Kerner | | Unknown |
| | | *E. sikkimensis* Boiss. | | Ingenol |
| | | *E. stricta* L. | | Ingenol: 5-Deoxy ingenol |
| | 11. *Esculae* Boiss. | *E. amygdaloides* L. | | Unknown |
| | | *E. androsaemifolia* Willd. | | Unknown ingenane |
| | | *E. characias* L. | | Unknown ingenane |
| | | *E. cyparissias* L. | | Unknown ingenane |
| | | *E. exigua* L. | | Unknown ingenane |
| | | *E. paralias* L. | Europe | Unknown ingenane |
| | | *E. peplus* L. | | Ingenol: 5-Deoxy ingenol |
| | | *E. portlandica* L. | | None |
| | | *E. sibthorpii* Boiss. | | Ingenol: 5-Deoxy ingenol |
| | | *E. wulfenii* Hoppe | | Unknown ingenane |
| | 12. *Myrsiniteae* Boiss. | *E. biglandulosa* Desf. | Mediterranean | Ingenol: 5-Deoxy ingenol |
| | | *E. myrsinites* L. | | Ingenol: 5-Deoxy ingenol |

Table 6 (continued)

| Genus *Euphorbia* L. section | Subsection and series | Species[1] | Geographical origin | Diterpenes present |
|---|---|---|---|---|
| Species not classified by Pax & Hoffman .......... | | *E. memoralis* R. A. Dyer | Africa | Ingenol |
| | | *E. paganorum* A. Chev. | | 12-Deoxy phorbol |
| | | *E. poissonii* Pax | W. Africa | 12-Deoxy phorbol Tinyatoxin Resiniferatoxin |
| | | *E. unispina* N. E. Br. | | 12-Deoxy phorbol Resiniferatoxin |
| | | *E. robbiae* Turrill | Europe | Ingenol unknown ingenanes |
| | | *E. kotschyana* Fenzl. | E. Africa | Ingenol and unknown ingenanes |
| | | *E. pseudograntii* Pax | E. Africa | Ingenol and unknown ingenanes |
| Genus *Elaeophorbia* Stapf .... | | *E. drupifera* Thonner | W. Africa | Ingenol |
| | | *E. grandifolia* Haw. | | Ingenol |

[1] Species classified according to Pax & Hoffman (1931) from (*25*).

*References, pp. 90—99*

Table 7. *Esters of Ingenol*

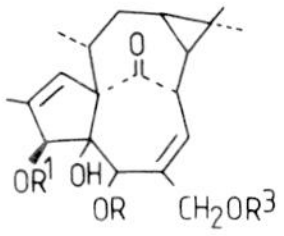

| $R^1$ | $R^2$ | $R^3$ | Botanical source | References |
|---|---|---|---|---|
| $-CO(CH_2)_{14}CH_3$ | H | H | E. ingens E. lathyris | (126–128, 90) |
| H | H | $-CO(CH_2)_{14}CH_3$ | E. ingens, E. lathyris, E. serrata | (126–128, 90, 137) |
| $-CO(CH{=}CH)_3(CH_2)_2CH_3$ | H | H | E. ingens, E. lathyris, E. tirucalli | (127, 90, 85) |
| $-CO(CH{=}CH)_4(CH_2)_2CH_3$ | H | H | E. tirucalli | (85) |
| $-CO(CH{=}CH)_5(CH_2)_2CH_3$ | H | H | E. tirucalli, E. lathyris, E. esula, E. jolkini | (90, 85, 138 - 140) |
| $-CO(CH{=}CH)_2(CH_2)_4CH_3$ | H | H | E. tirucalli, E. lathyris | (90, 85) |
| $-CO(CH{=}CH)_3(CH_2)_4CH_3$ | H | H | E. tirucalli, E. lathyris | (90, 85) |
| $-CO(CH{=}CH)_4(CH_2)_4CH_3$ | H | H | E. tirucalli | (85) |
| $-CO(CH{=}CH)_2(CH_2)_6CH_3$ | H | H | E. lathyris | (90) |
| $-CO(CH_2)_8CH_3$ | H | H | E. kamerunica | (136) |

| $R^1$ | $R^2$ | $R^3$ | Botanical source | References |
|---|---|---|---|---|
| H | $-COCH{=}CH(CH_2)_4CH_3$ | H | E. kamerunica | (136) |
| H | H | $-COCH(CH_3)-CH_3$ | E. cotinifolia | (132) |
| peptide | H | H | E. millii | (139, 140) |
| H | H | peptide | E. millii | (139, 143) |
| peptide | H | $-COCH_3$ | E. millii | (139, 143) |
| $-CO(CH{=}CH)_2(CH_2)_4CH_3$ | H | $-COCH_3$ | E. kansui | (139, 144) |
| $-COC_6H_5$ | H | $-COC_6H_5$ | E. esula | (9) |
| $-COCH_2CH_3$ | H | $-COCH(CH_3)CH_2CH_3$ | E. cotinifolia | (132) |
| $-COCH_2CH_3$ | H | $-COCH(CH_3)-CH_3$ | E. cotinifolia | (132) |
| $-COCH(CH_3)-CH_3$ | H | $-COCH(CH_3)-CH_3$ | E. cotinifolia | (132) |
| $-COCH{=}CH(CH_2)_4CH_3$ | H | $-COCH_3$ | E. kamerunica | (136) |
| $-CO-C(CH_3){=}C(CH_3)H$ | H | $-COCH_3$ | E. kamerunica | (136) |

Table 7 (continued)

| $R^1$ | $R^2$ | $R^3$ | Botanical source | References |
|---|---|---|---|---|
| $-CO(CH_2)_{10}CH_3$ | H | H | *E. esula*, *E. kamerunica* | (*136*, *138*) |
| $-COCH(CH_3)(CH_2)_3CH(CH_3)-CH_2CH_3$ | H | H | *Euphorbium* | (*64*) |
| $-COCH(CH_3)(CH_2)_6CH_3$ | H | H | *Euphorbium* | (*64*) |
| $-COCH(CH_3)(CH_2)_7CH_3$ | H | H | *Euphorbium* *E. virgata* | (*64*, *141*) |
| $-COCH(CH_3)(CH_2)_3CH(CH_3)(CH_2)_2CH_3$ | H | H | *Euphorbium* | (*64*) |
| $-COCH(CH_3)(CH_2)_3CH(CH_3)(CH_2)_2CH_3$ | H | H | *Euphorbium* | (*64*) |
| $-COCH(CH_3)(CH_2)_8CH_3$ | H | H | *Euphorbium* | (*64*) |
| $-COCH(CH_3)(CH_2)_3CH(CH_3)(CH_2)_4CH_3$ | H | H | *Euphorbium* | (*64*) |
| $-CO-C(CH_3)=C(CH_3)H$ | H | H | *E. paralias*, *E. virgata* | (*141*, *142*) |
| H | H | $-COC_6H_5$ | *E. virgata* | (*141*) |
| H | H | $-COCH_2C_6H_5$ | *E. virgata* | (*141*) |
| H | $-CO(CH=CH)_2CH_3$ | H | *E. kamerunica* | (*136*) |

Structure of peptide from *E. millii*

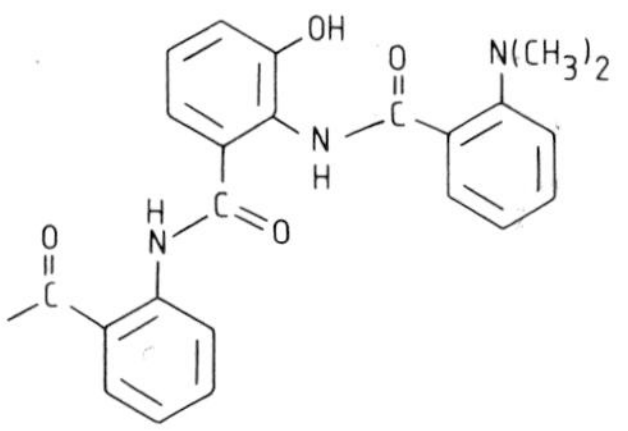

## VIII.2. Distribution of Ingenol Esters in Plants

A comparative phytochemical study of about sixty *Euphorbia* species (*104*) showed ingenol esters to be the most common irritant of the *Euphorbia* genus (see Table 6) and in a recent screening of 15 of the 19 *Euphorbia* species growing in the Azarbaijan province of Iran (*135*), the irritant principles were exclusively of the ingenol type. No ingenanes have been isolated to date from the Thymelaeaceae family.

Ingenol esters have been isolated as either monoesters (esterified at C-3, C-5 or C-20) or diesters (esterified at both C-3 and C-20). The naturally occurring ingenol esters isolated to date are shown in Table 7.

*References, pp. 90—99*

## VIII.3. 5-Deoxyingenol

Hydrolysis and subsequent acetylation of the acetone extracts of *Euphorbia biglandulosa* and *E. myrsinites* latices (*133*) yielded two products. The first was identified as ingenol triacetate, and the second less polar compound as 5-deoxyingenol diacetate. The structure and spectral characteristics of 5-deoxyingenol diacetate are shown in Fig. 55.

Empirical formula $C_{24}H_{32}O_6$

M.P. 205 - 207°C

I.R. ($CHCl_3$) $\nu_{max}$ 3430, 1740, 1705, 1640 $cm^{-1}$

e.i. MS fragmentation

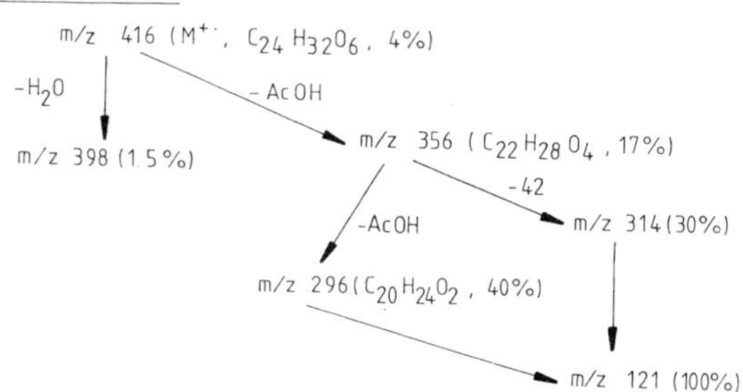


U.V. (MeOH) $\lambda_{max}$ 210, 282 nm

C.D (MeOH) ($\Delta$E): 204 (+3.01), 210 (0.00), 222 (-2.62), 252 (0.00) 268 (-0.10), 275 (0.00), 300 (+0.34), 315 (0.00) nm

$^1H$-NMR ($CDCl_3$, 60 MHz) - assigned according to (133)

Fig. 55. Structure and spectral characteristics of 5-deoxyingenol-diacetate

The structural similarity of 5-deoxyingenol diacetate and ingenol triacetate is evident from their similar UV, IR and CD spectra. Both compounds lack conjugation and have UV absorption maxima at 282 nm due to the $\beta,\gamma$-unsaturated keto groups present. The IR spectrum shows the same functional groups as ingenol triacetate, with hydroxyl absorbance at $3430\,cm^{-1}$, the ketone bridge between rings B and C at $1705\,cm^{-1}$ and the

$-C=C-$ groups at $1640\,cm^{-1}$. The *trans* junction between rings A and B of 5-deoxyingenol diacetate is evident from the CD spectrum (see Fig. 54) which shows Cotton effects similar to those of ingenol triacetate over the region 200 – 390 nm. The presence of the β,γ-unsaturated ketone group linking rings B and C is confirmed by the positive Cotton effect at ~300 nm.

In the mass spectrum, ingenol triacetate and 5-deoxyingenol diacetate show comparably intense fragment ions throughout their spectra, both having a base peak at m/z 121. Accurate mass determination gave the empirical formula $C_{24}H_{32}O_6$ for the molecular ion (*133*).

The structure of 5-deoxyingenol diacetate was fully established by comparing its $^1$H-NMR spectrum with that of ingenol triacetate (Fig. 52). The differences in the $^1$H-NMR spectra of 5-deoxyingenol diacetate and ingenol triacetate are attributable to removal of the C-5 acetoxy group in 5-deoxyingenol diacetate which as a result of a diamagnetic shift of H-7 to 5.82 ppm caused an inversion of the H-1 and H-7 signals. 5-Deoxyingenol diacetate also differs from ingenol triacetate in having a smaller coupling constant for H-20, the upfield shift of H-5 which is no longer geminal to an acetoxy group and the loss of one acetyl methyl signal.

5-Deoxyingenol has also been cited as the parent diterpene of the irritant principles of *Euphorbia peplus* and *E. sibthorpii* (*104*).

## VIII.4.16-Hydroxyingenol

A further derivative of ingenol was isolated from *Euphorbia ingens* (*127*) and identified as 16-hydroxyingenol-3-(2,4,6-decatrienoate)-16-angelate. Base catalysed transesterification yielded the parent alcohol 16-hydroxyingenol, which was isolated as its stable tetraacetate after acetylation. 16-Hydroxyingenol-3,5,16,20-tetraacetate was later obtained after hydrolysis and subsequent acetylation of the irritant fraction of *E. lactea* latex (*145*); its structure and spectroscopic data are shown in Fig. 56.

In the $^1$H-NMR spectrum the chemical shifts of the H-13 and H-14 signals appear downfield at 1.40 ppm – 0,90 ppm compared with the signals in ingenol triacetate (0.73 ppm), indicating that the additional OH group is positioned at C-16 of the ingenol nucleus (*127*). The additional acetyl moiety did not influence the chemical shifts of H-8, H-11 and H-12 in 16-hydroxyingenol tetraacetate, as would be expected for an acetate at position C-16 rather than at the alternative C-17 position. The c.i.-m.s. shows a fragmentation pattern typical of an ingenane derivative, with the exception that the base peak is at m/z 310, rather than at m/z 121 as for ingenol triacetate and 5-deoxyingenol diacetate. The circular dichroism curve is similar to that of ingenol triacetate, showing that the stereochemistry of the ring junctions is identical.

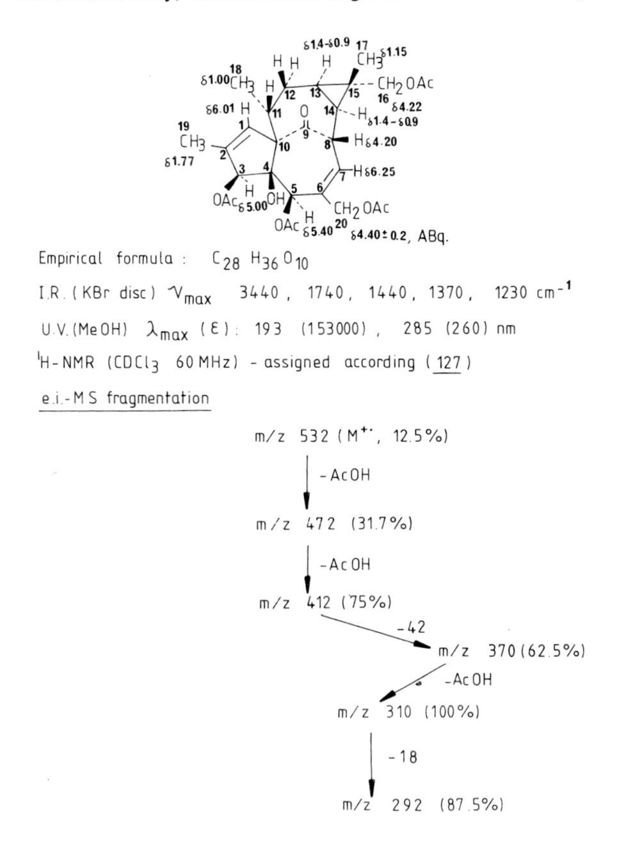


Fig. 56. Structure and spectroscopic data of 16-hydroxyingenol-3,5,16,20-tetraacetate

## VIII.5. 13-Hydroxyingenol

Several diesters of 13-hydroxyingenol have been isolated from *Euphorbia* species. The first of these was 20-O-hexanoyl-13-O-dodecanoyl-13-hydroxyingenol from *E. kansui* (*44, 139*). Hydrolysis of this compound with sodium methoxide in methanol gave the monoester, 13-O-dodecanoyl-13-hydroxyingenol, which was stable in alkali and resisted attempts to hydrolyse it further to the parent alcohol (*44, 139*). Final confirmation of the structure was obtained by a series of chemical reactions and by comparison of the spectral data with those of other ingenanes (see *44, 139* for details). The 13-hydroxyingenols isolated to date are shown in Table 8; Fig. 57 gives the structure and spectral data of one of these, 20-O-(2,3-dimethylbutyryl)-13-O-dodecanoyl-13-hydroxyingenol ($Cy_{12}$), which was isolated from *Euphorbia cyparissias* (*146*).

Table 8. *Esters of 13-Hydroxyingenol*

Esters of 13-hydroxyingenol

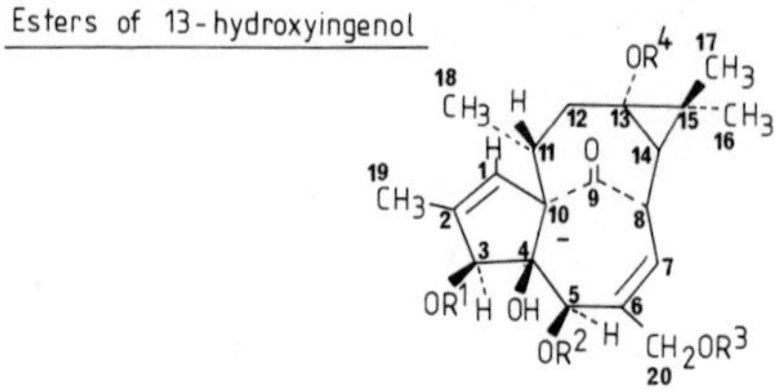

| $R^1$ | $R^2$ | $R^3$ | $R^4$ | Botanical source | Reference |
|---|---|---|---|---|---|
| $-CO(CH=CH)_2(CH_2)_5CH_3$ | H | H | $-CO(CH_2)_{10}CH_3$ | *E. kansui* | 139, 44 |
| H | H | $-CO(CH_2)_4CH_3$ | $-CO(CH_2)_{10}CH_3$ | *E. kansui* | 139, 44 |
| $-CO-CH(CH_3)-CH(CH_3)-CH_3$ | H | H | $-CO(CH_2)_{10}CH_3$ | *E. cyparissias* | 146 |
| H | $-CO-CH(CH_3)-CH(CH_3)-CH_3$ | H | $-CO(CH_2)_{10}CH_3$ | *E. cyparissias* | 146 |
| H | H | $-CO-CH(CH_3)-CH(CH_3)-CH_3$ | $-CO(CH_2)_{10}CH_3$ | *E. cyparissias* | 146 |
| $-CO-CH(CH_3)-CH(CH_3)-CH_3$ | H | H | $-CO(CH_2)_{10}CH_3$ | *E. cyparissias* | 146 |

$R^1 = -COCH(CH_3)-CH(CH_3)-CH_3$

$R^2 = -CO(CH_2)_{10}CH_3$

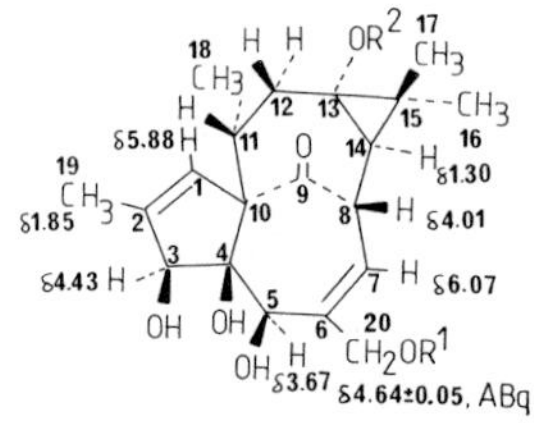

Empirical formula : $C_{38}O_8H_{60}$

I.R. ($CH_2Cl_2$): $\nu_{max}$ : 3590, 3540, 3470, 2925, 2855, 1720 $cm^{-1}$

U.V. (MeOH) $\lambda_{max}$ : 193, 291 nm.

M.S. m/z 644 ($M^{+\cdot}$), 626 ($M^+-18$), 510 ($M^+-134$), 426 ($M^+-218$), 310 (100%)

$^1$H-NMR ($CDCl_3$, 90 MHz) assigned according to (146)

Fig. 57. Structure and spectral data of $Cy_{12}$ isolated from *E. cyparissias* (*146*)

## VIII.6.13,19-Dihydroxyingenol

A second group of compounds isolated from *Euphorbia cyparissias* had Rf-values similar to the di-o-acyl-13-hydroxyingenols, but differed in partition behaviour and colour reactions with vanillin/sulphuric acid spray (*146*). They were identified as triesters of 13,19-dihydroxyingenol; the structure and spectral data of one of these compounds, $Cy_6$, is shown in Fig. 58. The UV and IR-spectra are similar to those of 13-hydroxyingenol esters, whereas the molecular formula, $C_{38}H_{58}O_{10}$, together with the mass fragmentation pattern, indicated $Cy_6$ to be a trihexanoate of a dihydroxyingenol (*146*).

R = -CO-CH(CH₃)-CH(CH₃)-CH₃

Empirical formula : $C_{38}H_{58}O_{10}$

U.V. ($CH_3OH$) : $\lambda_{max}$ ($\varepsilon$) 193 (17830), 285 (290) nm

I.R. ($CH_2Cl_2$) : $\nu_{max}$ 3850, 3550, 3500, 3430, 2960, 2925, 2870, 1725 $cm^{-1}$

e.i. m.s. fragmentation

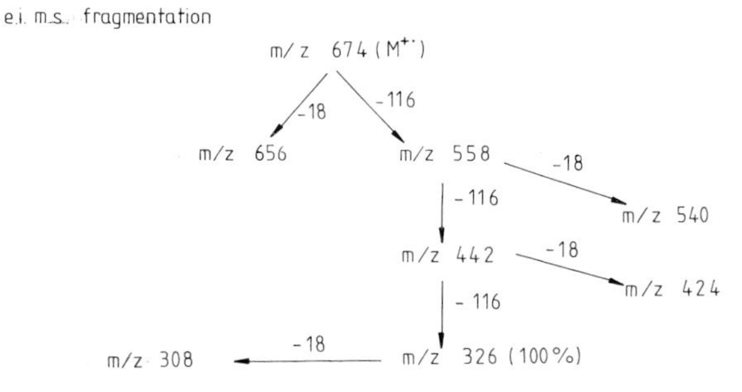


¹H-NMR (CDCl₃ , 90 MHz ) assigned according to (146)

Fig. 58. Structure and spectral characteristics of 3,13,19-tris-O-(2,3-dimethyl-butyryl)-13,19-dihydroxyingenol, $Cy_6$, isolated from *E. cyparissias* (*146*)

In the ¹H-NMR spectrum of $Cy_6$, as compared with the spectrum of a 13-hydroxyingenol ester, the H-1 signal is shifted downfield, a new signal for two geminal protons at 4.63 ppm replaces the H-19 vinyl methyl signal for 3H-19, and the H-14 signal appears as a doublet at 1.26 ppm. These findings showed the new parent compound to be 13,19-dihydroxyingenol;

the positions of the acyl groups were determined by double resonance experiments (*146*).

The structures of the three esters of this new parent compound are shown in Table 9.

Table 9. *13,19-Dihydroxyingenol Esters from E. cypariassias (146)*

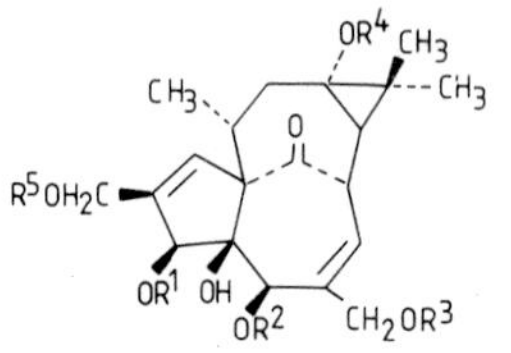

| | $R^1$ | $R^2$ | $R^3$ | $R^4$ | $R^5$ |
|---|---|---|---|---|---|
| Cy6 | $-COCH(CH_3)-CH(CH_3)-CH_3$ | H | H | $-COCH(CH_3)-CH(CH_3)-CH_3$ | $-CO-CH(CH_3)-CH(CH_3)-CH_3$ |
| Cy4 | H | $-COCH(CH_3)-CH(CH_3)-CH_3$ | H | $-CO-CH(CH_3)-CH(CH_3)-CH_3$ | $-CO-CH(CH_3)-CH(CH_3)-CH_3$ |
| Cy2 | H | H | $-CO-CH(CH_3)-CH(CH_3)-CH_3$ | $-CO-CH(CH_3)-CH(CH_3)-CH_3$ | $-CO-CH(CH_3)-CH(CH_3)-CH_3$ |

## VIII.7. 20-Deoxyingenol

Four ingenane monoesters based on 20-deoxyingenol have been isolated from *Euphorbia* species. The first of these were two isomers, 3-O-benzoyl-20-deoxyingenol and 5-O-benzoyl-20-deoxyingenol isolated from *E. kansui* (*144*). These two compounds were hydrolysed with sodium methoxide, followed by neutralisation with IRC-50, to yield 20-deoxyingenol, and methyl benzoate. The structure and spectral characteristics of 20-deoxyingenol are shown in Fig. 59.

The $^1$H-NMR spectrum showed the presence of a new vinyl methyl group at 1.98 ppm, and the absence of the two proton low-field signal attributed to H-20 of ingenol. Other signals in the spectrum were similar to those of ingenol. The structure of 20-deoxyingenol was established unequivocally by hydrogenation of its 3,5-diacetate with Pd-C in ethyl acetate which gave a compound identical with that formed by the hydrogenation of ingenol triacetate (*144*).

3-Angelyl-20-deoxyingenol was obtained as the major irritant from *Euphorbia paralias* L. latex (*142*), where it occurred together with smaller amounts of 3-hexanoyl-20-deoxyingenol. The $^1$H-NMR spectrum of 3-hexanoyl-20-deoxyingenol is shown diagrammatically in Fig. 60.

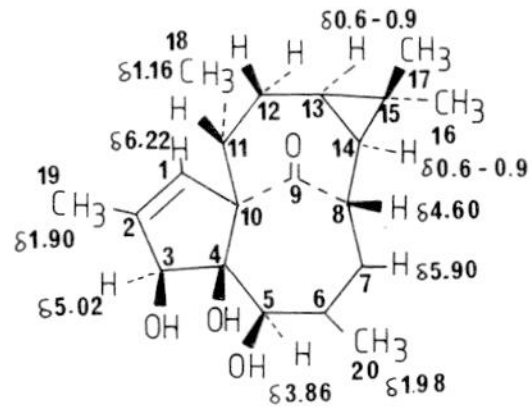

Empirical formula : $C_{20}O_4H_{28}$

M.P. 201 - 203° C (decomp.)

I.R. (KBr) $\nu_{max}$ 3550, 1705, 1665, 1640 $cm^{-1}$

M.S. m/z 332 ($M^{+\cdot}$), 314, 296, 278

$^1$H-NMR ($C_6D_6N$, 100 MHz): assigned according to (144)

Fig. 59. Structure and spectral characteristics of 20-deoxyingenol

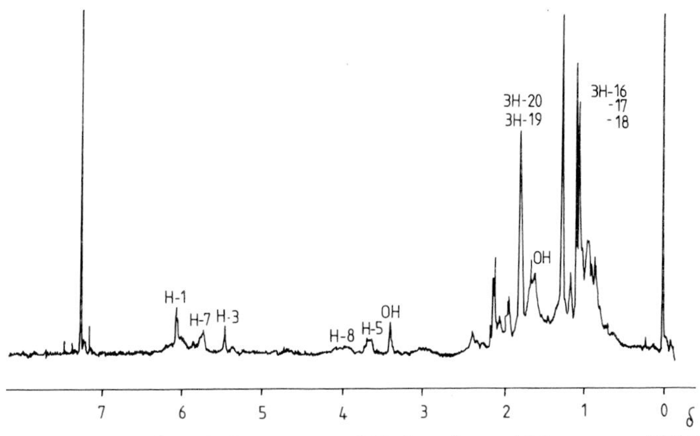


Fig. 60. 100 MHz $^1$H-NMR spectrum of 20-deoxyingenol-3-hexanoate ($CDCl_3$)

# IX. Daphnane Derivatives

## IX.1. Daphnetoxin Type

In 1970, a highly poisonous principle was isolated in 0.02% yield from "mezeron" bark which is a commercially available mixture of the barks of *Daphne mezereum, Daphne laureola* and *Daphne gnidium* (Thymelaeaceae) (*147*). This compound was named daphnetoxin, and represented the first of a series of related orthoester diterpenes to be isolated from the Euphorbiaceae and Thymelaeaceae.

The orthoester structure of daphnetoxin (Fig. 61) is reminiscent of the hemiacetal structure of tetrodotoxin (*148*). It is a tricyclic diterpene whose structure is closely related to the tigliane skeleton, with the cyclopropane ring opened so as to give an isopropenyl side chain at C-13. Present also are a 6 – 7 epoxide group, an extra secondary hydroxyl group at C-5 and the orthoester benzoate moiety which bridges C-9, C-13 and C-14 of ring C. The structure was elucidated by X-ray analysis of the bromoacetate (*147*) which demonstrated that the stereochemistry of daphnetoxin was similar to that of phorbol. The A/B ring junction is *trans*, with a β-orientated hydroxyl at C-4 and the 6 – 7 epoxide group in the α-configuration. The structure and spectroscopic properties of daphnetoxin are shown in Fig. 61.

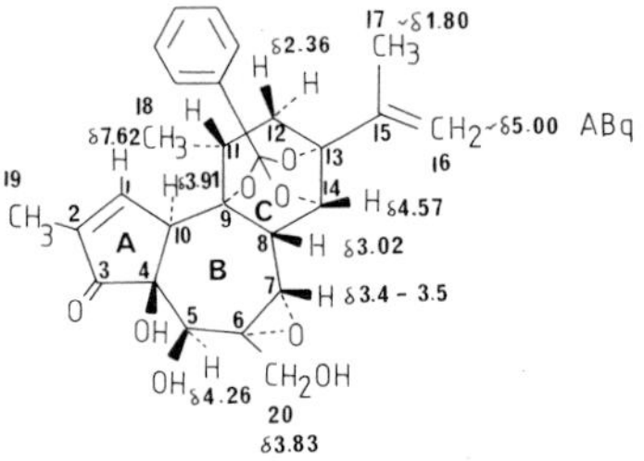

Empirical formula $C_{27}H_{30}O_8$

M.P. 194 - 196° C

U.V. (EtOH) : $\lambda_{max}$ (ε) : 243 (8950), 337 (84) nm

C.D. (MeOH) 342 ([θ] = + 3050), 243 ([θ] = - 8300) nm

$^1$H-NMR (CDCl$_3$, 100 MHz) assigned according to (149)

Fig. 61. Spectroscopic data of daphnetoxin

In the $^1$H-NMR spectrum of daphnetoxin, the C-17 methyl frequency occurs at about 1.80 ppm and the C-16 methylene signals at 5.0 ppm as an AB quartet, thereby confirming the presence of the isopropenyl group at C-13. The signal for the proton on C-14 is diagnostic of orthoesters, and is found as a sharp doublet at 4.57 ppm. Double resonance experiments have shown that it couples with H-8 at 3.02 ppm. The chemical shift of H-7 is influenced by the α-orientated epoxide group, and is observed at about 3.40 – 3.50 ppm. This is in contrast to phorbol esters, where the H-7 signal is downfield at about 5.8 ppm (see VII.1).

SAKATA *et al.* isolated a piscicidal constituent from *Hura crepitans* (Euphorbiaceae) (*150* – *152*) which was called huratoxin. Spectral studies, chemical degradation and X-ray crystallography showed the structure to be a non-aromatic daphnetoxin derivative, with a tetradeca-2,4-dienoic acid orthoester (see Table 10).

Table 10. [1]*H-NMR Data of Some Daphnetoxin-Type Diterpenes*

| Protons of carbon number | δ ppm compounds (**1**) – (**5**)* | | | | |
|---|---|---|---|---|---|
| | (**1**) (*150*) | (**2**) (*153*) | (**3**) (*108*) | (**4**) (*108*) | (**5**) (*116*) |
| 1 | 7.61 | 7.68 | 7.62 | 7.66 | 7.58 |
| 7 | 3.46 | 3.46 | 3.44 | 3.44 | 3.42 |
| 16 | 5.04 | 5.05 | 5.00, 4.88 | 5.04, 4.92 | 5.00, 4.88 |
| 20 | 3.84 | 3.84 | 3.82 | 3.81 ± 0.03 | 3.82 |
| 14 | 4.45 | 4.46 | 4.35 | 4.40 | 4.37 |
| 8 | 2.94 | 2.93 | 2.90 | 2.90 | 2.88 |
| 10 | 3.84 | 3.84 | 3.75 | 3.75 | n. r. |
| 5 | 4.23 | 4.28 | 4.23 | 4.22 | 4.23 |
| 19 | 1.80 | 1.74 | 1.80 | 1.80 | 1.77 |
| 17 | 1.80 | 1.74 | 1.80 | 1.80 | 1.77 |
| 18 | 0.90 | 0.86 | 1.15 | 1.14 | n. r. |

* Compound (**1**). Huratoxin ($CDCl_3$, 60 MHz).
(**2**). Excoecariatoxin ($CDCl_3$, 90 MHz).
(**3**). Simplexin (Pimela factor $P_1$) ($CDCl_3$, n. r.).
(**4**). Pimelea factor $P_4$ ($CDCl_3$, n. r.).
(**5**). Montanin ($CDCl_3$, 60 MHz).

n. r. = not recorded.

Huratoxin has also been isolated from *Hippomane mancinella* (Euphorbiaceae) (*132*), where it occurred together with a closely related compound, Hippomane factor $M_2$, and a complex mixture of corresponding diesters, the C-20 position being esterified with a series of long chain fatty acids. This mixture was termed Hippomane factor $M'_x$, and could not be separated into individual compounds by chromatographic techniques. The C-20 ester moieties were identified by quantitative GLC of the methyl esters produced by transesterification of $M'_x$ with $10^{-2}$ M. NaOMe/MeOH (*132*).

A piscicidal compound isolated from *Excoecaria agallocha* (Euphorbiaceae) (*153*) is closely related to huratoxin but has a shorter aliphatic orthoester moiety. Montanin, obtained from both *Cunuria spruceana* (Euphorbiaceae) (*154*) and *Baliospermum montanum* (Euphorbiaceae) (*116*) contains a saturated straight chain aliphatic orthoester.

Other daphnetoxin-type orthoesters are simplexin (or Pimelea factor $P_1$) and Pimelea factor $P_4$, isolated from *Pimelea* spp. (*108, 155, 156*). It has been suggested (*108, 156*) that these compounds are responsible for the condition known as St. George's disease in cattle.

[1]H-NMR data for some daphnetoxin-type orthoesters are listed in Table 10; the structures and distribution of this group are shown in Table 11.

Table 11. *Daphnetoxin-Type Orthoesters*

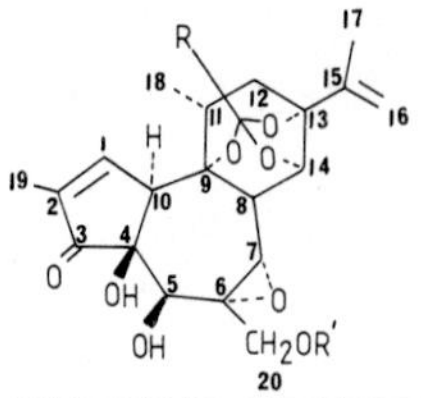

| Compound | R | R' | Botanical source | References |
|---|---|---|---|---|
| Daphnetoxin | $C_6H_5$ | H | Daphne mezereum, D. laureola, D. gnidium | (147) |
| Excoecariatoxin | $-(CH=CH)_2(CH_2)_4CH_3$ | H | Excoecaria agallocha | (153) |
| Simplexin (Pimelea factor P1) | $-C_9H_{19}$ | H | Pimelea simplex, P. prostrata | (155, 108) |
| Pimelea factor $P_4$ | $-C_{13}H_{27}$ | H | Pimelea prostrata | (108) |
| Huratoxin | $-(CH=CH)_2(CH_2)_8CH_3$ | H | Hippomane mancinella, Hura crepitans | (132, 150 - 152) |
| Montanin | $-(CH_2)_{10}CH_3$ | H | Baliosperm montanum<br>Cunuria spruceana | (116, 154) |
| Hippomane factor $M_2$ | $-(CH=CH)_3(CH_2)_8CH_3$ | H | Hippomane mancinella | (132) |
| Hippomane factor $M_x'$ | $-(CH=CH)_n(CH_2)_8CH_3$<br>n = 2, 3 | $-CO(CH_2)_nCH_3$<br>n = 14, 16, 18, 20, 22, 24 | Hippomane mancinella | (132) |

## IX.2. 12-Hydroxydaphnetoxin Type

In 1970, a novel compound, mezerein, was found to be the major toxic constituent of the seeds of *Daphne mezereum*, where it occurs together with small amounts of daphnetoxin (*149, 157, 158, 159*). The structure of mezerein was elucidated by RONLÁN and WICKBERG (*149*); it was found to be the 12-O-cinnamyl-dienacetyl ester of daphnetoxin. The absolute configuration was determined by X-ray crystallography in 1975 (*159*). The molecular structure is compact except for the side chain, and the fused ring systems are approximately spherical in shape. The six-membered ring is in a slightly twisted boat conformation and the seven-membered ring can be regarded as a basal plane of five carbon atoms, with the two bonds shared with the five-membered and six-membered rings at an angle of 110° to this plane (*159*). Structure and spectroscopic properties of mezerein are shown in Fig. 62.

The $^1$H-NMR spectrum is similar to that of daphnetoxin (see Fig. 61), the main difference being the chemical shift of the H-12 signal, which is a one proton singlet at 5.12 ppm in mezerein compared with a two proton multiplet at 2.36 ppm in daphnetoxin, and the appearance of signals corresponding to the 12-O-acyl moiety. RONLÁN and WICKBERG (*149*) performed extensive double resonance experiments in the $^1$H-NMR spectrum of mezerein; the results of this decoupling are reproduced in Table 12.

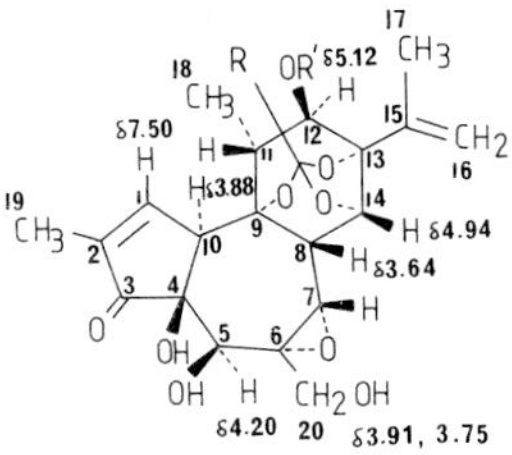

$R = C_6H_5$

$R' = OC(CH{=}CH)_2{-}C_6H_5$

Empirical formula: $C_{38}H_{38}O_{10}$

M. P. 265 - 269 °C (decomp.)

U.V. (EtOH) $\lambda_{max}$ (log ε) 227 (4.24), 234 (4.29), 241 (4.27) 314 (4.60) nm

I.R. (KBr) $\nu_{max}$ 3520, 1714, 1698, 1626 cm⁻¹

M.S. m/z 654.2482 ($M^+$), 157, 149, 105

$^1H$-NMR ($CDCl_3$, 100 MHz) assigned according to (149)

Fig. 62. Structure and spectral data of mezerein

Table 12. *Decoupling in the $^1H$-NMR Spectrum of Mezerein* (after RONLÁN and WICKBERG) (*149*)

| Irradiated at | ppm | Observed at | ppm | Change in multiplicity | Coupling removed Hz |
|---|---|---|---|---|---|
| H-1 | 7.48 | H-10 | 3.85 | sharpening | <0.5 |
| H-1 | 7.48 | H-19 | 1.74 | d. d.→d. | 1.2 |
| H-5 | 4.21 | OH-5 | 4.11 | d.→s. | 3.0 |
| H-8 | 3.61 | H-14 | 4.95 | d.→s. | 2.3 |
| H-10 | 3.85 | H-1 | 7.48 | sharpening | <0.5 |
| H-10 | 3.85 | H-19 | 1.78 | d. d.→d. | 2.6 |
| H-11 | 2.47 | H-18 | 1.36 | d.→s. | 7.0 |
| H-18 | 1.36 | H-11 | 2.47 | q.→s. | 7.0 |

The parent alcohol of mezerein, 12-hydroxydaphnetoxin, was isolated following hydrolysis of the toxic principles of *Lasiosiphon burchellii* (Thymelaeaceae) (*160*), the structure of the isolated 12-hydroxydaphnetoxin being determined by X-ray crystallography of its tribromoacetate.

KUPCHAN and co-workers isolated a series of antileukemic orthoesters from several *Gnidia* species (Thymelaeaceae). Three of these, gnididin, gniditrin and gnidicin (*161*), are structurally related to mezerein, with an aromatic orthoester and either an aliphatic or aromatic ester moiety at C-12. These three compounds were shown to have significant *in vivo* activity against P388 leukemia in mice (*161*). Two other antileukemic diterpenes, gnidilatin and gnidilatidin have an aliphatic orthoester and an aromatic ester at C-12. Their corresponding C-20 palmitate analogues were isolated from the same plant (*162*). Gnidilatidin has been isolated more recently from *Stillingia sylvatica* (*121*).

Odoracin, from *Daphne odora,* is closely related to gnidilatin and gnidilatidin, and has been shown to possess nematicidal properties (*163*).

Recently, from the root sap of *Stillingia sylvatica* (Euphorbiaceae), a series of interesting orthoesters has been isolated (*121*). These are Stillingia factors $S_1-S_5$, which have an aliphatic orthoester group and are esterified at C-12 with ω-hydroxy-decatrienoic acid ($S_1$) or the tetradecanoyl and 2-pentadecanoyl esters of ω-hydroxydecatrienoic acid ($S_2$, $S_3$, $S_4$ and $S_5$). $^1$H-NMR data of some 12-hydroxydaphnetoxin derivatives are shown in Table 13.

Table 13. *$^1$H-NMR Data of Some 12-Hydroxydaphnetoxin Derivatives [after (162)]*

| Protons of carbon number | δ ppm of compounds (**1**) – (**6**)* | | | | | |
|---|---|---|---|---|---|---|
| | (**1**) | (**2**) | (**3**) | (**4**) | (**5**) | (**6**) |
| 1 | n. r. | 7.46 | 7.46 | 7.50 | 7.59 | 7.47 |
| 5 | 4.27 | 4.11 | 4.12 | 4.13 | 4.24 | 4.17 |
| 7 | 3.63 | 3.53 | 3.40 | 3.56 | 3.64 | 3.47 |
| 8 | 3.63 | 3.53 | 3.51 | 3.56 | 3.71 | 3.39 |
| 10 | 3.92 | 3.78 | 3.70 | 3.80 | 3.93 | 3.70 |
| 11 | 2.50 | 2.48 | 2.47 | 2.50 | 2.50 | 2.28 |
| 12 | 5.11 | 5.10 | 5.10 | 5.12 | 5.23 | 4.85 |
| 14 | n. r. | 4.74 | 4.71 | 4.80 | 4.90 | 4.61 |
| 16 | 5.02 | 4.90 | 4.90 | 4.92 | 5.01 | 4.89 |
| 17 | 1.88 | 1.80 | 1.78 | 1.82 | 1.88 | 1.77 |
| 18 | 1.38 | 1.33 | 1.31 | 1.34 | 1.32 | 1.23 |
| 19 | 1.79 | 1.72 | 1.70 | 1.72 | 1.77 | 1.77 |
| 20 | 3.92 | 3.73, 3.55 ABq | 4.73, 3.77 ABq | 3.80 | 4.72, 3.87 ABq | 3.84, 3.70 ABq |

* Compound (**1**). Gnididin ($CDCl_3$, 100 MHz).
(**2**). Gnidilatin ($CDCl_3$, 100 MHz).
(**3**). Gnidilatin-20-palmitate ($CDCl_3$, 100 MHz).
(**4**). Gnidilatidin ($CDCl_3$, 100 MHz).
(**5**). Gnidilatidin-20-palmitate ($CDCl_3$, 100 MHz).
(**6**). Gnidiglaucin ($CDCl_3$, 100 MHz).

The naturally occurring 12-hydroxydaphnetoxin derivatives isolated to date are shown in Table 14.

Table 14. *Naturally Occurring 12-Hydroxydaphnetoxin Derivatives*

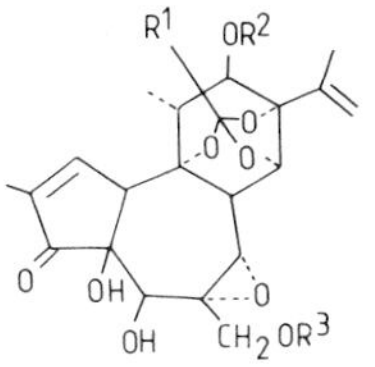


| Compound | $R^1$ | $R^2$ | $R^3$ | Botanical source | References |
|---|---|---|---|---|---|
| Mezerein | $-C_6H_5$ | H | H | Daphne mezereum | (149, 157-159) |
| Odoracin | $-(CH{=}CH)_2(CH_2)_4CH_3$ | $-COC_6H_5$ | H | Daphne odora | (163) |
| Gnididin | $-C_6H_5$ | $-CO(CH{=}CH)_2(CH_2)_4CH_3$ | H | Gnidia lamprantha | (161) |
| Gniditrin | $-C_6H_5$ | $-CO(CH{=}CH)_3(CH_2)_2CH_3$ | H | Gnidia lamprantha | (161) |
| Gnidicin | $-C_6H_5$ | $-CO-CH{=}CH-C_6H_5$ | H | Gnidia lamprantha | (161) |
| Gnidiglaucin | $-C_9H_{19}$ | $-COCH_3$ | H | Gnidia glaucus | (162) |

| Compound | $R^1$ | $R^2$ | $R^3$ | Botanical source | References |
|---|---|---|---|---|---|
| Gnidilatin | $-C_9H_{19}$ | $-COC_6H_5$ | H | Gnidia latifolia | (162) |
| Gnidilatin-20-palmitate | $-C_9H_{19}$ | $-COC_6H_5$ | $-COC_{15}H_{31}$ | Gnidia latifolia | (162) |
| Gnidilatidin | $-(CH{=}CH)_2(CH_2)_4CH_3$ | $-COC_6H_5$ | H | Gnidia latifolia<br>Stillingia syvatica | (162, 121) |
| Gnidilatidin-20-palmitate | $-(CH{=}CH)_2(CH_2)_4CH_3$ | $-COC_6H_5$ | $-COC_{15}H_{31}$ | Gnidia latifolia | (162) |
| Stillingia factor $S_1$ | $-(CH{=}CH)_2(CH_2)_2CH_3$ | $-CO(CH{=}CH)_3(CH_2)_3COOH$ | H | Stillingia sylvatica | (121) |
| Stillingia factor $S_2$ | $-(CH{=}CH)_2(CH_2)_2CH_3$ | $-CO(CH{=}CH)_3(CH_2)_3O$<br>$CH_3(CH_2)_{12}-C{=}O$ | H | Stillingia sylvatica | (121) |
| Stillingia factor $S_3$ | $-(CH{=}CH)_2(CH_2)_2CH_3$ | $-CO(CH{=}CH)_3(CH_2)_3O$<br>$CH_3(CH_2)_{12}-C{=}O$ | H | Stillingia sylvatica | (121) |
| Stillingia factor $S_4$ | $-(CH{=}CH)_2(CH_2)_4CH_3$ | $-CO(CH{=}CH)_3(CH_2)_3O$<br>$CH_3(CH_2)_{12}-C{=}O$ | H | Stillingia sylvatica | (121) |
| Stillingia factor $S_5$ | $-(CH{=}CH)_2(CH_2)_4CH_3$ | $-CO(CH{=}CH)_3(CH_2)_3O$<br>$CH_3(CH_2)_{12}-C{=}O$ | H | Stillingia sylvatica | (121) |

## IX.3. Resiniferonol Type

A novel orthoester was isolated from three *Euphorbia* species, *E. resinifera, E. poisonii* and *E. unispina* (*12, 33, 113*) and named resiniferatoxin. It differs from previous examples of orthoesters in that it has a $\Delta^{6-7}$ bond typical of the phorbol-type diterpenes and that the secondary hydroxy group at C-5 is absent. It has an aromatic orthoester and an aromatic ester moiety at C-20. Structure and spectral properties of resiniferatoxin are shown in Fig. 63, with the $^1$H-NMR spectrum represented diagrammatically in Fig. 64.

Empirical formula: $C_{37}H_{40}O_9$

I.R. (KBr) $\nu_{max}$: 3460, 2950, 1740, 1715, 1520, 1275, 1240, 1150, 1030, 910, 740, 705 $cm^{-1}$

U.V. (MeOH) $\lambda_{max}$ (log $\varepsilon$): 238 (3.9), 290, (3.5) (bathochromic shift with the addition of NaOH to 305) nm.

e.i. m.s. fragmentation (see also figure 12)

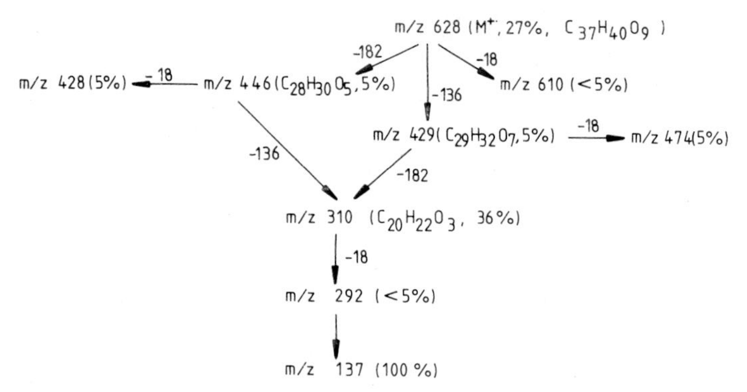


Fig. 63. Structure and spectral data of resiniferatoxin

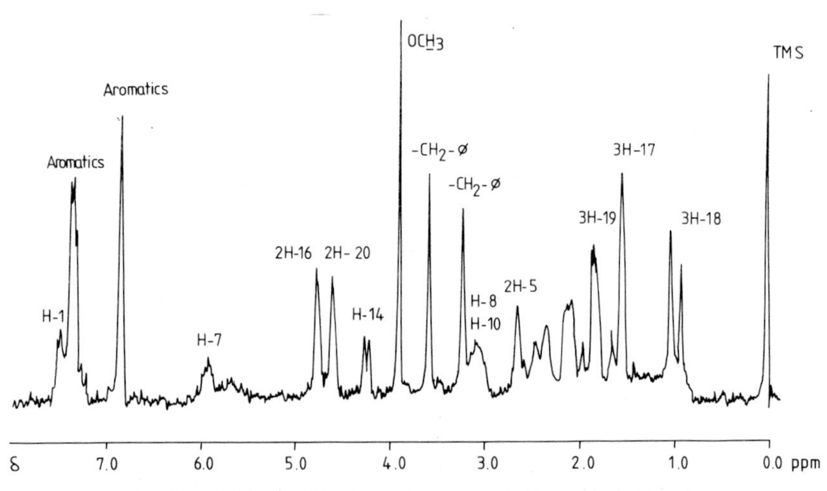


Fig. 64. 60 MHz $^1$H-NMR spectrum of resiniferatoxin in $CDCl_3$

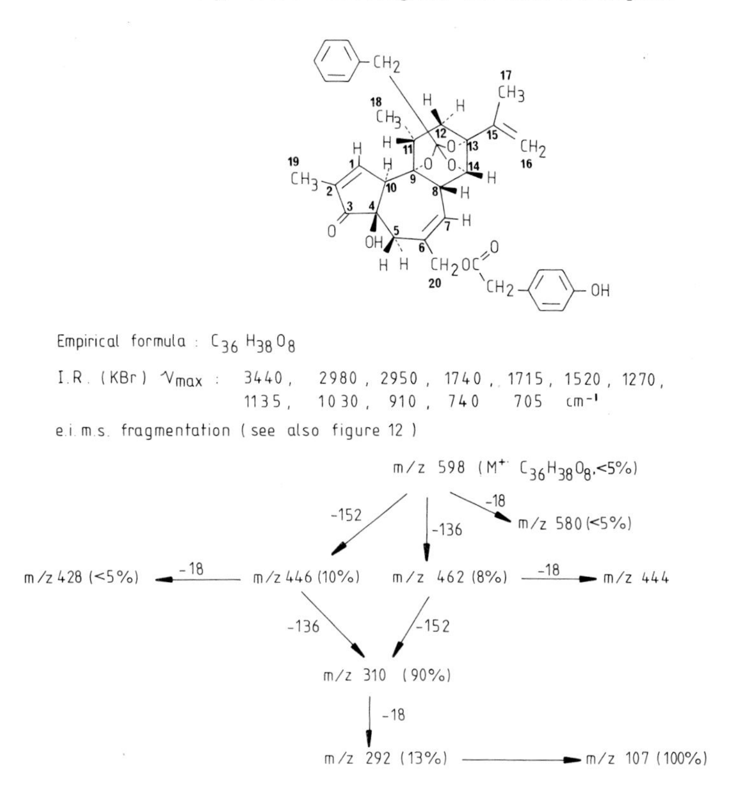


Fig. 65. Structure and spectral data of tinyatoxin

Resiniferatoxin occurred in the latex of *Euphorbia resinifera* together with its 20-desacyl analogue (*12*) and in *E. poisonii* together with tinyatoxin (*113, 164*). Tinyatoxin differs from resiniferatoxin only in the nature of its C-20 acyl group. The structure and spectral properties of tinyatoxin are shown in Fig. 65, its $^1$H-NMR in Fig. 66.

A difference in the $^1$H-NMR spectra of resiniferatoxin and tinyatoxin compared with other orthoesters is the chemical shift of the H-7 signal. As the resiniferonol esters possess a $\Delta^{6-7}$ bond, the H-7 is olefinic, its signal appearing at about 5.8 – 5.9 ppm. This is in contrast to the orthoesters possessing a 6 – 7 epoxide, where the H-7 signal has a chemical shift of about 3.4 – 3.5 ppm. Because the OH-5 is absent, the two proton H-5 signal in resiniferonol esters occurs at about 2.1 – 2.3 ppm as either a quartet or a multiplet. In orthoesters with a secondary OH group in this position, the H-5 resonance appears further downfield as a singlet at about 4.2 – 4.5 ppm.

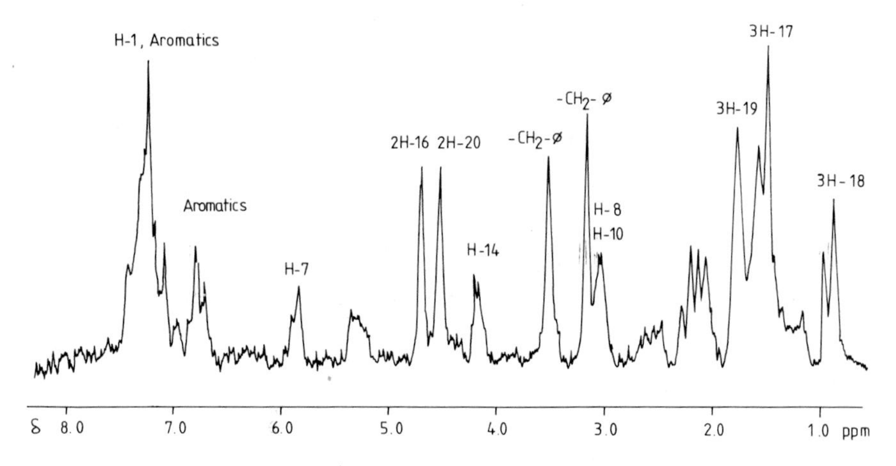


Fig. 66. 90 MHz $^1$H-NMR spectrum of tinyatoxin in $CDCl_3$

Structures of resiniferatoxin and tinyatoxin were proposed on the basis of chemical and spectroscopic evidence (*12, 120*). Mild acid-catalysed transesterification of these two compounds afforded the same product, 9,13,14-orthophenylacetylresiniferol, which indicated that the two compounds differed only in the nature of the acyl function on C-20. The structures of the esterifying acids were deduced by reference to mass spectral data for resiniferatoxin and tinyatoxin, and to the $^1$H-NMR spectra of 4-hydroxyphenylacetic acid and its methyl ester. The $^1$H-NMR and the high mass region of the e.i.-m.s. of the transesterification product,

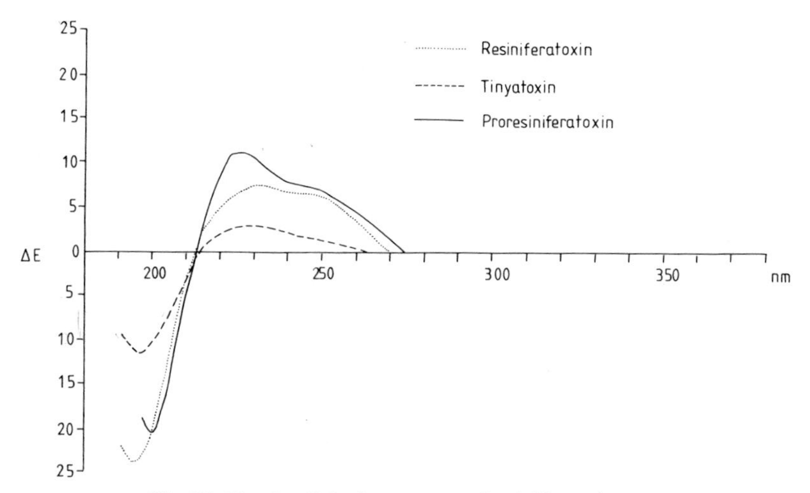


Fig. 67. Circular dichroism spectra of resiniferonol esters

*References, pp. 90—99*

9,13,14-orthophenylacetylresiniferonol, confirmed the presence of phenylacetate as the orthoester moiety. Confirmation of an α,β-unsaturated cyclopentenone, with a *trans* AB ring junction was provided by circular dichroism (Fig. 67) which shows Cotton effects comparable to those of phorbol esters.

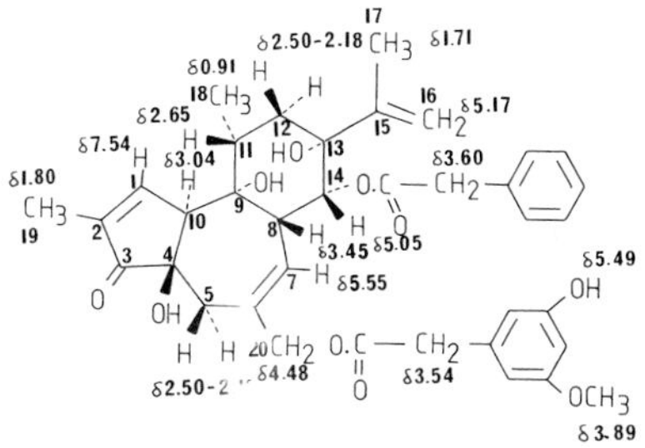

Empirical formula: $C_{37}H_{40}O_9$

I.R. (KBr): $\nu_{max}$ 3340, 2970, 1735, 1715, 1630 $cm^{-1}$

e.i. m.s: m/z 646 ($M^{+\cdot}$, $<2\%$) 628 (30%). Rest of spectrum similar to resiniferatoxin (see figure 63)

C.D. (MeOH): (Δ E): 224 (+6.80), 230 (+7.48), 195 (-11.29) nm.

$^1H$-NMR ($CDCl_3$, 60 MHz) assigned according to (33)

Fig. 68. Structure and spectral data of pro-resiniferatoxin

Pro-resiniferatoxin accompanies resiniferatoxin in *Euphorbia* species (*12, 33, 113*) and has been suggested as a possible biogenic precursor of the latter because it is readily converted to resiniferatoxin by mild acid catalysed dehydration (*12*). The structure and spectral properties of pro-resiniferatoxin are shown in Fig. 68. Its structure was deduced from its $^1H$-NMR spectrum by comparison with the spectra of resiniferatoxin and 12-deoxyphorbol (*12*). The e.i.-m.s. is identical with that of resiniferatoxin except for a weak ion 18 mass units higher than the molecular ion of resiniferatoxin. The CD curve was essentially similar (see Fig. 67), thus confirming the structure in the region of the chromophore. The two proton singlet arising from the methylene protons of the phenylacetate ester was, however, indicative of a normal ester (3.5 – 3.6 ppm) rather than an orthoester function (3.2 ppm).

The occurrence and structures of the resiniferonol type diterpenes are summarised in Table 15*.

Table 15. *Resiniferonol Esters Isolated from Euphorbia Species*

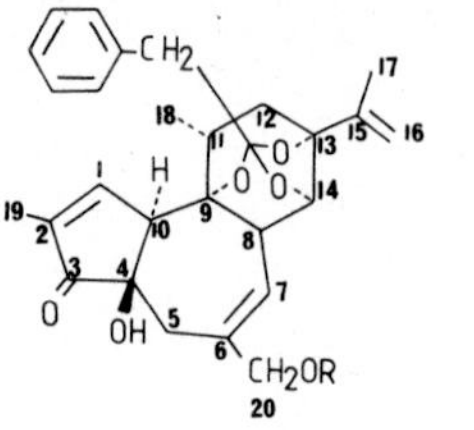

| Compound | R | Botanical source | References |
|---|---|---|---|
| Resiniferatoxin | 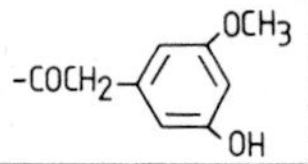 $-COCH_2-C_6H_3(OCH_3)(OH)$ | E. resinifera<br>E. unispina<br>E. poisonii | (12, 13, 164) |
| 20-desacylresiniferatoxin | H | E. resinifera | (12) |
| Tinyatoxin | 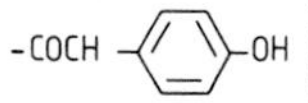 $-COCH-C_6H_4-OH$ | E. poisonii | (113, 164) |

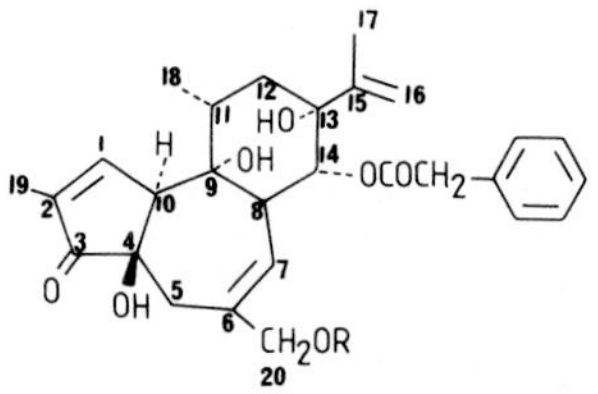

| Compound | R | Botanical source | References |
|---|---|---|---|
| Pro-resiniferatoxin | 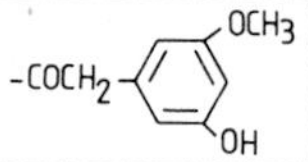 $-COCH_2-C_6H_3(OCH_3)(OH)$ | E. resinifera<br>E. poisonii | (12, 33) |

## IX.4. 1-Alkyldaphnane Type

The 1-alkyldaphnane derivatives represent the most recent group of toxic diterpene esters to be isolated from the plant family Thymelaeaceae. These compounds have not been found, as yet, in any member of the Euphorbiaceae family. They are similar in structure to daphnetoxin

* The structure of the C-20 acyl group in resiniferatoxin was recently amended to (4-hydroxy-3-methoxyphenyl)-acetic acid. See W. ADOLF, B. SORG, M. HERGENHAHN, and E. HECKER: Structure activity relationships of polyfunctional diterpenes of the daphnane type. I. Revised structure for resiniferatoxin and structure-activity relationships of resiniferonol – some of its esters. J. Nat. Prod. **45**, 347 (1982.

derivatives (see Section IX.1.), but with a carbon atom in the chain of the aliphatic orthoester attached to C-1 of ring A, so that a fourth, macrocyclic ring is formed. In addition, the $\Delta^{1-2}$ bond in ring A is absent, and at C-3 of ring A, the keto group is sometimes replaced by a benzoate ester. The orthoester bridging rings C and A may contain a substituent, either a hydroxyl or occasionally an ester function.

The first of this group of compounds to be encountered was a potent antileukemic compound, gnidimacrin, isolated by KUPCHAN and co-workers in 1976 from *Gnidia subcordata* (*165*). The structure and spectral properties of gnidimacrin are shown in Fig. 69.

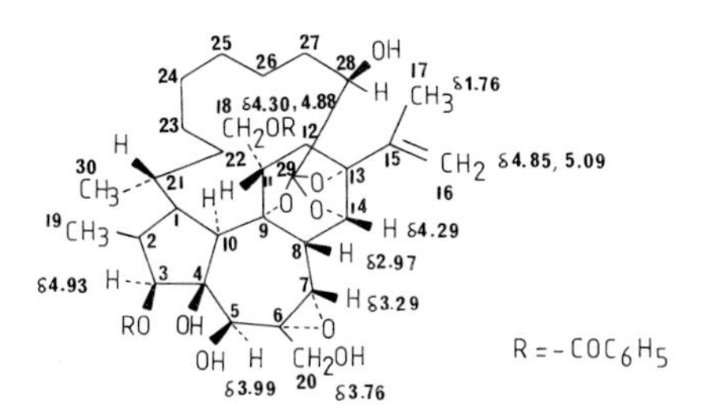


Empirical formula $C_{44}H_{54}O_{12}$

M.P. 172–174° C

I.R. ($CHCl_3$) $\nu_{max}$ 2.82, 5.82, 6.25, 6.32 μ

C.I.M.S. (methane) m/z 775.3679 ($M^+$+H, calc 775.3691), 739, 635, 513, 123, 105

$^1$H-NMR ($CDCl_3$) assigned according (165)

Fig. 69. Structure and spectral data of gnidimacrin

The $^1$H-NMR spectrum is similar in many respects to spectra of daphnetoxin-type derivatives (see Table 10), with characteristic signals for H-7, H-14 and H-16 at 3.29 ppm, 4.29 ppm and an AB system at 4.85 ppm and 5.09 ppm respectively. The major differences are the signals of the geminal protons on C-18, which occur a separately at 4.30 ppm and 4.88 ppm, and the resonance of H-3 geminal to the benzoyl ester moiety, which appears as a doublet at 4.93 ppm, while the vinylic H-1 resonance the signal of the secondary methyl group on C-11 of daphnetoxin were absent. The structure of gnidimacrin was determined unequivocally by X-ray crystallography of the 20-*p*-iodobenzoate (*165*), which showed the stereochemistry to be comparable with daphnetoxin derivatives, having a *trans* A/B ring junction, 4β- and 5β-hydroxyls and an α-orientated 6,7 epoxide group.

Gnidimacrin was accompanied in the plant by its 20-O-palmitate analogue, and both were demonstrated to have potent antileukemic activity *in vivo* (*165*).

Most 1-alkyldaphnanes have been isolated from *Pimelea* species. Pimelea factor $P_2$, from *P. prostrata* (*166*), has a secondary benzoyl group at C-3, with a methyl-substituted seven-carbon aliphatic orthoester joined to C-1 in ring A and is identical with Daphnopsis factor $R_1$, from *Daphnopsis* spp. (*166*) and linifolin B, isolated from both *Pimelea ligustrina* and *Pimelea linifolia* (*167*). Linifolin A, also from *Pimelea linifolia* (*167*), differs in having an acetate at C-12. In this respect, it is anomalous in being the only 1-alkyldaphnane to resemble a 12-hydroxydaphnetoxin derivative.

Pimelea factor $S_7$, from *Pimelea simplex* (*166*), differs from Pimelea factor $P_2$ in having a keto group at C-3, thus lacking an ester group in ring A. Other 1-alkyldaphnanes isolated to date are Pimelea factor $P_6$ from *Pimelea prostrata* (*166*) and Synaptolepsis factor $K_1$ from *Synaptolepsis* spp. (*166*).

The $^1$H-NMR data of some of these compounds are recorded in Table 16; their structures are listed in Table 17.

Table 16. $^1$*H-NMR Data of Some 1-Alkyldaphnane Derivatives*

| Protons attached to carbon number | δ ppm compounds (**1**)–(**6**)* | | | | | |
|---|---|---|---|---|---|---|
| | (**1**) (*165*) | (**2**) (*166*) | (**3**) (*166*) | (**4**) (*166*) | (**5**) (*166*) | (**6**) (*167*) |
| 7 | 3.27 | 3.32 | 3.34 | 3.42 | 3.42 | 3.43 |
| 16 | 4.95, 5.17 | 4.85, 4.95 | 5.05. 4.94 | 5.05, 4.92 | 5.00, 4.90 | 4.96, 4.88 |
| 20 | 3.81, 4.83 ABq | 3.76 ± 0.12 ABq | 3.80 ± 0.03 ABq | 3.80 ± 0.02 ABq | 3.82 | 3.84 |
| 14 | 4.36 | 4.24 | 4.24 | 4.35 | 4.32 | 4.60 |
| 8 | 3.06 | 2.88 | 2.93 | 2.94 | 2.90 | 3.48 |
| 10 | n. r. | 3.10 | 3.10 | 3.00 | 3.35 | 3.11 |
| 3 | 4.97 | 5.06 | — | — | — | 5.04 |
| 5 | 4.02 | 4.10 | 4.05 | 4.05 | 4.12 | 4.10 |
| 19 | n. r. | 1.04 | 1.13 | 1.14 | 1.08 | 1.03 |
| 17 | 1.82 | 1.75 | 1.71 | 1.80 | 1.76 | 1.77 |
| 18 | 4.30, 5.03 | 0.82 | 0.95 | 0.90 | 0.92 | 0.84 |
| 12 | n. r. | n. r. | n. r. | n. r. | n. r. | 4.93 |

* Compounds (**1**). Gnidimacrin-20-palmitate ($CDCl_3$).
(**2**). Pimelea factor $P_2$/Daphnopsis factor $R_1$/Linifolin B ($CDCl_3$).
(**3**). Pimelea factor $S_7$ ($CDCl_3$).
(**4**). Synoleptis factor $K_1$ ($CDCl_3$).
(**5**). Pimelea factor $P_6$ ($CDCl_3$).
(**6**). Linifolin A ($CDCl_3$).

n. r. = not recorded.

Table 17. *Naturally Occurring 1-Alkyl Daphnanes*

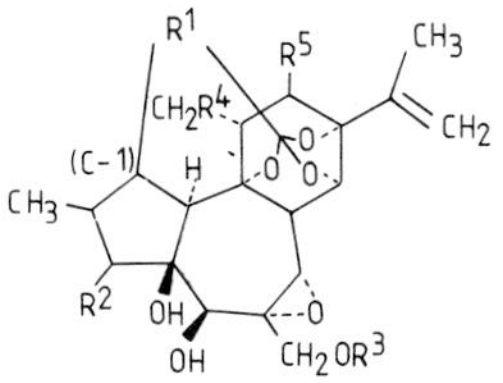

| Compound | $R^1$ | $R^2$ | $R^3$ | $R^4$ | $R^5$ | Botanical source | References |
|---|---|---|---|---|---|---|---|
| Gnidimacrin | $(C\text{-}1)\text{-}(CH_2)_7\text{-}CH(OH)\text{-}$ | $OCOC_6H_5$ | H | $OCOC_6H_5$ | H | *Gnidia subcordata* | (165) |
| Gnidimacrin - 20-palmitate | $(C\text{-}1)\text{-}(CH_2)_7\ CH(OH)\text{-}$ | $\text{-}OCOC_6H_5$ | $\text{-}COC_{15}H_{31}$ | $\text{-}OCOC_6H_5$ | H | *Gnidia subcordata* | (165) |
| Pimelea factor $P_2$<br>Daphnopsis factor $R_1$<br>Linifolin B | $(C\text{-}1)\text{-}CH(CH_3)\text{-}(CH_2)_7\text{-}$ | $\text{-}OCOC_6H_5$ | H | H | H | *Pimelea prostrata*<br>*Pimelea linifolia*<br>*Pimelea Ligustrina*<br>*Daphnopsis spp.* | (166, 167) |
| Pimelea factor $P_6$ | $(C\text{-}1\text{-}CH(CH_3)\text{-}(CH_2)_6\text{-}CH(OCOC_6H_5)\text{-}$ | =0 | H | H | H | *Pimelea prostrata* | (166) |
| Pimelea factor $S_7$ | $(C\text{-}1)\text{-}CH(CH_3)(CH_2)_7\text{-}$ | =0 | H | H | H | *Pimelea simplex* | (166) |
| Synaptolepsis factor $K_1$ | $(C\text{-}1)\text{-}C_{13}H_{26}\text{-}CH{=}CH\text{-}$ | =0 | H | H | H | *Synaptolepsis spp.* | (166) |
| Linifolin A | $(C\text{-}1)\text{-}CH(CH_3)\text{-}(CH_2)_7\text{-}$ | $\text{-}OCOC_6H_5$ | H | H | $OCOCH_3$ | *Pimelea linifolia* | (167) |

## X. Closing Remarks

The Euphorbiaceae and Thymelaeaceae produce a variety of diterpene esters. These compounds range in structural type from macrocyclic diterpenes, some of which have anti-tumour activity (*16*), to the tigliane diterpenes which are well known tumour-promoting phorbol esters of *Croton* oil (*34*). It has been suggested that all of these compounds are biosynthetically related (*31*) and may be useful chemotaxonomic agents (*104*) for these diverse plant families.

During the last 15 years the isolation of toxic diterpenes from these plants in ever increasing numbers, coupled with intensive biological evaluation of their pro-inflammatory and tumour-promoting activities (*6, 168*), has been of considerable importance from the point of view of environmental hygiene. Identification of second order carcinogenic risk factors of the tigliane, daphnane and ingenane types has considerably extended the availability of this group of substances for experimental use. Their involvement in the etiology of human cancer remains to be explained, but intensive biochemical research (*70*) has already been stimulated. On the other hand the macrocyclic diterpenes may eventually be responsible for the development of new types of chemotherapeutic agents (*10*).

Recent biological studies of the pro-inflammatory and tumour-promoting agents of the Euphorbiaceae has centered upon the elucidation of their mechanism of action and identification of the tigliane-ester receptor site on cells (*169*). Although these compounds were known at an early stage (*170, 171, 172*) to induce cellular division, stimulate RNA (*173, 174, 175*) DNA (*175, 176, 177*) and protein (*175, 178, 179, 180*) synthesis, it soon became evident that these activities might not be mediated *via* inhibition of DNA repair (*181, 182, 183, 184, 185*). Attention was focused on the possibility that there exists on cells a specific tigliane ester receptor site (*186*) which would be located at a membrane surface (*187, 188, 189, 190*).

A growing body of evidence supporting this view has been accumulated in recent years. Firstly, 12,13-diesters of phorbol are amphiphilic in nature. The acyl functions at C-12 or C-13 render the phorbol molecule active because increased lipid solubility allows penetration of membrane structures, whilst the shape and functional groups of the nucleus itself induce interaction with membrane proteins thus producing a disruption of normal membrane function. Concerning the nature of such interactions, it has been suggested that the $\alpha,\beta$-unsaturated ketone of phorbol could undergo Michael addition *in vivo*, thus making the esters alkylating agents. Accordingly it has been proposed that TPA acts by alkylating a membrane receptor (*191*). Secondly, the tumour-promoting action (*6*), pro-inflammatory action (*168*) and activity in various *in vitro* biochemical systems (*192, 193, 194, 195*) of tigliane esters has been shown to be structure-dependent. This structural specificity supports the concept of a tigliane-ester receptor. Furthermore the daphnane orthoesters have different structural requirements for their pro-inflammatory activity (*196*) and do not share many of the *in vitro* activities of tigliane esters (*197, 198*). The daphnanes are not tumour-promotors in that they do not act at the second or promotion stage of carcinogenesis, but may be involved with an hypothetical third stage (*70*). Thirdly, the tigliane and ingenane esters are active at nanomolar dose levels and their behaviour is reminiscent of hormones in that they induce a variety of effects depending upon the target cell (*169*). The fact that the earliest such response occurs at the cell surface membrane suggests that tigliane receptors are associated with plasma membranes (*169*). Finally, evidence for the existence of a membrane receptor for phorbol derivatives has been provided by the use of pharmacological inhibitors. The tumour-promoting activity of 12,13-diesters of phorbol has been inhibited by corticosteroids (*199*) and by synthetic protease inhibitors (*200*). In addition, membrane stabilising agents such as imipramine and desmethylimipramine inhibit both *in vivo* (*201*) and *in vitro* (*202*) actions. Also of significance is the observation that phospholipase-$A_2$ inhibitors (*201, 202*) and calmodulin inhibitors antagonise the activity of a number of tigliane derivatives. The use of a range of structurally related tigliane-esters, some which were tumour-promoters whilst others were non-promoters, to induce lymphocyte mitogenesis in the presence of

*References, pp. 90—99*

cyclosporin-A has led to the suggestion that promoting agents have an additional receptor site on human lymphocytes to the non-promoting analogues (*198*). Recently phorbol-diester receptors have been located on intact mammalian cells using tritiated phorbol-derivatives. These receptors were specific and saturatable (*203*).

The nature of tigliane-ester membrane receptors should become clearer in the near future. However several hypothetical suggestions have already been made. ROHRSCHNEIDER and BOUTWELL (*186*) proposed that tigliane esters mimic an endogenous hormone at the cell membrane in such a manner as to alter cellular communication and membrane control mechanisms. SMYTHIES and others (*204*) have indicated that phorbol esters can be shown to lie within a three-ring grid structure based upon a model for membrane prostaglandin receptors. If the tumour-promoting agents exert their effects *via* a prostaglandin receptor on membrane bound guanylcyclase, then some anti-tumour diterpene derivatives may act at the same receptor as antagonists. As yet, experimental evidence for the agonist activity of tigliane esters at a prostaglandin receptor has not been provided. However, recently it was demonstrated that in isolated gastric mucosae several phorbol and 12-deoxyphorbol esters could mimic the action of prostaglandin-$E_2$ by inhibiting histamine induced acid secretion (*205*).

The elucidation of the mechanism of action of these diterpenes is further complicated by the fact that tigliane esters have indirect effects upon cells due to the release of other biologically active materials (*206*). In this respect it is significant that TPA has been shown to release prostaglandins and arachadonic acid *in vitro* (*207, 208*). Stimulation of phospholipid turnover is therefore a feature of their interaction with living cells. In addition 12-deoxyphorbol phenylacetate has been shown to liberate biologically active substances from human platelets which induce further platelet aggregation (*202*). Future investigation of the nature of released mediators and their role in thrombosis, inflammation and cell division is obviously warranted.

The diterpenes of Euphorbiaceae and Thymelaeaceae are increasingly important tools for the investigation of diseased states and future work will involve the elucidation of their mechanism of action and the characterisation of their receptor sites on cell membrane surfaces.

**Acknowledgements**

We are indebted to Drs. A. D. KINGHORN, M. EL-MISSIRY, R. J. SCHMIDT and K. A. ABO for the isolation of some of the compounds used in obtaining spectral data, and we are particularly grateful to the Chemistry Department, King's College, London, and Queen Mary College, London, for 250 MHz $^1$H-NMR and $^{13}$C-NMR spectra respectively. We would also like to thank Mr. D. CARTER for mass-spectra, Mr. W. BALDEO for 60 MHz $^1$H-NMR spectra, and Dr. P. SCOPES for circular dichroism spectra. Finally we are grateful to Mrs. A. DUNCAN for producing the figures and to Mrs. E. COOPER for typing this manuscript.

## References


1. HECKER, E.: Co-carcinogenic principles from the seed oil of *Croton tiglium* and from other Euphorbiaceae. Cancer Res. **28,** 2338 (1968).
2. — Co-carcinogenesis and tumour promoters of the diterpene-ester type as possible carcinogenic risk factors. J. Cancer Res. Clin. Oncol. **99,** 103 (1981).
3. YAMAGIWA, K.: Collected papers on artificial production of cancer, 1. Tokyo: Maruzen Co. Ltd. 1965.
4. LIKHACHER, A. Y.: Combined effect of the carcinogenic substances. Vopr. Oncol. **14,** 114 (1968).
5. KIDD, J. G., and P. ROUS: The carcinogenic effect of a papilloma virus on the tarred skin of rabbits. J. Exp. Medicine **68,** 529 (1938).
6. HECKER, E.: Structure-activity relationships in diterpene esters irritant and co-carcinogenic to mouse skin. In: Carcinogenesis Vol. 2. Mechanisms of tumour promotion and co-carcinogenesis (SLAGA, T. J., A. SIVAK, and R. K. BOUTWELL, eds.). New York: Raven Press. 1978.
7. BERENBLUM, I., and P. SHUBIK: The role of *Croton* oil applications, associated with a single painting of a carcinogen in tumour induction of the mouse skin. Brit. J. Cancer **1,** 379 (1947).
8. SCHMIDT, R. J., and F. J. EVANS: Skin irritant effects of esters of phorbol and related polyols. Arch. Toxicol. **44,** 279 (1980).
9. KUPCHAN, S. M., I. UCHIDA, A. R. BRANFMAN, R. G. DAILEY, and B. YEFEI: Antileukemic principles isolated from Euphorbiaceae plants. Science **191,** 571 (1976).
10. SAHAI, R., R. P. RASTOGI, J. JAKUPOVIC, and F. BOHLMANN: A diterpene from *Euphorbia maddeni*. Phytochem. **20,** 1665 (1981).
11. EVANS, F. J., and C. J. SOPER: The tigliane, daphnane and ingenane diterpenes, their chemistry, distribution and biological activities. J. Nat. Prod. (Lloydia) **41,** 193 (1978).
12. HERGENHAHN, M., W. ADOLF, and E. HECKER: Resiniferatoxin and other novel polyfunctional diterpenes from *Euphorbia resinifera* and *E. unispina*. Tetrahedron Letters **1975,** 1595.
13. BURKE, B. A., W. R. CHAN, K. O. PASCOE, J. F. BLOUNT, and P. S. MANCHAND: The structure of crotonitenone, a novel casbane diterpene from *Croton nitens* SW. (Euphorbiaceae). J. Chem. Soc. Perkin I **1981,** 2666.
14. ADOLF, W., E. HECKER, A. BALMAIN, M. F. LOHMME, Y. NAKATANI, G. OURISSON, G. PONSINET, R. J. PRYCE, T. S. SANTHANAKRISHNAN, L. G. MATYUKHINA, and I. A. SALTIKOVA: Euphorbiasteroid (epoxylathyrol), a new tricyclic diterpene from *Euphorbia lathyris*. Tetrahedron Letters **1970,** 2241.
15. GHISALBERTI, E. L., P. R. JEFFERIES, T. G. PAYNE, and G. K. WORTH: Bertyadionol – an unusual diterpene. Tetrahedron Letters **1970,** 4599.
16. KUPCHAN, S. M., C. W. SIGEL, M. J. MATZ, J. A. S. RENAULD, R. C. HALTIWANGER, and R. F. BRYAN: Jatrophone, a novel macrocyclic diterpenoid tumour inhibitor from *Jatropha gossypiifolia*. J. Amer. Chem. Soc. **92,** 4476 (1970).
17. UEMURA, D., and Y. HIRATA: The structure of kansuinine A, a new multi-oxygenated diterpene. Tetrahedron Letters **1975,** 1697.
18. TORRANCE, S. J., R. M. WIEDHOPF, J. R. COLE, S. K. ARORA, R. B. BATES, W. A. BEAVER, and R. S. CUTLER: Anti-tumour agents from *Jatropha macrorhiza* (Euphorbiaceae) II. Isolation and characterisation of jatrophatrione. J. Org. Chem. **41,** 1855 (1976).
19. CHAN, W. R., E. C. PRINCE, P. S. MANCHARD, J. P. SPRINGER, and J. CLARDY: The structure of crotofolin-A, a diterpene with a new skeleton. J. Amer. Chem. Soc. **97,** 4437 (1975).
20. PUROSHATHAMAN, K. K., S. CHANDRASEKHAREN, A. F. CAMERON, J. D. CONNOLLY, C. LABBE, A. MALTZ, and D. S. RYCROFT: Jatropholones A and B, new diterpenoids from

the roots of *Jatropha gossypiifolia*. Crystal structure analysis of jatropholone B. Tetrahedron Letters **1979,** 979.

21. STUART, K. L., and M. BARRETT: A phorbol derivative from *Croton rhamnifolius*. Tetrahedron Letters **1969,** 2399.
22. WEBSTER, G. L.: The genera of the Euphorbiaceae in the South-Eastern United States. J. Arnold Arbor. **48,** 303 (1967).
23. FARNSWORTH, N. R., R. N. BLOMSTER, W. M. MESSMER, J. C. KING, G. J. PERSINOS, and J. D. WILKES: A phytochemical and biological review of the genus *Croton*. Lloydia **32,** 1 (1969).
24. EVANS, F. J., and R. J. SCHMIDT: Plants and plant products that induce contact dermatitis. Planta Medica **38,** 289 (1980).
25. PAX, F., and H. HOFFMAN: Euphorbiaceae. In: Die natürlichen Pflanzenfamilien, 2nd ed. 19C, p. 208 (ENGLER, A., and K. PRANTL eds.). Leipzig: Engelmann. 1931.
26. WHITE, A., R. A. DYER, and B. L. SLOANE: The succulent Euphorbiëae. Vol. I and II, p. 1. Pasadena, California: Abbey Garden Press. 1941.
27. AIRY-SHAW, H. K.: The Euphorbiaceae of Borneo. Kew Bull. Additional Series IV. London, H.M.S.O. iv (1975).
28. BOISSIER, E.: Euphorbieae. In: *Prodromus Systematis Naturalis Regni Vegetabilis* (DE CANDOLLE, A. P., ed.). **15,** 3 (1862).
29. NEVLING, L. I.: The Thymelaeaceae in the South Eastern United States. J. Arnold Arbor. **43,** 428 (1962).
30. ROBINSON, D. R., and C. A. WEST: Biosynthesis of cyclic diterpenes in extracts from seedlings of *Ricinus communis* L. I. Identification of diterpene hydrocarbons formed from mevalonate. Biochemistry **9,** 70 (1970).
31. ADOLF, W., and E. HECKER: Diterpenoid irritants and co-carcinogens in Euphorbiaceae and Thymelaeaceae, structural relationships in view of their biogenesis. Isr. J. Chem. **16,** 75 (1977).
32. UEMURA, D., K. NOBUHARA, Y. NAKAYAMA, Y. SHIZURI, and Y. HIRATA: The structure of new lathyrane diterpenes, jolkinols A, B, C and D from *Euphorbia jolkini* Boiss. Tetrahedron Letters **1976,** 4593.
33. SCHMIDT, R. J., and F. J. EVANS: A new aromatic ester diterpene from *Euphorbia poissonii*. Phytochem. **15,** 1778 (1976).
34. HECKER, E., and R. SCHMIDT: Phorbol esters, the irritants and co-carcinogens of *Croton tiglium* L. Fortschr. Chem. Organ. Naturstoffe **31,** 377 (1974).
35. TAYLOR, S. E., and F. J. EVANS: Unpublished results (1980).
36. EVANS, F. J.: The irritant toxins of the blue *Euphorbia (E. coerulescens)*. Toxicon **16,** 51 (1978).
37. KINGHORN, A. D., and F. J. EVANS: Skin irritants of *Euphorbia fortissima*. J. Pharm. Pharmacol. **27,** 329 (1975).
38. OCKEN, P. R.: Dry column chromatographic isolation of fatty acid esters of phorbol from *Croton* oil. J. Lipid. Res. **10,** 460 (1969).
39. OHIGASHI, H., and T. MITSUI: Studies on the biologically active substances of *Sapium japonicum*. Bull. Inst. Chem. Res. Kyoto University **50,** 239 (1972).
40. EDWARDS, M. E., S. E. TAYLOR, E. M. WILLIAMSON, and F. J. EVANS: New phorbol and deoxyphorbol esters, isolation and relative potencies in inducing platelet aggregation *in vitro* and erythema *in vivo*. Acta Pharmacol. et Toxicol. In press (1982).
41. MARSHALL, G. T., and A. D. KINGHORN: Isolation of phorbol and 4α-phorbol from *Croton* oil by droplet counter current chromatography. J. Chromatogr. **206,** 421 (1981).
42. EVANS, F. J., R. J. SCHMIDT, and A. D. KINGHORN: A microtechnique for the identification of diterpene ester inflammatory toxins. Biomed. Mass Spectr. **2,** 126 (1975).
43. EVANS, F. J., and A. D. KINGHORN: Thin layer chromatographic behaviour of the

acetates of some polyfunctional diterpene alcohols of toxicological interest. J. Chromatogr. **87,** 443 (1973).

44. UEMURA, D., and Y. HIRATA: New diterpene. 13-Oxy-ingenol isolated from *Euphorbia kansui.* Tetrahedron Letters **1974,** 2529.
45. DRIEDGER, P. E., and P. M. BLUMBERG: Structure-activity relationships in chick embryo fibroblasts for phorbol-related diterpene esters showing anomalous activities *in vivo.* Cancer Res. **40,** 339 (1980).
46. KINGHORN, A. D., and F. J. EVANS: A quantitative GLC method for phorbol and related diterpenes as their acetates. J. Pharm. Pharmacol. **26,** 408 (1974).
47. EVANS, F. J., and A. D. KINGHORN: A screening method for co-carcinogens. J. Pharm. Pharmacol. **25,** 145P (1973).
48. EVANS, F. J., and R. J. SCHMIDT: An assay procedure for the comparative irritancy testing of esters in the tigliane and daphnane series. Inflammation **3,** 215 (1978).
49. DUBLYANSKAYA, N. F.: Über den biologischen Effekt der die Toxizität von *E. lathyris* bedingenden Fraktionen. Pharmazie Pharmakol. **11—12,** 50 (1937).
50. ABO, K., and F. J. EVANS: Macrocyclic diterpene esters of the cytotoxic fraction from *Euphorbia kamerunica.* Phytochem. **20,** 2535 (1981).
51. CROMBIE, L., G. KNAEN, G. PATTANDEN, and D. WHYBROW: Total synthesis of the macrocyclic diterpene (−)-casbene, the putative biogenetic precursor of the lathyrane, tigliane, ingenane and related terpenoid structures. J. Chem. Soc. **8,** 1711 (1980).
52. KUPCHAN, S. M., C. W. SIGEL, M. J. MATZ, C. J. GILMORE, and R. F. BRYAN: Structure and stereochemistry of jatrophone, a novel macrocyclic diterpenoid tumour inhibitor. J. Amer. Chem. Soc. **98,** 2295 (1976).
53. UEMURA, D., C. KATAYAMA, E. UNO, K. SASAKI, and Y. HIRATA: Kansuinine B. A novel multi-oxygenated diterpene from *Euphorbia kansui* Liou. Tetrahedron Letters **21,** 1703 (1975).
54. UEMURA, D., and Y. HIRATA: Stereochemistry of kansuinine A. Tetrahedron Letters **21,** 1701 (1975).
55. ZECHMEISTER, K., M. RÖHRL, F. BRANDL, S. HECHTFISCHER, W. HOPPE, E. HECKER, and H. KUBINYI: Röntgenstrukturanalyse eines neuen makrozyklischen Diterpen-Ester aus der Springwolfsmilch (*Euphorbia lathyris* L.). Tetrahedron Letters **1970,** 3071.
56. HUI, W.-H., and K. K. NG: Terpenoids and steroids from *Macaranga tanarius.* Phytochem. **14,** 816 (1975).
57. NARAYANAN, P., M. RÖHRL, K. ZECHMEISTER, D. W. ENGEL, W. HOPPE, E. HECKER, and W. ADOLF: Structure of 7-hydroxylathyrol. A further diterpene from *Euphorbia lathyris* L. Tetrahedron Letters **1971,** 1325.
58. ADOLF, W., and E. HECKER: Further new diterpenes from the irritant and co-carcinogenic seed oil and latex of the Caper spurge (*Euphorbia lathyris* L.). Experientia **27,** 1393 (1971).
59. GHISALBERTI, E. L., P. R. JEFFERIES, T. G. PAYNE, and G. K. WORTH: Structure and stereochemistry of bertyadionol. Tetrahedron Letters **1973,** 403.
60. GHISALBERTI, E. L., P. R. JEFFERIES, R. F. TOIA, and G. K. WORTH: Stereochemistry of bertyadionol and related compounds. Tetrahedron **30,** 3269 (1974).
61. LOTTER, H., H. J. OPFERKUCH, and E. HECKER: X-ray crystallography of ingol-3,7,8,12-tetra-acetate obtained from naturally occurring esters by mild hydrolysis and acetylation procedures. Tetrahedron Letters **1979,** 77.
62. OPFERKUCH, H. J., and E. HECKER: Über ein neues Diterpen aus *Euphorbia ingens.* XVI Colloquium Spectroscopium Internationale. Heidelberg 1971. Vol. I, pp. 282.
63. —— Ingol – a new macrocyclic diterpene alcohol from *Euphorbia ingens.* Tetrahedron Letters **1973,** 3611.
64. HERGENHAHN, M., S. KUSUMOTO, and E. HECKER: Diterpene esters from Euphorbium and their irritant and co-carcinogenic activity. Experientia **30,** 1438 (1974).

65. UPADHYAY, R. R., and E. HECKER: Diterpene esters of the irritant and co-carcinogenic latex of *Euphorbia lactea*. Phytochem. **14,** 2514 (1975).
66. ABO, K., and F. J. EVANS: The composition of a mixture of ingol-esters from *Euphorbia kamerunica*. Planta Medica **43,** 392 (1981).
67. — — A triester of ingol from the latex of *Euphorbia kamerunica* Pax. J. Nat. Prods. **45,** 365 (1981).
68. SCHROEDER, G., M. ROHMER, J. P. BECK, and R. ANTON: Cytotoxic activity of 6,20-epoxylathyrol and its aliphatic diesters. Planta Medica **35,** 235 (1979).
69. MARSH, N., M. ROTHSCHILD, A. SCUTT, F. J. EVANS, and A. TEDSTONE: A cytotoxin from the pupa of the spurge Hawk Moth (*Hyles euphorbiae* L. Sphingidae: LEP), inhibiting the growth of malignant cells *in vivo* and *in vitro*. Unpublished results (1982).
70. BLUMBERG, P. M.: *In vitro* studies on the mode of action of the phorbol esters, potent tumour promoters. Parts I and II. C. R. C. Critical Reviews in Toxicology **8,** 153 (1980 – 1981).
71. BOHM, R., B. FLASCHENTRAGER, and L. LENDLE: Über die Wirksamkeit von Substanzen aus dem Kroton-Öl. Arch. exp. Pathol. Pharmakol. **177,** 212 (1935).
72. HECKER, E., CH. V. SZCZEPANSKI, H. KUBINYI, H. BRESCH, E. HÄRLE, H. U. SCHAIRER and H. BARTSCH: Über die Wirkstoffe des Crotonöls, VII. Phorbol. Z. Naturforsch. **21b,** 1204 (1966).
73. ARROYO, E. R., and J. HOLCOMB: Isolation and structure elucidation of a highly active principle from Croton oil. Chem. and Ind. **1965,** 350.
74. — — Structural studies of an active principle from *Croton tiglium* L. J. Med. Chem. **8,** 672 (1965).
75. HECKER, E., E. KUBINYI, CH. V. SZCZEPANSKI, E. HÄRLE, and H. BRESCH: Phorbol – ein neues tetracyclisches Diterpen aus Crotonöl. Tetrahedron Letters **1965,** 1837.
76. HECKER, E., H. BARTSCH, H. BRESCH, M. GSCHWENDT, E. HÄRLE, G. KREIBICH, H. KUBINYI, H. U. SCHAIRER, CH. V. SZCZEPANSKI, and H. W. THIELMANN: Structure and stereochemistry of the tetracyclic diterpene phorbol from *Croton tiglium* L. Tetrahedron Letters **1967,** 3165.
77. HOPPE, W., F. BRANDL, I. STRELL, M. RÖHRL, I. GASSMANN, E. HECKER, H. BARTSCH, G. KREIBICH, and CH. V. SZCZEPANSKI: X-ray structure analysis of neophorbol. Angew. Chem. internat. edit. **6,** 809 (1967).
78. PETTERSEN, R. C., G. I. BIRNBAUM, G. FERGUSON, K. M. S. ISLAM, and J. G. SIME: X-ray investigation of several phorbol derivatives. The crystal and molecular structure of phorbol bromofuroate chloroform solvate at $-160°$ C. J. Chem. Soc. (B) **1968,** 980.
79. HOPPE, W., K. ZECHMEISTER, M. RÖHRL, F. BRANDL, E. HECKER, G. KREIBICH, and H. BARTSCH: The structure determination of a solvate of phorbol, the diterpene parent of the tumour promoters from *Croton* oil. Tetrahedron Letters **1969,** 667.
80. CROMBIE, L., M. L. GAMES, and D. J. POINTER: Chemistry and structure of phorbol, the diterpene parent of the co-carcinogens of *Croton* oil. J. Chem. Soc. (C) **1968,** 1347.
81. PACK, G. R.: The molecular structure and charge distribution of phorbol, parent compound of the tumour-promoting phorbol-diesters. Cancer Biochem. Biophys. **5,** 183 (1981).
82. JACOBI, P., E. HÄRLE, H. U. SCHAIRER, and E. HECKER: Chemie des Phorbols, XVI. 4$\alpha$-Phorbol. Liebigs Ann. Chem. **741,** 13 (1970).
83. HECKER, E.: Isolation and characterisation of the co-carcinogenic principles of *Croton* oil. In: Methods of Cancer Research (BUSCH, H., ed.), Vol. vi. London: Academic Press. 1971.
84. UPADHYAY, R. R., and E. HECKER: A new cryptic irritant and co-carcinogen from the seeds of *Croton sparciflorus*. Phytochem. **15,** 1070 (1976).
85. FÜRSTENBERGER, G., and E. HECKER: New highly irritant Euphorbia factors from the latex of *Euphorbia tirucalli* L. Experientia **33,** 986 (1977).

*86.* Evans, F. J.: A new phorbol triester from the latices of *Euphorbia franckiana* and *E. coerulescens*. Phytochem. **16,** 395 (1977).

*87.* Szczepanski, Ch. V., H. U. Schairer, M. Gschwendt, and E. Hecker: Zur Chemie des Phorbols III. – Mono- und Diacetate des Phorbols. Liebigs Ann. Chem. **705,** 199 (1967).

*88.* Bartsch, H., and E. Hecker: Circular dichroism in phorbol derivatives. Angew. Chem. internat. edit. **6,** 974 (1967).

*89.* Fürstenberger, G., and E. Hecker: The new diterpene 4-deoxyphorbol and its highly unsaturated irritant diesters. Tetrahedron Letters **1977,** 925.

*90.* — — Zum Wirkungsmechanismus co-carcinogener Pflanzeninhaltsstoffe. Planta Medica **22,** 241 (1972).

*91.* Kinghorn, A. D.: Characterisation of an irritant 4-deoxyphorbol diester from *Euphorbia tirucalli*. J. Nat. Prod. **42,** 112 (1979).

*92.* Falsone, G., and A. E. G. Crea: Drei neue 4-deoxyphorbol-tri-esters aus *Euphorbia biglandulosa*. Liebigs Ann. Chem. **1979,** 1116.

*93.* Taylor, S. E., M. A. Gafur, A. K. Choudhury, and F. J. Evans: Sapintoxin A, a new biologically active nitrogen containing phorbol ester. Experientia **37,** 681 (1981).

*94.* Taylor, S. E., F. J. Evans, M. A. Gafur, and A. K. Choudhury: Sapintoxin D, a new phorbol ester from *Sapium indicum*. J. Nat. Prod. **44,** 729 (1981).

*95.* Taylor, S. E., M. A. Gafur, A. K. Choudhury, and F. J. Evans: Sapatoxins, aliphatic ester tigliane diterpenes from *Sapium indicum*. Phytochem. **21,** 405 (1982).

*96.* Miana, G. A., R. Schmidt, E. Hecker, M. Shamma, J. L. Moniot, and M. Kiamuddin: 4α-Sapinine – a novel diterpene ester from *Sapium indicum*. Z. Naturforsch. **326,** 727 (1977).

*97.* Weber, J., and E. Hecker: Co-carcinogens of the diterpene ester type from *Croton flavens* and esophageal cancer in Curaçao. Experientia **34,** 1595 (1978).

*98.* Hecker, E., and J. Weber: Co-carcinogens from *Croton flavens* and the high incidence of esophageal cancer in Curaçao. In: Proc. 7th Internat. Symp. on the Biological Characterisation of Human Tumours (Davis, W., ed.). Excerpta Medica (1977).

*99.* Taylor, S. E., M. A. Gafur, A. K. Choudhury, and F. J. Evans: Nitrogen containing phorbol derivatives of *Sapium indicum*. Phytochem. **20**, 2749 (1981).

*100.* — — — — 4-Deoxyphorbol and 4α-deoxyphorbol aldehydes, new diterpenes and their esters. Tetrahedron Letters **22,** 3321 (1981).

*101.* Gschwendt, M., and E. Hecker: Tumour promoting compounds from *Euphorbia triangularis*. Mono- and diesters of 12-deoxyphorbol. Tetrahedron Letters **1969,** 3509.

*102.* — — Tumorpromovierende Diterpenfettsäureester aus *Euphorbia triangularis* und *E. cooperi*. Fette, Seifen, Anstrichmittel **73,** 221 (1971).

*103.* — — Hautreizende und co-carcinogene Fraktionen aus *E. triangularis*. Z. Krebsforsch. **81,** 193 (1974).

*104.* Evans, F. J., and A. D. Kinghorn: A comparative phytochemical study of the diterpenes of some species of the genera *Euphorbia* and *Elaeophorbia* (Euphorbiaceae). Bot. J. Linn. Soc. **74,** 23 (1977).

*105.* Kinghorn, A. D.: Some biologically active constituents of the genus *Euphorbia*. Ph. D. thesis, University of London (1975).

*106.* Cashmore, A. R., R. N. Seelye, B. F. Cairn, H. Mack, R. Schmidt, and E. Hecker: The structure of prostratin, a toxic tetracyclic diterpene ester from *Pimelea prostrata*. Tetrahedron Letters **1976,** 1737.

*107.* McCormick, I. R. N., P. E. Nixon, and T. N. Waters: On the structure of prostratin, an X-ray study. Tetrahedron Letters **1976,** 1735.

*108.* Zayed, S., A. Hafez, W. Adolf, and E. Hecker: New tigliane and daphnane derivatives from *Pimelea prostrata* and *P. simplex*. Experientia **33,** 1554 (1977).

*109.* Evans, F. J., and A. D. Kinghorn: New diesters of 12-deoxyphorbol. Phytochem. **14,** 1669 (1975).

110. EVANS, F. J., A. D. KINGHORN, and R. J. SCHMIDT: Some naturally occurring skin irritants. Acta Pharmacol. et Toxicol. **37,** 250 (1975).
111. SCHMIDT, R. J., and F. J. EVANS: The succulent *Euphorbias* of Nigeria. Part II. Aliphatic diterpene esters from *Euphorbia poissonii* Pax and *E. unispina* N. E. Br. Lloydia **40,** 225 (1977).
112. — — Two minor diterpenes from *Euphorbia* latex. Phytochem. **17,** 1436 (1978).
113. EVANS, F. J., and R. J. SCHMIDT: Two new toxins from the latex of *Euphorbia poissonii.* Phytochem. **15,** 333 (1976).
114. — — The succulent *Euphorbias* of Nigeria. Part III. Structure and potency of the aromatic diterpenes of *Euphorbia poissonii* Pax. Acta Pharmacol. et Toxicol. **45,** 181 (1979).
115. SCHMIDT, R. J., and F. J. EVANS: Skin irritants of the sun spurge (*Euphorbia helioscopia* L.). Contact dermatitis **6,** 204 (1980).
116. OGURA, M., K. KOIKE, G. A. CORDELL, and N. R. FARNSWORTH: Potential anticancer agents. VIII. Constituents of *Baliospermum montanum* (Euphorbiaceae). Planta Medica **33,** 128 (1978).
117. GSCHWENDT, M., and E. HECKER: Tumour-promoting compounds from *Euphorbia cooperi.* Di- and triesters of 16-hydroxy-12-deoxyphorbol. Tetrahedron Letters **1970,** 567.
118. — — Hautreizende und Co-carcinogene aus *Euphorbia cooperii.* Z. Krebsforsch. **80,** 335 (1973).
119. SCHMIDT, R. J., and F. J. EVANS: Candletoxins A and B. Two new aromatic esters of 12-deoxy-16-hydroxyphorbol from the latex of *Euphorbia poissonii.* Experientia **33,** 1197 (1977).
120. SCHMIDT, R. J.: Chemical and biological studies on the tigliane and daphnane diterpenes of three *Euphorbia* species. Ph. D. thesis, University of London (1978).
121. ADOLF, W., and E. HECKER: New irritant diterpene esters from the roots of *Stillingia sylvatica* L. (Euphorbiaceae). Tetrahedron Letters **21,** 2887 (1980).
122. — — On the irritant and co-carcinogenic principles of *Hippomane mancinella.* Tetrahedron Letters **1975,** 1587.
123. OKUDA, T., T. YOSHIDA, S. KOIKE, and N. TOH: The toxic constituents of the fruits of *Aleurites fordii.* Chem. Pharm. Bull. **22,** 971 (1974).
124. — — — — New diterpene esters from *Aleurites fordii.* Phytochem. **14,** 509 (1975).
125. NEEMAN, M., and O. D. SIMMONS: Carbon-13 nuclear magnetic resonance spectroscopy of phorbol. Canad. J. Chem. **57,** 2071 (1979).
126. ADOLF, W., H. J. OPFERKUCH, and E. HECKER: Über die diterpenoiden Inhaltsstoffe des Samenöls von *Euphorbia lathyris* und ihre tumorpronovierende Wirkung. Fette, Seifen, Anstrichmittel **70,** 850 (1968).
127. OPFERKUCH, H. J., and E. HECKER: New diterpenoid irritants from *Euphorbia ingens.* Tetrahedron Letters **1974,** 261.
128. ZECHMEISTER, K., F. BRANDL, W. HOPPE, E. HECKER, H. J. OPFERKUCH, and W. ADOLF: Structure determination of the new tetracyclic diterpene ingenol-triacetate with triple product methods. Tetrahedron Letters **1970,** 4075.
129. EVANS, F. J., and A. D. KINGHORN: Ingenol from *Euphorbia desmondi.* Phytochem. **13,** 1011 (1974).
130. — — The succulent *Euphorbias* of Nigeria. Part I. Lloydia **38,** 363 (1975).
131. UPADHYAY, R. R., M. H. ZARINTAN, and M. ANSARIN: Irritant constituents of Iranian plants. Ingenol from *Euphorbia seguieriana.* Planta Medica **30,** 196 (1976).
132. HIROTA, M., H. OHIGASHI, Y. OKI, and K. KOSHIMIZU: New ingenol esters as piscicidal constituents of *Euphorbia cotinifolia* L. Agric. Biol. Chem. **44,** 1351 (1980).
133. EVANS, F. J., and A. D. KINGHORN: A new ingenol type diterpene from the irritant fraction of *Euphorbia myrsinites* and *Euphorbia biglandulosa.* Phytochem. **13,** 2324 (1974).

*134.* BIEMAN, K.: Mass spectrometry – organic chemical applications, p. 121. London: McGraw-Hill. 1962.

*135.* UPADHYAY, R. R., F. BAKHTAVAR, M. MOHSENI, A. M. SATER, N. SALEH, A. TAFAZULI, F. N. DIZAJI, and G. MOHADDES: Screening of *Euphorbia* from Azarbaijan for skin irritant activity and for diterpenes. Planta Medica **38,** 151 (1980).

*136.* ABO, K. A., and F. J. EVANS: Ingenol esters from the pro-inflammatory fraction of *Euphorbia kamerunica.* Phytochem. **21,** 725 (1981).

*137.* UPADHYAY, R. R., M. ANSARIN, M. H. ZARINTAN, and P. SHAKUI: Tumour promoting constituent of *Euphorbia serrata* L. latex. Experientia **32,** 1196 (1976).

*138.* UPADHYAY, R. R., F. BAKHTAVAR, M. GHAISARZEDEH, and J. TILABI: Co-carcinogenic and irritant factors of *Euphorbia esula* L. latex. Tumori **64,** 99 (1978).

*139.* HIRATA, Y.: Toxic substances of Euphorbiaceae. Pure Appl. Chem. **41,** 175 (1975).

*140.* UEMURA, D., and Y. HIRATA: Isolation and structures of irritant substances obtained from *Euphorbia* species (Euphorbiaceae). Tetrahedron Letters **1973,** 881.

*141.* UPADHYAY, R. R., R. SAMIYEH, and A. TAFAZULI: Tumor promoting and skin irritant diterpene esters of *Euphorbia virgata* latex. Neoplasma **28,** 555 (1981).

*142.* SAYED, M. D., A. RISZK, F. M. HAMMOUDA, M. M. EL-MISSIRY, E. M. WILLIAMSON, and F. J. EVANS: Constituents of Egyptian Euphorbiaceae IX. Irritant and cytotoxic ingenane esters from *Euphorbia paralias* L. Experientia **36,** 1206 (1980).

*143.* UEMURA, D., and Y. HIRATA: The isolation and structures of two new alkaloids, milliamines A and B obtained from *Euphorbia millii.* Tetrahedron Letters **1971,** 3673.

*144.* UEMURA, D., H. OHWAKI, Y. HIRATA, Y.-P. CHEN, and H.-Y. HSU: Isolation and structures of 20-deoxyingenol, new diterpene derivatives and ingenol derivative obtained from "kansui". Tetrahedron Letters **1974,** 2527.

*145.* UPADHYAY, R. R., and E. HECKER: Diterpene esters of the irritant and co-carcinogenic latex of *Euphorbia lactea.* Phytochem. **14,** 2514 (1975).

*146.* OTT, H. H., and E. HECKER: Highly irritant ingenane-type diterpene esters from *Euphorbia cyparissias* L. Experientia **37,** 88 (1981).

*147.* STOUT, G. H., W. G. BALKENHOL, M. POLING, and G. L. HICKERNELL: The isolation and structure of Daphnetoxin, the poisonous principle of *Daphne* species. J. Amer. Chem. Soc. **92,** 1070 (1970).

*148.* GOTO, T., Y. KISHI, S. TAKAHASHI, and Y. HIRATA: Tetrodotoxin. Tetrahedron **21,** 2059 (1965).

*149.* RONLÁN, A., and B. WICKBERG: The structure of mezerein, a major toxic principle of *Daphne mezereum* L. Tetrahedron Letters **1970,** 4261.

*150.* SAKATA, K., K. KAWAZU, and T. MITSUI: Studies on the piscicidal constituent of *Hura crepitans.* Part I: Isolation and characterisation of huratoxin and its piscicidal activity. Agric. Biol. Chem. **35,** 1084 (1971).

*151.* — — — Studies on the piscicidal constituent of *Hura crepitans.* Part II: Chemical structure of huratoxin. Agric. Biol. Chem. **35,** 2113 (1971).

*152.* SAKATA, K., K. KAWAZU, T. MITSUI, and N. MASAKI: The structure and stereochemistry of huratoxin: a piscicidal constituent of *Hura crepitans.* Tetrahedron Letters **1971,** 1141.

*153.* OHIGASHI, H., H. KATSUMATA, K. KAWAZU, K. KOSHIMIZU, and T. MITSUI: A piscicidal constituent of *Excoecaria agallocha.* Agric. Biol. Chem. **38,** 1093 (1974).

*154.* GUNASEKERA, S. P., G. A. CORDELL, and N. R. FARNSWORTH: Potential anticancer agents. XIV. Isolation of spruceanol and montanin from *Cunuria spruceana* (Euphorbiaceae). J. Nat. Prod. **42,** 658 (1979).

*155.* ROBERTS, H. B., T. H. MCCLURE, E. RITCHIE, W. C. TAYLOR, and P. W. FREEMAN: The isolation and structure of the toxin of *Pimelea simplex* responsible for St George's disease of cattle. Austral. Vet. J. **51,** 325 (1975).

*156.* KELLY, W. R., and A. A. SEAWRIGHT: *Pimelea* spp. poisoning of cattle. In: Effects of poisonous plants on livestock, p. 293. New York-San Francisco-London: Academic Press Inc. 1978.

157. SCHILDKNECHT, H., G. EDELMAN, and R. MAURER: Zur Chemie des Mezereins, des entzündlichen und co-carcinogenen Giftes aus dem Seidelbast *Daphne mezereum*. Chem. Ztg. **94,** 347 (1970).

158. KUPCHAN, S. M., and R. L. BAXTER: Mezerein: antileukemic principle isolated from *Daphne mezereum* L. Science **187,** 652 (1974).

159. NYBORG, J., and T. LA COUR: X-ray diffraction study of molecular structure and conformation of mezerein. Nature **257,** 824 (1975).

160. COERTZER, J., and M. J. PIETERSE: The isolation of 12-hydroxydaphnetoxin, a degradation product of a constituent of *Lasiosiphon burchellii*. J. van die Suid-Afrikaanse Chem. Inst. **24,** 241 (1974).

161. KUPCHAN, S. M., J. G. SWEENY, R. L. BAXTER, T. MURAE, V. A. ZIMMERLY, and B. R. SICKLES: Gnididin, Gniditrin and Gnidicin, novel potent antileukemic diterpenoid esters from *Gnidia lamprantha*. J. Amer. Chem. Soc. **97,** 672 (1975).

162. KUPCHAN, S. M., Y. SHIZURI, W. C. SUMNER, H. R. HAYNES, A. P. LEIGHTON, and B. R. SICKLES: Isolation and structural elucidation of a new potent antileukemic diterpenoid esters from *Gnidia* species. J. Org. Chem. **41,** 3850 (1976).

163. KOGISO, S., K. WADA, and K. MUNAKATA: Odoracin, a nematicidal constituent from *Daphne odora*. Agric. Biol. Chem. **40,** 2119 (1976).

164. SCHMIDT, R. J., and F. J. EVANS: The structure and potency of the Tinyatoxins. J. Pharm. Pharmacol. **27,** 50P (1975).

165. KUPCHAN, S. M., Y. SHIZURI, T. MURAE, J. G. SWEENY, H. R. HAYNES, M.-S. SHEN, J. C. BARRICK, R. F. BRYAN, D. VAN DER HELM, and K. K. WU: Gnidimacrin and gnidimacrin-20-palmitate, novel macrocyclic antileukemic diterpenoid esters from *Gnidia subcordata*. J. Amer. Chem. Soc. **98,** 5719 (1976).

166. ZAYED, S., W. ADOLF, A. HAFEZ, and E. HECKER: New highly irritant l-alkyldaphnane derivatives from several species of Thymelaeaceae. Tetrahedron Letters **1977,** 3481.

167. TYLER, M. I., and M. E. H. HOWDEN: Piscicidal constituents of *Pimelea* species. Tetrahedron Letters **22,** 689 (1981).

168. SCHMIDT, R. J., and F. J. EVANS: Skin irritant effects of esters of phorbol and related polyols. Arch. Toxicol. **44,** 279 (1980).

169. LEE, L. S., and I. B. WEINSTEIN: Tumour-promoting phorbol esters inhibit binding of epidermal growth factor to cellular receptors. Science **202,** 313 (1978).

170. FREI, J. V., and P. STEVENS: The correlation of promotion of tumour growth and of induction of hyperplasia in epidermal two stage carcinogenesis. Brit. J. Cancer **22,** 63 (1968).

171. RAICK, A. N.: Cell proliferation and promoting action in skin carcinogenesis. Cancer Res. **34,** 920 (1974).

172. DIAMOND, L., S. O'BRIEN, C. DONALDSON, and Y. SHIMIZU: Growth stimulation of human diploid fibroblasts by tumour-promoter 12-O-tetradecanoyl-phorbol-13-acetate. Int. J. Cancer **13,** 721 (1974).

173. SIVAK, A., and B. L. VAN DUUREN: RNA synthesis induction in cell cultures by a tumour-promoter. Cancer Res. **30,** 1203 (1970).

174. BAIRD, W. M., P. W. MELERA, and R. K. BOUTWELL: Acrylamide gel electrophoresis studies of the incorporation of cytidine-$^3$H into mouse skin RNA at early times after treatment with phorbol-esters. Cancer Res. **32,** 781 (1972).

175. BAIRD, W. H., J. A. SEDGWICK, and R. K. BOUTWELL: Effects of phorbol and four diesters of phorbol on the incorporation of tritiated precursors into DNA, RNA and protein in mouse epidermis. Cancer Res. **31,** 1434 (1971).

176. SHINOZUKA, H., and A. C. RITCHIE: Pretreatment with *Croton* oil, DNA synthesis and carcinogenesis by carcinogen followed by *Croton* oil. Int. J. Cancer **2,** 77 (1967).

177. YUSPA, S. H., H. HENNINGS, and T. BEN: Stimulated DNA synthesis in mouse epidermal cell structures treated with 12-O-tetradecanoylphorbol-13-acetate. Proc. Am. Assoc. Cancer Res. **16,** 69 (1975).
178. HENNINGS, H., G. T. BOWDEN, and R. K. BOUTWELL: The effect of *Croton* oil pretreatment on skin tumour initiation in mice. Cancer Res. **29,** 1773 (1969).
179. SCRIBNER, J. D., and R. K. BOUTWELL: Inflammation and tumour-promotion: Selective protein induction in mouse skin by tumour-promoters. Eur. J. Cancer **8,** 617 (1972).
180. SCRIBNER, J. D., and T. J. SLAGA: Multiple effects of dexamethasone on protein synthesis and hyperplasia caused by a tumour-promoter. Cancer Res. **33,** 542 (1973).
181. VAN DUUREN, B. L.: Tumour promoting agents in two stage carcinogenesis. Progr. Exptl. Tumour Res. **11,** 31 (1964).
182. LANGENBACK, R., and C. KUSYYNSKI: Non-specific inhibition of DNA repair by promoting and non-promoting phorbol esters. J. Natl. Cancer Inst. **55,** 801 (1975).
183. POIRIER, M. C., and M. W. LIEBERMAN: Non-specific inhibition of DNA repair synthesis by tumour-promoters. Proc. Am. Assoc. Cancer Res. **16,** 43 (1975).
184. TROSKO, J. E., C. CHANG, L. P. YOTTI, and E. H. Y. CHU: Effect of phorbol myristate acetate on the recovery of spontaneous and ultra-violet light induced 6-thioguanine and oubain resistant Chinese hamster cells. Cancer Res. **37,** 188 (1977).
185. SOPER, C. J., and F. J. EVANS: Investigations into the mode of action of the co-carcinogen 12-O-tetradecanoylphorbol-13-acetate using auxotrophic bacteria. Cancer Res. **37,** 2487 (1977).
186. ROHRSCHNEIDER, L. R., and R. K. BOUTWELL: Phorbol-esters, fatty acids and tumour-promotion. Nature New Biol. **243,** 212 (1973).
187. SIVAK, A., and B. L. VAN DUUREN: Cellular interactions of phorbol myristate acetate in tumour-promotion. Chem. Biol. Interact. **3,** 401 (1971).
188. KUBINSKI, H., M. A. STRANGSTALIEN, W. BAIRD, and R. K. BOUTWELL: Interaction of phorbol-esters with cellular membranes *in vitro*. Cancer Res. **33,** 3103 (1973).
189. WITZ, G., B. L. VAN DUUREN, and S. BANERJEE: The interaction of phorbol myristate acetate, a tumour-promoter, with rat liver plasma membrane – a fluorescence study. Proc. Am. Assoc. Cancer Res. **16,** 30 (1975).
190. WENNER, C. E., J. HACKNEY, H. K. KIMELBERG, and E. MAYHEW: Membrane effects of phorbol esters. Cancer Res. **34,** 1731 (1974).
191. WILSON, S. R., and J. C. HUFFMAN: The structural relationship of phorbol and cortisol, a possible mechanism for the tumour-promoting activity of phorbol. Experientia **32,** 1489 (1976).
192. EWARDS, M. C., S. E. TAYLOR, A. M. E. NOURI, D. GORDON, and F. J. EVANS: Tumour-promoting and non-promoting pro-inflammatory phorbol esters act as human lymphocyte mitogens. J. Pharm. Pharmacol. **33,** 55P (1981).
193. WESTWICK, J., E. M. WILLIAMSON, and F. J. EVANS: Structure activity relationships of 12-deoxyphorbol esters on human platelets. Thrombosis Res. **20,** 683 (1980).
194. BLUMBERG, P. M., P. E. DRIEDGER, and P. W. ROSSOW: Effects of phorbol esters on a transformation sensitive surface protein of chick fibroblasts. Nature **264,** 446 (1976).
195. DRIEDGER, P. E., and P. M. BLUMBERG: Quantitative correlation between *in vitro* and *in vivo* activities of phorbol esters. Cancer Res. **39,** 714 (1979).
196. SCHMIDT, R. J., and F. J. EVANS: Investigations into the skin-irritant properties of resiniferonol ortho-esters. Inflammation **3,** 273 (1979).
197. WILLIAMSON, E. M., J. WESTWICK, and F. J. EVANS: The effect of daphnane esters on platelet aggregation and erythema of the mouse ear. J. Pharm. Pharmacol. **32,** 373 (1980).
198. EDWARDS, M. C., A. M. E. NOURI, D. GORDON, and F. J. EVANS: Tumour-promoting and non-promoting pro-inflammatory phorbol esters act as human lymphocyte mitogens with different sensitivities to inhibition by cyclosporin-A. Molecular Pharmacol. In press (1982).

*199.* Slaga, T. G., and J. D. Scribner: Inhibition of tumour initiation and promotion by anti-inflammatory agents. J. Natl. Cancer Inst. **51,** 1723 (1974).
*200.* Troll, W., A. Klassen, and A. Janoff: Tumourgenesis in mouse skin: inhibition by synthetic inhibitors of proteases. Science **169,** 1211 (1970).
*201.* Williamson, E. M., and F. J. Evans: Inhibition of erythema induced by pro-inflammatory Esters of 12-deoxyphorbol. Acta Pharmacol. et Toxicol. **48,** 47 (1981).
*202.* Williamson, E. M., J. Westwick, V. V. Kakkar, and F. J. Evans: Studies on the mechanism of action of 12-deoxyphorbolphenyl acetate, a potent platelet aggregating tigliane ester. Biochem. Pharmacol. **30,** 2691 (1981).
*203.* Horowitz, A. D., E. Greenebaum, and I. B. Weinstein: Identification of receptors for phorbol-ester tumour promoters in intact mammalian cells and of an inhibitor of receptor binding in biologic fluids. Proc. Natl. Acad. Sci. U.S.A. **78,** 2315 (1981).
*204.* Smythies, J. R., F. Bennington, and R. D. Morin: On the molecular structures of receptors for co-carcinogens and some anti-cancer drugs. Psychoneuroendocrinol. **1,** 123 (1975).
*205.* Pearce, J. B., A. W. Baird, E. M. Williamson, and F. J. Evans: Inhibition of histamine-induced acid secretion in rat isolated gastric mucosa by esters of phorbol and 12-deoxyphorbol. J. Pharm. Pharmacol. **33,** 737 (1981).
*206.* Williams, T. J., J. Westwick, E. M. Williamson, and F. J. Evans: Vascular changes in rabbit skin induced by pro-inflammatory phorbol and 12-deoxyphorbol esters. Inflammation **5,** 29 (1981).
*207.* Ohuchi, K., and L. Levine: Stimulation of prostaglandin production, deacylation of lipids and morphological changes by tumour-promoting phorbol, 12,13-diesters and adriamycin in MDCK cells. Effect of cycloheximide, indomethacin and hydrocortisone on these stimulations. Prostaglandins **15,** 723 (1978).
*208.* Trashjian, A. H., L. I. Joel, and B. Declos: Stimulation of prostaglandin production in bone by phorbol-diesters and melittin. Prostaglandins **16,** 221 (1978).

*(Received July 6, 1982)*

# Bitter Principles of Cneoraceae

By A. Mondon and B. Epe, Institute of Organic Chemistry, University of Kiel, Federal Republic of Germany

With 2 coloured Figures

**Contents**

I. Introduction . . . . . . . . . . 102

II. Nomenclature and Classification . . . . . . . . . . 104

III. Constitution and Configuration of Cneorins **A**, **B**, **C** and **D** . . . . . . . . . . 106

1. Structural Studies by Ozonolysis . . . . . . . . . . 106
2. Establishment of Relative and Absolute Configuration . . . . . . . . . . 108
3. Carbon Skeleton of Cneorins **A** – **D**: Numbering of C-Atoms . . . . . . . . . . 112
4. Configuration of Cneorin-$\mathbf{C}_{\mathrm{I}}$ . . . . . . . . . . 112
5. Configuration of Cneorin **B**, $\mathbf{B}_{\mathrm{I}}$ and $\mathbf{B}_{\mathrm{III}}$ . . . . . . . . . . 114
   a) Elucidation of the Configuration at Carbon Atoms 7 and 9 . . . . . . . . . . 114
   b) Elucidation of the Configuration at Carbon Atoms 5, 10, 13 and 17 . . . . 116
   c) Hydrogenolysis of Cneorins $\mathbf{B}_{\mathrm{III}}$ and $\mathbf{C}_{\mathrm{III}}$ . . . . . . . . . . 118
   d) The Steric Series **B** and **C** with (17*R*)- and (17*S*)-Configuration . . . . . . . 118
6. Intramolecular Cyclisation to Cneorins $\mathbf{B}_{\mathrm{II}}$ and $\mathbf{C}_{\mathrm{II}}$ . . . . . . . . . . 119
7. Constitution and Configuration of Cneorins **A** and **D** and Pyrolysis of Cneorin-**D** . . . . . . . . . . 122

IV. The Series of Stereoisomeric $\Delta^{8(30)}$-Olefins . . . . . . . . . . 124

V. Stereoisomeric Alcohols with an 8-OH Group and (5*S*, 10*R*)-Configuration . . . 126

VI. Cneorins and Tricoccins with a C-7 Carbonyl Group . . . . . . . . . . 128

1. Cneorin-**F** and Tricoccin-$\mathbf{S}_{14}$ . . . . . . . . . . 128
2. The Tricoccins $\mathbf{R}_9$, $\mathbf{R}_{12}$ and Their Epi-compounds . . . . . . . . . . 131

VII. C-7 Hemiacetals and Methylacetals with (5*S*, 10*R*, 17*R*)-Configuration . . . . . . . 133

1. Cneorin **K** and $\mathbf{K}_1$ . . . . . . . . . . 133
2. Tricoccin $\mathbf{R}_1$ and $\mathbf{R}_{10}$ . . . . . . . . . . 134

VIII. C-9 Hemiacetals with (5*S*, 8*S*, 10*R*)-Configuration . . . . . . . . . . 136

IX. Cyclic Peroxides ........ 141
X. Bitter Principles with a γ-Lactol Ring (C-15 Hemiacetals) ........ 144
1. Precursors of Known Cneoroids ........ 144
2. C-15 Cycloacetals ........ 147
3. γ-Lactols with the Partial Structure of Ring A of Obacunone ........ 150
XI. Tetranortriterpenoids from Cneoraceae ........ 155
1. 3,4-*seco*-Meliacans ........ 155
2. 7,8-*seco*-Meliacans ........ 162
3. 3,4-16,17-*seco*-Meliacans (Limonin Group) ........ 166
4. 3,4-7,8-16,17-*seco*-Meliacans ........ 168
XII. Protolimonoids ........ 168
1. Tirucallan-(20*S*)-Triterpenoids ........ 168
2. Apotirucallan-(20*S*)-Triterpenoids ........ 172
XIII. Comments on Biosynthesis of Cneorins and Tricoccins ........ 173
XIV. Tables of Natural Bitter Principles from Cneoraceae ........ 178
1. Cneorins from *Neochamaelea pulverulenta* (Vent.) Erdtm. of Known Constitution and Configuration ........ 178
2. Cneorins of Unknown Structure ........ 179
3. Tricoccins from *Cneorum tricoccon* L. of Known Constitution and Configuration ........ 179
4. Tricoccins of Unknown Structure ........ 181
References ........ 181

## I. Introduction

The Cneoraceae, a plant family of considerable interest to the natural products chemist, comprises two genera with a total of only three species, two of which we have studied in considerable detail. These two species are *Neochamaelea pulverulenta* (Vent.) Erdtman (= *Cneorum pulverulentum* Vent.), a xerophytic shrub native to the Canary Islands, with silver-haired leaves, yellow blossoms and hard stone fruits (Fig. 1), and *Cneorum tricoccon* L., a shrub native to coastal areas of the western Mediterranean, with hairless leaves, yellow blossoms and red fruits (Fig. 2). We had no means of access to the third species, *Cneorum trimerum* (Urban) Chodat, which belongs to the flora of Cuba. Taxonomically, the family of Cneoraceae is assigned by A. TAKHTAJAN (*1*) to the Rutales, and by A. CRONQUIST (*2*) to the Sapindales; the botanically interested reader is referred to recent articles by H. STRAKA *et al.* (*3*), and by D. LOBREAU-CALLEN *et al.* (*4*).

Fig. 1

Fig. 2

We owe our thanks to Mr. WALTER DAEBEL (†), Kappeln/Schlei, both for suggesting the Cneoraceae as an object of study and for making available the first samples of *Neochamaelea pulverulenta* from Gomera. We undertook subsequent field trips to Teneriffa and Gran Canaria to collect further material. *Cneorum tricoccon* was investigated using material collected on Mallorca.

Non-specific constituents of the Cneoraceae include n-alkanes, wax esters, n-alkanols, phytosterols, and triterpenes such as lupeol and lupeone (*5, 6*), flavonoids (*6, 7*), tocotrienols (*8*), phenols and carboxylic acids (*9, 10*). Among the species-specific constituents we encountered novel diterpenes, the so-called cneorubins (*11*), and a great variety of chromones (*5, 6, 12–15*), which were also studied by A. G. GONZALEZ *et al.* (*16–21*); in contrast to the chromones, the coumarins identified totalled only three (*12, 13, 22*).

The polyterpenoid bitter principles take pride of place among the constituents characteristic of the Cneoraceae and are dealt with as a coherent whole for the first time in this review. Although related to the bitter principles of the Rutaceae and Meliaceae, even sharing with the latter some common structures, the majority of the Cneoraceae bitter principles differ in a characteristic fashion from those of the other two families.

The bitter principles were isolated by standard column and thin-layer chromatographic techniques. Benzene, or benzene/methanol extracts of plant material were separated on silica gel after removal of chlorophyll by treatment with activated charcoal. The compounds could be detected on thin-layer chromatograms by spraying with sulphuric acid followed by heating, causing red or blac spots to develop. Use of acids or bases during work-up of plant extracts must be avoided, owing to the sensitivity of many constituents; solvents such as chloroform or dichloromethane are also unsuitable.

## II. Nomenclature and Classification

The nomenclature of the bitter principles reflects the order of their isolation. In his 1972 thesis, H. CALLSEN (*23*) designated the first four bitter principles isolated from *Neochamaelea pulverulenta* as "terpenes **A, B, C** and **D**" respectively. B. EPE (*24*) continued this sequence to **L**, and later as far as **T**, the remaining letters of the alphabet having been already assigned to the Cneorubins (*11*). Later compounds were assigned a number affixed to the letters "**NP**", starting with $\mathbf{NP}_{27}$ and subsequently continued by U. OELBERMANN as far as $\mathbf{NP}_{38}$.

In later investigations of Cneorum tricoccon carried out for a diploma (1975) and a doctoral thesis (1977), D. TRAUTMANN (*25, 26*) subdivided the bitter principles into an **R**-series and an **S**-series according to their characteristic colour reaction. Compounds isolated in collaboration with U. OELBERMANN were numbered according to this system, the two series reaching $\mathbf{R}_9$ and $\mathbf{S}_{12}$ respectively. More recent studies have shown compounds of the **S**-series to be particularly numerous, this series having currently reached $\mathbf{S}_{43}$. This abundance has led us to abandon the attempt to follow the customary practice in natural products chemistry of assigning an individual trivial name to each. We have instead restricted ourselves to two collective names, "Cneorin" (*27*) for all bitter principles from *Neochamaelea pulverulenta* and "Tricoccin" (*28, 29*) for all from *Cneorum tricoccon,* the name being followed by the appropriate determinant as described above. We shall use the abbreviation "*Np*" for *Neochamaelea pulverulenta* and "*Ct*" for *Cneorum tricoccon.*

The sequence in which the bitter principles were isolated from the Cneoraceae and their classification according to structural features show a striking parallelism. The four compounds referred to above as Cneorins **A**, **B**, **C**, and **D**, but later renamed to conform to the revised nomenclature, have a common 25-carbon atom skeleton. Because of their empirical formulae they were at first considered to be sesterterpenes (*5*), until more thorough structural elucidation showed them to be pentanortriterpenes, a class of which at that time only two naturally occurring examples were known, namely simarolide (*30*) and picrasin-A (*31*) found in Simaroubaceae. Three further bitter principles of this type were discovered later (*32, 33*). Most recently, individual pentanortriterpenes such as clausenolide (*34*) and nimbinene (*35*) have been identified in Rutaceae and Meliaceae.

On the other hand, the pentanortriterpenes of the Cneoraceae form a new class of natural products, characterized by a common carbon skeleton. We suggest the name "cneoroids" for this class, in analogy to the limonoids (*36 – 38, 40*) of the Rutaceae and Meliaceae, or to the quassinoids (*37, 39*) of the Simaroubaceae.

The Cneoraceae also contain $C_{26}$-tetranortriterpenes, some of which, such as the familiar compounds obacunone, α-obacunol and α-obacunol acetate, belong to the limonin group, although the majority are otherwise unkown compounds based on the meliacane skeleton but lacking one C-C bond of the latter. Tetranortriterpenoids of this type are closely related to both the Rutaceae and the Meliaceae, confirming the chemotaxonomic assignment of the Cneoraceae to the Rutales or Sapindales respectively.

The most recently discovered group of bitter principles comprises $C_{30}$-triterpenes with the carbon framework of the tirucallanes or apotirucallanes and an oxidised side chain. One noteworthy example is melianone (*41*) with the tirucallan skeleton, a compound which we have also isolated from *Np*.

# III. Constitution and Configuration of Cneorins A, B, C and D

## 1. Structural Studies by Ozonolysis

The elucidation of the constitution and configuration of the cneorins **A**, **B**, **C** and **D** forms the foundation on which all subsequent research on the cneoroids is based (*42*–*45*).

Cneorins **B** and **C** have the same empirical formulae, $C_{25}H_{26}O_7$, and are stereoisomers with matching fragments in their mass spectra. Cneorin-**D** is a tertiary alcohol $C_{25}H_{28}O_8$, and cneorin-**A** is a tertiary methyl ether, $C_{26}H_{30}O_8$, which is also formed when cneorin-**D** is methylated.

A striking property of cneorins **B** and **C** is their extreme sensitivity to acid, observed at an early stage by H. CALLSEN (*23*) when recording their spectra in $CDCl_3$. As the highly complex acid-catalysed isomerization will be discussed in detail later, it is sufficient to note at this point that **B** and **C** are rearranged by traces of acid to the labile cneorins $\mathbf{B}_{\mathrm{I}}$ and $\mathbf{C}_{\mathrm{I}}$, and under more drastic acid conditions to the stable cneorins $\mathbf{B}_{\mathrm{III}}$ and $\mathbf{C}_{\mathrm{III}}$. Elimination of water from cneorin-**D** by treatment with phosphoryl chloride in pyridine yields cneorin-$\mathbf{C}_{\mathrm{I}}$. Cneorins **A**, **C** and **D** thus all belong to a single steric series.

$$\text{Cneorin-}\mathbf{A} \xleftarrow[\text{NaH}]{CH_3I} \text{Cneorin-}\mathbf{D} \xrightarrow[\text{Py}]{POCl_3} \text{Cneorin-}\mathbf{C}_{\mathrm{I}}$$

All four cneorins have a β-furyl substituent linked to a methine group, which in turn is adjacent to an oxygen and a tertiary carbon atom; an additional common feature is a γ-substituted butenolide ring bearing a γ-methyl group, corresponding to the partial structures (**1**) and (**2**). Cneorins **B** and **C** have a trisubstituted double bond with an allylic proton, which in its turn is adjacent to an oxygen atom and a methylene group, as in (**3**). In cneorins **A** and **D** the elements of methanol or water respectively are added across the trisubstituted double bond, resulting in partial structures (**4**) and (**5**).

Ozonolysis of cneorin-**C** followed by perhydrol oxidation yields a crystalline dilactone $C_{11}H_{14}O_4$ with the structure (**6**), corresponding to partial structure (**7**). In contrast, ozonolysis of cneorin-**A** followed by oxidation and methylation with diazomethane yields a crystalline methyl ester $C_{24}H_{30}O_9$; in this sequence only the furan ring is cleaved and the β-carbon atom oxidised to the carboxyl group. In neither case is the double bond of the butenolide ring attacked under the mild ozonolysis conditions.

Ozonolysis of cneorin-$\mathbf{C}_{\mathrm{III}}$ affords further insight into cneorin molecular structure. Oxidation of the ozonolysis mixture followed by methylation

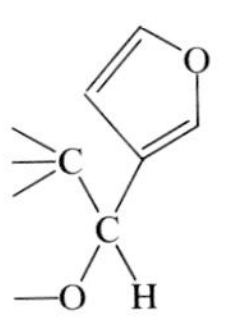

(1) 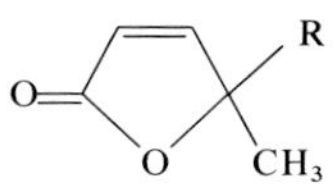 (2) 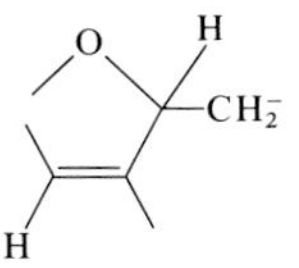 (3) 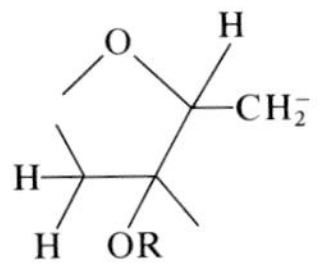 (4) R = $CH_3$
(5) R = H

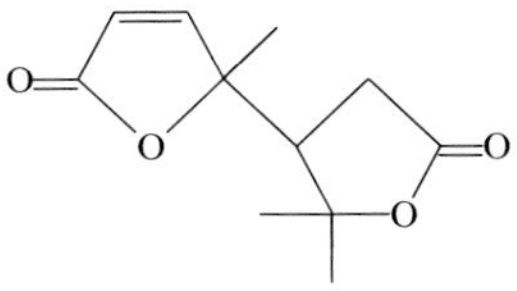

(6) 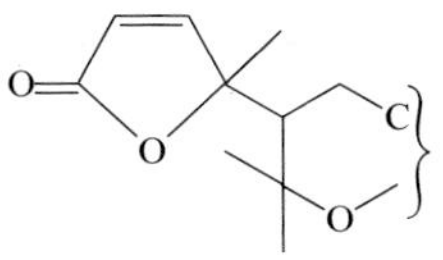 (7) 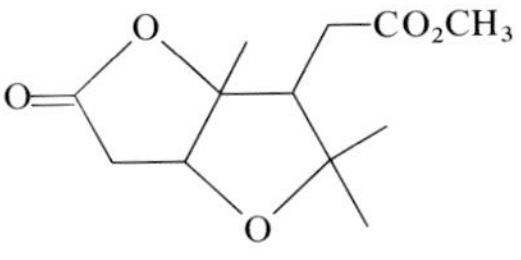 (8) 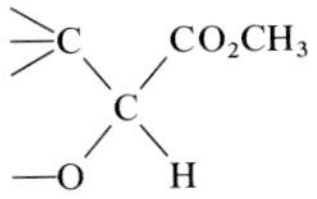 (9)

with diazomethane yields a crystalline monomethyl ester $C_{12}H_{18}O_5$ with structure (**8**) and an oily although pure tricarboxylic acid trimethyl ester $C_{13}H_{16}O_8$. Six of these eight oxygen atoms are accounted for by the ester groups, and one by partial structure (**9**), which is formed from (**1**) on oxidative degradation. The eight oxygen atom can be assigned to the carbonyl group of a saturated γ-lactone on IR spectral grounds, allowing (**9**) to be expanded to partial structure (**10**). Of the empirical formula there remains a difference of $C_3H_6$ to be accounted for, by incorporating three methylene groups and a ring in the structural formula. Of the structures compatible with the evidence, those with a cyclopropane ring, namely (**11**) and (**12**), are to be preferred. Degradation of cneorin-$\mathbf{C}_{III}$ is accompanied by the loss of a carbon atom, clearly the unsubstituted β-carbon atom of a trisubstituted furan ring generated from partial structure (**3**) on acid-catalysed rearrangement.

On piecing together the structural formula of cneorin-$\mathbf{C}_{III}$ from the evidence, structure (**13**) was preferred both on steric grounds and from a consideration of possible biosynthetic pathways. The preliminary structural studies on the cneorins culminated in structure (**14**), proposed for cneorin-**C** (*43*).

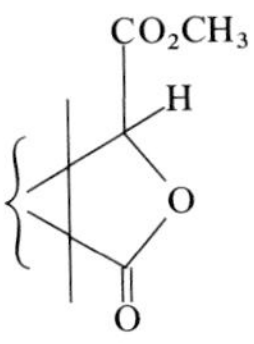

(10) 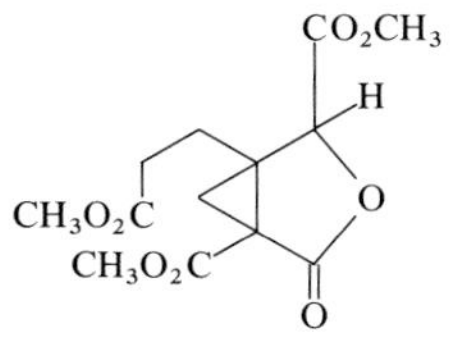 (11) 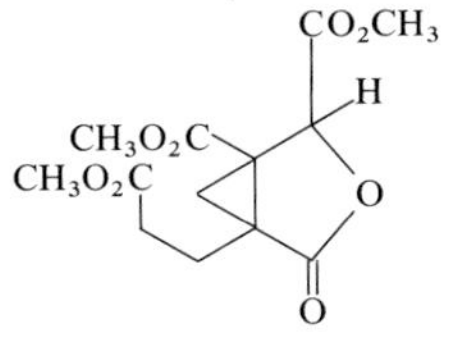 (12)

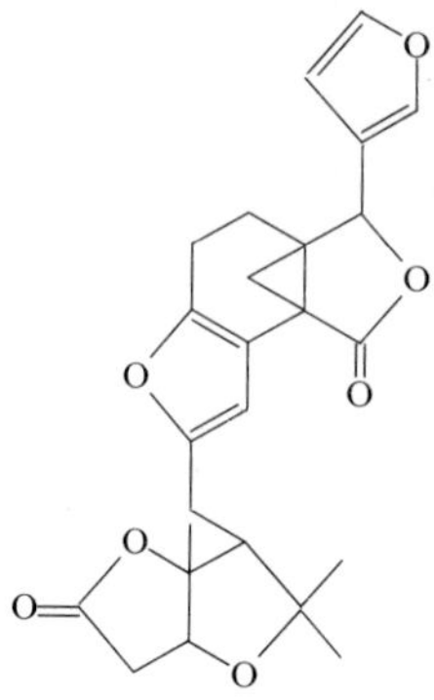

(13)

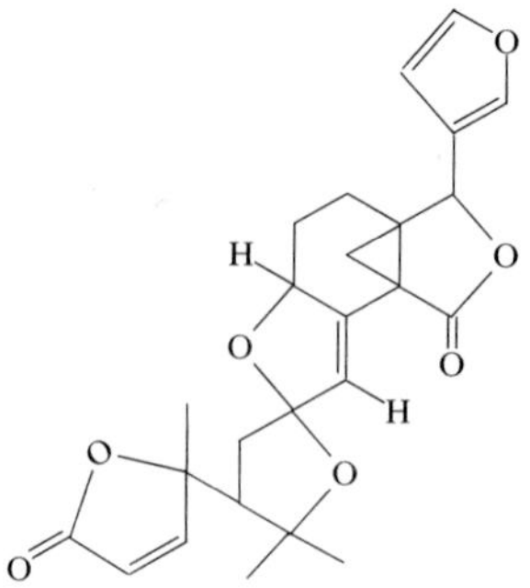

(14)

## 2. Establishment of Relative and Absolute Configuration

X-ray diffraction analysis was undertaken, not only to confirm or refute the structure proposed for cneorin-**C**, but also to elucidate the complex stereochemistry of cneorins **A**–**D** with either six or seven chiral centres respectively.

Treatment with osmium tetroxide in dioxane converts cneorin-**C** into an osmate ester, which on cleavage with $H_2S$ yields the high-melting *cis*-diol-**C**, a compound which forms compact crystals. This hydroxylation takes place at the trisubstituted double bond, as the spectrum contains a singlet at $\delta_H = 4.20$ (DMSO-$d_6$) instead of the olefinic proton signal (*47*). The spectrum of the diacetate has a singlet at 6.63 ppm. As the identical diol is formed on treatment with osmium tetroxide in pyridine, a configurational change at the spiro centre can be excluded.

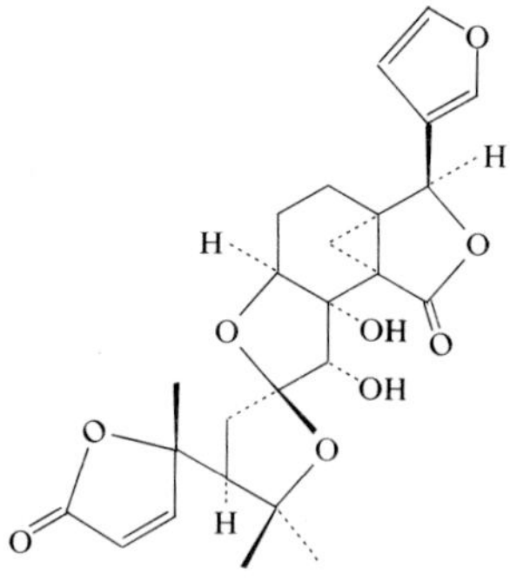

(15)

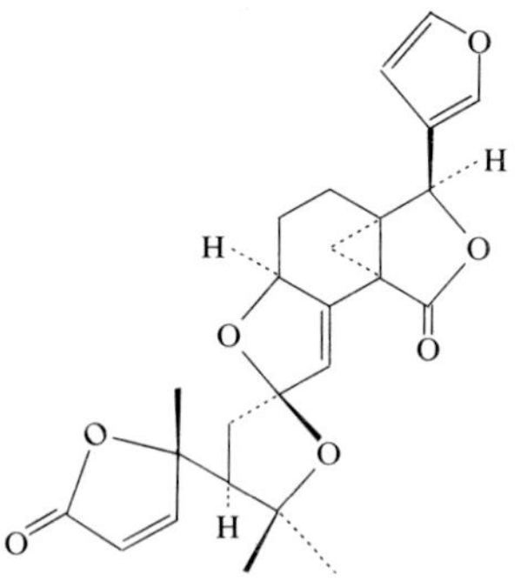

(16)

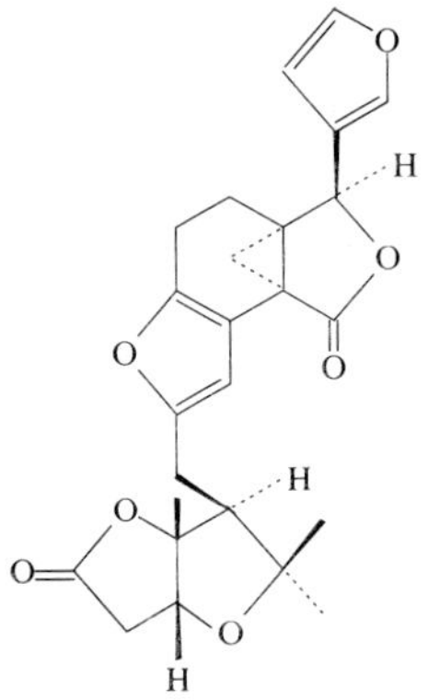

**(17)** 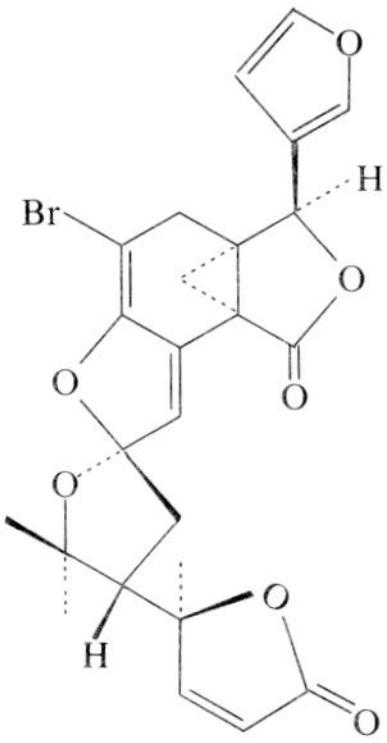 **(18)**

The X-ray diffraction analysis, carried out in collaboration with G. HENKEL and H. DIERKS (*44*), confirmed the structure proposed for cneorin-**C** and shown in formula (**14**), and also the relative configuration for all chiral carbon atoms of *cis*-diol-**C**, corresponding to formula (**15**) with hydroxy groups at the α-positions (*48*). Cneorin-**C** and cneorin-$\mathbf{C}_{III}$ can now be assigned the structures shown in formulae (**16**) and (**17**).

Introduction of a heavy atom into the optically active compound enables the absolute configuration to be determined, taking the anomalous dispersion of the atoms into account (*49*). Addition of bromine to the reactive double bond of cneorin-**C** seemed the most obvious approach to the preparation of a suitable compound. A solution of cneorin-**C** in chloroform reacts with bromine after a short induction period; the experiment was terminated after addition of 1.2 equivalents of bromine. Cneorin-$\mathbf{C}_{III}$ (**17**) was isolated from the reaction mixture, together with a bromoterpene $C_{25}H_{23}O_7Br$, which crystallized from methanol in long needles. Comparison of this empirical formula with that of cneorin-**C** shows that incorporation of one bromine atom is accompanied by the loss of three hydrogen atoms, in other words the reaction cannot be interpreted as a simple addition or substitution; the bromine atom is firmly bound and does not react with alcoholic silver nitrate solution. In addition to the absorption due to the furan and butenolide rings at 210 nm, the UV spectrum shows an additional maximum at 267 nm ($\varepsilon = 12100$) due to a conjugated system. The $^1$H-NMR spectrum has no signal which can be assigned to the allylic proton of partial structure (**3**), but all signals due to the olefinic protons of the starting material are present, shifted to a minor degree.

*Diagram 1*

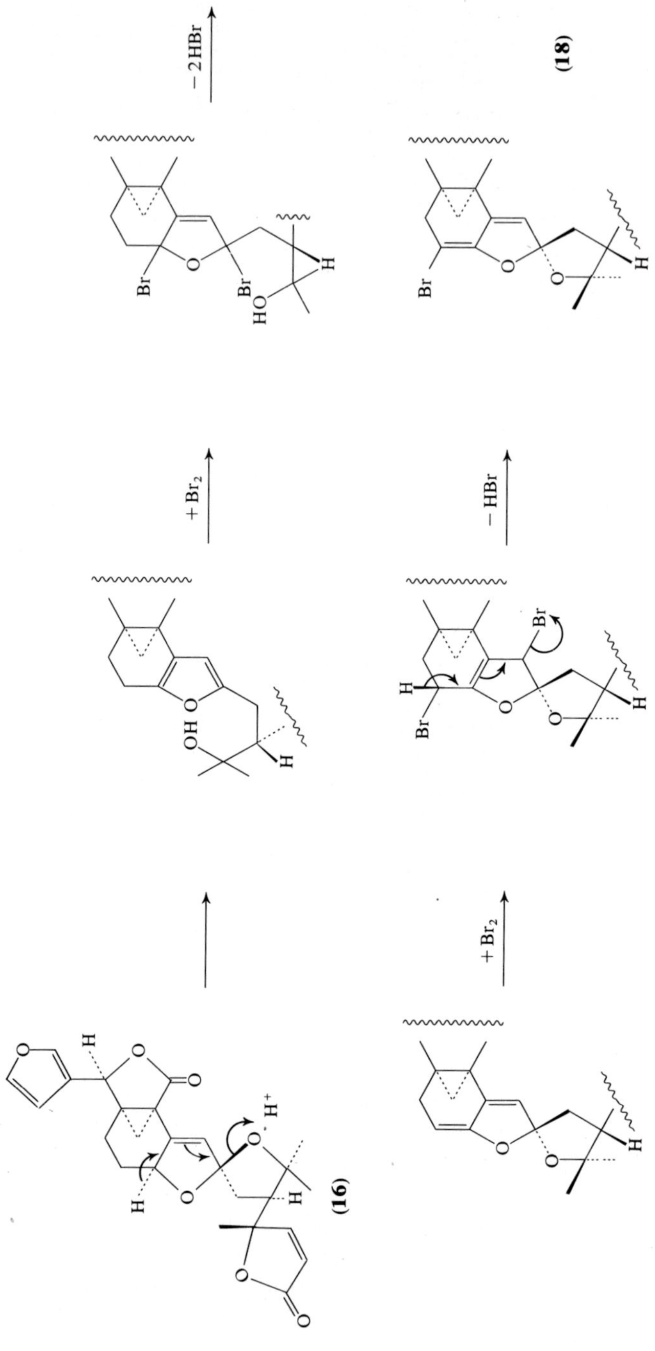

Formula (**18**) shows the result of the X-ray diffraction analysis of bromoterpene-**C** (*8*). The absolute configuration shown here is correct with a statistical probability greater than 99%. By reference to formula (**18**), the arbitrarily drawn enantiomer in formula (**15**), (**16**) and (**17**) can be seen to correspond to the absolute configuration. This proves the β-orientation of the furan ring in cneorin-**C** (*50*).

Formation of (**18**) from cneorin-**C** requires two equivalents of bromine; a possible mechanism is presented in diagram 1.

Traces of acid catalyse cleavage of the acetal structure and formation of the furan ring, the latter undergoing 1,4-addition of bromine. The acetal is reformed by nucleophilic substitution, with inversion of configuration at the spiro centre. Subsequent steps are elimination of HBr, renewed 1,4-addition of bromine, and 1,4-elimination to yield (**18**). The hydrogen bromide formed accelerates the first stage of the reaction and also catalyses the rearrangement of cneorin-**C** to cneorin-$\mathbf{C}_{III}$, isolated with the bromoterpene from the reaction mixture.

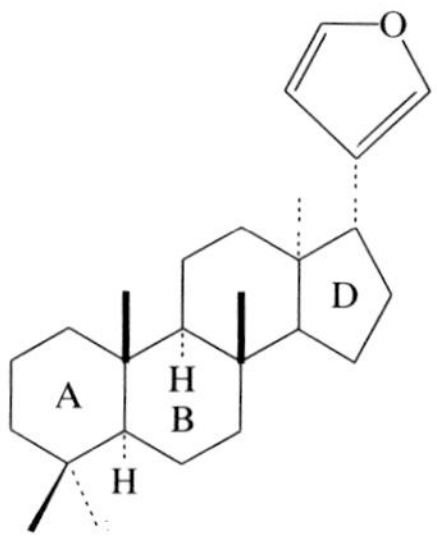

(**19**)

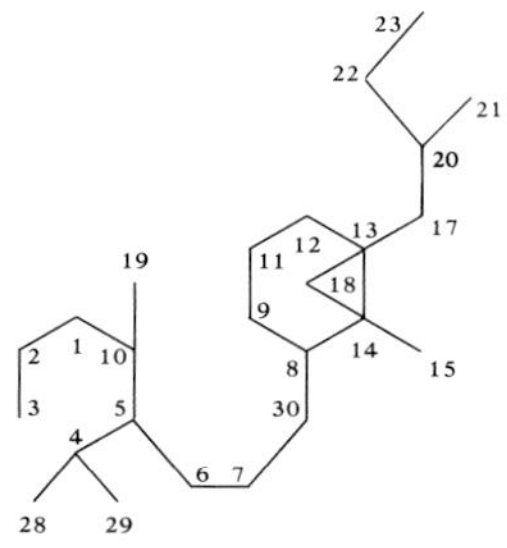

(**20**)

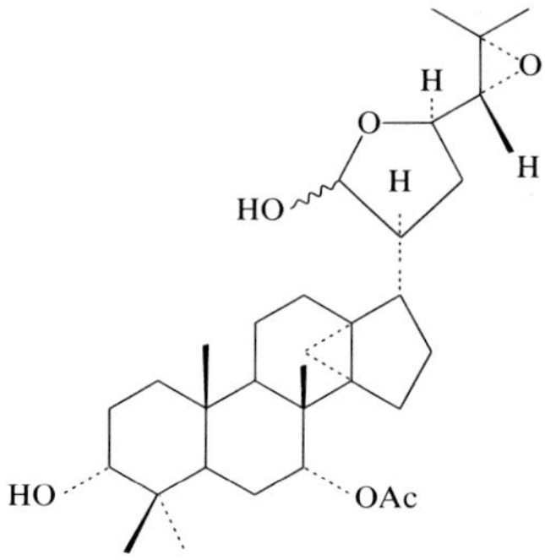

(**21**)

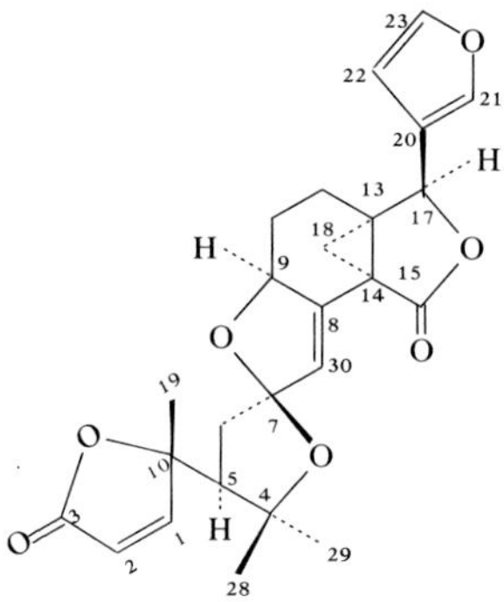

(**22**)

### 3. Carbon Skeleton of Cneorins A—D: Numbering of the C-Atoms

The carbon skeleton (**20**) of cneorins **A**–**D**, which we have named "cneoran" (*3*), derives from meliacan (*51*) (**19**), the parent skeleton of the limonoids and meliacins. Diverging from familiar biosynthetic patterns, ring B is formally enlarged to a seven-membered ring incorporating the angular methyl group at C-30 and then opened between carbon atoms 9 and 10. Incorporation of the methyl group at C-13 into a cyclopropane ring with point of attachment at C-14 is accompanied by cleavage of ring D with loss of C-16; on the other hand, an opened ring A between carbon atoms 3 and 4 is a feature familiar from the many examples known. The β-substituted furan ring of (**19**) is retained as a structural element in cneorins **A**–**D**. The carbon atoms of the cneoran skeleton are numbered by following the system used for triterpenoids, as shown for cneorin-**C** in formula (**22**). The α-orientation of the cyclopropane ring established by X-ray analysis is of biogenetic origin, and is a feature common to all cneoroids. Glabretal (**21**), isolated from *Guarea glabra* (Meliaceae) (*52*), also incorporates an α-oriented cyclopropane ring at the equivalent site, but has the apotirucallan skeleton with an oxidised side chain.

### 4. Configuration of Cneorin-$C_I$

H. CALLSEN (*23*) was the first to prepare cneorin-$\mathbf{C_I}$ which he obtained as a mixture with a second component, referred to as $\mathbf{C_{II}}$, on treating cneorin-**C** with dilute acid. B. EPE (*24*) later isolated the same compound as a natural product of *Np*, and called it cneorin-**E**. It is possible to suppress the acid-catalysed formation of $\mathbf{C_{II}}$ completely by carrying out the rearrangement of cneorin-$\mathbf{C_I}$ in glycol dimethyl ether with one drop of aqueous ammonium nitrate solution, under virtually neutral conditions. It is noteworthy however, that the reaction does not proceed to completion no matter how long the reaction time, but that an equilibrium mixture containing **C** and $\mathbf{C_I}$ in the approximate ratio 1 : 10, i.e. containing a large excess of cneorin-$\mathbf{C_I}$, is formed. The same equilibrium mixture is of course obtained from pure $\mathbf{C_I}$ as starting material:

$$\text{Cneorin-}\mathbf{C} \rightleftharpoons \text{Cneorin-}\mathbf{C_I}$$

Both this equilibrium and the mass spectra of **C** and $\mathbf{C_I}$ support the stereoisomeric relationship between the two. The $^1$H-NMR spectra of **C** and $\mathbf{C_I}$ differ to only a very minor degree. On the other hand, conversion of one to the other is accompanied by a drastic change in specific rotation, from +66.4° to −47.3°. The nature of this interconversion is an inversion of configuration at the spiro centre C-7 as shown below:

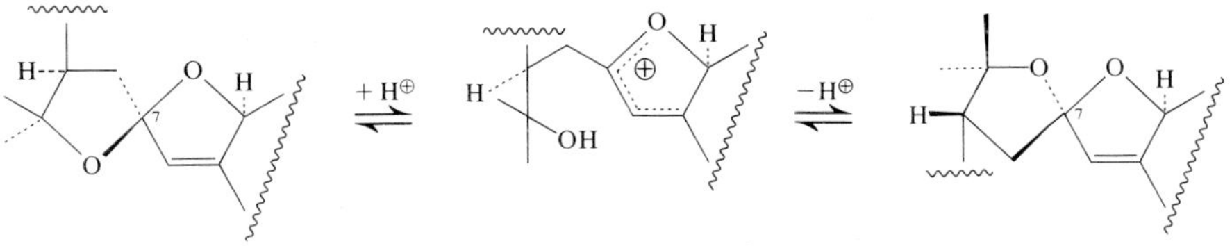

Epimerization of cneorin **C** and $\mathbf{C}_\mathrm{I}$

This facile opening of the cycloacetal structure to the carbonium ion is evidently assisted by mesomeric stabilization with the olefinic double bond. Cneorin-**C** (**16**) [or (**22**)] has the (7*R*)- and cneorin-$\mathbf{C}_\mathrm{I}$ (**23**) the (7*S*)-configuration.

The stereoisomers display a most striking difference in their reactivities. Reaction of cneorin-$\mathbf{C}_\mathrm{I}$ with osmium tetroxide in dioxane requires an increase in reaction time from 3 to 18 hours. After treatment of the reaction mixture with $H_2S$ and work-up, the same *cis*-diol-**C** (**15**) as previously described, and not the epimeric diol, is isolated. It is clear from this circumstance that the reaction requires an acid concentration sufficient to promote inversion of configuration at C-7. If the same reaction is carried out in pyridine as solvent, starting material can still be detected after 48 hours. As expected, no *cis*-diol-**C** can be isolated after work-up; instead, a structural isomer, a diol (**24**) formed by hydroxylation of the double bond of the butenolide ring is obtained.

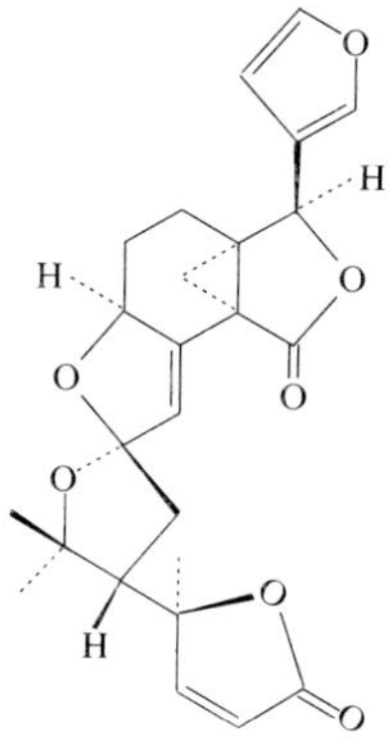

(**23**) 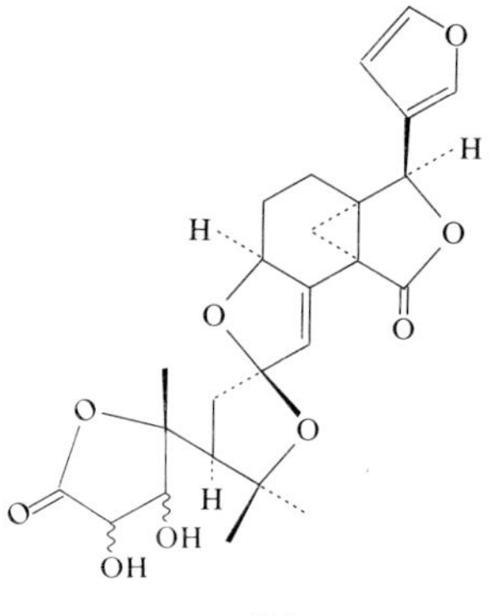 (**24**) 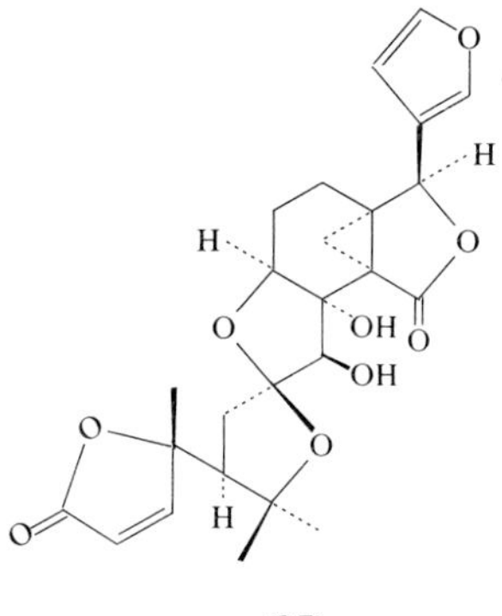 (**25**)

Similarly, the attempted preparation of *cis*-diol-$\mathbf{C}_\mathrm{I}$ by acid catalysed epimerization of *cis*-diol-**C** is unsuccessful, as the diol (**15**) is unexpectedly stable towards acid. Acid-catalysed transformation of *cis*-diol-**C** requires the action of concentrated HCl for days. Although the molecular weight and the spectra of the reaction product point to a stereoisomeric diol structure, the compound is not the expected C-7 epimer. While *cis*-diol-**C** is rapidly cleaved by periodate, the new diol fails to react within a comparable length of time, this lack of reactivity being evidence for the *trans*-diol-**C** structure (**25**).

## 5. Configuration of Cneorin B, $B_I$ and $B_{III}$ (*45*)

### *a) Elucidation of the Configuration at Carbon Atoms 7 and 9*

As already mentioned, cneorin-**B** and cneorin-**C** are stereoisomers. Cneorin-**B** is also rearranged by traces of acid to a stereoisomeric compound cneorin-$\mathbf{B_I}$ (*23*). $\mathbf{B_I}$ was also isolated as a natural product from *Np,* receiving the designation cneorin-**P**. The compound occasionally crystallizes as a low-melting modification with an abnormal IR spectrum, when the latter is recorded in KBr. $\mathbf{B_{III}}$ has been detected in *Ct* extracts, but whether as a genuine natural product or as an artefact remains an open question.

The obvious suspicion that cneorin-**B** and cneorin-$\mathbf{B_I}$ might be a further pair of epimers, differing only in their configuration at C-7, is readily confirmed. Stirring a solution of cneorin-$\mathbf{B_I}$ in glycol dimethyl ether with a drop of ammonium nitrate solution leads to an equilibrium mixture with cneorin-**B**, again in the approximate ratio $\mathbf{B_I} : \mathbf{B} = 10 : 1$. Like cneorin-$\mathbf{C_I}$, cneorin-$\mathbf{B_I}$ is the thermodynamically more stable epimer.

$$\text{Cneorin-}\mathbf{B} \rightleftharpoons \text{Cneorin-}\mathbf{B_I}$$

Comparison of spectra and specific rotations (see Table 1) provides additional support for the analogy of **B** and $\mathbf{B_I}$ with **C** and $\mathbf{C_I}$. Referring to the numbered cneorin-**C** formula (**22**), in which the proton on C-9 is *trans* with respect to the acetal oxygen between C-7 and C-4, the 9-H signal is observed at $\delta_H = 4.24$ ppm. The proximity of this oxygen atom in cneorin-$\mathbf{C_I}$ causes a low-field shift of 0.2 to $\delta_H = 4.44$ ppm. As the same effect accompanies the transition from cneorin-**B** to $\mathbf{B_I}$, it is safe to assume that here again the less stable cneorin-**B** is the epimer with 9-H *trans* with respect to the acetal oxygen between C-4 and C-7.

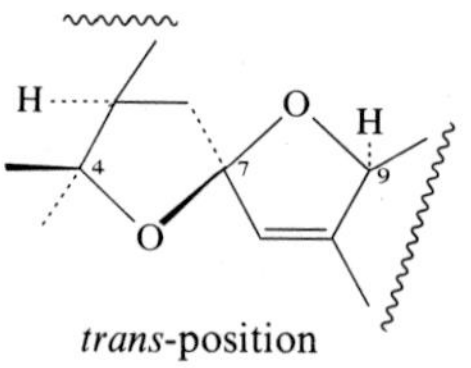

*trans*-position

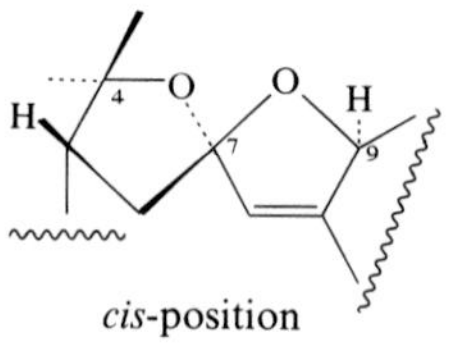

*cis*-position

The weakened 9-H shielding in cneorins $\mathbf{B}_I$ and $\mathbf{C}_I$ leads to an enhanced shielding of C-9, observed in the $^{13}$C-NMR spectrum as a high-field shift (cf. Table 1). In the other five-membered spiro ring, an analogous interaction should lead to a low-field shift of 5-H and a high-field shift of C-5. The C-5 signal position in the $^{13}$C-NMR spectrum of cneorin-**C** bears out this prediction (*53*).

As previously stated, the configuration at C-7 strongly influences the rotatory power of the C-7 epimeric cneorins. The observation that the transitions from cneorin-**B** to $\mathbf{B}_I$ and from cneorin-**C** to $\mathbf{C}_I$ are accompanied by approximately the same alteration in specific rotation with respect to both direction and magnitude can be taken as an indication that in both cases the transition takes place from the (7*R*)- to the (7*S*)-configuration.

Table 1. $\Delta^{8(30)}$-*Cneorin Olefins with* (5*S*,9*S*,10*R*)-*Configuration*

| Cneorin- | $[\alpha]_D^{20}$ Acetone | Config. C-7 | C-17 | $^1$H-NMR[a] 5-H | 9-H | $^{13}$C-NMR[b] C-5 | C-9 | MS/rel. Int.% m/z 273 | 268 | 166 |
|---|---|---|---|---|---|---|---|---|---|---|
| **B** (33) | +93° | *R* | *R* | <2.5 | 4.29 | — | — | 100 | 45 | 75 |
| $\mathbf{B}_I$ (34) | + 1° | *S* | *R* | 2.70 | 4.47 | 53.4 | 80.9 | 36 | 100 | 35 |
| **C** (16) | +66° | *R* | *S* | <2.5 | 4.24 | 54.0 | 82.5 | 100 | 45 | 73 |
| $\mathbf{C}_I$ (23) | −47° | *S* | *S* | 2.70 | 4.44 | 53.4 | 80.9 | 4 | 100 | 2 |

[a] Multiplets, δ (ppm), $CDCl_3$.
[b] Bruker HX-90 R, δ (ppm), $CDCl_3$.

In combination with the results obtained by comparing NMR spectra one may conclude that the configuration of cneorins **B** and $\mathbf{B}_I$ at C-9 corresponds with that of cneorins **C** and $\mathbf{C}_I$.

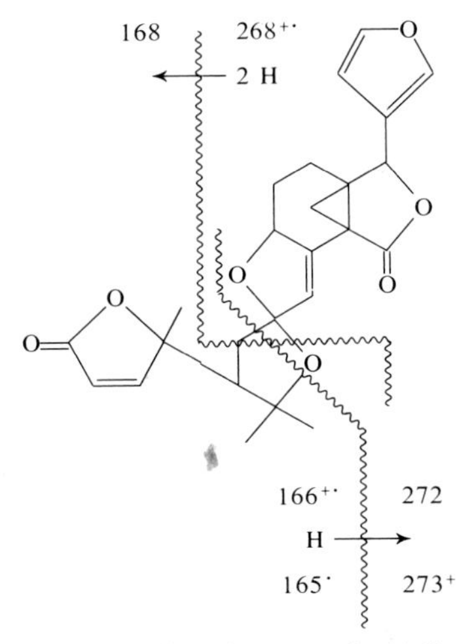


Fragmentation of the cneorin Olefins

A final point to be mentioned is the effect of C-7 epimerization on the fragmentation patterns of the cneorin olefins. Characteristic fragments are at mass nos. 273, 268 and 166, the relative intensity of which show a marked dependence on the configuration at C-7 (cf. Table 1). As the relevant fragmentations involve hydrogen transfers between the spiro-linked parts of the molecule, a stereochemical influence on these processes is hardly surprising. The mechanisms of these rearrangements are not known with certainty, although the exact composition of the fragments has been confirmed by high resolution spectra.

### *b) Elucidation of the Configuration at Carbon Atoms 5, 10, 13 and 17*

Once the configuration at carbon atoms 7 and 9 of cneorins **B** and $\mathbf{B}_{\mathrm{I}}$ was known, the next task was to establish the configuration at carbon atoms 5, 10, 13, and 17 in relation to the corresponding atoms in cneorin-**C**. For this purpose, *cis*-diol-**B** was prepared as described above by treating cneorin-$\mathbf{B}_{\mathrm{I}}$ with osmium tetroxide in dioxan, with inversion of configuration at C-7, the stereoisomeric *cis*-diols then being cleaved with periodate in the presence of acid. Crystalline compounds were obtained from the neutral fractions of each cleavage reaction mixture. These compounds were identical both with each other and with the dilactone (**6**) previously obtained by ozonolysis of cneorin-**C** by all criteria, melting points, specific rotations and spectra. This identity proves that cneorins **B** and **C** have the same configuration at carbon atoms 5 and 10.

The dilactone formula (**6**) can now be replaced by the stereoformula (**26**). Similarly, formula (**8**) for the bicyclic lactone ester resulting from ozonolysis of cneorin-$\mathbf{C}_{\mathrm{III}}$ is superseded by (**27**). Acidification of an alkaline solution of (**26**) yields the optically active acid (**28**) which is methylated with diazomethane to the methyl ester (**27**).

H. HEIDENREICH (*54*) synthesized the racemate (**27**) which on optical resolution of the racemic acid with brucine yielded a dextrorotatory carboxylic acid corresponding in melting point and specific rotation to the degraded natural product (**28**).

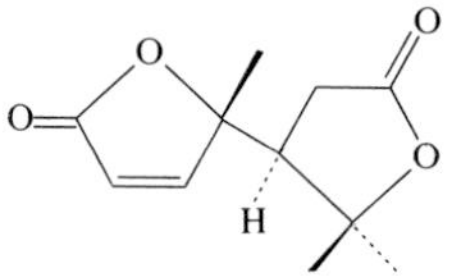

O O O H

(**26**)

O $CO_2R$ O H H O

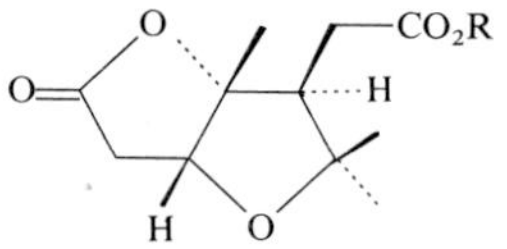

(**27**) R = $CH_3$
(**28**) R = H

Periodate oxidation also gives rise to an acidic fragment which incorporates the furan part of the parent molecules and is isolated as the methyl ester after methylation with diazomethane. The cleavage can be formulated as shown below, using *cis*-diol-**C** (**15**) as an example:

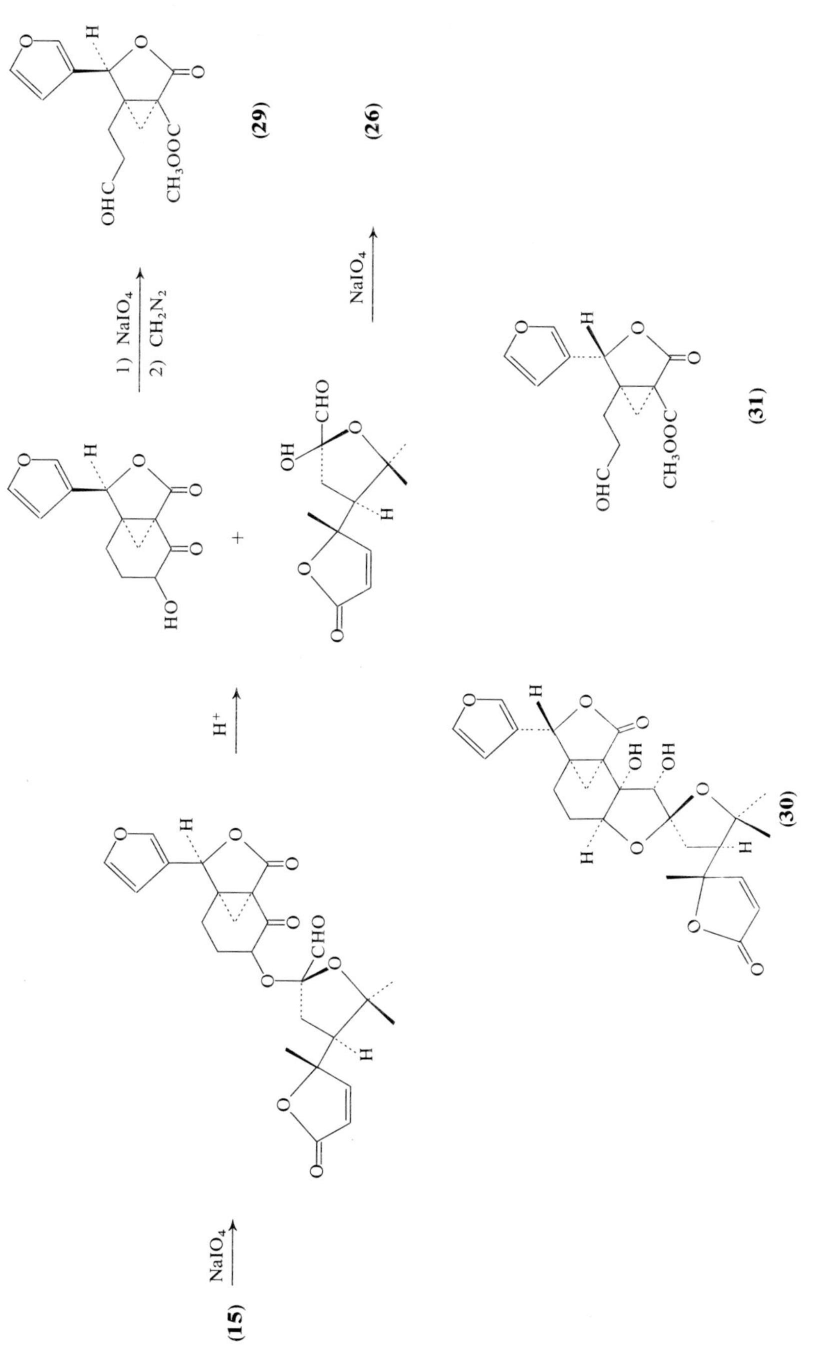

(15) $\xrightarrow{NaIO_4}$ (30)

$\xrightarrow{H^+}$

$\xrightarrow[2)\ CH_2N_2]{1)\ NaIO_4}$ (29)

$\xrightarrow{NaIO_4}$ (26)

(31)

The $^{1}$H-NMR spectrum of the methyl ester obtained from *cis*-diol-**C** reveals, in addition to the signals for the protons of the 3-furylmethylene group and the methyl ester, a triplet at $\delta = 9.75$ with $J = 1$ Hz for an aldehydic proton. Doublets at $\delta = 2.00$ and 1.55, each with $J = 5$ Hz, corresponding to the cyclopropane protons which are largely obscured in the spectra of the natural cneorins, are also visible. The aldehyde ester-**C** has the structure (**29**).

The methyl ester obtained on cleavage of *cis*-diol-**B** has a very similar $^{1}$H-NMR spectrum, but thin-layer chromatographic properties and an IR spectrum which show it to differ from (**29**). Assuming that the cyclopropane ring is invariably $\alpha$-oriented, the only remaining site which can account for this difference is at C-17; the aldehyde ester-**B** is accordingly assigned structure (**31**), in which the cyclopropane and furan rings are mutually *cis*, while *cis*-diol-**B** has the structure (**30**) with $\alpha$-hydroxy groups.

### *c) Hydrogenolysis of Cneorins* $\mathbf{B}_{\mathrm{III}}$ *and* $\mathbf{C}_{\mathrm{III}}$ *(45)*

Hydrogenolysis of $\beta$-furyl-substituted $\gamma$-lactones yields carboxylic acids (*55*), which lose their chiral properties on ring opening. If the supposition is correct that cneorins of the **B** and **C** series differ solely in their configuration at C-17, identical hydrogenolysis products should be obtained from either series. The most suitable starting materials for such experiments are the rearranged products $\mathbf{B}_{\mathrm{III}}$ and $\mathbf{C}_{\mathrm{III}}$ owing to the anticipated low reactivity of the highly substituted furan ring in these compounds.

Hydrogenolysis of $\mathbf{B}_{\mathrm{III}}$ and $\mathbf{C}_{\mathrm{III}}$ over palladium in methanol is stopped in each case after 1.5 molar equivalents of hydrogen have been taken up, and the product methylated with diazomethane. The major products of both reactions are identical by all chromatographic and spectroscopic criteria, and have the same specific rotation of $+5.5°$. The $^{1}$H-NMR spectrum and the molecular weight of 454 are in agreement with structure (**32**) for the esterified hydrogenolysis product.

Cneorins **B**, $\mathbf{B}_{\mathrm{I}}$ and $\mathbf{B}_{\mathrm{III}}$ can thus be assigned the stereo-formulae (**33**), (**34**) and (**35**).

### *d) The Steric Series* **B** *and* **C** *with (17R)- and (17S)-Configuration (45)*

As shown by degradation studies, the natural cneorins occur in two steric series with opposite configurations at C-17. The furan ring is $\alpha$-oriented in the **B**-series and $\beta$-oriented in the **C**-series, so that the **B**-series has the (17*R*)-, and the **C**-series the (17*S*)-configuration. This occurrence of two steric series represents a major difference between the cneoroids and the

*References, pp. 181 – 187*

limonoids, as the furan ring at C-17 of the latter is, with one exception (*56*), invariably α-oriented (*36* – *38*, *40*, *57*).

Cneorins $\mathbf{B}_{III}$ and $\mathbf{C}_{III}$ serve as reference standards for stereochemical assignments to the two series.

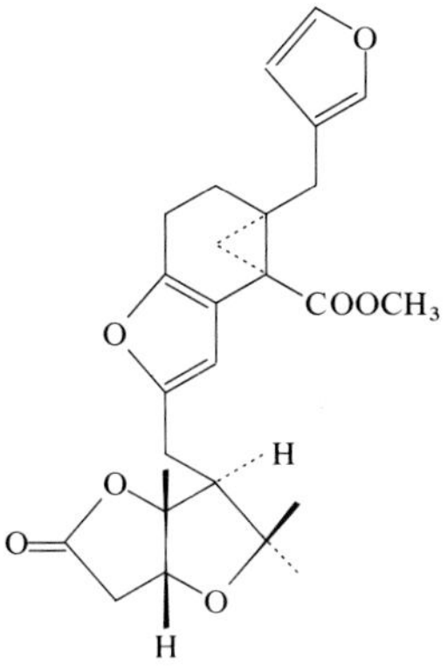

**(32)** 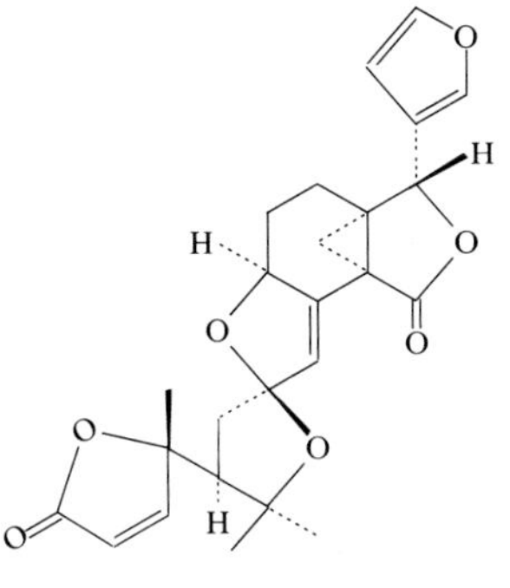 **(33)**

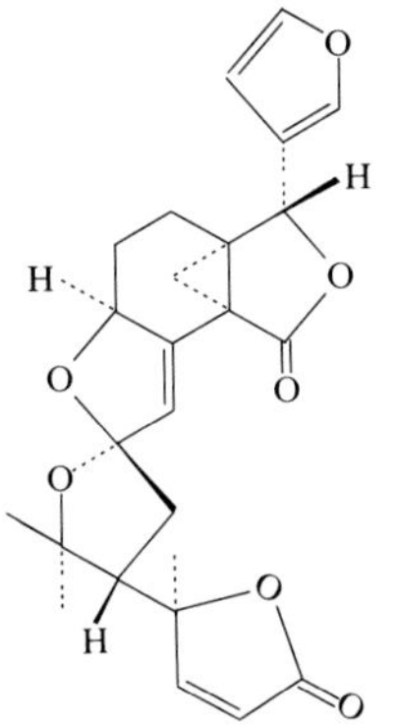

**(34)** 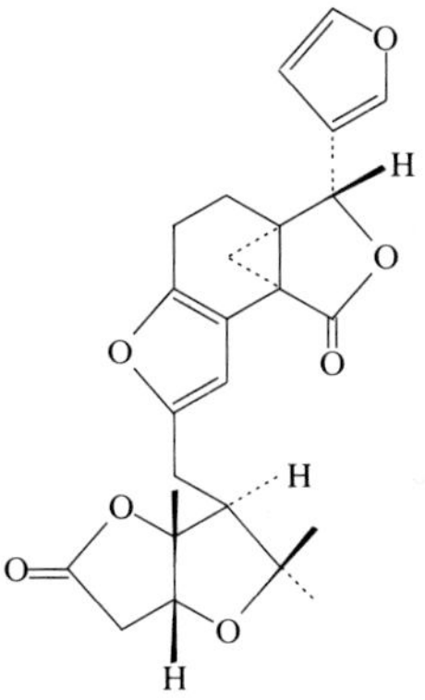 **(35)**

## 6. Intramolecular Cyclisation to Cneorins $\mathbf{B}_{II}$ and $\mathbf{C}_{II}$

Besides the acid-catalysed rearrangements of cneorins **B** and **C** to $\mathbf{B}_{I}$ and $\mathbf{C}_{I}$, or $\mathbf{B}_{III}$ and $\mathbf{C}_{III}$ respectively, H. Callsen (*23*) mentioned two rearranged products $\mathbf{B}_{II}$ and $\mathbf{C}_{II}$ but did not investigate them further. He assumed them to be intermediates of acid-catalysed rearrangement and to be finally transformed into the stable end products of this reaction.

Cneorins $\mathbf{B}_{II}$ and $\mathbf{C}_{II}$ can be prepared by stirring cneorin **B** (or $\mathbf{B}_{I}$) and **C** (or $\mathbf{C}_{I}$) respectively with 0.2 n HCl in glycol dimethyl ether. At this low acid concentration, $\mathbf{B}_{II}$ and $\mathbf{C}_{II}$ are formed as major products in high yields, rearrangement to $\mathbf{B}_{III}$ and $\mathbf{C}_{III}$ taking place only slowly under these conditions. The two new compounds are isomers of the starting material and are mutual stereoisomers.

$\mathbf{B}_{II}$ and $\mathbf{C}_{II}$ are tertiary alcohols. Their $^{1}$H-NMR spectra show neither the doublets corresponding to the olefinic butenolide ring protons nor signals for an ABX system as in $\mathbf{B}_{III}/\mathbf{C}_{III}$. The only signal above 3 ppm apart from those of the 3-furylmethylene group protons is a sharp, one-proton singlet at $\delta = 6.88$ and 6.90 respectively. The mass spectra have a molecular ion at $m/z = 438$ and the same base peak at $m/z = 268$ as the cneorins $\mathbf{B}_{I}$ and $\mathbf{C}_{I}$ (cf. Table 1); intense peaks correspond to fragments formed by elimination of water from the molecular ion.

In the $^{13}$C-NMR spectra are found two lactone carbonyl group singlets. In the olefinic carbon atom region, at $\delta = 143$ and 135, and at $\delta = 141$ and 133 ppm respectively, a singlet and a doublet corresponding to a trisubstituted double bond are seen. Making allowance for the double bonds, an eight-ring skeleton can be deduced from the empirical formula for $\mathbf{B}_{II}$ and $\mathbf{C}_{II}$. Reaction with osmium tetroxide in dioxane as previously described leads to the high-melting *cis*-diols $\mathbf{B}_{II}$ and $\mathbf{C}_{II}$, with the hydroxy groups added across the trisubstituted double bond. Elimination of water from $\mathbf{C}_{II}$ with phosphoryl chloride in pyridine yields the anhydrocneorins $\mathbf{C}_{II,1}$ and $\mathbf{C}_{II,2}$ with a ditertiary, or a terminal double bond respectively. Consideration of their spectroscopic and chemical properties permits assignment of formulae (**37**) and (**38**) to cneorins $\mathbf{B}_{II}$ and $\mathbf{C}_{II}$, and (**41a** and **b**) to the anhydro compounds.

An X-ray analysis of *cis*-diol-$\mathbf{B}_{II}$ was undertaken to prove the constitution and configuration of $\mathbf{B}_{II}$. The result, which provides confirmation of the structures formulated above, appears in formula (**39**) which shows a diol with β-oriented hydroxy groups. The atomic position indicate considerable strain in the ring system: for example, the bond angle at the bridging oxygen atom is 97.5°.

The X-ray analysis also supplies retrospective proof of the α-orientation of the furan ring in the B-series, which had been deduced previously from other evidence.

*cis*-Diol-$\mathbf{C}_{II}$ has formula (**40**), as inspection of a Dreiding model shows that the hydroxy groups can only be β-oriented. A feature of the $^{1}$H-NMR spectra of (**37**/**38**) and (**41a**/**41b**) is the unusual shift of the 30-H singlet to ca. 6.9 ppm, ascribed to the anisotropic effect of the C-15 carbonyl group and the high degree of strain in the rigid ring system.

A mechanism to rationalize the formation of cneorins $\mathbf{B}_{II}$ and $\mathbf{C}_{II}$ from cneorins **B** and **C** (or $\mathbf{B}_{I}$ and $\mathbf{C}_{I}$) respectively can be formulated via the non-

*References, pp. 181—187*

isolable intermediate (**36**). While subsequent reaction to $\mathbf{B}_{III}/\mathbf{C}_{III}$ requires strong acid catalysis, $\mathbf{B}_{II}/\mathbf{C}_{II}$ are formed spontaneously by intramolecular Diels-Alder ring closure between the butenolide and trisubstituted furan rings. Although intramolecular cycloadditions to furan rings are familiar (*58*), the example under discussion is surprising in view of the facility of cyclization to such a strained ring system.

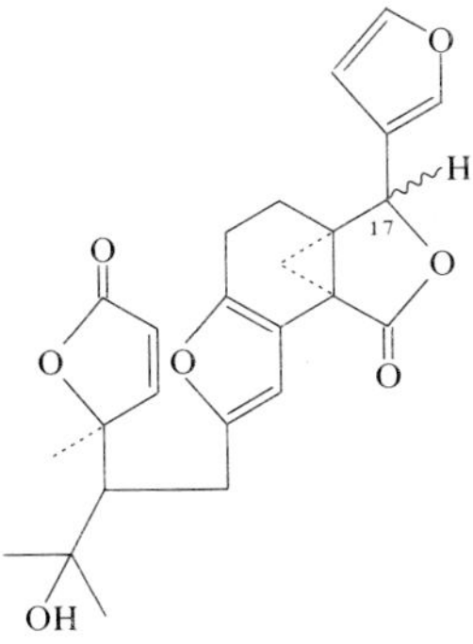

(**36**)

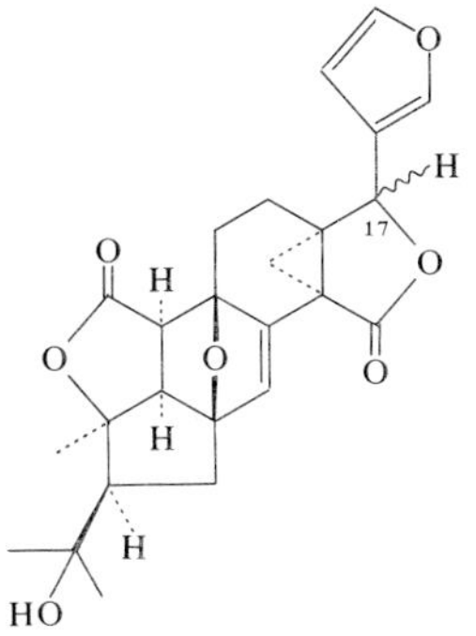

(**37**) 17β-H
(**38**) 17α-H

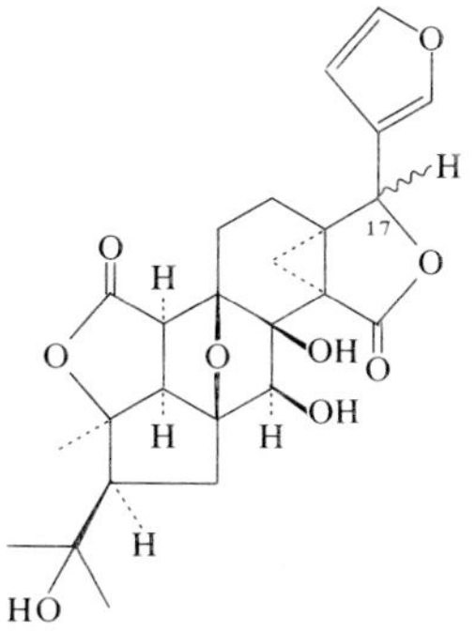

(**39**) 17β-H
(**40**) 17α-H

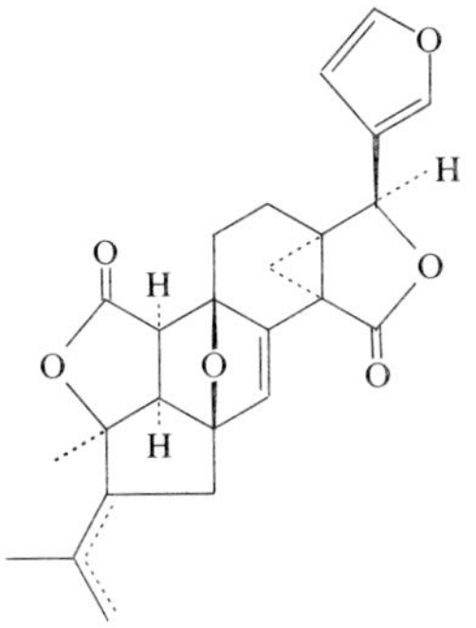

(**41a**), (**41b**)

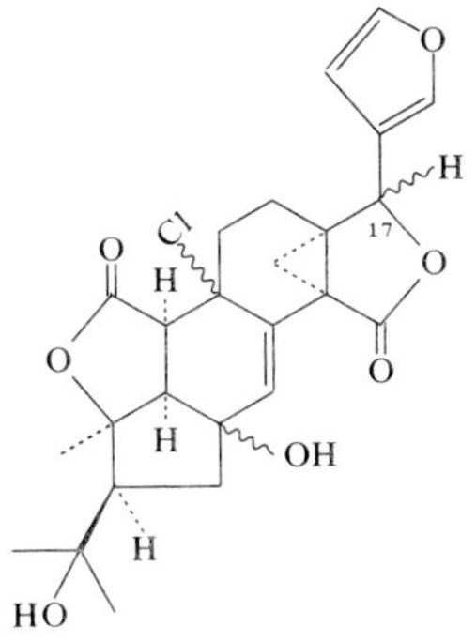

(**42**)
(17α- and β-H)

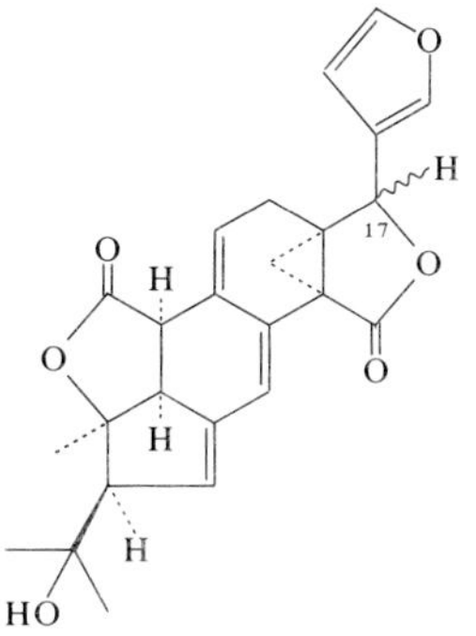

(**43**)
(17α- and β-H)

When $\mathbf{B}_{II}$ and $\mathbf{C}_{II}$ are stirred with conc. hydrochloric acid in glycol dimethyl ether, only small amounts of $\mathbf{B}_{III}$ and $\mathbf{C}_{III}$ are formed by retrograde scission via (**36**). The major products, formed by opening of the oxygen bridge, are the vinylogous chlorohydrins $\mathbf{B}_{IIa}$ and $\mathbf{C}_{IIa}$ formulated as (**42**); these compounds decompose with evolution of gas at the melting point, and even during measurement of the $^{13}$C-NMR spectrum. Elimination of HCl and $H_2O$ leads to the unsaturated compounds $\mathbf{B}_{IIb}$ and $\mathbf{C}_{IIb}$, with the structures shown in (**43**) and $\lambda_{max}$ 284 nm ($\varepsilon = 12600$) (calc. 285 nm).

The acid-catalysed rearrangements of cneorins **B** and **C** are summarized in Diagram 2.

*Diagram 2*

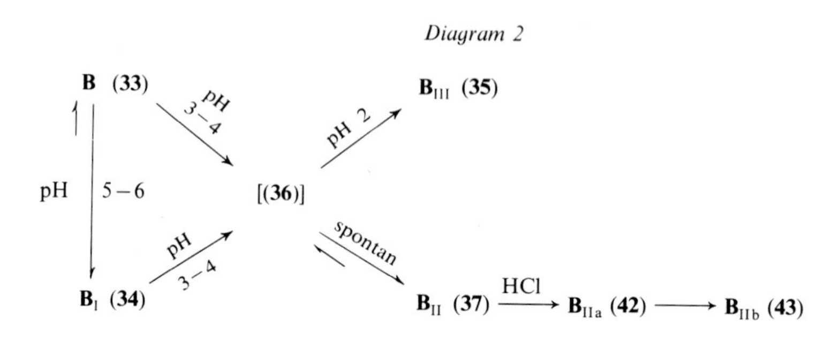


## 7. Constitution and Configuration of Cneorins A and D (*45*) and Pyrolysis of Cneorin-D (*8*)

As previously mentioned, the tertiary alcohol cneorin-**D** yields cneorin-$\mathbf{C}_I$ (**23**) on elimination of water; the constitution and absolute configuration of cneorin-**D** are thus known with the exception of the hydroxyl group at C-8.

The configuration at C-8 is deduced from the $^1$H-NMR spectrum, in which the 9-H signal is seen as a triplet at $\delta = 3.84$ ppm with $J = 5$ Hz, reflecting the equatorial position of the proton at the cyclohexane ring. The hydroxy group at C-8 must therefore be axial, as the dihydrofuran ring cannot be *trans*-diaxially fused with the cyclohexane ring. Formula (**44**) conforms to this requirement and shows cneorin-**D** with 8-OH and 9-H mutually *cis*.

Methylation of cneorin-**D** yields the methyl ether (**45**), and acetylation the acetate (**46**). As previously stated, the methyl ether (**45**) is identical in all respects with cneorin-**A**. Unlike cneorins **B** and **C**, cneorins **A** and **D** are not sensitive to dilute acid.

The acetate (**46**) does not yield an olefin with a ditertiary double bond when pyrolysed. In contrast, pyrolysis of cneorin-**D** at 300° C causes fragmentation, with formation of the dilactone (**26**) and the unsaturated methyl ketone (**49**), the structure of which is deduced from its spectral and chemical characteristics. The ketone has a maximum at 231 nm ($\varepsilon = 19700$), which results from the overlapping of the acetylcyclohexene and vinylfuran chromophore at 232 nm ($\varepsilon = 12500$) (*59*) and 233 nm ($\varepsilon = 9000$) (*60*). The 2,4-dinitrophenylhydrazone of (**49**) has $\lambda_{max} = 376$ nm ($\varepsilon = 13200$), cor-

*References, pp. 181—187*

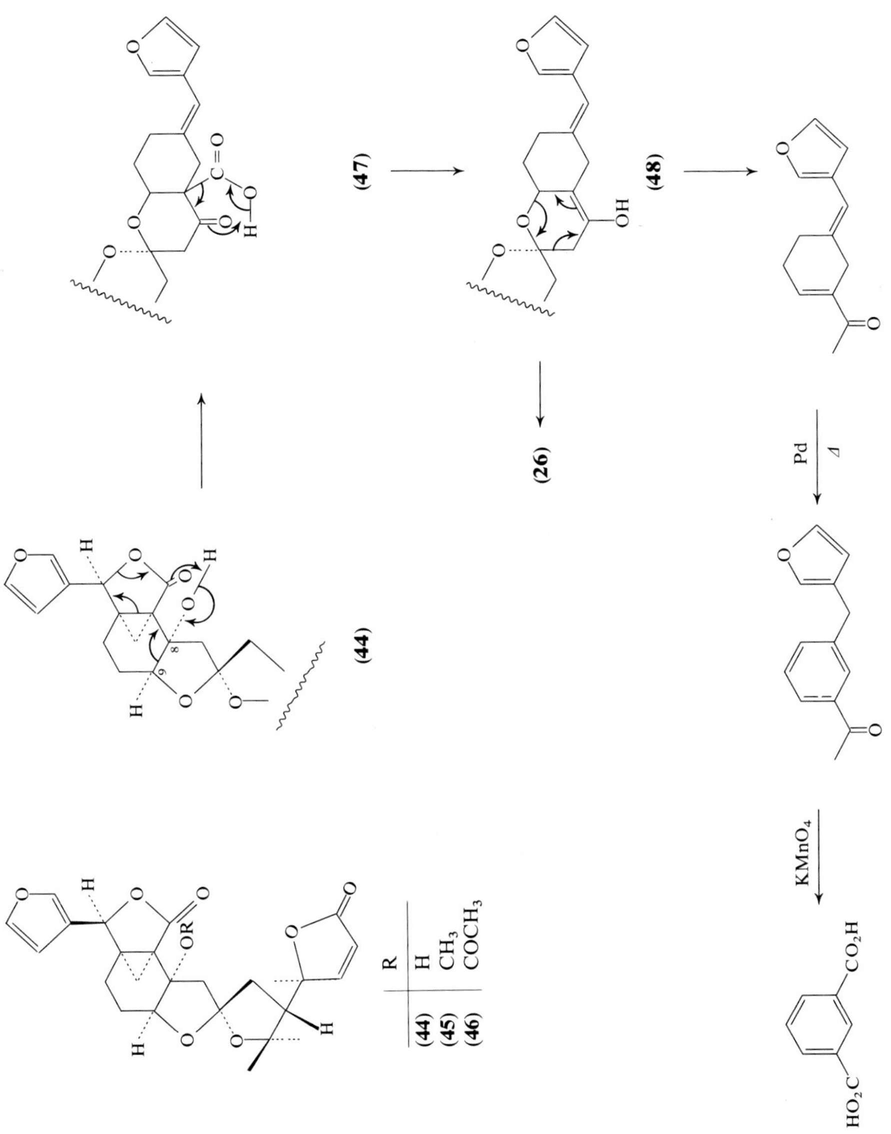

| | R |
|---|---|
| (44) | H |
| (45) | $CH_3$ |
| (46) | $COCH_3$ |

responding to an α,β-unsaturated ketone. Proof of its constitution was obtained by dehydration of (**49**) with palladium to 3-(3-furylmethyl)-acetophenone (**50**). Oxidation with potassium permanganate yields isophthalic acid, confirming the substitution in (**49**).

The pyrolysis probably proceeds by a mechanism initiated by migration of the C-8/C-9 bond, opening of the cyclopropane and γ-lactone rings and formation of the β-keto acid (**47**); this undergoes decarboxylation to the enol (**48**), from which the dilactone (**26**) and, via the enol form, the methyl ketone (**49**) are formed by a retro-Diels-Adler reaction.

## IV. The Series of Stereoisomeric $\Delta^{8(30)}$-Olefins (*28*)

The cneorins **B**, **C**, $\mathbf{B_I}$ and $\mathbf{C_I}$ are stereoisomeric olefins with a $\Delta^{8(30)}$ double bond and the same (5*S*, 9*S*, 10*R*)-configuration. D. TRAUTMANN (*26*) described the stereoisomeric tricoccins $\mathbf{R_0}$, $\mathbf{R_2}$, $\mathbf{R_3}$ and $\mathbf{R_5}$ in his thesis. $\mathbf{R_0}$ and $\mathbf{R_3}$ are identical in all respects with cneorins **B** and $\mathbf{B_I}$.

The constitution and configuration (**51**) and (**53**) for the tricoccins $\mathbf{R_2}$ and $\mathbf{R_5}$ are elucidated by periodate degradation of the corresponding *cis*-diols, as described above. The same dilactone (**26**) is obtained from both, but the aldehyde esters formed differ. These aldehyde esters correspond to (**29**) and (**31**) from *cis*-diols **C** and **B** respectively; this proves that $\mathbf{R_2}$ has the (17*S*)-, and $\mathbf{R_5}$ the (17*R*)-configuration.

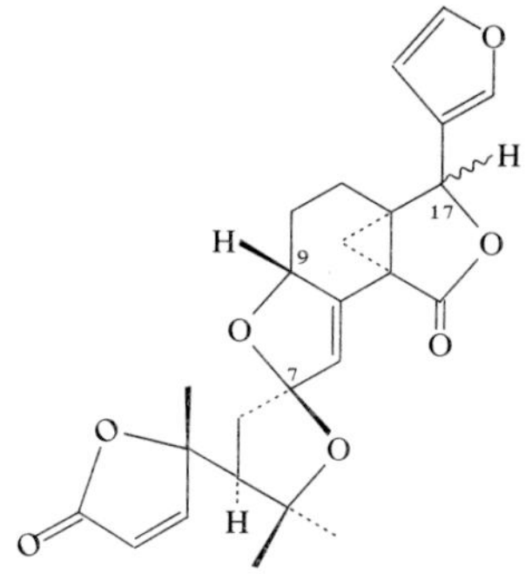

(**51**) 17α-H
(**52**) 17β-H

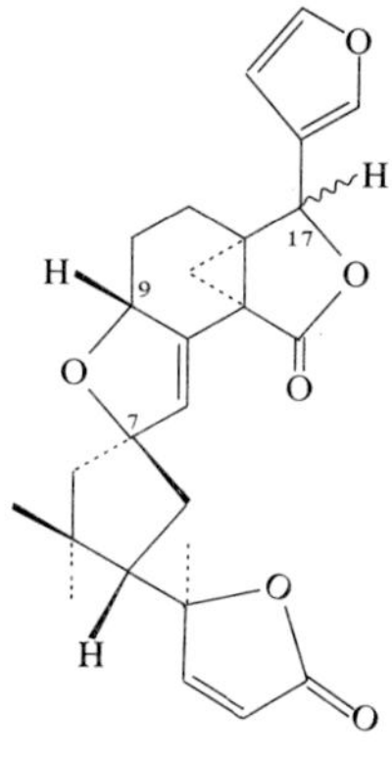

(**53**) 17β-H
(**54**) 17α-H

The stereoisomeric olefins differ markedly in reactivity: the formation of *cis*-diols **B** and **C** is complete in a few hours, but $\mathbf{R_2}$ and $\mathbf{R_5}$ require about a week for the reaction to go to completion. The configuration of the diols at carbon atoms 7, 8 and 30 is not known, but is not important for their degradation. $\mathbf{R}_2$ and $\mathbf{R}_5$ also prove inert when attempts are made to epimerize them at C-7. Both remain unchanged in the presence of ammonium nitrate or dilute hydrochloric acid, although a stabilized mesomeric cation can be formed; this unusual behaviour is thus due solely to the (9*R*)-configuration, which is the opposite of that in the reactive cneorins **B** and **C**. Rearrangement of $\mathbf{R}_2$ and $\mathbf{R}_5$ does not occur unless they are treated with strong acid, when rearrangement to the cneorins $\mathbf{C}_{III}$ and $\mathbf{B}_{III}$ takes place, thus confirming the steric series already established above.

Inversion at C-7 is achieved by altering the experimental conditions. When a suspension of $\mathbf{R}_2$ or $\mathbf{R}_5$ in absolute ether is stirred with $BF_3$-etherate, $\mathbf{R}_2$ affords 7-*epi*-$\mathbf{R}_2$ in relatively good yields after 12 hours, together with starting material; the equilibrium lies here on the side of the *epi*-compound. $\mathbf{R}_5$ reacts considerably more slowly, 7-*epi*-$\mathbf{R}_5$ being obtained in moderate yield after 36 hours, together with starting material and cneorin-$\mathbf{B}_{II}$ (**37**), the latter being formed via the intermediate (**36**) and withdrawn from the equilibrium.

The respective configurations at C-7 remain to be determined for the four new $\Delta^{8(30)}$ olefins with (9*R*)-configuration. Referring back to Table 1, the assignments deduced from the NMR spectra, specific rotations and mass spectra are to be found in Table 2.

Table 2. $\Delta^{8(30)}$-*Tricoccin Olefins with (5S,9R,10R)-Configuration*

| Tricoccin- | $[\alpha]_D^{20}$ Acetone | Config. C-7 | C-17 | $^1$H-NMR[a] 5-H | 9-H | $^{13}$C-NMR[b] C-5 | C-9 | MS/rel. Int.%m/z 273 | 268 | 166 |
|---|---|---|---|---|---|---|---|---|---|---|
| $\mathbf{R}_2$ (**51**) | −25° | *R* | *S* | <2.5 | 4.30 | 54.6 | 80.9 | 15 | 100 | 27 |
| *epi*-$R_2$ (**54**) | −134° | *S* | *S* | 2.62 | 4.00 | 52.6 | 80.7 | 57 | 3 | 100 |
| *epi*-$\mathbf{R}_5$ (**52**) | −10° | *R* | *R* | <2.5 | 4.49 | — | — | — | — | — |
| $\mathbf{R}_5$ (**53**) | −108° | *S* | *R* | 2.70 | 4.21 | 52.8 | 80.9 | 52 | 24 | 100 |

[a] Multiplets, δ (ppm), $CDCl_3$.
[b] δ (ppm), $CDCl_3$.

In the $^1$H-NMR spectra of all olefins with (7*S*)-configuration, a signal with four lines at $\delta = 2.62$ to 2.70 is seen for the 5-H proton. In the spectra of all olefins with (7*R*)-configuration, this signal lies below 2.5 ppm and is no longer recognizable. This difference is paralleled by the signal for 9-H, which appears at lower field in the (9*S*)-series (Table 1), and at higher field in the (9*R*)-series (Table 2). In both series, inversion at C-7 is accompanied by a change in specific rotation of about minus 100°. The mass spectra, too, show a fragmentation dependent in characteristic fashion on the molecular geometry. Tricoccin $\mathbf{R}_2$ has the (7*R*)-, while $\mathbf{R}_5$ has the (7*S*)-configuration: the artificially prepared tricoccins *epi*-$\mathbf{R}_2$ and *epi*-$\mathbf{R}_5$ have correspondingly inverted configurations, as shown in the formulae (**54**) and (**52**).

## V. Stereoisomeric Alcohols with an 8-OH Group and (5*S*, 10*R*)-Configuration (*28*)

The most prominent member of the 8-OH series is cneorin-**D**, the constitution and absolute configuration (**44**) of which were previously established by elimination of water and by analysis of the $^{1}$H-NMR spectrum. Further stereoisomeric alcohols are the cneorins **H**, **N** and **O** from *Np*, and the tricoccins $\mathbf{R}_6$ and $\mathbf{R}_8$ from *Ct*. Table 3 summarizes the conversions effected by eliminating water from the alcohols with phosphoryl chloride in pyridine, followed by acid-free work-up. The olefins formed are the familiar cneorin olefins of the steric series **B** and **C**, with uniform configuration at carbon atoms 5 and 10.

Table 3. *Elimination of Water from the 8-OH-Cneorins and Tricoccins with Phosphoryl Chloride in Pyridine*

| | | |
|---|---|---|
| Cneorin-**D** | ⟶ | Cneorin-$\mathbf{C}_{\mathrm{I}}$ (**23**) |
| Cneorin-**H** | ⟶ | Cneorin-**C** (**16**) (**22**) |
| Cneorin-**N** | ⟶ | Cneorin-$\mathbf{B}_{\mathrm{I}}$ (**34**) |
| Cneorin-**O** | ⟶ | Cneorin-**B** (**33**) |
| Tricoccin-$\mathbf{R}_6$ | ⟶ | Cneorin-$\mathbf{B}_{\mathrm{I}}$ (**34**) |
| Tricoccin-$\mathbf{R}_8$ | ⟶ | Tricoccin-$\mathbf{R}_2$ (**51**) |

Table 4. *8-OH-Cneorins and Tricoccins with* (5*S*,10*R*)-*Configuration*

| Name/ formula | $[\alpha]_D^{20}$ Aceton | Configuration C-7 | C-8 | C-9 | C-17 | $^{1}$H-NMR[a] 9-H | 17-H | MS/rel. Int. % m/z[b] 456 | 235 | 97 |
|---|---|---|---|---|---|---|---|---|---|---|
| **D** (**44**) | − 67° | *S* | *S* | *S* | *S* | 3.84 | 5.38 | 5 | 100 | 90 |
| **H** (**55**) | + 1° | *R* | *S* | *S* | *S* | 3.61 | 5.36 | 10 | 100 | 58 |
| **N** (**59**) | − 6° | *S* | *S* | *S* | *R* | 3.92 | 5.22 | 11 | 100 | 65 |
| **O** (**57**) | + 35° | *R* | *S* | *S* | *R* | 3.75 | 5.25 | 10 | 91 | 100 |
| $\mathbf{R}_6$ (**60**) | +117° | *S* | *R* | *S* | *R* | 3.50 | 5.37 | 5 | 70 | 100 |
| $\mathbf{R}_8$ (**58**) | −131° | *R* | *S* | *R* | *S* | 3.50 | 5.38 | 3 | 53 | 100 |

[a] δ (ppm) in $CDCl_3$; 9-H: **D**, **H**, **N** = t, $J = 5$ Hz, **O** = m (small), $\mathbf{R}_6$ and $\mathbf{R}_8$ = m (broad).
[b] 456 = $C_{25}H_{28}O_8$, 235 = $C_{13}H_{15}O_4$ and 97 = $C_5H_5O_2$.

As can be seen from Table 4, the $^{1}$H-NMR spectra of cneorins **D**, **H**, **N** and **O** each have a triplet with $J = 5$ Hz for the proton at C-9, or a narrow multiplet, which, based on the interpretation given above for cneorin-**D**, is diagnostic for the *cis*-orientation of 8-OH and 9-H relative to each other; the (8*S*)-, (9*S*)-configuration is common to all four alcohols which have the structures shown in formulae (**44**), (**55**), (**59**) and (**57**). In contrast, the C-9 proton signal in the spectra of tricoccins $\mathbf{R}_6$ and $\mathbf{R}_8$ is a broad multiplet at higher field, corresponding to its *trans*-orientation with respect to the OH group. As elimination of water to form $\mathbf{B}_I$ establishes the α-, and to form $\mathbf{R}_2$ the β-orientation for 9-H, the 8-OH group must be β-oriented in $\mathbf{R}_6$ and α-oriented in $\mathbf{R}_8$. Tricoccin-$\mathbf{R}_6$ thus has the (8*R*,9*S*)-configuration (**60**) and $\mathbf{R}_8$ the (8*S*,9*R*)-configuration (**58**).

The hydroxy group influences the chemical shift of the proton at C-17. If 8-OH and 17-H are both α-oriented, the 17-H singlet is seen at $\delta_H = 5.36$ to 5.38 ppm, this shift being the same for all relevant structures. If, on the other hand, they are oriented α- and β-, as is the case in the cneorins **N** and **O**, the 17-H signal is shifted upfield by about 0.13 ppm.

The mass spectra of the stereoisomeric alcohols have fragments with the same mass values, but with different intensities, reflecting the stereochemistry of the individual alcohols; the relative intensities are given for the molecular ion peak and the key fragments m/z = 235 and 97 in Table 4.

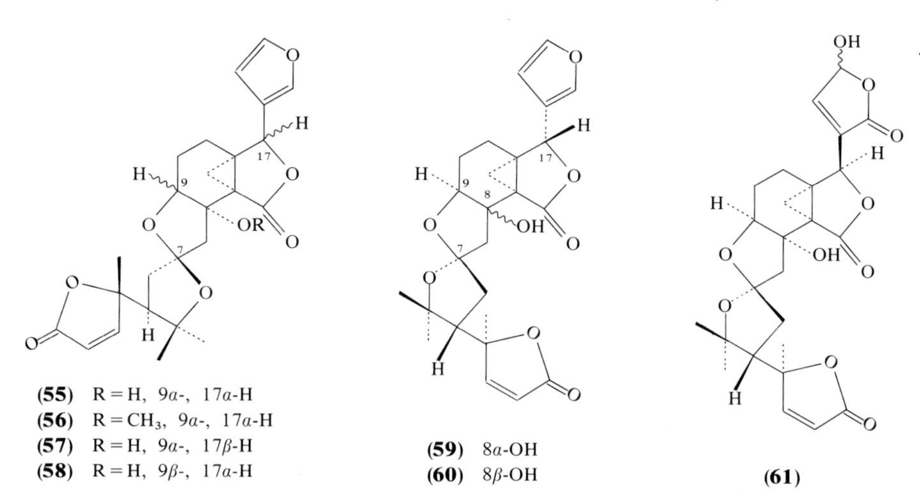

(**55**) R = H, 9α-, 17α-H
(**56**) R = $CH_3$, 9α-, 17α-H
(**57**) R = H, 9α-, 17β-H
(**58**) R = H, 9β-, 17α-H

(**59**) 8α-OH
(**60**) 8β-OH

(**61**)

The cneorins **D**, **H**, **N** and **O** epimerize at the spiroacetal centre C-7 in the presence of acid. To achieve an acceptable reaction rate, a higher than the minimum concentration of acid is required: an equilibrium is established which remains unchanged over considerable periods of time as it is not disturbed by other competing reactions as are observed with the cneorin olefins. In the **C**-series, cneorin-**D** is thermodynamically favoured, in the **B**-series, contrary to expectation, cneorin-**O**.

$$\text{Cneorin-}\mathbf{D}\ (\mathbf{44}) \rightleftharpoons \text{Cneorin-}\mathbf{H}\ (\mathbf{55})$$

$$\text{Cneorin-}\mathbf{N}\ (\mathbf{59}) \rightleftharpoons \text{Cneorin-}\mathbf{O}\ (\mathbf{57})$$

As in the case of the cneorin olefins, the configuration at C-7 influences the position of the 9-H signal in the $^1$H-NMR spectrum. As can be seen from Tables 1 and 4, lower field signals are always associated with the (7*S*)-configuration, and higher field signals with the (7*R*)-configuration. The specific rotations are also affected by configuration in the same manner as previously observed, the change taking place in the same direction but to a lesser degree.

The methyl ethers of cneorin **D** and **H** also establish an equilibrium, which lies on the side of cneorin-**A**.

$$\text{Cneorin-}\mathbf{A}\ (\mathbf{45}) \rightleftarrows \text{Cneorin-}\mathbf{H}\text{-methyl ether}\ (\mathbf{56})$$

In contrast to the cneorins **D**, **H**, **N** and **O**, no equilibrium due to inversion at C-7 in the presence of acid is detectable for the tricoccins $\mathbf{R}_6$ and $\mathbf{R}_8$, as both these alcohols undergo immediate rearrangement to the cneorins $\mathbf{B}_{III}$ and $\mathbf{C}_{III}$.

Cneorin-$\mathbf{NP}_{27}$ (**61**), with a γ-hydroxybutenolide ring instead of the furan ring, is occasionally isolated from *Np*. This variation is frequently encountered among the limonoids and meliacines (*61*), and also known in the case of tricoccins $\mathbf{S}_8$ and $\mathbf{S}_{19}$ from the Cneoraceae (*62*). $\mathbf{NP}_{27}$ is also formed artificially when a methanolic solution of cneorin-**D** is exposed to sunlight (*63*).

## VI. Cneorins and Tricoccins with a C-7 Carbonyl Group

### 1. Cneorin-F (*45*) and Tricoccin-$S_{14}$ (*64*)

The first example of a bitter principle from the Cneoraceae with a keto group was cneorin-**F**, which was isolated from the leaves and fruits of *Np*. Cneorin-**F** has the empirical formula $C_{25}H_{28}O_8$ and the constitution and absolute configuration shown in (**62**) (*24*). Besides the γ-lactone carbonyl band at 1750 cm$^{-1}$, the IR spectrum has a ketone carbonyl band at 1715 cm$^{-1}$, confirmed by the singlet at $\delta = 205.3$ ppm in the $^{13}$C-NMR spectrum. The $^1$H-NMR spectrum shows the protons of the β-furyl group, and the methine proton at C-17, and also an olefinic proton multiplet at $\delta = 5.42$ and the typical ABX-system centred at $\delta = 2.70$ and 4.19 ppm corresponding to the partial structure of the bicyclic lactone ether of (**27**), or of cneorins $\mathbf{B}_{III}$ and $\mathbf{C}_{III}$ respectively. The double bond with $\delta_C = 127.9$ s and 124.4 d is trisubstituted, and not conjugated with the keto group, as can be seen from the UV spectrum with $\lambda_{max}$ 281 nm ($\varepsilon = 130$). An isolated methylene group between the keto group and the double bond is recognized by the broadened doublets at $\delta_H = 3.86$ and 3.46 ppm with $J = 16$ Hz.

The mass spectrum of the cneorin-**F** ethylene acetal (**63**), with the prominent ions m/z=255 and 301 for the α-cleavage, eliminates the possibility that the keto group is incorporated in a ring. Cneorin-**F** is converted by osmium tetroxide into the *cis*-diol (**64**), with the hydroxy groups on the β-side of the molecule, the α-side being severely hindered by the cyclopropane and furan rings. As a β,γ-dihydroxyketone, the *cis*-diol (**64**) undergoes ring closure and loss of water when treated with conc. hydrochloric acid in glycol dimethyl ether, rearranging to cneorin-$\mathbf{B}_{III}$ (**35**). Cneorin-**F** thus belongs to the **B**-series, with (17*R*)-configuration as shown in (**62**).

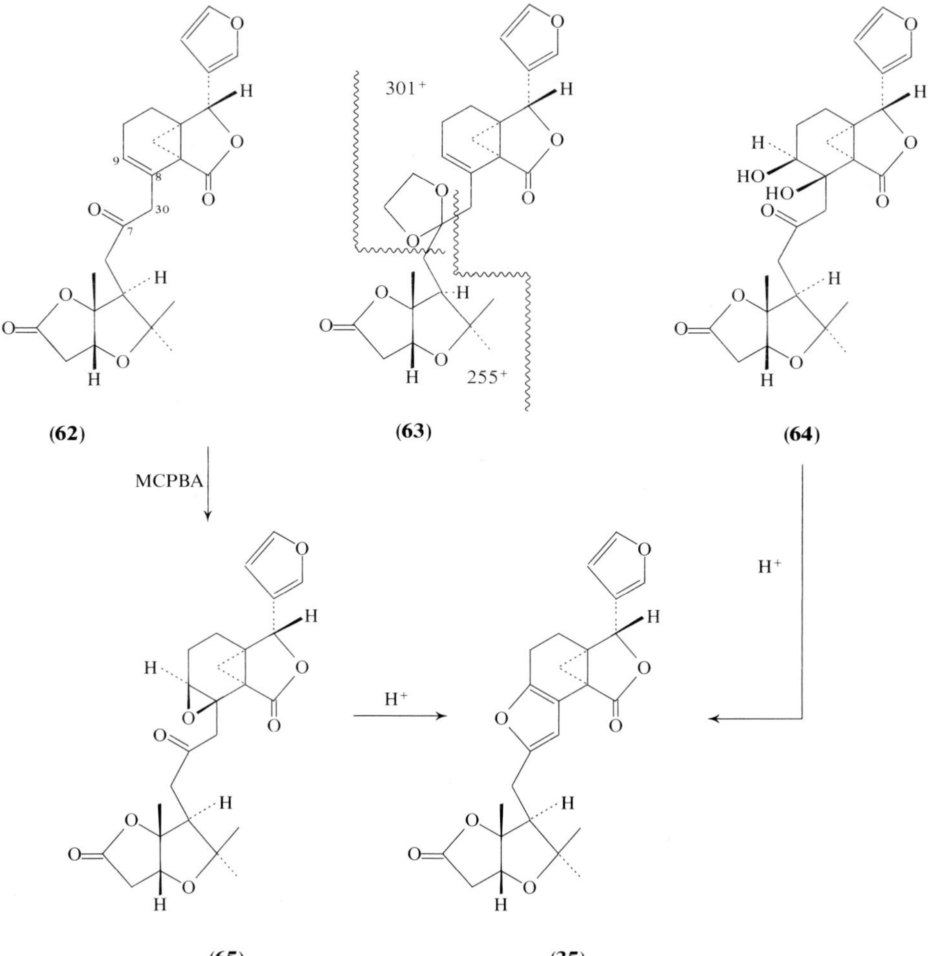

(**62**) (**63**) (**64**)

MCPBA

H⁺

H⁺

(**65**) (**35**)

In the [1]H-NMR spectrum of cneorin-**F**, the signals for 30-$H_2$ with the calculated value 3.25 ppm for the grouping $-CO-CH_2-\overset{|}{C}=\overset{|}{C}-$ are shifted downfield by a maximum of 0.6 ppm owing to the anisotropic effect of the C-15 carbonyl group.

Tricoccin-$\mathbf{S}_{14}$ from *Ct* has the empirical formula $C_{25}H_{28}O_8$ and the stereochemistry shown in (**65**). This compound, too, has a keto group with $\nu_{CO}=1730\,cm^{-1}$ and $\delta_C=205.3$ s (DMSO-$d_6$), and instead of the double bond in (**62**) an epoxide bridge between the carbon atoms 8 and 9, with $\delta_H=3.08$ m and $\delta_C=54.2$ s and 57.4 d (DMSO-$d_6$). Just as (**64**), tricoccin-$\mathbf{S}_{14}$ is rearranged by acid to cneorin-$\mathbf{B}_{III}$.

Epoxidation of cneorin-**F** with m-chloroperbenzoic acid in methylene chloride yields cneorin-**F** epoxide, identical with tricoccin-$\mathbf{S}_{14}$ in all respects, including the β-orientation taken by the incoming epoxide group owing to steric hindrance on the other side.

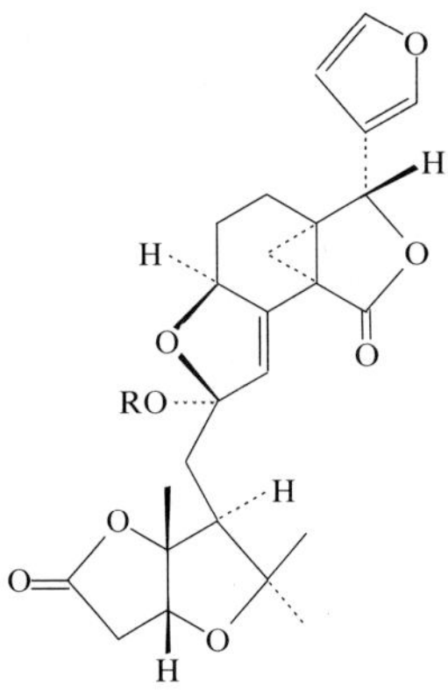

(**66**) R = H
(**67**) R = $CH_3$

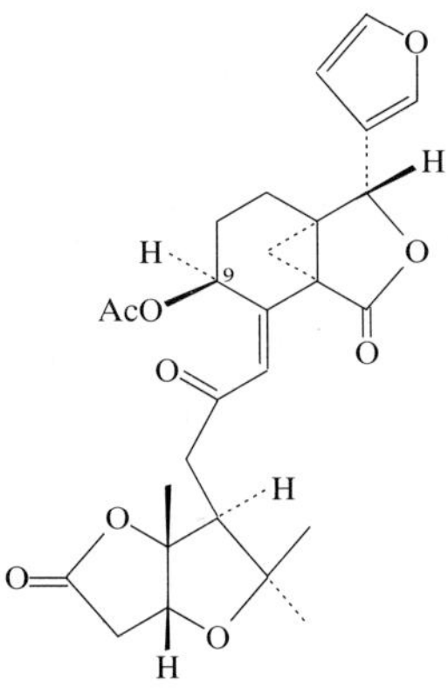

(**68**)

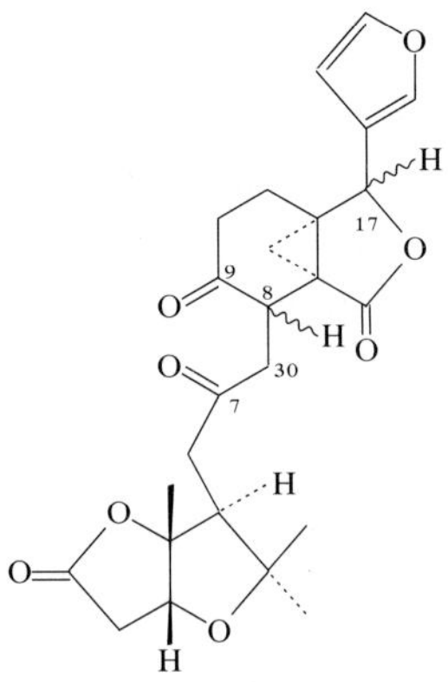

(**69**) 8*α*-, 17*α*-H
(**70**) 8*β*-, 17*α*-H
(**71**) 8*α*-, 17*β*-H
(**72**) 8*β*-, 17*β*-H

Tricoccin-$\mathbf{S}_{14}$ rearranges to the hemiacetal (**66**) in methanol containing a trace of KOH; the [1]H-NMR spectrum (acetone-$d_6$) shows 9-H as a broad, 8-line signal at $\delta=4.54$, the olefinic proton 30-H as a doublet at 6.06 with $J=2$ Hz and the proton of the 7-OH group as a singlet at 4.83 ppm which disappears on $D_2O$ exchange. The methyl ether (**67**) is formed quantitatively from the hemiacetal in methanol. Acetylating conditions cause opening of the dihydrofuran ring, with formation of the conjugated

*References, pp. 181—187*

ketoacetate (**68**) with $\lambda_{max}$ 210 and 248 nm ($\varepsilon = 9800$ and 11000) for the chromophore of the furan ring and the conjugated ketone. The IR spectrum (KBr) has three carbonyl bands at 1775 ($\gamma$-lactone), 1735 (acetate) and 1690 $cm^{-1}$ (conjugated ketone). In the $^{1}$H-NMR spectrum (acetone-$d_6$) the 9-H appears as a multiplet at $\delta = 6.11$, and in the $^{13}$C-NMR the conjugated carbonyl group as a singlet at $\delta = 197.4$ ppm.

## 2. The Tricoccins $R_9$, $R_{12}$ and Their Epi-compounds (*64–66*)

D. Trautmann (*26*) describes the bitter principle tricoccin-$\mathbf{R_9}$ from *Ct* in his thesis as a compound which crystallizes well, has the empirical formula $C_{25}H_{28}O_8$, and the constitution and configuration shown in (**69**). The spectra are similar to those of cneorin-**F** (**62**), from which $\mathbf{R_9}$ differs by virtue of an additional oxygen atom. In the mass spectrum of this C-7 ketone, the ions associated with $\alpha$-cleavage are found at m/z = 211, as in the case of cneorin-**F**, and at m/z = 273 for an ion heavier by 16 mass units than the corresponding ion from cneorin-**F**. Compared with the spectrum of cneorin-**F**, the proton spectrum of $\mathbf{R_9}$ shows an additional splitting of the 30-$H_2$ doublet with $J = 4$ Hz, the result of coupling with a neighbouring proton on C-8; the exact data are $\delta = 3.90$ dd (1 H) and 3.43 dd (1 H) with $J_1 = 18$ and $J_2 = 4$ Hz. The striking shift of the signal for the $-CO-CH_2-$ group, with the calculated value 2.45 ppm, to lower field points to the anisotropic effect of an additional carbonyl group, which can be only at C-9; by doubling the cneorin-**F** increment, $2 \times 0.6$ ppm, the mean value of 3.65 ppm for the chemical shift of 30-$H_2$ in $\mathbf{R_9}$ is obtained.

In methanol containing a trace of alkali, or in chloroform with trifluoroacetic acid, cneorin-$\mathbf{R_9}$ (**69**) establishes an equilibrium with 8-*epi*-cneorin-$\mathbf{R_9}$ (**70**) which the NMR-spectrum shows to lie on the side of the *epi*-compound. The epimers have virtually the same $R_F$-values in all eluants and cannot be separated chromatographically. However, the poorly soluble cneorin-$\mathbf{R_9}$ crystallizes quantitatively from ether solution. The *epi*-compound remains in solution, from which it is precipitated as an amorphous powder by adding pentane; its specific rotation differs only marginally from that of $\mathbf{R_9}$.

The mass spectra of the C-8 epimers have fragments common to both but of different intensities. In the $^{1}$H-NMR spectrum of 8-*epi*-$\mathbf{R_9}$ the 30-$H_2$ signals are shifted to higher field, $\delta = 3.47$ m (1 H) and 3.13 m (1 H).

Inspection of a model leads to the conclusion that in tricoccin-$\mathbf{R_9}$, with the greater low-field shift for 30-$H_2$, the proton at C-8 must be axial and $\alpha$-oriented, and conversely in the *epi*-compound equatorial and $\beta$-oriented. Proof that $\mathbf{R_9}$ belongs to the **C**-series with (17*S*)-configuration will be given later (see below).

Tricoccin-$\mathbf{R}_{12}$, a stereoisomer of tricoccin-$\mathbf{R}_9$, has not been identified as a natural product to date, but has probably been merely overlooked, as the compound has properties most unfavourable to its isolation from a complex mixture. $\mathbf{R}_9$ and $\mathbf{R}_{12}$ have the same constitution and differ solely in the configuration at C-17; $\mathbf{R}_{12}$ (**71**) belongs to the B-series with (17*R*)-configuration.

$\mathbf{R}_{12}$ is obtained from tricoccin-$\mathbf{S}_{14}$ or cneorin-**F** epoxide by rearrangement with $BF_3$-etherate in ether. Cneorin-$\mathbf{B}_{III}$ (**35**), formed as a by-product, crystallizes readily, $\mathbf{R}_{12}$ remaining in the mother liquor. TLC shows a double spot for the equilibrium

$$\text{Tricoccin-}\mathbf{R}_{12}\ (\mathbf{71}) \rightleftharpoons \text{Tricoccin-8-}epi\text{-}\mathbf{R}_{12}\ (\mathbf{72})$$

which is also set up from either side in solutions containing a trace of alkali. This pair of epimers can be separated by preparative layer chromatography, but form only amorphous precipitates from ether solution on addition of pentane. Table 5 shows that the $^1$H-NMR spectra of both epimer pairs are similar in the critical region with the exception of the signal for 22-H, which is shifted to lower field in the spectrum of $\mathbf{R}_9$. This effect reflects the conformation of the molecule with (8*S*)-configuration, disappears in the spectrum of the 8-*epi*-compound (**70**) and is of course absent in the spectra of the epimer pair with an α-oriented furan ring. Only marginal differences are found between the values for 17α-H and 17β-H.

Table 5. *Proton Spectra of the Tricoccins $R_9$ and $R_{12}$ and Their 8-epi-Compounds*[a]

| | $\mathbf{R}_9$ (**69**) | *epi*-$\mathbf{R}_9$ (**70**) | $\mathbf{R}_{12}$ (**71**) | *epi*-$\mathbf{R}_{12}$ (**72**) |
|---|---|---|---|---|
| | C-Series | | **B**-Series | |
| 30-$H_1$ | 3.90 dd | 3.47 m | 3.80 dd | 3.55 m |
| 30-$H_2$ | 3.43 dd | 3.13 m | 3.41 dd | 3.12 m |
| | $J_1$ = 18 Hz | | $J_1$ = 18 Hz | |
| | $J_2$ = 4 Hz | | $J_2$ = 4 Hz | |
| 22-H | 6.91 m | 6.40 m | 6.45 m | 6.41 m |
| 17-H | 5.38 s | 5.40 s | 5.43 s | 5.46 s |

[a] 90 MHz, δ (ppm) in $CDCl_3$.

## VII. C-7 Hemiacetals and Methylacetals with (5*S*, 10*R*, 17*R*)-Configuration

### 1. Cneorin K and $K_I$ (*64*)

Cneorin-**K**, with the empirical formula $C_{25}H_{28}O_7$, was first isolated from the hard fruits of the "Canarian stoneberry" and later also from the leaves of *Np*, but in very small amounts whatever the source (*24*). The compound has a tertiary hydroxy group and is converted quantitatively into the methyl ether cneorin-$\mathbf{K_I}$ when a methanolic solution is allowed to stand; $\mathbf{K_I}$ can also be isolated from plant extracts, but probably as an artefact.

The proton spectra of **K** and $\mathbf{K_I}$ show the usual signals for the butenolide and β-substituted furan rings, the methine proton at C-17, and also a multiplet at $\delta = 5.50$ ppm for the single proton of a trisubstituted double bond. The ion of greatest mass in the mass spectra of **K** and $\mathbf{K_I}$ is at m/z = 422, corresponding to $M^+ - H_2O$ and $M^+ - CH_3OH$ respectively. The base peak at m/z = 325 results from the further loss of the butenolide ring as a radical with mass 97.

The constitution and configuration were determined by X-ray analysis of cneorin-$\mathbf{K_I}$ (*8*), the result of which is shown in the stereoformula (**74**), which also gives the absolute configuration. As no inversion of configuration is observed when a hemiacetal is etherified, cneorin-**K** has structure (**73**), the hemiacetal group having the (7*R*)-configuration. **K** and $\mathbf{K_I}$ belong to the **B**-series with (17*R*)-configuration.

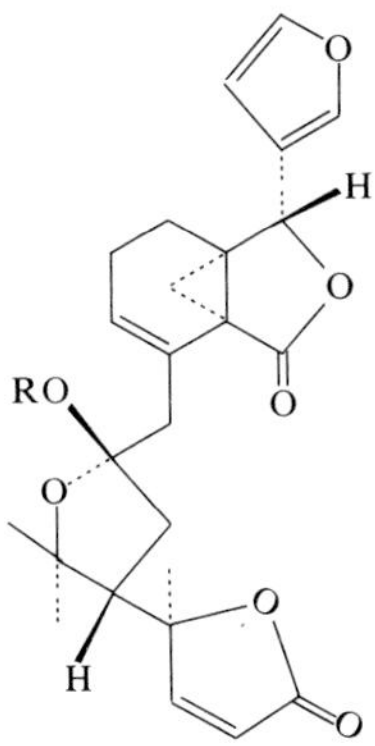

(**73**) R = H
(**74**) R = $CH_3$

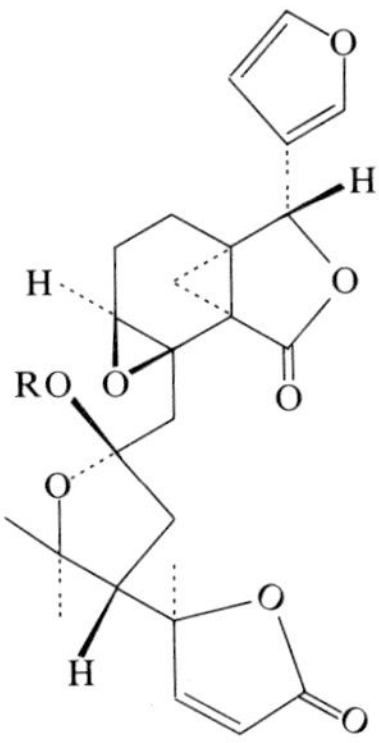

(**75**) R = $CH_3$
(**76**) R = H

## 2. Tricoccin $R_1$ and $R_{10}$ (*26*)

Tricoccin-$\mathbf{R}_1$, with the constitution and absolute configuration shown in formula (**75**), was the first bitter principle isolated from the leaves of *Ct* (*25*). With the empirical formula $C_{26}H_{30}O_8$, the compound is an isomer of cneorin-**A** (**45**) and also a methyl ether, with $\delta_H = 3.22$ s (3H); the same compound was later also isolated from *Np* and designated cneorin-**M**.

In contrast to cneorin-**A**, tricoccin-$\mathbf{R}_1$ is very sensitive to acid, resembling cneorins **B** and **C** in this respect. In chloroform solution, $\mathbf{R}_1$ is very rapidly transformed into cneorin-$\mathbf{B}_I$ (**34**), $\mathbf{B}_{II}$ (**37**) being identified as a by-product. Stronger acid at once converts $\mathbf{R}_1$ to cneorin-$\mathbf{B}_{III}$ (**35**).

$$\text{Tricoccin-}\mathbf{R}_1\ (\mathbf{75}) \xrightarrow{CHCl_3} \text{Cneorin-}\mathbf{B}_I\ (\mathbf{34}) \text{ and } \mathbf{B}_{II}\ (\mathbf{37})$$

$$\text{Tricoccin-}\mathbf{R}_1\ (\mathbf{75}) \xrightarrow{HCl} \text{Cneorin-}\mathbf{B}_{III}\ (\mathbf{35})$$

These characteristics of tricoccin-$\mathbf{R}_1$ were at first wrongly interpreted. The structure originally proposed (*28*) was revised on $^{13}C$-NMR spectral grounds (*64*). The spectrum shows an acetal centre with $\delta_C = 104.5$ s and an oxirane ring at a trisubstituted double bond with $\delta_C = 55.8$ s and 57.1 d. The single proton of the latter appears as multiplet at $\delta_H = 3.53$ ppm.

X-ray analysis of cneorin-$\mathbf{K}_I$, which has a novel structure compared with cneorins **A**, **B**, **C** and **D**, indicated a close relationship between $\mathbf{K}_I$ and $\mathbf{R}_1$. This assumption was confirmed by the preparation of cneorin-$\mathbf{K}_I$ epoxide, which proved identical with tricoccin-$\mathbf{R}_1$ in all respects; cneorin-$\mathbf{B}_I$ (**34**) is formed as a by-product of epoxidation.

$$\text{Cneorin-}\mathbf{K}_I\ (\mathbf{74}) \xrightarrow{\text{Peracid}} \text{Tricoccin-}\mathbf{R}_1\ (\mathbf{75}) + \text{Cneorin-}\mathbf{B}_I\ (\mathbf{34})$$

The extremely facile transformation of tricoccin-$\mathbf{R}_1$ into cneorin-$\mathbf{B}_I$ by the mechanism shown below proves the β-orientation of the epoxide ring, i.e. the (9*S*)-configuration:

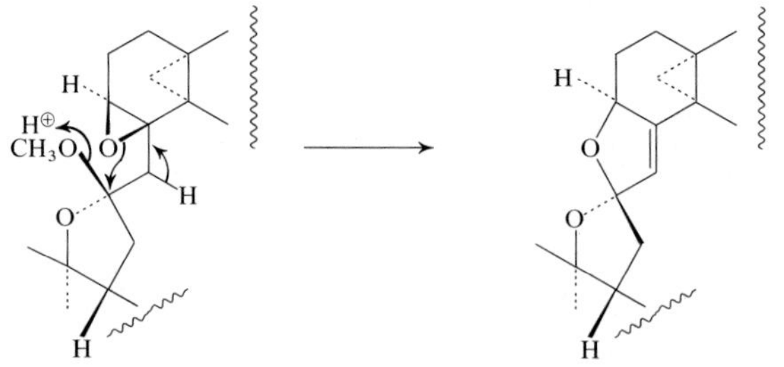

H
$H^{\oplus}$
$CH_3O$ O
H
O
H
$\mathbf{R}_1$ (**75**)

H
O
O
H
$\mathbf{B}_I$ (**34**)

Tricoccin-$\mathbf{R}_{10}$ has the empirical formula $C_{25}H_{28}O_8$ and is also converted by traces of acid into cneorin-$\mathbf{B}_{\mathrm{I}}$. On the other hand, tricoccin-$\mathbf{R}_{10}$ methyl ether is formed quantitatively in methanolic solution, the methyl ether proving identical to tricoccin-$\mathbf{R}_1$ by all criteria. $\mathbf{R}_{10}$ is obtained from cneorin-**K** by epoxidation. Tricoccin-$\mathbf{R}_{10}$ thus has the stereo-formula (**76**) with (7*R*,9*S*,17*R*)-configuration.

$$\text{Tricoccin-}\mathbf{R}_{10}\ (\mathbf{76}) \xrightarrow{H^+} \text{Cneorin-}\mathbf{B}_{\mathrm{I}}\quad (\mathbf{34})$$

$$\text{Tricoccin-}\mathbf{R}_{10}\ (\mathbf{76}) \xrightarrow{CH_3OH} \text{Tricoccin-}\mathbf{R}_1\quad (\mathbf{75})$$

$$\text{Cneorin-}\mathbf{K}\quad (\mathbf{73}) \xrightarrow{\text{Peracid}} \text{Tricoccin-}\mathbf{R}_{10}\ (\mathbf{76})$$

$\mathbf{R}_{10}$ occurs abundantly in the leaves of *Ct*; here too, the methyl ether $\mathbf{R}_1$ may be merely a secondary product formed on work-up.

The close relationships between the C-7 hemiacetals cneorin-**K** and tricoccin-$\mathbf{R}_{10}$ on the one hand, and the C-7 carbonyl compounds cneorin-**F** and tricoccin-$\mathbf{S}_{14}$ on the other is illustrated by the base-catalysed rearrangement of cneorin-**K** to cneorin-**F**, and of tricoccin-$\mathbf{R}_{10}$ via $\mathbf{S}_{14}$ to the hemiacetal (**66**) described above.

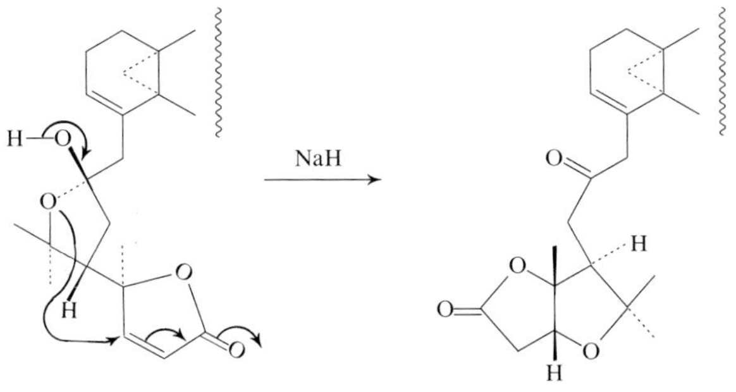

NaH

(**73**) (**62**)

$OH^\ominus$ $OH^\ominus$

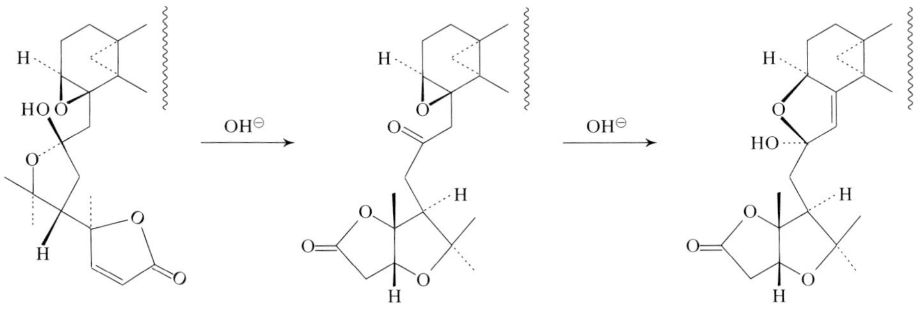

(**76**) (**65**) (**66**)

## VIII. C-9 Hemiacetals with (5*S*, 8*S*, 10*R*)-Configuration (*29*)

In his thesis D. TRAUTMANN (*26*) describes tricoccin-$\mathbf{S_2}$, the main bitter principle of the Cneoraceae species native to the Mediterranean area. This compound has the empirical formula $C_{25}H_{28}O_8$ and is isomeric with the above-mentioned cneorin and tricoccin alcohols bearing a hydroxy group at C-8. Spectral data show $\mathbf{S_2}$ to have the cneoroid skeleton with a tertiary hydroxy group, the position of which can be deduced from the NMR spectra. In the $^{13}$C off-resonance spectrum the singlet at $\delta = 108.5$ for the spiroacetal centre C-7 is accompanied by a further singlet at 103.5 ppm for an additional $-O-\overset{|}{\underset{|}{C}}-O-$ grouping; as the $^1$H-NMR spectrum lacks a C-9 proton signal, the hydroxy group must be bound to this carbon atom. Tricoccin-$\mathbf{S_2}$ is a hemiacetal and readily forms a methyl or ethyl acetal.

Neither cneorin-$\mathbf{B_{III}}$ nor cneorin-$\mathbf{C_{III}}$ is formed by acid-catalysed rearrangement of tricoccin-$\mathbf{S_2}$ which resembles the cneorin alcohols **D**, **H**, **N** and **O** in this respect. However, catalytic amounts of acid cause rearrangement of $\mathbf{S_2}$ to tricoccin-$\mathbf{S_1}$ which accompanies $\mathbf{S_2}$ naturally in the plant, although in smaller quantities. In the presence of acid, $\mathbf{S_2}$ and $\mathbf{S_1}$ form an equilibrium, reached rapidly from either side and analogous to the equilibria between **B** and $\mathbf{B_I}$, or between **C** and $\mathbf{C_I}$. This analogy prompts the conclusion that $\mathbf{S_2}$ and $\mathbf{S_1}$ are a further pair of stereoisomers distinguished from each other by the configuration at the spiroacetal centre C-7.

An X-ray analysis of tricoccin-$\mathbf{S_2}$ was carried out to assign it to the correct steric series and to confirm its constitution (*8*). Formula (**77**) shows the result and also the absolute configuration, bearing in mind the assumption, that the $\alpha$-orientation of the cyclopropane ring is biogenetically imposed. $\mathbf{S_2}$ belongs to the **C**-series with (17*S*)-configuration; 8-H and 9-OH are $\beta$-oriented and mutually *cis*, while the spiroacetal centre has the (7*R*)-configuration. The C-7 epimer tricoccin-$\mathbf{S_1}$ has the structure shown in (**82**).

*Diagram 3*

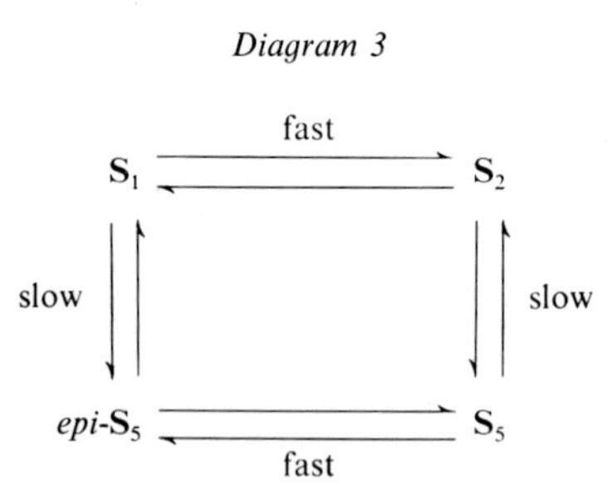


A total of four stereoisomers are possible through inversion at the acetal carbon atoms 7 and 9. D. TRAUTMANN (*26*) discovered tricoccin-$\mathbf{S_5}$ as a natural product, while the fourth stereoisomer, designated *epi*-$\mathbf{S_5}$, was later prepared artificially from $\mathbf{S_2}/\mathbf{S_1}$ (*8*). All four stereoisomers equilibrate with each other in the presence of acid, as shown in Diagram 3, although with great variations in the speed with which the individual equilibria are established.

A thin-layer chromatogram of the equilibrium mixture always shows only three spots, as $\mathbf{S_2}$ and *epi*-$\mathbf{S_5}$ have the same $R_F$-value. If, on the other hand, a catalytic amount of ammonium hydroxide is used for equilibration, in each case only two spots are visible, corresponding to the two rapidly reached equilibria $\mathbf{S_1} \rightleftharpoons \mathbf{S_2}$ and $\mathbf{S_5} \rightleftharpoons$ *epi*-$\mathbf{S_5}$ respectively. Exploiting these differences, *epi*-$\mathbf{S_5}$ was obtained via $\mathbf{S_5}$ from $\mathbf{S_1}/\mathbf{S_2}$.

Assignment of configuration at the carbon atoms 7 and 9 of the stereoisomers is possible by a study of the $^{13}$C-NMR spectra; these show characteristic signal positions for C-5, depending on whether the proton on this carbon atom is on the same or opposite side as the oxygen atom between carbon atoms 7 and 9. In the former case, the $\delta_C$ value is less, in the latter case greater than 54 ppm. $\mathbf{S_2}$ (**77**) serves as a reference compound in which the proton at C-5 is on the same side as the oxygen atom between carbon atoms 7 and 9. The specific rotations, the $\delta$ values for C-5 in the $^{13}$C-NMR spectra with the corresponding configurations, and some mass spectroscopic data for the stereoisomers are listed in Table 6. Formulae (**77**) and (**82**) for $\mathbf{S_2}$ and $\mathbf{S_1}$ can now be supplemented by the stereoformulae (**86**) and (**90**) for $\mathbf{S_5}$ and *epi*-$\mathbf{S_5}$.

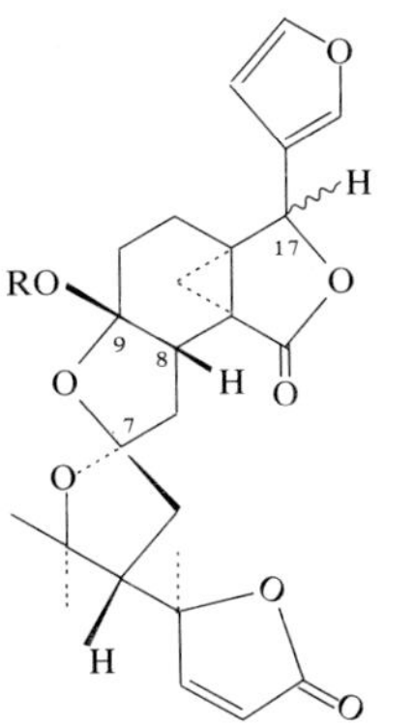

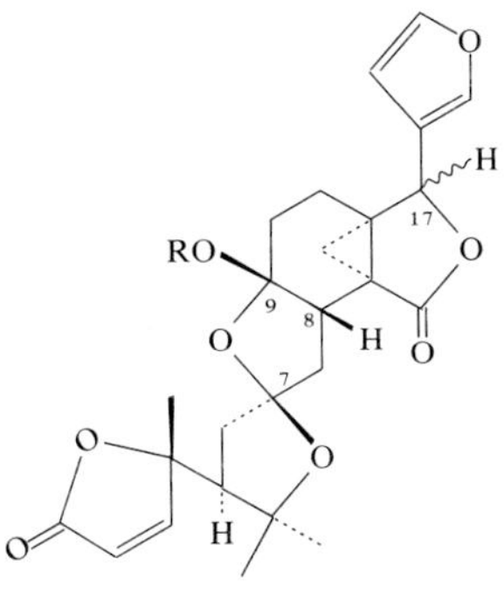

| | R | 17α- | 17β-H |
|---|---|---|---|
| $S_2$ (**77**) | H | + | – |
| $S_{10}$ (**78**) | $CH_3$ | + | – |
| — (**79**) | $C_2H_5$ | + | – |
| $S_{15}$ (**80**) | H | – | + |
| — (**81**) | $CH_3$ | – | + |

| | R | 17α- | 17β-H |
|---|---|---|---|
| $S_1$ (**82**) | H | + | – |
| $S_{11}$ (**83**) | $CH_3$ | + | – |
| $S_{20}$ (**84**) | H | – | + |
| — (**85**) | $CH_3$ | – | + |

| | | R | 17α- | 17β-H |
|---|---|---|---|---|
| $S_5$ | **(86)** | H | + | − |
| $S_{12}$ | **(87)** | $CH_3$ | + | − |
| — | **(88)** | $C_2H_5$ | + | − |
| *epi*-$S_{26}$ | **(89)*** | H | − | + |

| | | 17α- | 17β-H |
|---|---|---|---|
| *epi*-$S_5$ | **(90)** | + | − |
| $S_{26}$ | **(91)** | − | + |

* only DC-evidence

Table 6. *Data on the C-9 Hemiacetals with* (5*S*,8*S*,10*R*)-*Configuration*

| Tricoccin/ formula | | $[\alpha]_D^{20}$ Acetone | $^{13}$C-NMR[a] C-5 | Configuration C-7 | C-9 | C-17 | MS/rel. Int. % m/z 456 | 341 | 97 |
|---|---|---|---|---|---|---|---|---|---|
| $S_1$ | **(82)** | + 1° | 54.6 | *S* | *R* | *S* | — | 100 | 75 |
| $S_2$ | **(77)** | −52° | 53.3 | *R* | *R* | *S* | — | 100 | 87 |
| $S_5$ | **(86)** | −17° | 53.5 | *R* | *S* | *S* | 2 | 60 | 100 |
| *epi*-$S_5$ | **(90)** | +10° | 54.3 | *S* | *S* | *S* | 1 | 72 | 100 |
| $S_{20}$ | **(84)** | 0° | 54.3 | *S* | *R* | *R* | — | 100 | 66 |
| $S_{15}$ | **(80)** | −56° | 53.6 | *R* | *R* | *R* | 2 | 97 | 100 |
| *epi*-$S_{26}$ | **(81)**[b] | — | — | *R* | *S* | *R* | — | — | — |
| $S_{26}$ | **(91)** | +22° | 54.2 | *S* | *S* | *R* | 1 | 100 | 65 |

[a] δ (ppm) in Acetone-$d_6$.
[b] Identified by TLC, substance not yet isolated.

The four stereoisomers display noticeable differences in chromatographic behaviour on silica gel plates developed with ether: **$S_1 > S_2$** and **$S_5$** > *epi*-**$S_5$**. Inspection of a DREIDING model supplies an explanation for the less polar character of tricoccin **$S_1$** and **$S_5$**, in which an intramolecular hydrogen bond 2.1 to 2.2 Å in length can form a bridge between the proton of the axial hydroxy group at C-9 and the oxygen atom between the carbon atoms 4 and 7.

*References, pp. 181—187*

In the plant itself, the tricoccins $S_2$, $S_1$ and $S_5$ are accompanied by their corresponding methyl acetals (**78**), (**83**) and (**87**), designated $S_{10}$, $S_{11}$ and $S_{12}$ respectively (*26*); the ethyl acetals (**79**) and (**88**) are products of chemical transformation.

The series of stereoisomeric C-9 hemiacetals is continued by the tricoccins $S_{15}$, $S_{20}$ and $S_{26}$ (*8*) which belong to the stereochemical **B**-series with (17*R*)-configuration; the missing stereoisomer *epi*-$S_{26}$ has been identified in thin layer chromatograms but not yet isolated in substance. Here too, the stereoisomers establish an equilibrium as shown in Diagram 4, the rapid equilibration between $S_{15}$ and $S_{20}$ corresponding to that between $S_2$ and $S_1$. Configurations are assigned on the basis of the $\delta_C$ values for C-5 and comparison of the specific rotations in the two steric series (cf. Diagram 4 and Table 5). Tricoccins $S_{15}$, $S_{20}$ and $S_{26}$ have the structures shown in (**80**), (**83**) and (**91**) respectively, while (**89**) is the structure of the still unknown *epi*-$S_{26}$. $S_{15}$ and $S_{20}$ readily form the methyl acetals (**81**) and (**85**).

*Diagram 4*

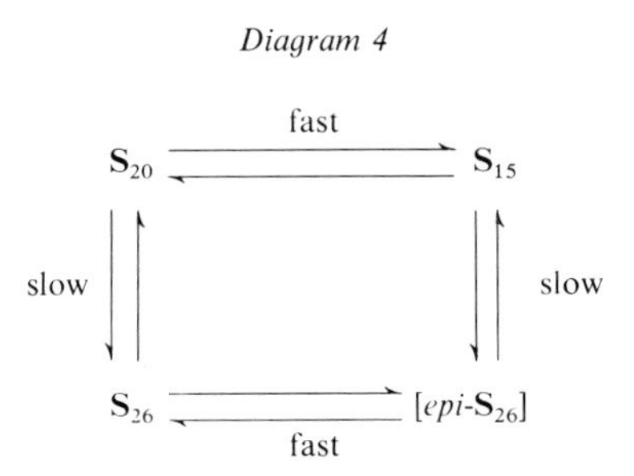


The natural stereoisomers $S_{26}$ and $S_5$ in the **B**- and **C**-series are conspicuous by their mutually opposed configurations.

The C-9 hemiacetals of the **C**-series each give three unsaturated compounds when treated with phosphoryl chloride in pyridine, elimination of water being followed by work-up with careful exclusion of acid. It is of no importance whether $S_1$, $S_2$ or $S_5$ is chosen as starting material (*8, 29*). $S_2$ has been most closely studied, elimination of water giving rise to the two enol ethers (**92**) and (**93**) with a $\Delta^{9(11)}$ double bond and (7*S*)- or (7*R*)-configuration respectively. A by-product of this reaction is the isomer (**94**) with a ditertiary double bond. In the $^1$H-NMR spectra of (**92**) and (**93**) the enol ether proton 11-H appears as a 6-line signal at $\delta = 4.50$ which is replaced in the spectrum of (**94**) by a pair of doublets for 30-$H_2$ at $\delta = 3.18$ and 2.88 with $J = 14$ Hz. The enol ether (**92**) with $\delta_C = 54.6$ ppm for C-5 has the (7*S*)-configuration and (**93**) thus the (7*R*)-configuration. When a moist solution of (**92**) or (**93**) in chloroform or methylene chloride is left in contact with an acid ion exchanger, water adds readily to re-form a mixture of tricoccins $S_1$, $S_2$ and $S_5$.

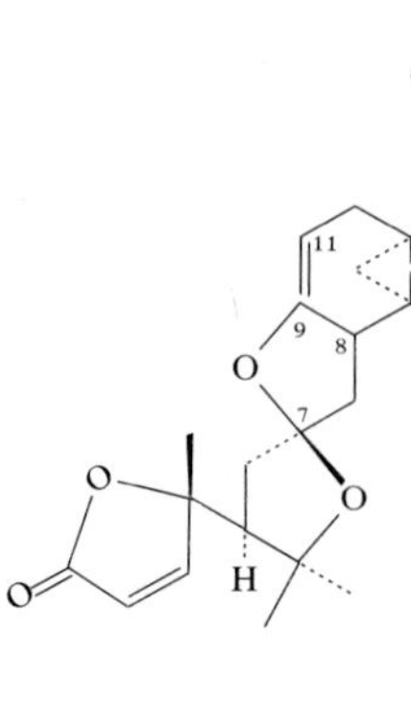

**(92)** 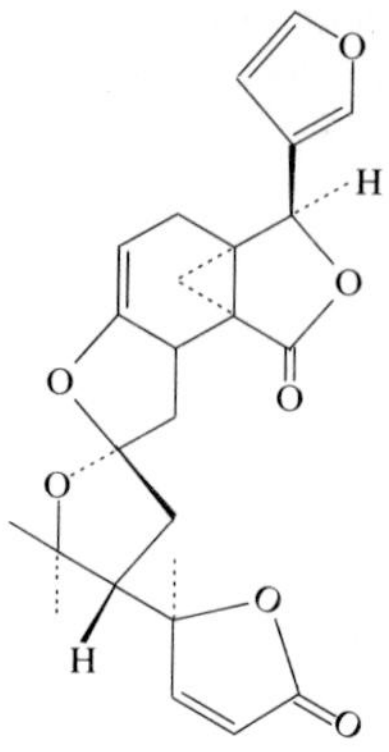 **(93)** 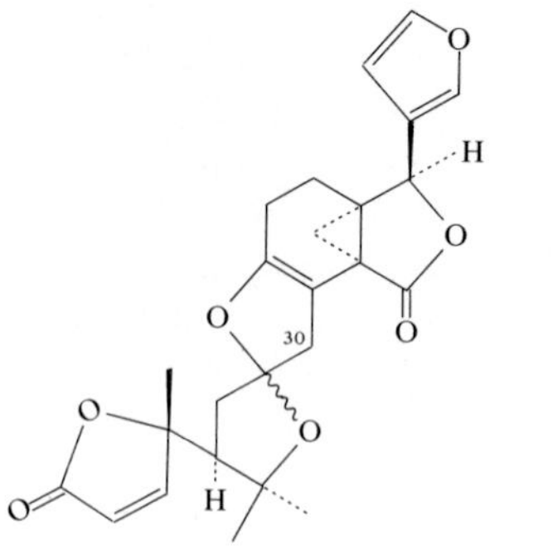 **(94)**

Tricoccins $\mathbf{S}_1$ and $\mathbf{S}_2$ undergo rearrangement to 1,4-dicarbonyl compounds when treated with sodium hydride in glycol dimethyl ether. When the reaction mixture is acidified, the crystalline tricoccin-$\mathbf{R}_9$ **(69)** described above is isolated, while 8-*epi*-$\mathbf{R}_9$ **(70)** can be recovered from the mother liquor. This rearrangement takes place *via* the anion formed by removal of the 8-OH group proton, the primary product being 8-*epi*-$\mathbf{R}_9$ from which $\mathbf{R}_9$ is formed by equilibration. This supplies the still missing proof of (17*S*)-configuration for the pair of epimers (see above).

Tricoccins $\mathbf{R}_{12}$ **(71)** and 8-*epi*-$\mathbf{R}_{12}$ **(72)** were prepared in just the same way from $\mathbf{S}_{15}/\mathbf{S}_{20}$.

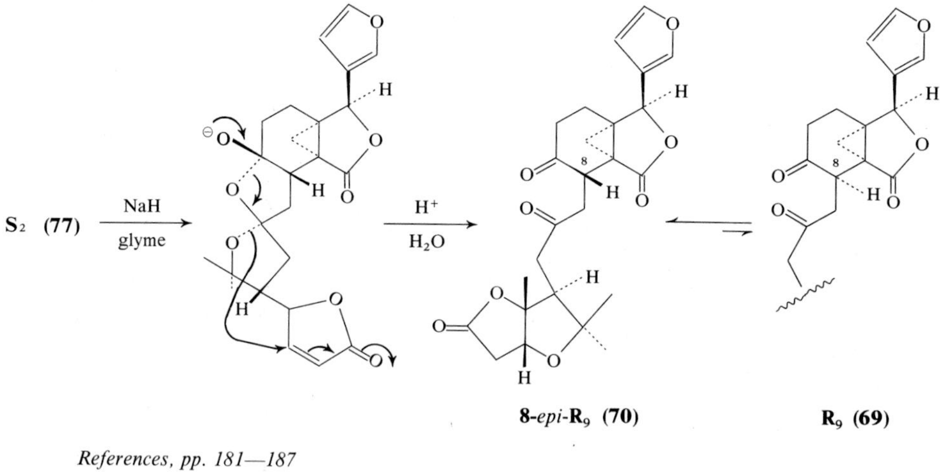

**S₂ (77)** NaH, glyme → H⁺, H₂O → **8-*epi*-R₉ (70)** ⇌ **R₉ (69)**

*References, pp. 181—187*

## IX. Cyclic Peroxides (*67*)

The cyclic peroxides cneorin **Q** and **NP**$_{29}$ are interesting bitter principles, found only in Cneoraceae of the Canaries. The two compounds have the same empirical formula, differing from that of the $\Delta^{8(30)}$ olefins by one additional oxygen atom in an ether-type linkage.

Equivalent fragments with equal intensities are seen in the mass spectra of **Q** and **NP**$_{29}$, from which they are recognized as stereoisomers. A characteristic feature is the splitting off of $O_2$ from the molecular ion m/z=454 to give the ion m/z=422 for $C_{25}H_{26}O_6$; this was the first indication of the peroxide structure of these two bitter principles. The loss of water always observed in the spectra of cneorins **B**, **B**$_I$ and **C**, **C**$_I$ is also found with **Q** and **NP**$_{29}$, giving rise to the ion m/z=436, from which either a methyl radical is lost to give m/z=421 or, the more important process, the butenolide ring is split off to give the base peak at m/z=339 for $C_{20}H_{19}O_5$.

Study of the $^{13}$C-NMR spectrum of cneorin-**Q** permits identification of the site of the peroxide bridge. By comparison with the data for cneorins **B**$_I$, **C**$_I$ and **C** in Table 8, it can be seen that the signals for C-7 to C-9 in the spectrum of **Q** are shifted to higher field, and those for C-30 and also C-10 to lower field. This identifies carbon atoms 7 and 9 as the atoms linked by the peroxide bridge, corresponding to formulae (**95**) and(**96**).

Table 7. *Selected Data on Cneorins* **Q** and **NP**$_{29}$ *with Reference Data on Cneorins and Tricoccins* **B**, **B**$_I$, **R**$_5$ *and epi-***R**$_5$

| Compound | Formula | $^1$H-NMR[a] 5-H | 9-H | 30-H |
|---|---|---|---|---|
| **Q** (**95**) | $C_{25}H_{26}O_8$ | <2.5 | 4.52 m | 6.30 d |
| **NP**$_{29}$ (**96**) | $C_{25}H_{26}O_8$ | 2.60 m | 4.45 m | 6.38 d |
| **B** (**33**) | $C_{25}H_{26}O_7$ | <2.5 | 4.29 m | 6.10 d |
| **B**$_I$ (**34**) | $C_{25}H_{26}O_7$ | 2.70 m | 4.47 m | 6.11 d |
| **R**$_5$ (**53**) | $C_{25}H_{26}O_7$ | 2.70 m | 4.21 m | 5.98 d |
| *epi*-**R**$_5$ (**52**) | $C_{25}H_{26}O_7$ | <2.5 | 4.49 m | 5.98 d |

[a] 90 MHz, δ (ppm) in $CDCl_3$.

Table 8. $^{13}$*C-NMR Data on Cneorin-***Q** *and Reference Data on Cneorins* **B**$_I$, **C**$_I$ *and* **C**

| Cneorin- | C-7 | C-9 | C-8 | C-30 | C-5 | C-17 | C-10 | C-4 |
|---|---|---|---|---|---|---|---|---|
| **Q** (**95**) | 105.8 | 75.8 | 131.2 | 126.9 | 52.9 | 77.6 | 84.8 | 87.0 |
| **B**$_I$ (**34**) | 115.1 | 79.7 | 138.6 | 120.2 | 51.9 | 76.4 | 81.8 | 87.4 |
| **C**$_I$ (**23**) | 115.3 | 79.5 | 138.6 | 120.3 | 51.9 | 77.1 | 82.0 | 87.4 |
| **C** (**16**) | 116.7 | 80.9 | 138.6 | 121.7 | 53.2 | 77.1 | 81.8 | 87.4 |

[a] Bruker HX-90 R, δ (ppm) in DMSO-$d_6$.

The $^1$H-NMR spectra of **Q** and **NP$_{29}$** also show a low-field shift, for the 30-H doublet at $\delta = 6.30$ and 6.38 ppm respectively ($J = 2$ Hz), when compared with the corresponding data for the $\Delta^{8(30)}$ olefins (Table 7). Conversely, the position of the 8-line signal for 9-H centred at $\delta = 4.52$ and 4.45 ppm respectively remains unchanged.

Chemical detection of the peroxide structure fails with acid potassium iodide solution, but is readily achieved with $Fe^{2+}/SCN^-$ spray reagent (*68*). Unlike cneorins **B** and **C**, the peroxides do not undergo rearrangement in acid solution.

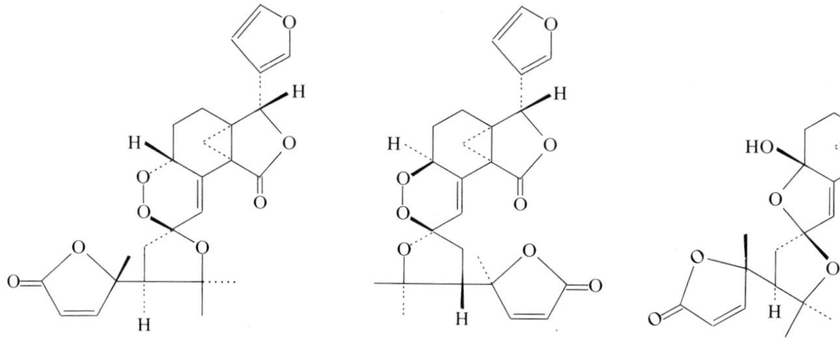

**(95)** —$Ph_3P$→ **(53)**

**(96)** —$Ph_3P$→ **(34)**

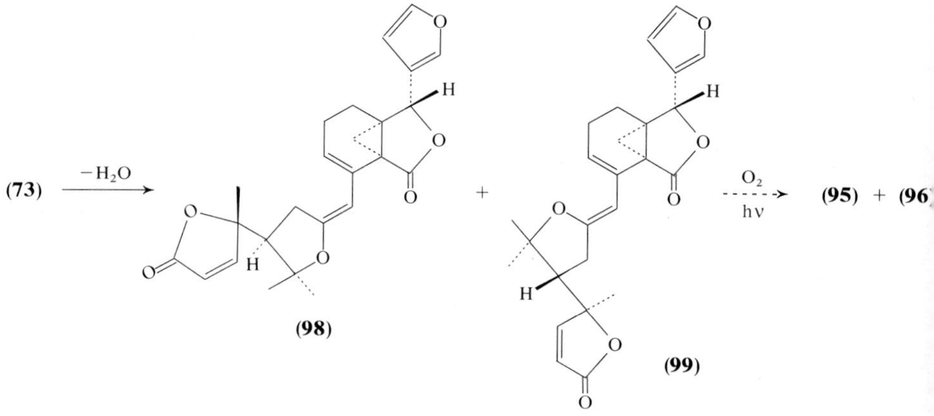

**(97)**

**(73)** —$-H_2O$→ **(98)** + **(99)** ⇢($O_2$, $h\nu$) **(95)** + **(96)**

J. A. TURNER and W. HERZ (*69*) described the $FeSO_4$-catalysed rearrangement of cyclic peroxides to hemiacetals. When applied to cneorin-**Q**, this reaction leads smoothly and in high yield to a compound with mp. 187° C (from ether) which has an OH band at 3400 $cm^{-1}$ in the IR spectrum; in the $^1$H-NMR spectrum, the signal for the olefinic proton at 30-H now appears as a singlet at $\delta = 6.23$ ppm, and the 9-H multiplet is missing. The mass spectrum, with the molecular ion peak at $m/z = 454$, has no peak corresponding to loss of $O_2$, but a base peak at $m/z = 339$ as seen in the spectrum of cneorin-**Q**. All the evidence confirms the hemiacetal structure (**97**) with the hydroxy group at C-9, thus providing additional proof that the peroxide bridge in cneorin-**Q** is between carbon atoms 7 and 9 (**95**).

**$NP_{29}$** conspicuously fails to react with $FeSO_4$ in THF in a comparable time, but as has been pointed out repeatedly above, reactivities are known to be often highly sensitive to stereochemical influences.

**Q** and **$NP_{29}$** form a further pair for which it is possible to deduce the configuration from the C-5 proton signal in the $^1$H-NMR spectrum: if a doubled doublet is seen above 2.5 ppm, this proton and the oxygen atom between carbon atoms 7 and 9 are on the same side of the molecule. As seen from Table 7, **Q** and **$NP_{29}$** thus have opposite configurations at C-7.

The configuration at the other centres is identified by deoxygenating the peroxides with triphenylphosphine in hot toluene as described by L. HORNER and W. JURGELEIT (*70*). Under these conditions, cneorin-**Q** yields tricoccin-**$R_5$** (**53**) with 9β-H; inversion at C-7 proceeds according to the HORNER mechanism via a cation at this carbon atom. This confirms the (7*S*,9*R*,17*R*)-configuration of cneorin-**Q**. Cneorin-**$NP_{29}$**, which is deoxygenated to cneorin-**$B_I$** (**34**) with 9α-H, has the (7*R*,9*S*,17*R*)-configuration (**96**).

The hemiacetal (**97**), with the (7*R*)- and (17*R*)-configuration of cneorin-**B** (**33**), probably has a β-oriented OH group with a hydrogen bond to the oxygen atom between carbon atoms 5 and 7. Both the IR spectrum and the non-polar nature of the compound are evidence for this assumption.

We consider it plausible that the peroxides are formed by photo-oxidation (*71*) of the postulated dienes (**98**) and (**99**), which can arise as products of water elimination from cneorin-**K** (**73**). The ions found in the mass spectra of cneorins **K**, **Q** and **$NP_{29}$** at $m/z = 422$ clearly correspond to $C_{25}H_{26}O_6$. This assumption is supported by a qualitative test, in which pyrolysis of cneorin-**K** at 350° C/$10^{-4}$ mm Hg yields a distillate with $\lambda_{max}$ 264 nm. In addition inspection of models shows that the oxygen molecule will approach the conjugated dienes from either the α- or the β-side respectively, owing to steric hindrance by the butenolide ring, such an approach necessarily leading to cneorins **Q** or **$NP_{29}$** respectively.

## X. Bitter Principles with a γ-Lactol Ring (C-15 Hemiacetals)

### 1. Precursors of Known Cneoroids *(65)*

In Cneoraceae species native to the Mediterranean area we found a great variety of γ-lactols, characterized by a hemiacetal group at C-15 in ring D instead of a carbonyl group. Four variations of this new structural type are theoretically possible and are illustrated by partial structures (**100**) to (**103**). Oxidation of these leads to partial structures (**104**) and (**105**) which correspond to the two steric series **B** and **C** of the cneoroids described so far.

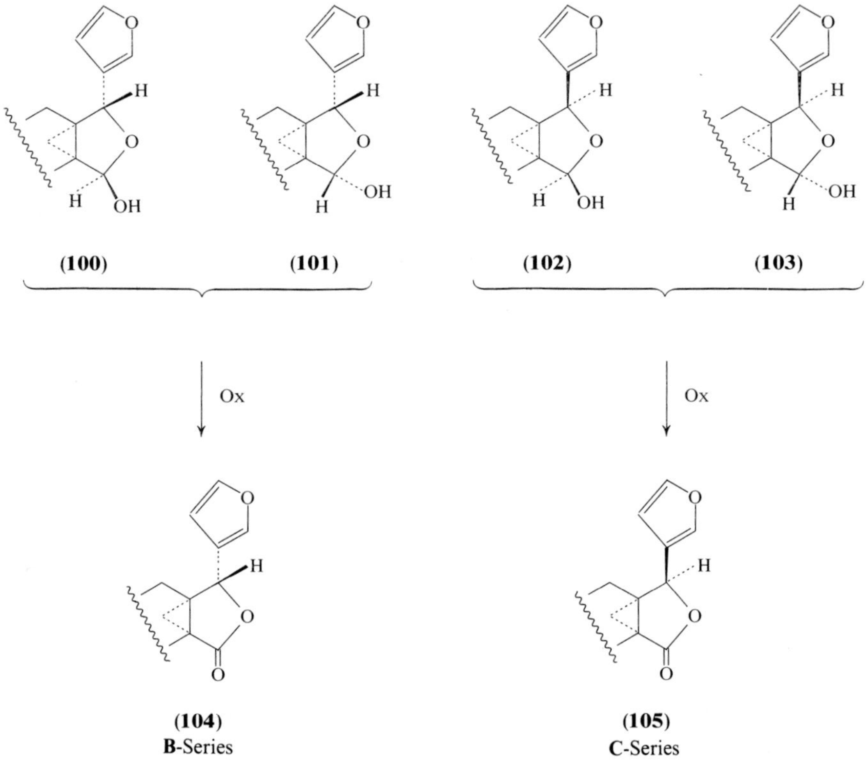

(**100**) (**101**) (**102**) (**103**)

Ox Ox

(**104**) **B**-Series

(**105**) **C**-Series

Apart from a single exception not yet definitely confirmed as such, all tricoccins found with the γ-lactol structure belong sterically to the **B**-series with an α-oriented furan ring; configurational assignment of the compounds discussed below is thus restricted to the elucidation of the configuration at C-15.

The labile γ-lactols usually occur as mixtures of epimers which are difficult to separate and are frequently obtained in an amorphous state. Oxidation with pyridinium chlorochromate (*72*) converts them into the corresponding γ-lactones. For all the tricoccins (**106**) to (**112**) listed in Table 9, the γ-lactone obtained from them by oxidation are known. The constitution and configuration of the lactols, except that at C-15, are deduced from the structures of the oxidation products.

Table 9. *Tricoccins with a γ-Lactol Ring and Their Oxidation Products*

| Tricoccin- | Formula | Oxidation Product |
|---|---|---|
| $S_6$ (**106**) | $C_{25}H_{30}O_8$ | Tricoccin-$\mathbf{R}_6$ (**60**) |
| $S_{17}$ (**107**) | $C_{25}H_{28}O_7$ | Cneorin-$\mathbf{B}_I$ (**34**) |
| $S_{18}$ (**108**) | $C_{25}H_{28}O_7$ | Cneorin-$\mathbf{B}_{III}$ (**35**) |
| $S_{24}$ (**109**)[a] | $C_{25}H_{30}O_8$ | Tricocccin-$\mathbf{R}_{12}$ (**71**) |
| $S_{28}$ (**110**)[a] | $C_{25}H_{30}O_7$ | Cneorin-**F** (**62**) |
| $S_{35}$ (**111**) | $C_{26}H_{32}O_8$ | Tricoccin-$\mathbf{R}_1$ (**75**) |
| $S_{36}$ (**112**) | $C_{25}H_{30}O_8$ | Cneorin-**N** (**59**) |

[a] Characterized as acetate, cf. Table 11.

The hydroxy group at C-15 of the γ-lactols prefers the β-orientation. In the proton spectra, the signal for 17β-H is shifted upfield to approx. 5.0 ppm compared with the related lactones, while the 15α-H appears as a singlet at about 5.5 ppm after $D_2O$ exchange. The $^1$H-NMR data are listed in Table 10, although it should be noted that the use of different solvents does not allow a rigorous comparison. In the single case of $\mathbf{S}_{36}$ a 1 : 1 mixture of the epimeric lactols with unambiguous assignments exists in $CDCl_3$ solution.

Table 10. *Special $^1$H-NMR Data on the γ-Lactols (δ ppm for $CDCl_3$)*

| Tricoccin- | $S_6$[a] (**106**) 15α-OH | $S_{17}$ (**107**) 15β-OH | $S_{18}$ (**108**) 15β-OH | $S_{24}$[b] (**109**) 15β-OH | $S_{28}$[b] (**110**) 15β-OH | $S_{35}$ (**111**) 15β-OH | $S_{36}$ (**112a**) 15α-OH | $S_{36}$ (**112b**) 15β-OH |
|---|---|---|---|---|---|---|---|---|
| 15-H[c] | 5.61 | 5.50 | 5.60 | 5.25 | 5.45 | 5.40 | 5.39 | 5.50 |
| 17β-H | 4.62 | 5.00 | 5.08 | 5.15 | 5.10 | 4.93 | 4.63 | 4.98 |

[a] DMSO-$d_6$. [b] Acetone-$d_6$. [c] After $D_2O$ exchange.

Separation and characterization of the epimeric lactols is often facilitated by preparing their crystalline acetates. The following rule is applicable to the $^1$H-NMR data in Table 11: if the value for 17β-H lies above 4.9 ppm, the acetate group at C-15 is β-oriented.

Table 11. *Properties of the γ-Lactol Acetates*

| Tricoccin- | Schmp. °C | $[\alpha]_D^{20}$ Acetone | Formula | $^1$H-NMR data[a] | | |
|---|---|---|---|---|---|---|
| | | | | 15α-H | 15β-H | 17β-H |
| $S_6$-15α-Ac | amorph | − 81.3° | $C_{27}H_{32}O_9$ | — | 6.57[b] | 4.82 |
| $S_6$-15β-Ac | 161 | −147.3° | $C_{27}H_{32}O_9$ | 6.30 | — | 5.05 |
| $S_{18}$-15β-Ac | amorph | — | $C_{27}H_{30}O_8$ | 6.57 | — | 5.09 |
| $S_{24}$-15β-Ac | 168 | + 60.0° | $C_{27}H_{32}O_9$ | 6.23 | — | 5.20 |
| $S_{28}$-15β-Ac | 175 | +107.2° | $C_{27}H_{32}O_8$ | 6.38 | — | 5.04 |
| $S_{36}$-15α-Ac | 169 | — | $C_{27}H_{32}O_9$ | — | 6.35 | 4.80 |

[a] 90 MHz, δ (ppm) in $CDCl_3$.
[b] Shift caused by interaction with 8β-OH.

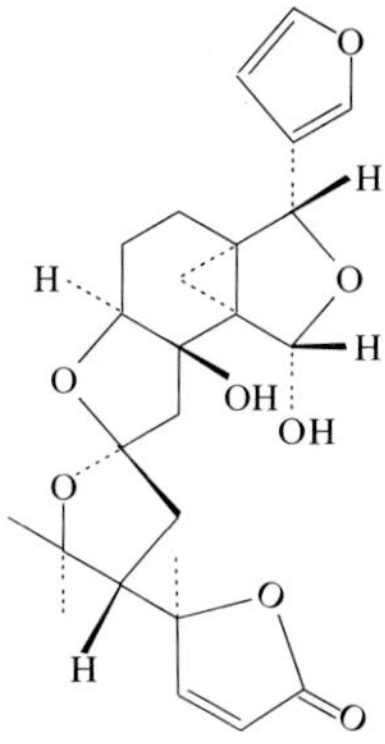

**(106)** 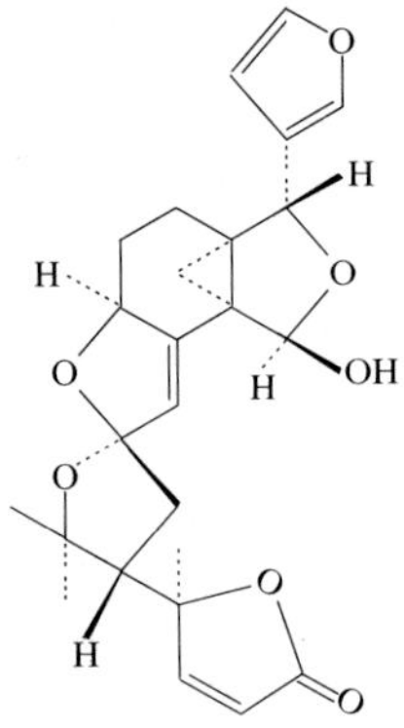 **(107)** 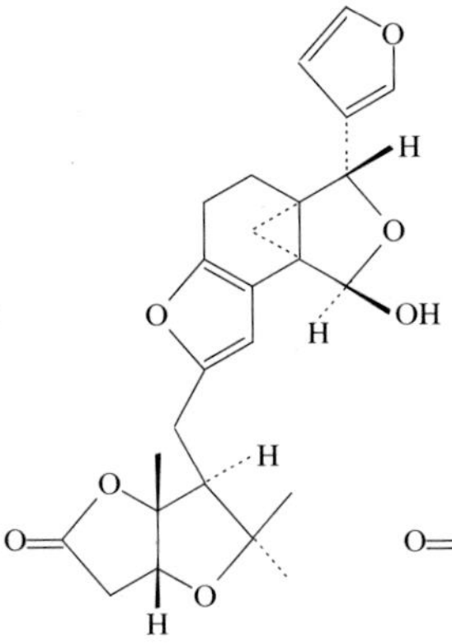 **(108)** 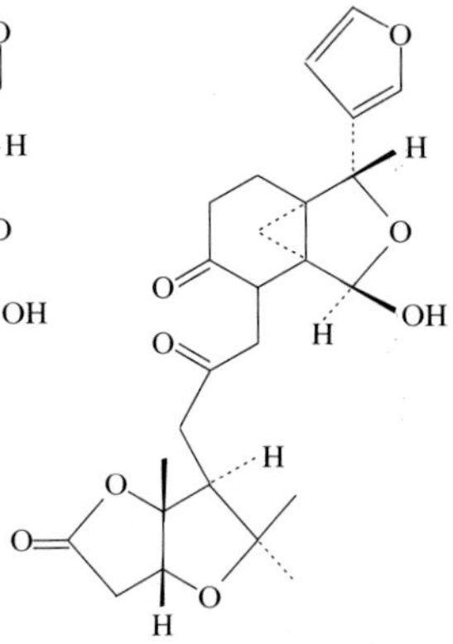 **(109)**

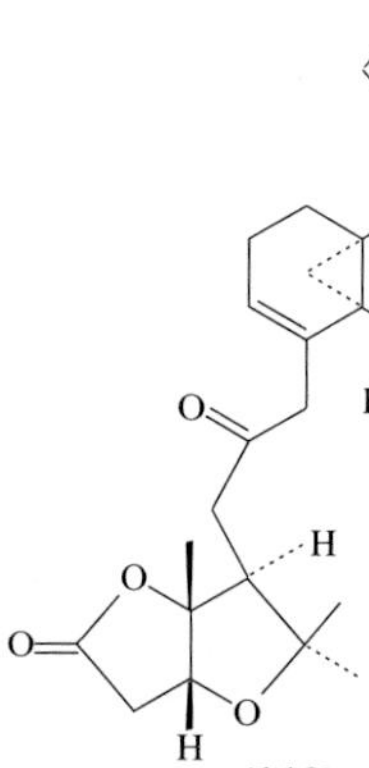

**(110)** 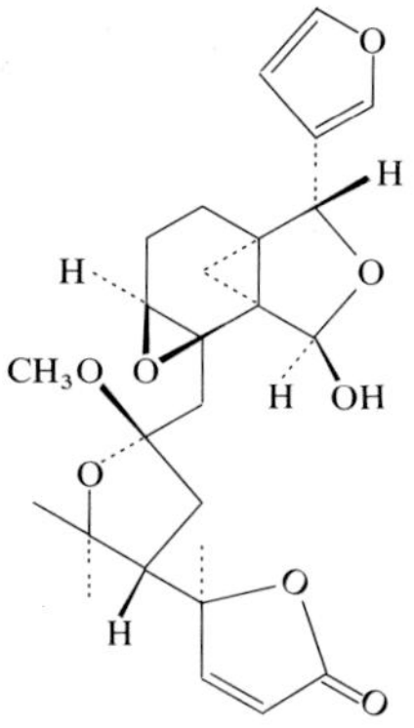 **(111)** 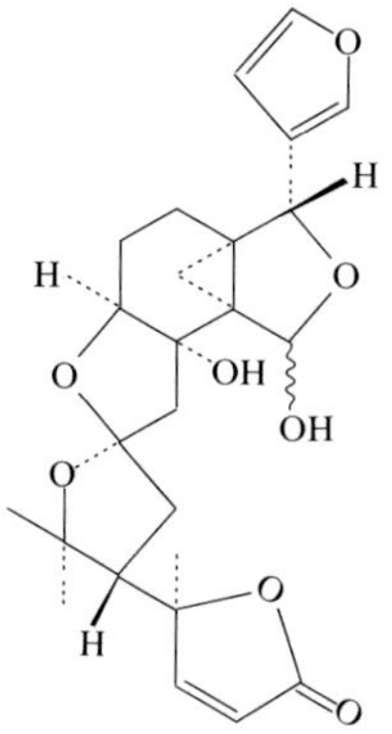 **(112 a)** 15α-OH
**(112 b)** 15β-OH

The authenticity of cneorin-$\mathbf{B}_{III}$ (**35**) as a natural product was stated above to be open to doubt. Following discovery of the bitter principle tricoccin-$\mathbf{S}_{18}$ (**108**), cneorin-$\mathbf{B}_{III}$ can be definitely included among naturally-occurring compounds.

## 2. C-15 Cycloacetals (*65,73*)

C-15 Cycloacetals are formed in the plant by intramolecular acetalization of the γ-lactols. The first representative of this new structural type, described by D. TRAUTMANN (*26*) in his thesis, was tricoccin-$S_4$ with the empirical formula $C_{25}H_{28}O_6$. TRAUTMANN was able to establish the constitution and, with the exception of the spiroacetal centre at C-7, the configuration of the molecule.

A fortunate circumstance later led to our isolation of tricoccin-$\mathbf{S}_{42}$, the stereoisomer of $\mathbf{S}_4$ with the opposite configuration at C-7. By comparing the spectra of the two, the epimer pair $\mathbf{S}_4$ and $\mathbf{S}_{42}$ could be assigned the stereoformulae (**113**) and (**114**) respectively.

The proton spectra of $\mathbf{S}_4$ and $\mathbf{S}_{42}$ show the familiar signals for the furan and butenolide rings and singlets for three methyl groups. As seen from Table 12, the olefinic proton 9-H appears as a multiplet at $\delta = 5.40$ ppm, while the signal for 17β-H, shifted to higher field at $\delta = 5.09$ and 5.10 respectively, indicates the reduced ring D of the γ-lactols. Also listed in Table 12 are the $^{13}$C-NMR signals for the carbon atoms 8 and 9 of the trisubstituted double bond, and for the spiroacetal centres at C-7 and C-15. As there are no OH bands in the IR spectra, the hydroxy group originally present at C-15 must have been incorporated in an inner acetal; the corresponding α-oriented acetal proton at 15-H appears as a singlet.

Confirmation of the spiroacetal structure is found in the mass spectra, with a base peak at m/z = 214 for an ion formulated as (**115**), which is formed when the bislactone (**26**) is split off as a neutral fragment.

Table 12. *Selected $^{13}$C- and $^{1}$H-NMR Data on the Cycloacetals, δ (ppm) in $CDCl_3$*

| Tricoccin- | $^{13}$C-NMR | | | | $^{1}$H-NMR | | | |
|---|---|---|---|---|---|---|---|---|
| | C-7 | C-8 | C-9 | C-15 | 5-H | 9-H | 15α-H | 17β-H |
| $S_4$ (**113**) | 105.0 s | 132.6 s | 116.8 d | 99.2 d | <2.5 | 5.40 | 5.45 | 5.09 |
| $S_{42}$ (**114**) | 103.9 s | 131.6 s | 118.1 d | 99.9 d | 2.75 | 5.40 | 5.11 | 5.10 |
| $S_{33}$ (**116**) | — | — | — | — | <2.6 | 4.29 | 5.52 | 5.15 |
| $S_{33}$-Ox (**117**) | — | — | — | — | <2.5 | — | 5.60 | 5.23 |
| $S_{39}$ (**122**) | 105.6 s | 72.4 s | 77.8 d | 100.2 d | <2.5 | 3.60 | 5.36 | 5.00 |
| $S_{39}$-Ac (**123**) | — | — | — | — | <2.5 | 4.74 | 5.41 | 5.05 |
| $S_{39}$-Ox (**124**) | — | — | — | — | <2.5 | — | 5.50 | 5.13 |

The assignment of configuration at C-7 is deduced from the positions of the signals for 5-H and 15-H as follows (cf. Table 12): if the proton at C-5 and the oxygen atom between C-7 and C-15 project from the same side of the molecule, a 4-line signal above 2.5 ppm appears for 5-H, and a singlet for 15-H at higher field; if the reverse is true, the 5-H signal lies below 2.5 ppm, while the 15α-H singlet is shifted to lower field under the influence of the oxygen atom between C-4 and C-7.

The data obtained for the cycloacetals permit tricoccin-$S_4$ to be assigned structure (**113**) with (7*S*,15*S*,17*R*)-configuration, and tricoccin-$S_{42}$ the structure (**114**) with (7*R*,15*S*,17*R*)-configuration.

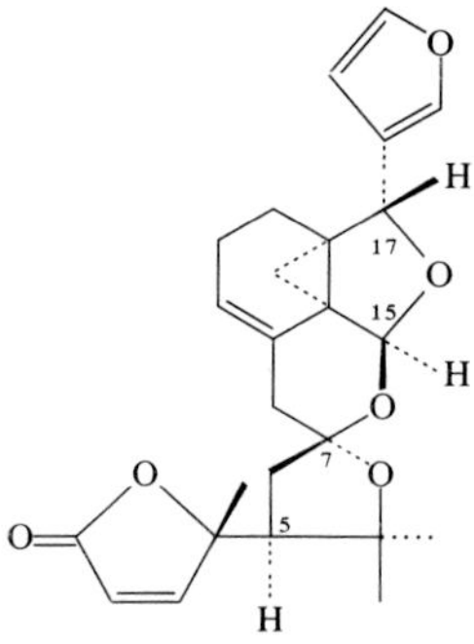

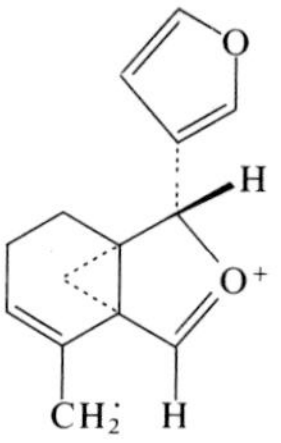

m/z = 214

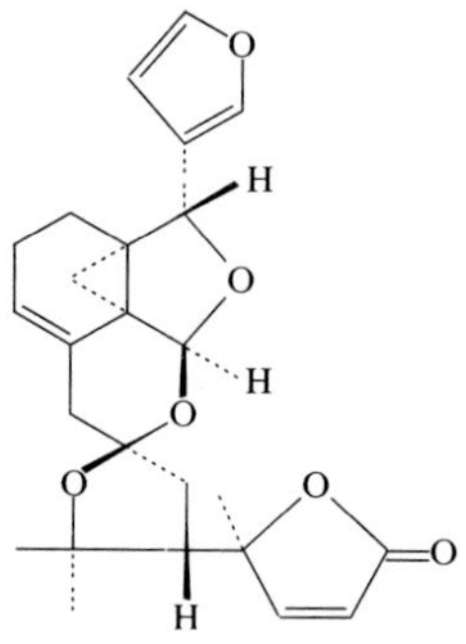

(**115**)

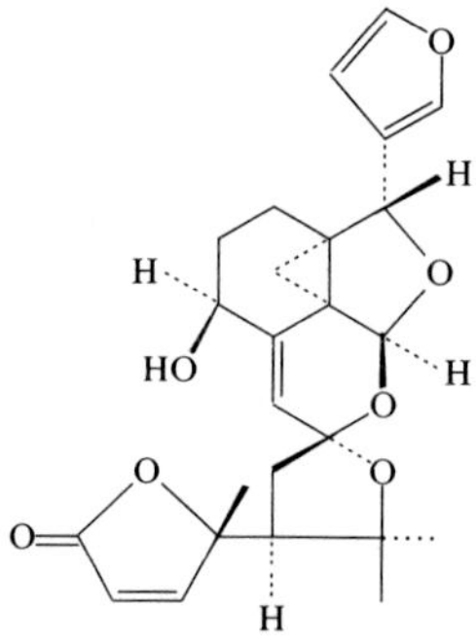

(**113**) (**114**)

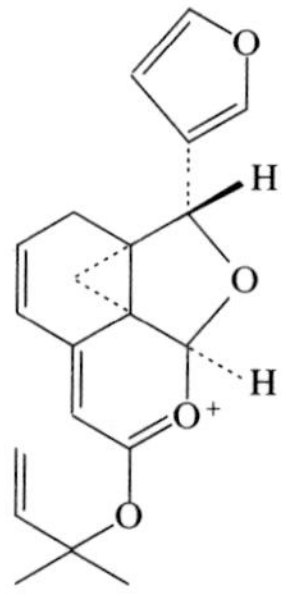

m/z = 325

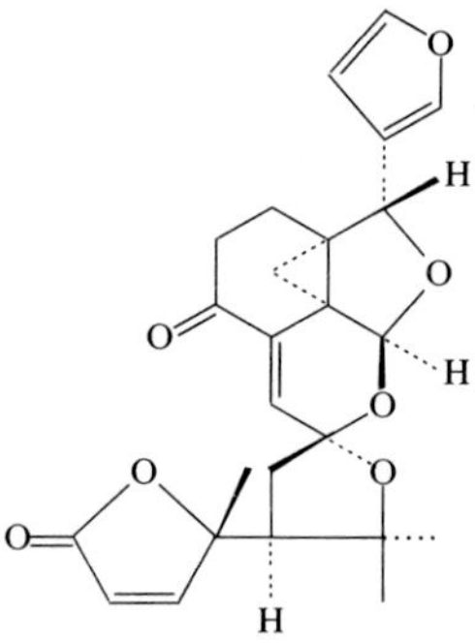

(**116**) (**117**) (**118**)

*References, pp. 181—187*

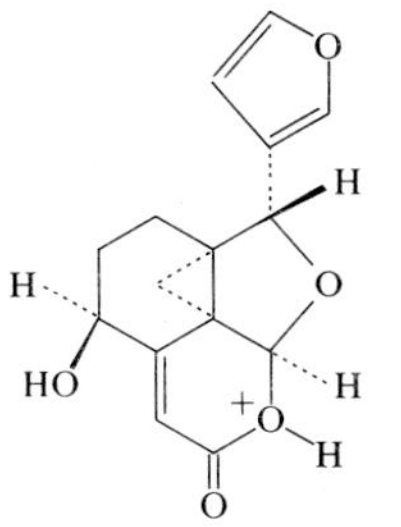

m/z = 275

**(119)**

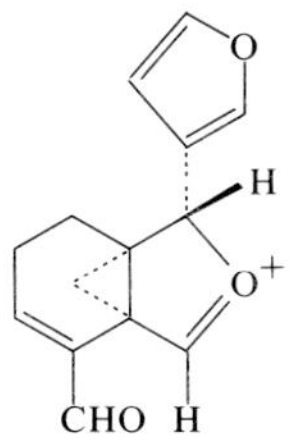

m/z = 229

**(120)**

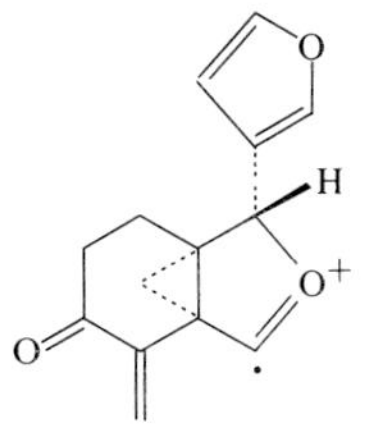

m/z = 228

**(121)**

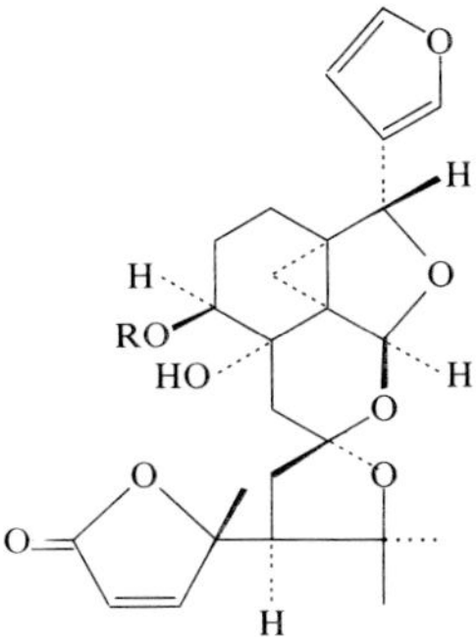

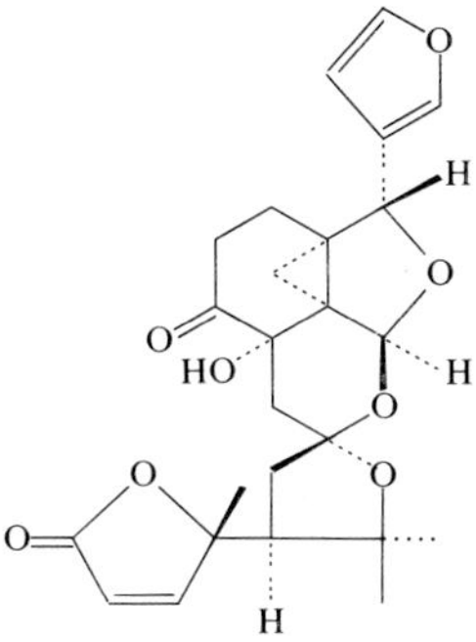

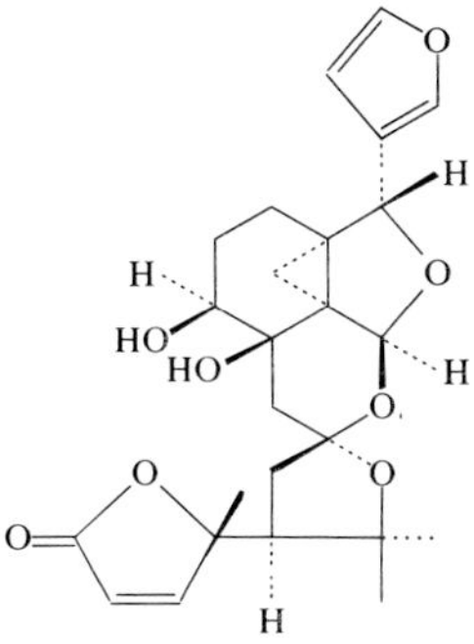

**(122)** R = H
**(123)** R = Ac

**(124)**

**(125)**

Tricoccin-$S_{33}$ (**116**) with the empirical formula $C_{25}H_{28}O_7$ also belongs to this new structural type. The proton spectrum again shows the familiar signals for the furan and butenolide rings and, as listed in Table 12, singlets for 17β-H at $\delta = 5.15$ and for 15α-H at 5.52 ppm; new additional signals are a multiplet for 9-H at $\delta = 4.29$ and a doublet for the C-30 olefinic proton at 5.57 ppm with $J = 2$ Hz. The proton of the secondary OH group appears as a doublet at 1.70 ppm with $J = 5$ Hz, exchangeable with $D_2O$. The allylic alcohol partial structure is confirmed by the IR spectrum which shows $\nu_{OH}$ 3460 and $\nu_{C=C}$ 1680 $cm^{-1}$.

When $\mathbf{S_{33}}$ is oxidized with pyridinium chlorochromate, the α,β-unsaturated ketone $\mathbf{S_{33}}$-Ox (**117**) is obtained. The IR spectrum of (**117**) has bands for a conjugated ketone at 1700 and 1635 $cm^{-1}$. The proton spectrum has a singlet at $\delta = 6.30$ ppm for the olefinic proton at C-30. The UV spectrum is noteworthy for a maximum at the strikingly long wavelength of 259 nm ($\varepsilon = 12000$), this red shift being ascribed to the high strain in the tetracyclic ring system.

Support for the proposed constitution of $S_{33}$ is also found in the mass spectrum, with a molecular ion peak at m/z=440, the base peak at m/z=97 for $C_5H_5O_2$ and the prominent ions at m/z=325 $C_{20}H_{21}O_4$ (**118**), 275 $C_{15}H_{15}O_5$ (**119**) and 229 $C_{14}H_{13}O_3$ (**120**). In the spectrum of the unsaturated ketone (**117**), the molecular ion peak at m/z=438 is at the same time the base peak, while an intense fragment ion is found at m/z=228 $C_{14}H_{12}O_3$ (**121**).

The figures given in Table 12 are again those on which the assignment of configuration is based, and according to which tricoccin-$S_{33}$ has the (7*S*,15*S*,17*R*)-configuration (*74*).

As a final example of the natural cycloacetals, mention may be made of tricoccin-$S_{39}$, the empirical formula of which is that of $S_{33}+H_2O$. Tricoccin-$S_{39}$ is assigned the *trans*-diol structure (**122**) on spectral and biogenetic grounds.

The $^{13}$C-NMR spectral data listed in Table 12 confirm the presence of acetal groups at C-7 and C-15 and the orientation of the OH groups at carbon atoms 8 and 9. In the $^1$H-NMR spectrum, 9α-H appears as a multiplet centred at 3.60 ppm and in the acetate (**118**) as a split doublet at 4.74 ppm with $J$=11 and 7 Hz.

Oxidation yields ketol (**124**) with the base peak at m/z=438 for $M^+-H_2O$ corresponding to the molecular ion peak of (**117**) and m/z=228 representing structure (**121**). $S_{39}$ is not identical with $S_4$-β-*cis*-diol (**125**), previously prepared by D. TRAUTMANN (*26*), although their mass spectra are similar. Based on the data in Table 12, tricoccin-$S_{39}$ also has the (7*S*)-configuration.

### 3. γ-Lactols with the Partial Structure of Ring A of Obacunone (*73*)

Tricoccins $S_{16}$ and $S_{27}$ with the stereoformulae (**126**) and (**129**) are biosynthetic intermediates derived from the cneoran skeleton (**20**) but still with the partial structure of ring A of obacunone (**181**). These bitter principles have a γ-lactol ring and a C-7 hemiacetal group and are potential precursors of cycloacetals.

Tricoccin-$S_{16}$, with the empirical formula $C_{25}H_{30}O_7$ (cf. Table 13), was first isolated from *Ct*. Its spectra reveal a β-substituted furan ring, and besides a tertiary, a secondary hydroxy group on the γ-lactol ring. As the compound is unstable in $CDCl_3$, the proton spectrum was measured in acetone-$d_6$. Characteristic features of the spectrum include the familiar singlets for 17β-H at $\delta_H$=4.95 and 15α-H at 5.45 ppm after exchange with $D_2O$, and also the doublets for the protons of the 18-$CH_2$ group of the cyclopropane ring, shifted to high field at δ=0.90 and 0.65 with $J$=5.5 Hz. The 9-H multiplet and the 30-$H_2$ doublets correspond to those of cneorin-**K** (**73**).

A novel feature for tricoccins with the cneoran skeleton is the conjugated ε-lactone ring, confirmed by the UV spectrum with $\lambda_{max}$ 208 nm ($\varepsilon = 11800$), the IR bands of the conjugated lactone at 1700 and 1640 $cm^{-1}$ and the $^1$H-NMR doublets for 1-H and 2-H at $\delta = 6.22$ and 5.67 ppm with $J = 13$ Hz. The 1-H signal shows a fine separation of 1.5 Hz due to coupling with the proton at C-5 (see below).

$^{13}$C-NMR spectral data support the proposed constitution as does the mass spectrum with a molecular ion peak at m/z = 442 and the base peak, characteristic for γ-lactols, at m/z = 214 for $C_{14}H_{14}O_2$ corresponding to structure (**115**), followed by the ions at m/z = 186 and 185 resulting through loss of CO and CHO respectively.

With acetic anhydride in pyridine, $\mathbf{S_{16}}$ forms the secondary monoacetate (**127**) $C_{27}H_{32}O_8$ with 15α-H at $\delta = 6.42$ s and 15-OAc at 2.10 s (3H). The singlet for 17β-H at 5.00 ppm is evidence that the acetate group is β-oriented (see above).

As already mentioned, tricoccin-$\mathbf{S_{16}}$ is extremely sensitive to acid, undergoing ring closure with loss of water to form the spirocyclic acetal (**128**). As summarized in Table 13, the 17β-H singlet is found in the proton spectrum at $\delta = 5.08$, the 15α-H singlet at 5.45, and the doublets of the 18-$CH_2$ group at 1.22 and 0.99 ppm with $J = 5.5$ Hz. The 9-H multiplet is centred at 5.39 and the 1-H and 2-H doublets at 6.18 and 5.88 ppm with $J = 13$ Hz. Here too, the 1-H signal shows a fine separation of 1.5 Hz. The molecular ion peak in the mass spectrum is at m/z = 424 and the base peak at m/z = 214, followed by the ions m/z = 186 and 185 as in $\mathbf{S_{16}}$. The assignments are confirmed by the $^{13}$C-NMR spectrum.

Table 13. *Selected Data on Tricoccins* $\mathbf{S_{16}}$, $\mathbf{S_{27}}$*-Diacetate and the Cycloacetals* (**128**) *and* (**131**)

| Compound | Formula | $^1$H-NMR, 90 MHz in δ (ppm)[a] 5-H | 15α-H | 17β-H | 18-$H_1$[b] | 18-$H_2$[b] |
|---|---|---|---|---|---|---|
| Tricoccin-$S_{16}$ (**126**) | $C_{25}H_{30}O_7$ | — | 5.45 | 4.95 | 0.90 | 0.65 |
| Tricoccin-$S_{27}(Ac)_2$ (**130**) | $C_{29}H_{34}O_{10}$ | — | 6.42 | 4.93 | 0.99 | 0.81 |
| Cycloacetal[d] (**128**) | $C_{25}H_{28}O_6$ | 3.00[c] | 5.45 | 5.08 | 1.22 | 0.99 |
| Cycloacetal[e] (**131**) | $C_{25}H_{28}O_7$ | 2.61[c] | 5.46 | 5.14 | 1.28 | 0.99 |

[a] Compound (**126**) in acetone-$d_6$, all others in $CDCl_3$.
[b] d, $J = 5.5$ Hz.
[c] 8-line signal through coupling with 1-H (1.5 Hz).
[d] M. p. 160° C, $[\alpha]_D^{20} + 123.9°$.
[e] M. p. 205° C.

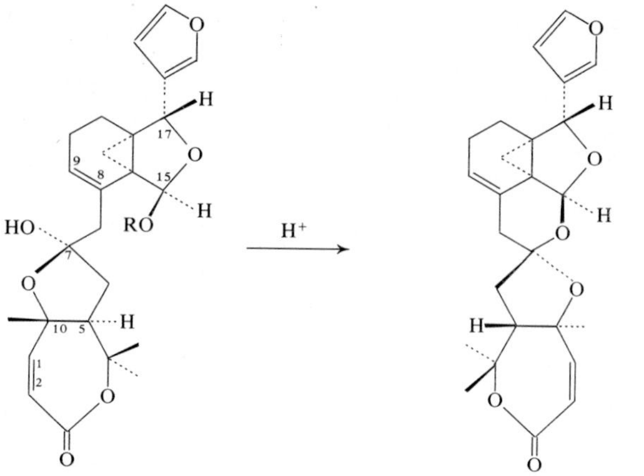

(**126**) R = H
(**127**) R = Ac

(**128**)

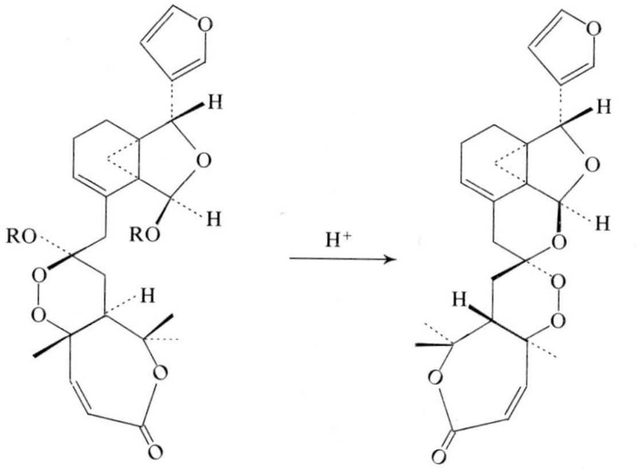

(**129**) R = H
(**130**) R = Ac

(**131**)

The data in Table 13 show that the 5- and 15-H signals of spiroacetal (**128**) are shifted to lower field and that both protons are *cis* to an oxygen atom. (**128**) thus has the (7*S*, 15*S*, 17*R*)- and tricoccin-$S_{16}$ the (7*R*, 15*S*, 17*R*)-configuration. The unusual coupling between 1- and 5-H with $J = 1.5$ Hz is confirmed by double resonance and results in an 8-line signal for 5-H.

Tricoccin-$S_{27}$ can be seen from Table 13 to differ from $S_{16}$ through having an extra oxygen atom. In the mass spectrum of this amorphous compound, a molecular ion peak at m/z = 458 can just be discerned. The compound is characterized by preparing the diacetate (**130**) which crystallizes readily. No molecular ion peak is seen in the mass spectrum of the

*References, pp. 181—187*

diacetate; following loss of acetic acid, the fragment of greatest mass is found at m/z=482 corresponding to $C_{27}H_{30}O_8$. The IR spectrum of (**130**) has no OH band from which may be concluded that the additional oxygen is incorporated in an ether-type bond.

$\mathbf{S}_{27}$ displays the same sensitivity to acid as $\mathbf{S}_{16}$, readily forming the spirocyclic acetal (**131**) with a molecular ion peak at m/z=440 for $C_{25}H_{28}O_7$; the ion at m/z=408, i.e. at $M^+ - 32$, indicates the presence of a peroxide function, which can also be detected chemically with $Fe^{2+}/SCN^-$ spray reagent in $\mathbf{S}_{27}$, its diacetate and cycloacetal. As the base peak of the cycloacetal is at m/e=214, the hemiacetal structure of $\mathbf{S}_{16}$ must in $\mathbf{S}_{27}$ be expanded to the peroxy-hemiacetal structure (**129**).

As the signals for 5- and 15-H are shifted to lower field in the proton spectrum of (**131**) (Table 13), the cycloacetal (**131**) and tricoccin-$\mathbf{S}_{27}$ have the (7*R*, 15*S*, 17*R*)-configuration. Transition from the γ-lactol to the cyclic structure is accompanied by a low-field shift of the 18-$CH_2$ group signals.

Precursors of the hemiacetals $\mathbf{S}_{16}$ and $\mathbf{S}_{27}$ are the 10-hydroxy- and 10-hydroperoxyketones (**132**) and (**133**) respectively, which are formed in the course of the biosynthesis of the cneoran skeleton (**20**) following the oxidative opening of ring B between carbon atoms 9 and 10 (see below).

Further compounds probably derived from the postulated precursor (**132**) are the epimeric tricoccins $\mathbf{S}_{30}$ and $\mathbf{S}_{31}$, with the stereoformulae (**134**) and (**135**); a possible sequence for their formation is allylic rearrangement, oxidation of the secondary hydroxy group at C-2 to a keto group, and reductive cyclization between carbon atoms 2 and 7.

The bitter principles $\mathbf{S}_{30}/\mathbf{S}_{31}$ are obtained as an amorphous mixture. Their monoacetates $C_{27}H_{32}O_8$ (**136**) and (**137**) can be separated chromatographically, but are also isolated as amorphous powders when precipitated from ether with hexane. Fragments in the mass spectra of the monoacetates coincide; a molecular ion peak is not found, or is scarcely visible, the fragment of greatest mass being seen at m/z=424 for $C_{25}H_{28}O_6$ equivalent to $M^+ -$HOAc, and the base peak at m/z=60. Taking the 17α-acetate (**136**) as an example, Diagram 5 illustrates the formation of further prominent fragments, in particular the RDA cleavage to m/z=256 for $C_{16}H_{16}O_3$, followed by loss of CHO to m/z=227 for $C_{15}H_{15}O_2$ which is typical of γ-lactols. Fragmentation to m/z 214 and 213 corresponds to the cleavage between carbon atoms 7 and 30, in the first case accompanied by a hydrogen atom transfer. Loss of water gives a low-intensity fragment at m/z=406.

Besides the familiar signals of the β-substituted furan ring, the IR spectra of the acetates have OH bands at 3530 and 3440 $cm^{-1}$ for the α-acetate, and at 3450 for the β-acetate, consistent with the broadened singlets, exchangeable with $D_2O$, at $\delta_H$=4.85 and 4.45, and at 4.50 and 2.85 ppm respectively. The carbonyl band at 1765 and 1700 in the former and at 1755 and 1705 $cm^{-1}$ in the latter case are assigned to lactone and acetate groups respectively, whose presence is confirmed by the $^{13}$C-NMR spectrum of the α-acetate; this has signals at δ = 171.6 s and 168.6 s, but no signals for the otherwise usual keto or hemiacetal group at C-7, only the C-15 hemiacetal group doublet at δ = 98.5 being present. Signals at δ = 127.6 d/141.2 s and 124.2 d/131.6 s confirm the presence of two trisubstituted double bonds.

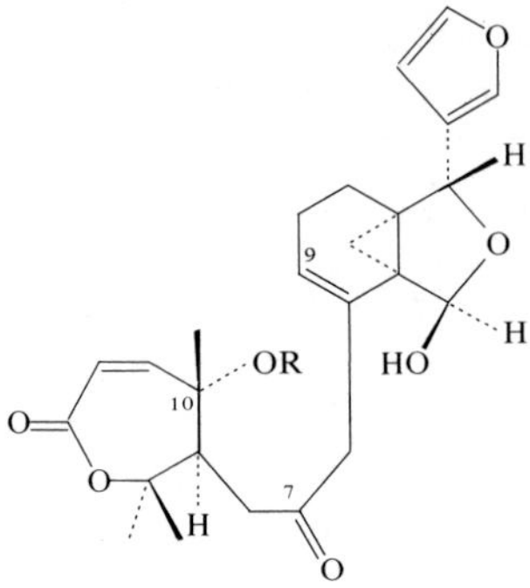

**(132)** R = H
**(133)** R = OH

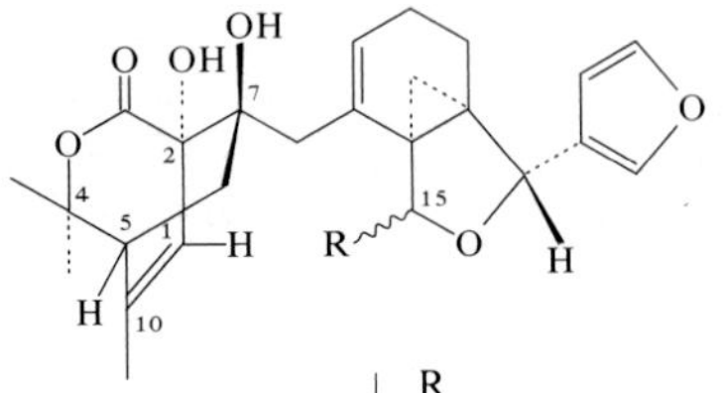

| | R |
|---|---|
| **(134)** | 15α-OH |
| **(135)** | 15β-OH |
| **(136)** | 15α-OAc |
| **(137)** | 15β-OAc |

*Diagram 5*

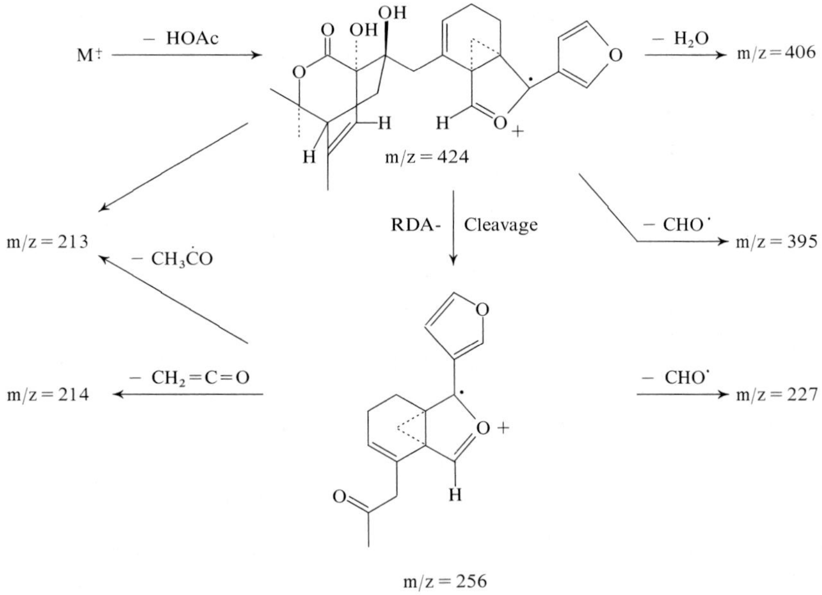

M⁺ — HOAc → m/z = 424 — $H_2O$ → m/z = 406

m/z = 424 — CHO˙ → m/z = 395

m/z = 424 — RDA-Cleavage → m/z = 256

m/z = 256 — $CH_3\dot{C}O$ → m/z = 213

m/z = 256 — $CH_2{=}C{=}O$ → m/z = 214

m/z = 256 — CHO˙ → m/z = 227

A feature of the proton spectra of both acetates is a doublet at $\delta = 1.88$ ppm with $J = 2$ Hz, corresponding to a methyl group linked to a carbon atom bearing two further substituents. The same coupling constant is found in the 4-line signals of the olefinic proton at $\delta = 5.52$ ppm. Irradation at either frequency causes the other to appear as a singlet. This defines the partial structure at carbon atoms 1, 10 and 19. The singlets of the geminal dimethyl groups on C-4 lie at about 1.5 ppm and confirm the lactone grouping; the broad multiplet centred at $\delta = 5.45$ and 5.35 ppm respectively is assigned to the olefinic proton 9-H.

In the $^{13}$C-NMR spectrum of the α-acetate, the signals for the carbon atoms 4 and 17 at $\delta = 81.2$ s and 77.9 d are accompanied by two others at 87.2 s and 80.0 s, which are assigned to carbon atoms 2 and 7.

As presented above, the $^1$H-NMR signals afford information on the steric arrangement of the lactol ring; the spectrum of the α-acetate has singlets for 15β-H at $\delta = 6.20$ and for 17β-H at 4.85, while the corresponding signals for the β-acetate are seen at 6.60 and 5.02 ppm.

Of those configurations possible at C-7, inspection of models leads us to prefer the (7*R*)-configuration with the OH group axial to the ε-lactone ring, other configurations being associated with considerable steric hindrance. Proof by elimination of water to form the cycloacetals must await renewed isolation of the bitter principles.

## XI. Tetranortriterpenoids from Cneoraceae

### 1. 3,4-seco-Meliacans (*75, 76*)

As mentioned at the outset, the Cneoraceae also contain tetranortriterpenoids of the limonoid and meliacin type, which are widespread in the Rutaceae and Meliaceae families. The 26-carbon atom bitter principles belonging to this class are derived from the carbon skeleton of meliacan (**19**), but not a single natural product with an intact carbon skeleton has as yet been isolated from the Cneoraceae. Oxidatively cleaved carbon-carbon bonds of the ring system are invariably found, leading us to follow the example of J. D. Connolly (*38*), who adopted this criterion as a classification principle.

Ring A of α,β-unsaturated ketones is frequently found to be modified in a manner resembling the Baeyer-Villiger oxidation, the bond between carbon atoms 3 and 4 being broken, with enlargement of the ring to a conjugated ε-lactone.

The first bitter principle of this type was tricoccin-**S**$_{13}$, isolated from *Ct*, which has the empirical formula $C_{26}H_{36}O_5$ and the constitution and configuration shown in (**138**). The partial structure of ring A is indicated by the following spectroscopic data: the UV spectrum with $\lambda_{max}$ 206 nm ($\varepsilon = 17800$), the IR bands at 1693 and 1620 $cm^{-1}$ for the carbonyl group and double bond, the $^{1}$H-NMR doublets for 1-H and 2-H at $\delta = 6.42$ and 5.81 ppm with $J = 12$ Hz and finally the $^{13}$C-NMR signals for C-1 at $\delta = 155.4$ d, C-2 119.5 d, C-3 167.5 s and C-4 85.1 s.

In common with all meliacan derivatives, $\mathbf{S_{13}}$ is derived from apotirucallol with (20*S*)- or apoeuphol with (20*R*)-configuration, and possesses the typical 7α-hydroxy- and 8β-methyl groups and also the $\Delta^{14}$ double bond. In the $^{1}$H-NMR spectrum, the 7β-H and 15-H give rise to triplets at $\delta = 3.90$ with $J = 3$ Hz and 5.48 ppm with $J = 2.5$ Hz.

Acetylation yields the monoacetate (**139**) with 7β-H at $\delta_H = 5.22$ t with $J = 3$ Hz, while oxidation with m-chloroperbenzoic acid leads selectively to the 14,15-β-epoxide (**140**), the epoxide proton 15-H of which is split only to a doublet at $\delta_H = 3.61$ ppm with $J = 3$ Hz; the 8β-methyl group singlet of $\mathbf{S_{13}}$ at $\delta = 1.12$ experiences a low-field shift to 1.33 ppm.

In the mass spectrum, the molecular ion peak at m/z = 428 first loses acetone and then the neutral fragment (**143**), to give the base peak m/z = 138 for $C_8H_{10}O_2$ formulated as (**144**); subsequent loss of the OH radical accompanied by H-transfer yields m/z = 121 corresponding to (**145**).

In contrast to (**19**), the β-substituted furan ring at C-17 is in $\mathbf{S_{13}}$ replaced by a saturated γ-lactone ring, characterized in the IR spectrum by $\nu_{CO} = 1782$ $cm^{-1}$ and in the $^{1}$H-NMR spectrum by $\delta_H = 4.42$ dd for 21-$H_a$ with $J = 8$ and 9 Hz, and 3.90 t for 21-$H_b$ with $J = 9$ Hz. The new partial structure poses a stereochemical problem in the assignment at C-20. This is resolved by X-ray analysis of $\mathbf{S_{13}}$ with the result shown in (**138**).

The (20*S*)-configuration found for $\mathbf{S_{13}}$ stimulated the conjecture that tirucallol-(20*S*) might be the biogenetic precursor of the Cneoraceae tetranortriterpenoids.

Closely related bitter principles which have the β-substituted furan ring of (**19**) instead of the γ-lactone ring of $\mathbf{S_{13}}$ are proceranone (*77*) (**141**) from *Carapa procera* (Meliaceae) and 7-deacetyl proceranone (*78*) (**142**) from *Teclea grandifolia* (Rutaceae).

Other relatives of $\mathbf{S_{13}}$ are the more highly oxidised tricoccins $\mathbf{S_{38}}$ and $\mathbf{S_{40}}$ with the formulae (**146**) and (**150**), both of which have an additional oxygen atom located in a hydroxy group.

$\mathbf{S_{38}}$, with a tertiary OH group at C-20, has broadened singlets at $\delta = 4.22$ and 2.62 ppm in the $^{1}$H-NMR spectrum for 21- and 22-$H_2$. The following signals in the $^{13}$C-NMR spectrum confirm the β-substituted β-hydroxy-γ-lactone ring: $\delta = 77.7$ s C-20, 78.0 t C-21, 29.8 t C-22 and 176.4 s C-24. The UV spectrum is unchanged with respect to $\mathbf{S_{13}}$, and the mass spectrum once again shows the base peak at m/z = 138 for (**144**). The configuration at C-20 is unknown.

*References, pp. 181—187*

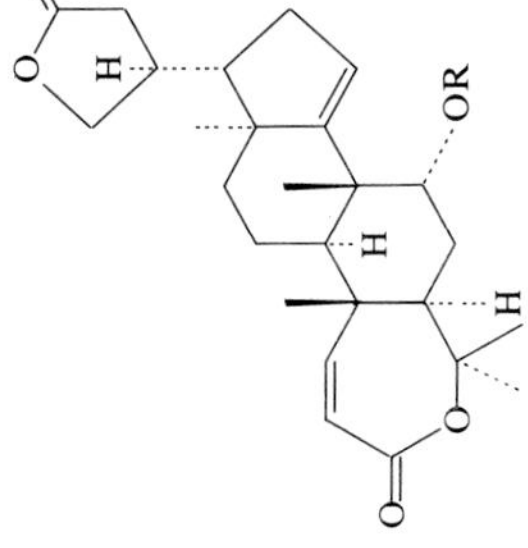

(138) R = H
(139) R = Ac

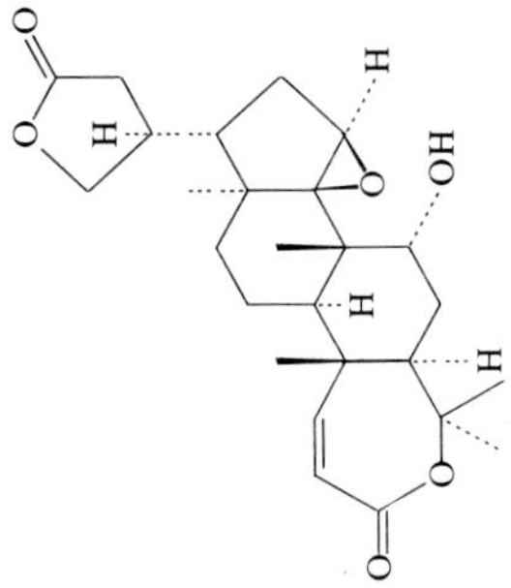

(140)

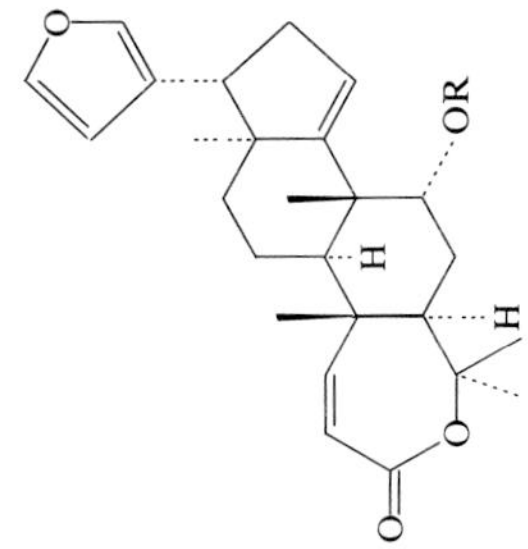

(141) R = Ac
(142) R = H

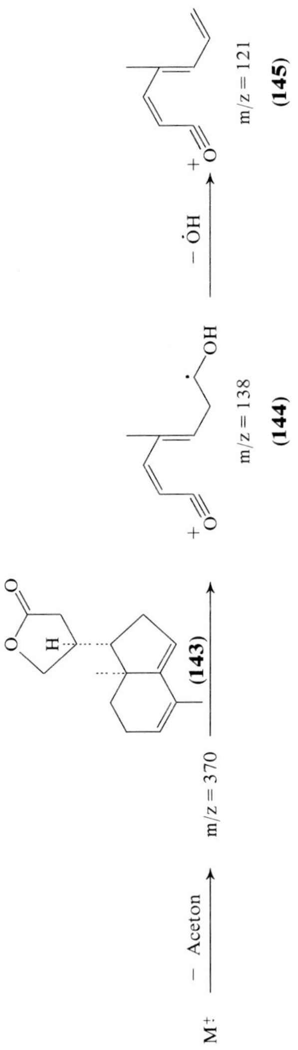

$M^{+\cdot}$ —(− Aceton)→ m/z = 370 (143) → m/z = 138 (144) —(− ·OH)→ m/z = 121 (145)

Acetylation of $\mathbf{S}_{38}$ yields the monoacetate (**147**) from which the anhydro compounds (**148**) and (**149**) are obtained by elimination of water with phosphoryl chloride in pyridine; (**148**) has a molecular ion peak at m/z=468. The butenolide ring is characterized by the double band in the IR spectrum at 1780/1755 $cm^{-1}$, and in the $^{1}H$-NMR spectrum by a triplet for 21-$H_2$ at $\delta=4.71$ with $J=2$ Hz, and a doublet for 22-H at 5.89 ppm, also with $J=2$ Hz. The coupling of the protons was confirmed by a double resonance experiment. In the UV spectrum, $\lambda_{max}$ is at 210.5 nm ($\varepsilon=32000$).

The molecular ion peak of (**149**) is at m/z=468 and the IR double bands at 1780/1765 $cm^{-1}$. The extinction of the UV maximum at 211 nm is distinctly lower at $\varepsilon=17400$. The $^{1}H$-NMR spectrum which has no olefinic proton signal shows a broadened singlet for 21-$H_2$ at $\delta=4.93$ and a quasi-triplet for 15-H at 5.39 ppm with $J=3$ Hz. Irradiation at these frequencies alters the 16-$H_2$ multiplet centred at 3.00 ppm (2H). In the anhydro derivatives the 18-$CH_3$ singlet is shifted to higher and to lower field respectively, appearing in (**148**) at $\delta=0.93$ and in (**149**) at 1.20 ppm.

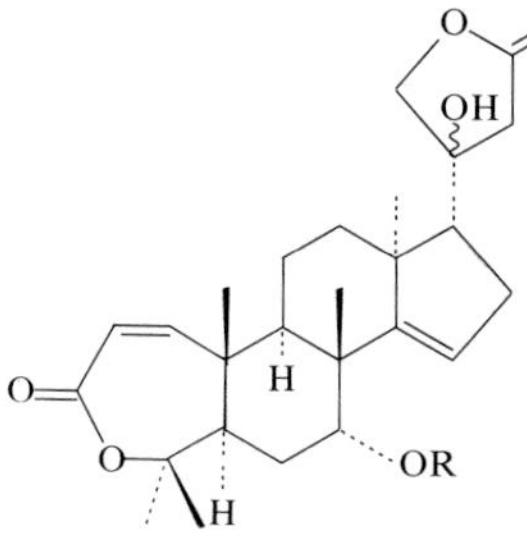

(**146**) R=H
(**147**) R=Ac

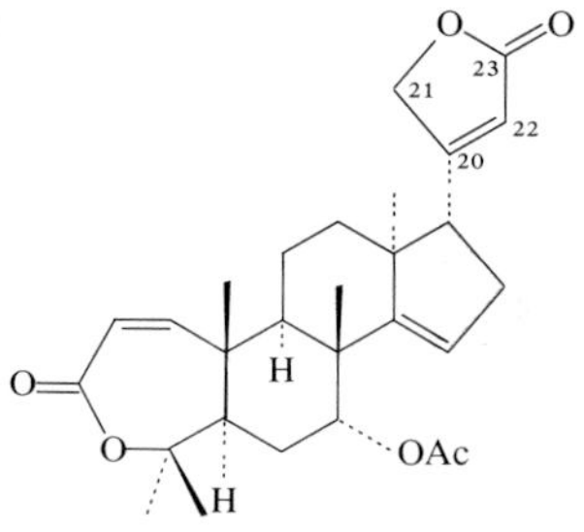

(**148**)

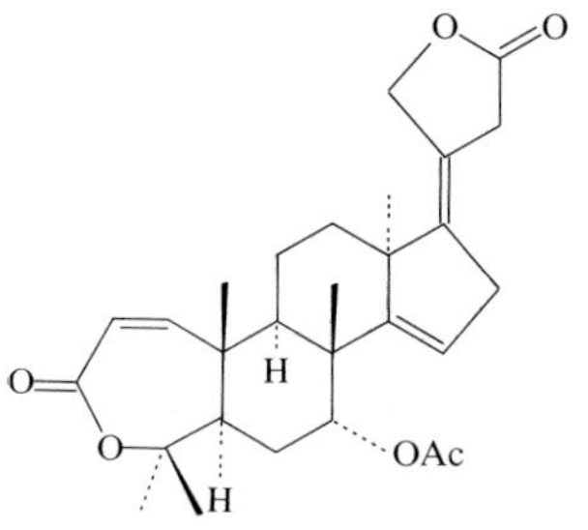

(**149**)

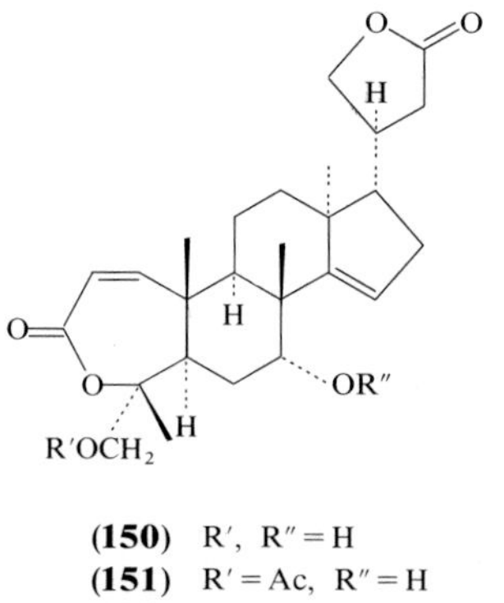

(**150**) R′, R″=H
(**151**) R′=Ac, R″=H
(**152**) R′, R″=Ac

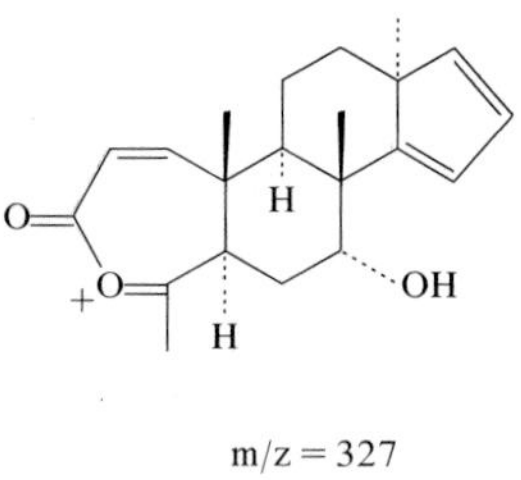

m/z=327

(**153**)

The proton spectrum of tricoccin-$S_{40}$ (**150**) shows only four instead of the usual five methyl groups, but in addition a tertiary $CH_2OH$ recognized by the doublets at $\delta=3.88$ and 3.55 ppm with $J=12$ Hz. The spectrum otherwise corresponds exactly to that of $S_{13}$, with the exception of the missing methyl singlet at $\delta=1.41$ ppm. Acetylation yields both a monoacetate (**151**) in the spectrum of which the above-mentioned doublets are shifted downfield to $\delta_H=4.39$ and 4.13 ppm with $J=12$ Hz and a diacetate (**152**).

It is clear from a study of the methyl signals listed in Table 14 that in $S_{40}$ one of the two geminal methyl groups at carbon atom 4 must be oxidized to the $CH_2OH$ group. As both the molecular model and the X-ray analysis of $S_{13}$ show that the *quasi*-equatorial 4α-methyl group is involved in no immediate interaction with the axial methyl groups at C-8, C-10 and C-13, it must be the 4α-methyl group of $S_{13}$ which in $S_{40}$ is oxidized.

The $^{13}C$-NMR spectrum also corresponds to that of $S_{13}$ with the exception of the missing methyl group quartet, replaced by a triplet at $\delta=67.8$. The data for the methyl group carbon atoms of $S_{13}$ and $S_{40}$ are listed in Table 15.

The constitution is confirmed by the mass spectrum, in which the molecular ion peak is at m/z=444, $M^+ - CH_2OH$ at m/z=413 and an ion at m/z=370 corresponds to $M^+ - CH_3COCH_2OH$, replacing $M^+$ − acetone in the spectrum of $S_{13}$. The neutral fragment (**143**), mentioned earlier in the discussion of $S_{13}$, is found here also as an ion with m/z=232: once again, loss of (**143**) results in the base peak at m/z=138. The key fragment at m/z=327 for $C_{21}H_{27}O_3$ (**153**) arises by loss of a methyl radical and of the γ-lactone ring with hydrogen transfer.

Table 14. *$^1H$-NMR Singlet Data for the Tricoccins $S_{13}$ and $S_{14}$ and Their Acetates [90 MHz, δ (ppm) in $CDCl_3$]*

| Tricoccin- | 4α,4β-$CH_3$ | 10β-$CH_3$ | 8β-$CH_3$ | 13α-$CH_3$ |
|---|---|---|---|---|
| $S_{13}$ (**138**) | 1.41/1.41 | 1.24 | 1.12 | 1.02 |
| $S_{13}$-Ac (**139**) | 1.35/1.42 | 1.28 | 1.18 | 1.00 |
| $S_{40}$ (**150**) | 1.41 | 1.29 | 1.10 | 1.05 |
| $S_{40}$-Ac (**151**) | 1.48 | 1.28 | 1.11 | 1.03 |
| $S_{40}$-$Ac_2$ (**152**) | 1.39 | 1.29 | 1.19 | 1.02 |

Table 15. *$^{13}C$-NMR Data for the C-Atoms of the Methyl Groups of Tricoccins $S_{13}$ and $S_{40}$ [δ (ppm) in $CDCl_3$]*

| Tricoccin | C-28 | C-29 | C-19 | C-30 | C-18 | |
|---|---|---|---|---|---|---|
| $S_{13}$ (**138**) | 31.9 | 27.1* | 26.2* | 19.7 | 15.7 | * exchangeable |
| $S_{40}$ (**150**) | — | 26.7** | 26.6** | 20.5 | 16.4 | ** exchangeable |

Correlation of the spectra permits assignment of the (20*S*)-configuration to $\mathbf{S}_{40}$ as well as $\mathbf{S}_{13}$.

The tricoccins $\mathbf{S}_{22}$ and $\mathbf{S}_{32}$, also isolated from *Ct*, again feature the β-substituted furan ring of (**19**) at C-17 and differ from the bitter principles already discussed by virtue of a new oxidation pattern in ring D of the meliacan skeleton. Rings A, B and C of $\mathbf{S}_{22}$ are identical with those of $\mathbf{S}_{13}$, while $\mathbf{S}_{32}$ has a saturated ε-lactone ring.

Tricoccin-$\mathbf{S}_{22}$ has the empirical formula $C_{26}H_{34}O_6$ and the constitution and configuration shown in (**154**). The proton spectrum shows the signals expected for the unsaturated ε-lactone ring A, the β-substituted furan ring and the secondary OH group at C-7. In the UV spectrum, $\lambda_{max}$ is at 214 nm ($\varepsilon = 17000$). The molecular ion peak is found at m/z=422 in the mass spectrum, in which the base peak is once again at m/z=138 for $C_8H_{10}O_2$ corresponding to (**144**).

The $^1$H-NMR spectrum (acetone-$d_6$) supplies information on the environment of the two remaining oxygen atoms. The relevant signals are at $\delta = 3.63$ s (1 H) and 4.22 ppm m (1 H), the latter simplifying to a doublet with $J = 9$ Hz after $D_2O$-exchange. The singlet is assigned to the epoxide proton 15-H and the multiplet (doublet after $D_2O$-exchange) to the proton at C-16 with the OH group at the same carbon atom. The absence of coupling between the protons at C-15 and C-16 is confirmed by comparison with $\mathbf{S}_{13}$ epoxide (**140**). In the spectrum of the latter, the epoxide proton is split simply to a doublet at $\delta = 3.60$ ppm with $J = 3$ Hz, with no further multiplicity.

$\mathbf{S}_{22}$ is readily acetylated to a diacetate (**155**), together with the C-16 monoacetate (**156**). In the $^1$H-NMR spectrum of the monoacetate, the 16-H doublet is shifted to $\delta = 5.27$ ppm, and the proton at C-17 with which it couples is also seen as a doublet at $\delta = 2.90$, both signals with $J = 9$ Hz. Acetylation of the 16-OH group causes a high-field shift of the 21-H signal to $\delta = 7.16$ ppm, so that the effect can be interpreted as an indication of the α-orientation of the 16-OAc group, assuming the biogenetically regulated stereochemistry at C-17.

Tricoccin-$\mathbf{S}_{22}$ is sensitive to acids, being rearranged by 0.1 n HCl in glycol dimethyl ether to an isomeric ketone $C_{26}H_{34}O_6$ with $\nu_{CO} = 1730\,cm^{-1}$. This ketone is assigned structure (**157**), based on the $^1$H-NMR spectrum (acetone-$d_6$) and mechanistic considerations. Double resonance experiments enable the signals at $\delta = 2.90$ d with $J = 2$ Hz, 4.27 dd after $D_2O$-exchange with $J = 2$ and 11, and 3.16 ppm d with $J = 11$ Hz to be assigned to the vicinal protons 15-, 16- and 17-H. Acetylation of the ketone yields the diacetate (**158**), in the $^1$H-NMR spectrum of which the 7-H and 16-H signals are shifted to $\delta = 5.08$ and 5.70 ppm. Once again, the base peak of the mass spectrum is found at m/z=138; it is noteworthy that this ion is missing in the mass spectra of the diacetates (**155**) and (**158**).

*References, pp. 181—187*

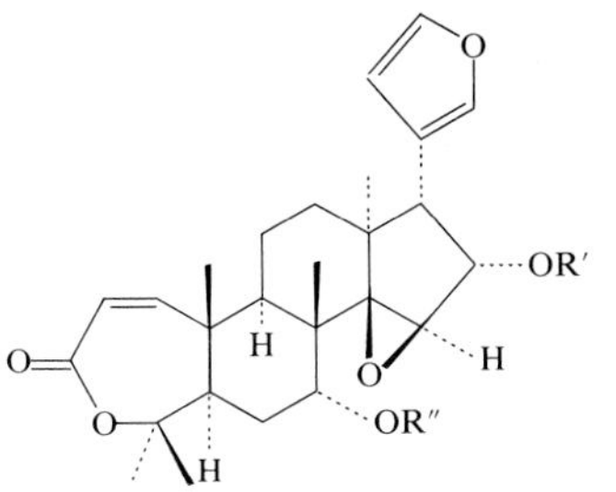

**(154)** R′, R″ = H
**(155)** R′, R″ = Ac
**(156)** R′ = Ac, R″ = H

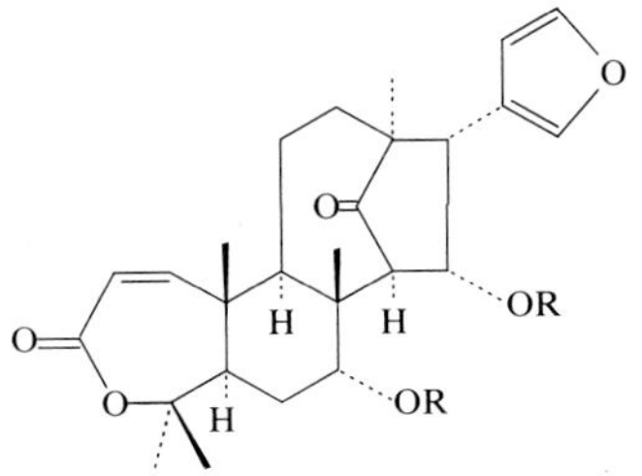

**(157)** R = H
**(158)** R = Ac

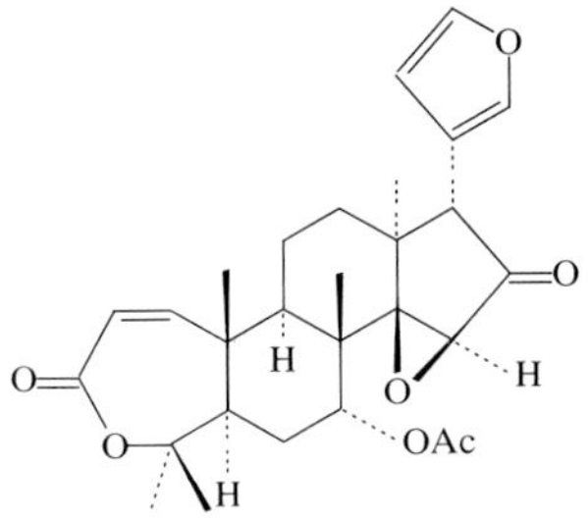

**(159)**

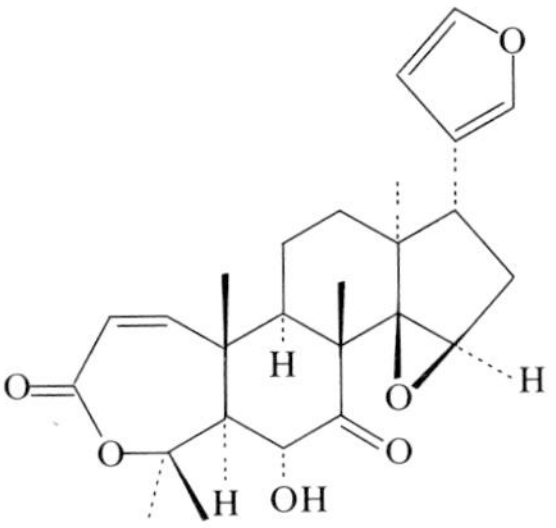

**(160)**

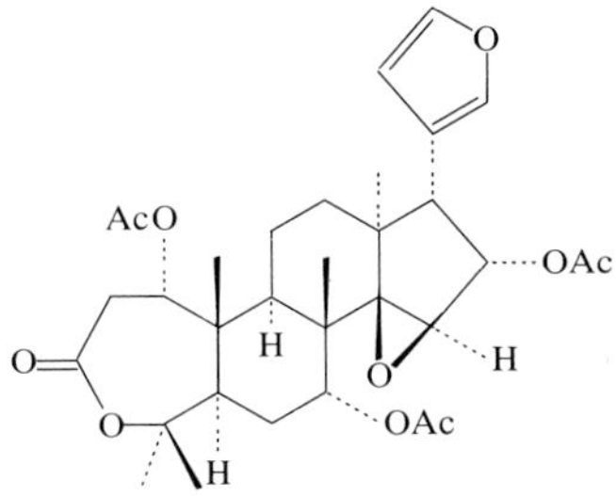

**(161)**

A close relative of $\mathbf{S}_{22}$ is evodulone (*79*) (**159**) from *Carapa procera*, with 7α-OAc and 16-keto groups. To the same group belong surenone (**160**) from *Toona sureni* (Meliaceae) and surenine with 6α- and 7α-OAc groups in place of the ketol grouping (*80*).

Tricoccin-$S_{32}$ is a natural diacetate which must be further acetylated before impurities can be removed. The pure triacetate thus obtained (**161**) has the empirical formula $C_{32}H_{42}O_{10}$ and is amorphous.

In the $^1$H-NMR spectrum of the triacetate the signals for the protons of ring D have the same shifts as the corresponding signals in the spectrum of $S_{22}$ diacetate (**155**). The spectrum also shows the signal pattern, familiar from cneorin-**G** [see below (**184**)], of a 1,7-diacylated meliacin with ring A expanded to an ε-lactone (1-H: mc 4.90 and 2-$H_2$: mc 3.10 ppm). The close similarity between the triacetate and cneorin-**G** is especially apparent when their $^{13}$C-NMR spectra are compared, that of the triacetate showing only two significant differences, namely in the positions of the signals for C-16 and C-17 at $\delta = 76.6$ and 44.1 ppm. The acetylated tricoccin-$S_{32}$ is thus assigned structure (**161**). Selected $^1$H-NMR data on the acetates of $S_{22}$ and $S_{32}$ are presented in Table 16.

Table 16. *Selected $^1$H-NMR Data for the Acetates of Tricoccins* $S_{22}$ *and* $S_{32}$ *[90 MHz, δ (ppm) in $CDCl_3$]*

| Tricoccin- | 7-H | 15-H | 16-H | 17-H | 21-H |
|---|---|---|---|---|---|
| $S_{22}$-Ac (**156**) | 3.55 m | 3.75 s | 5.27 d | 2.90 d | 7.16 s |
| $S_{22}$-$Ac_2$ (**155**) | 4.65 m | 3.58 s | 5.22 d | 2.85 d | 7.17 s |
| $S_{32}$-Ac (**161**) | 4.60 m | 3.53 s | 5.20 d | 2.85 d | 7.13 s |

## 2. 7,8-seco-Meliacans (*62*)

In his thesis D. TRAUTMANN (*26*) described the tricoccins $S_7$ and $S_8$ with the structures shown in (**162**) and (**173**), the first tetranortriterpenoids from Cneoraceae with the 7,8-*seco*-meliacan skeleton. Shortly afterwards, B. EPE found tricoccin-$S_{19}$ (**175**), a compound with similarities to $S_8$, and later U. OELBERMANN found tricoccin-$S_{43}$ (**177**), a structure previously predicted by D. TRAUTMANN, which in a departure from the usual procedure was isolated from fresh plant material.

Tricoccin-$S_7$ (**162**), with the empirical formula $C_{27}H_{32}O_6$, has a UV spectrum ($CH_3OH$) with $\lambda_{max}$ at 252 nm ($\varepsilon = 18000$) and IR bands (KBr) at 3130, 1502 and 870 (furan), 1735 and 1165 (ester), 1710 (cyclopentenone) and 1635 $cm^{-1}$ (enol ether of a 1,3-diketone). The signals of the $^{13}$C-NMR spectrum are important structural clues, including singlets at $\delta = 204.5$ and 202.3 for keto groups, 173.2 for ester carbonyl and 89.5 ppm for a saturated tertiary carbon atom linked to oxygen. Including the furan ring, all oxygen atoms are now accounted for. Assignment of two singlets at $\delta = 183.6$ and 180.3 ppm to the β-C-atoms of two α,β-unsaturated ketones finds support

*References, pp. 181—187*

through the model compounds (**163**) and (**164**); doublets at $\delta = 127.7$ and 100.7 ppm for the $\alpha$-carbon atoms show that both double bonds are trisubstituted. Taking these data together, the partial structures (**165**) and (**166**) can be drawn for $\mathbf{S_7}$.

The $^1$H-NMR spectrum of $\mathbf{S_7}$ has two one-proton singlets at $\delta = 5.44$ and 6.36 for the olefinic protons in (**165**) and (**166**), and a singlet at 3.53 ppm for the methine proton in (**166**); six three-proton singlets are seen, at $\delta = 3.75$ for the methyl ester, and at 1.05, 1.08, 1.23, 1.54 and 1.87 for five tertiary methyl groups; the signal with the greatest low-field shift belongs to the methyl group of a tertiary ether.

For the purpose of comparison with the data for the partial structure (**166**), azadiradione (**167**) from *Melia azadirachta* (Meliaceae) (*81*) serves as a suitable example. The compound has singlets at $\delta_H = 5.83$ and 3.43 ppm for 15- and 17-H and an IR band at 1710 $cm^{-1}$ for the conjugated five-membered cyclic ketone.

Tricoccin-$\mathbf{S_7}$ with a 26-carbon atom skeleton, five tertiary methyl groups and a methyl ester group, is a tetranortriterpene with a cleaved ring. Other possibilities having been excluded, only the bond between carbon atoms 7 and 8 can be the cleavage site, C-7 being transformed into the ester group and C-8 into a tertiary alcohol. Elimination of water between the latter and the enol group at C-1 gives rise to the tertiary ether formulated as (**162**).

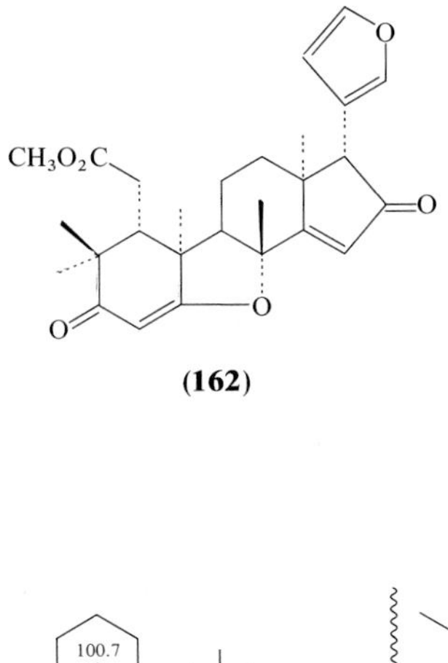

$CH_3O_2C$ O O O

(**162**)

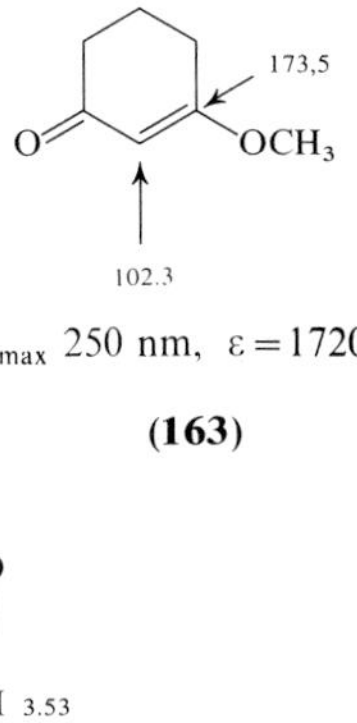

173,5 O $OCH_3$ 102.3

$\lambda_{max}$ 250 nm, $\varepsilon = 17200$

(**163**)

O 179.4

(**164**)

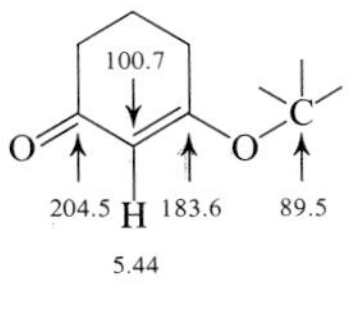

100.7 O O C 204.5 H 183.6 89.5 5.44

(**165**)

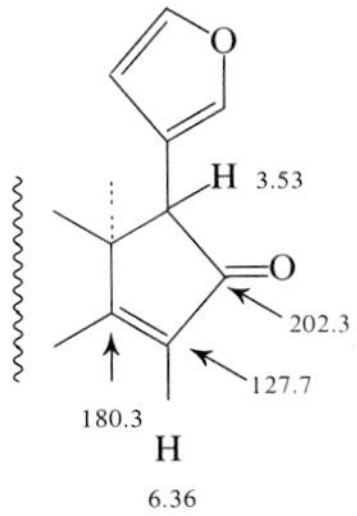

O H 3.53 O 202.3 127.7 180.3 H 6.36

IR : 1710

(**166**)

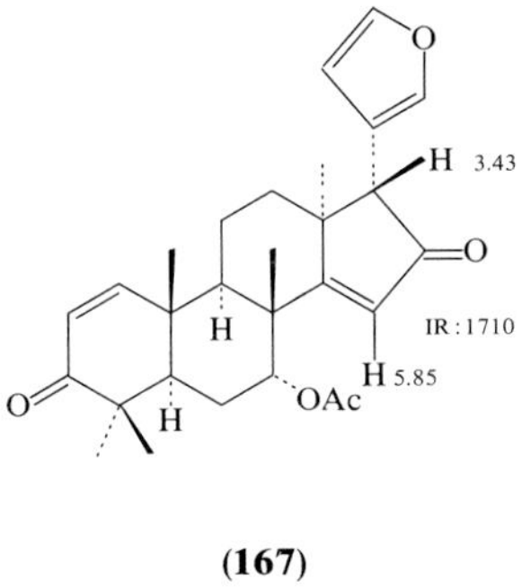

O H 3.43 O IR : 1710 H H 5.85 OAc O H

(**167**)

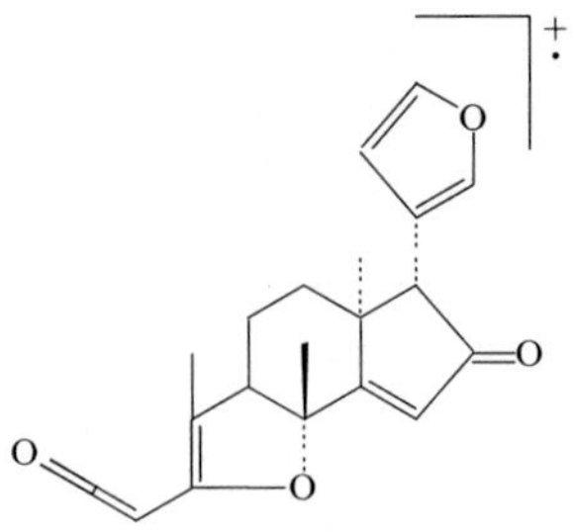

m/z = 324

**(168)**

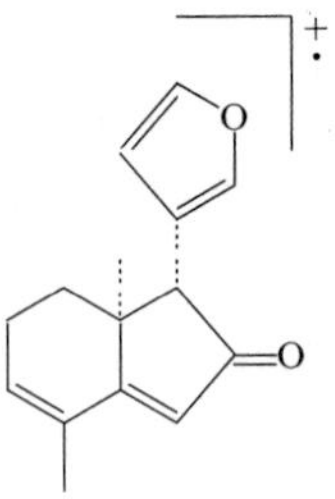

m/z = 228

**(169)**

The mass spectroscopic disintegration of (**162**) is also characteristic, with the molecular ion peak at m/z = 452. This undergoes retro-Diels-Alder cleavage in ring A to form the base peak at m/z = 324 for $C_{20}H_{20}O_4$ with formula (**168**), from which subsequent expulsion of the neutral fragment $C_5H_4O_2$ gives rise to the ion at m/z = 228 with structure (**169**).

It has not yet been possible, with the small quantity of this natural product at our disposal, to cleave the ether ring of $\mathbf{S_7}$ with formation of the β-diketone (**170**). The latter, when ring D is subjected to Baeyer-Villiger oxidation either before or following the ether cleavage as described by J. D. CONNOLLY, I. M. S. THORNTON and D. A. H. TAYLOR (*82*), represents the potential precursor for mexicanolide (*83*) (**171**) and other bicyclononanolides. A pyrolysis experiment gave only starting material and a stereoisomeric compound, probably the C-8 epimeric ether (**172**).

Tricoccins $\mathbf{S_8}$ and $\mathbf{S_{19}}$ (**175**) are isomers with the empirical formula $C_{27}H_{34}O_7$, differing from that of $\mathbf{S_7}$ by having the extra elements of water. Both compounds still have the $\mathbf{S_7}$ UV chromophore with higher extinction, but on the other hand their spectra have no signals corresponding to the β-substituted furan ring and the carbonyl group of the cyclopentenone ring.

The IR spectrum of $\mathbf{S_8}$ has additional bands at 3320 and 1760 $cm^{-1}$ for a γ-hydroxybutenolide ring, while the $^1$H-NMR spectrum shows the olefinic 15-H proton signal changed into a triplet at $\delta = 5.83$ with $J = 3$ Hz. Also seen in the $^1$H-NMR spectrum are the signals for 22- and 23-H as characteristically broadened singlets at $\delta = 7.00$ and 6.17 ppm of an α-substituted γ-hydroxybutenolide ring (*84*). The structure shown in formula (**173**) is supported by the mass spectrum with the base peak at m/z = 342; $\mathbf{S_8}$ also forms an acetate (**174**) which gives a mass spectrum with the molecular ion peak at m/z = 512 and the base peak at m/z = 384. Both $\mathbf{S_8}$ and its acetate occur as a C-23 epimeric mixture, recognizable by the splitting of the 18-$CH_3$ singlet.

Principal bands in the IR spectrum of $\mathbf{S_{19}}$ are at 3395 and 1743 $cm^{-1}$, the CO bands of the γ-hydroxybutenolide and the methyl ester coinciding. The

*References, pp. 181–187*

$^1$H-NMR spectrum (*85*) has, in addition to the 15-H triplet at $\delta = 5.85$ with $J = 3$ Hz, a typical singlet at 5.94 for the olefinic proton 22-H of a β-substituted γ-hydroxybutenolide ring, and two doublets at 6.04 and 4.67 ppm with $J = 8$ Hz for 21-H and 21-OH, the latter being exchangeable with $D_2O$. In contrast to $\mathbf{S_8}$, tricoccin-$\mathbf{S_{19}}$ (**175**) and its acetate (**176**) are stereochemically uniform, with unknown configuration at C-21.

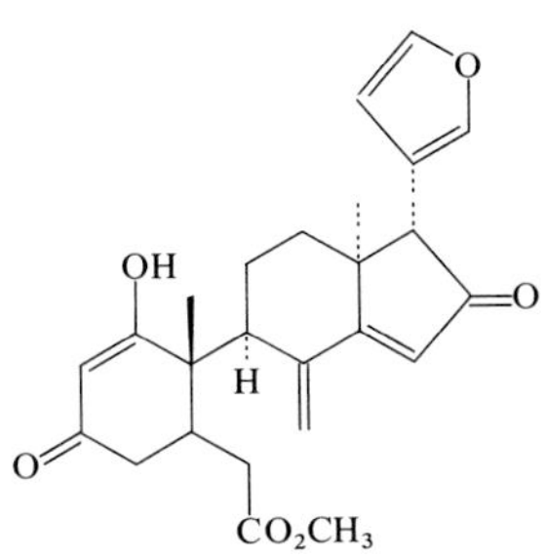

(**170**)

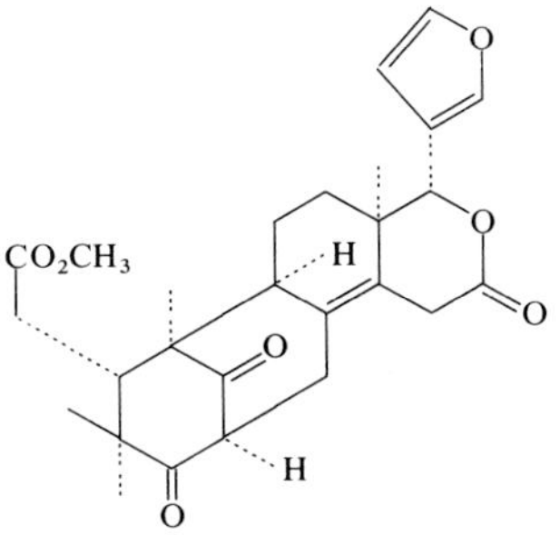

(**171**)

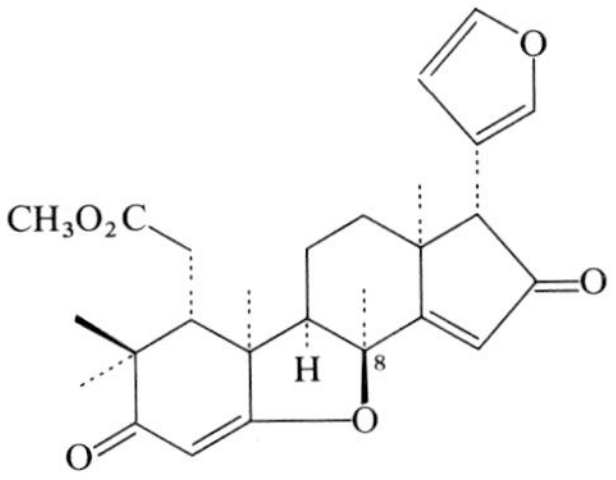

(**172**)

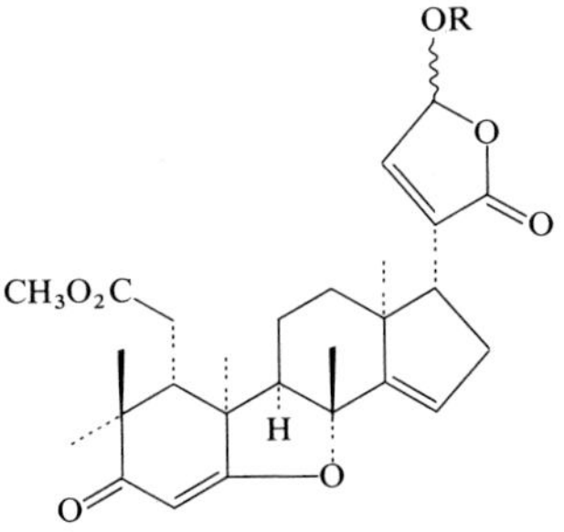

(**173**) R = H
(**174**) R = Ac

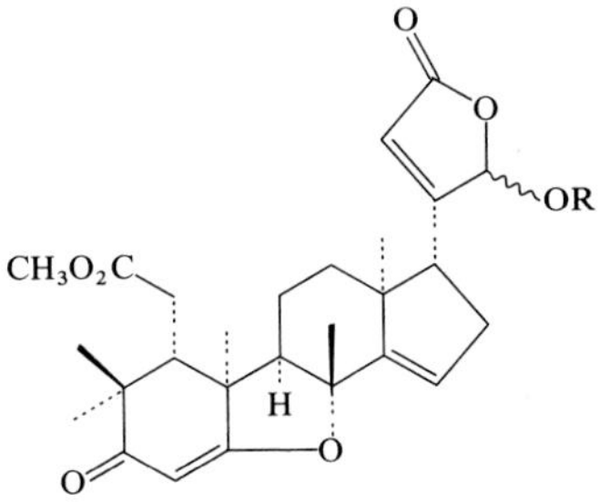

(**175**) R = H
(**176**) R = Ac

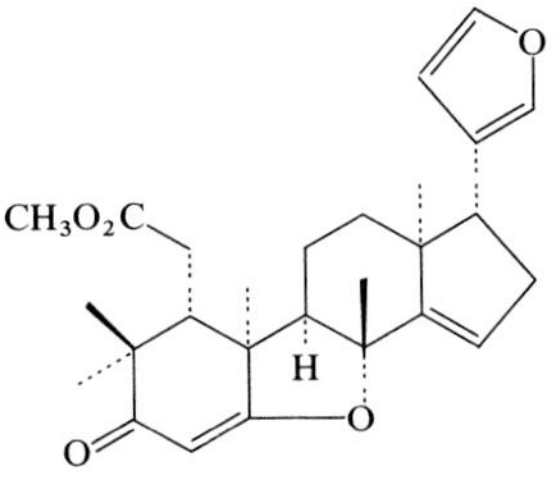

(**177**)

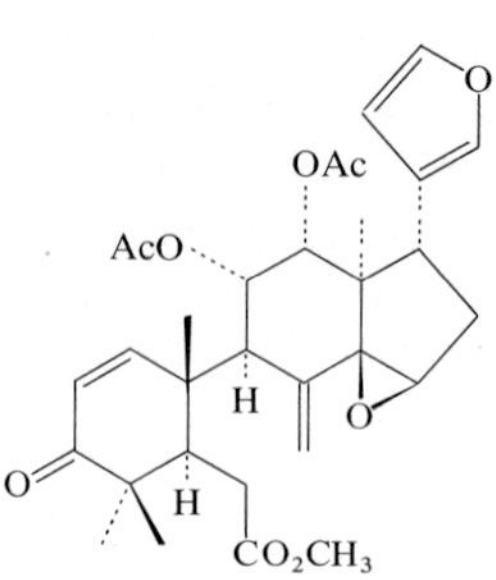

(178) 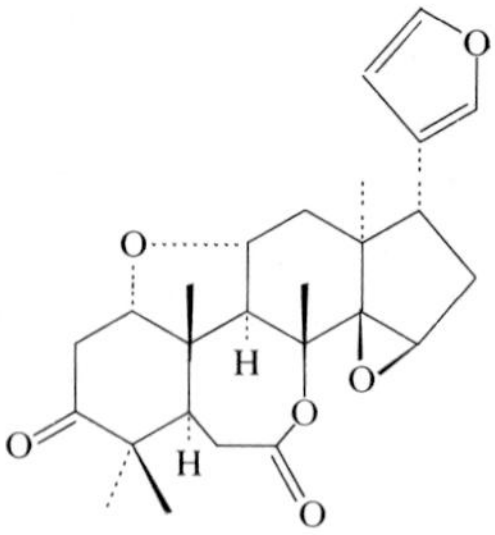 (179)

It is possible that $S_8$ and $S_{19}$ may, like cneorin-$NP_{27}$ (**61**), be simply artefacts of the most recently isolated tricoccin $S_{43}$, which has the empirical formula $C_{27}H_{34}O_5$ and the structure shown in (**177**). Like $S_7$, $S_{43}$ has a β-substituted furan ring, but both IR and $^{13}C$-NMR spectra lack the signals of a C-16 carbonyl group. As a direct result, the $^1H$-NMR spectrum has a triplet for 15-H at $\delta = 5.85$ ppm with $J = 2$ Hz, while the mass spectrum has a molecular ion peak at $m/z = 438$ and the base peak at $m/z = 310$.

W. KRAUS *et al.* (*86*, *87*) have isolated selectively cleaved 7,8-*seco*-meliacans from *Toona ciliata* (Meliaceae). Examples of this type are toonacilin (**178**) and the especially interesting toonafolin (**179**), the first natural tetranortriterpenoid in which ring B is expanded to an ε-lactone. Analogous structures are thought also to occur in the Cneoraceae as biosynthetic intermediates (see below).

### 3. 3,4-16,17-seco-Meliacans (Limonin Group)

Associated with the classical bitter principle limonin (*88*), (**180**) from citrus seeds is obacunone (**181**), isolated by O. H. EMERSON (*89*). The constitution and configuration of obacunone were established by D. H. R. BARTON *et al.* (*90*). In this compound ring D is also expanded, in a manner reminiscent of the Baeyer-Villiger oxidation, to the δ-lactone. D. TRAUTMANN (*26*) described this bitter principle under the designation tricoccin-$S_3$ from *Ct* (*3*, *62*). The same compound was later also isolated as cneorin-S from *Np*. Obacunone as a biogenetic precursor of limonin is probably also an intermediate on the biosynthetic pathway to the pentanor-triterpenoids with the cneoran skeleton.

D. L. DREYER (*36*) regarded obacunone with its expanded ring A as a typical Rutaceae bitter principle, although exceptions have since been found. At this point it is pertinent to mention the isolation of obacunone from *Harrisonia abyssinica* (Simaroubaceae) by K. NAKANISHI *et al.* (*91*). Identification of a bitter principle of the Rutaceae has taxonomic consequences, lending support as it does to the classification by A. TAKHTAJAN of the Cneoraceae as a family within the order Rutales.

Related compounds accompanying obacunone are α-obacunol acetate (*92, 93*) (**182**), isolated as cneorin-**I** from *Np* (*3, 24, 62*) and tricoccin-$S_{41}$ from *Ct* (*8*). The free alcohol α-obacunol (*92—95*) (**183**) has also been detected, as a constituent of *Ct* designated tricoccin-$\mathbf{S}_{23}$.

Cneorin-**G** (**184**) from *Np* (*24, 62*), at the time of these studies still of unknown constitution, was later described by F. R. AHMED *et al.* (*96*) as 7α-acetoxydihydronomilin (**184**); we had established the constitution and configuration, with the exception of that at C-1, by partial removal of acetic acid with formation of α-obacunol acetate (**182**) (*3*).

A close relative is 7α,11β-diacetoxydihydronomilin from *Cedrela mexicana* (Meliaceae), recently described by G. B. MARCELLE and B. S. MOOTOO (*97*).

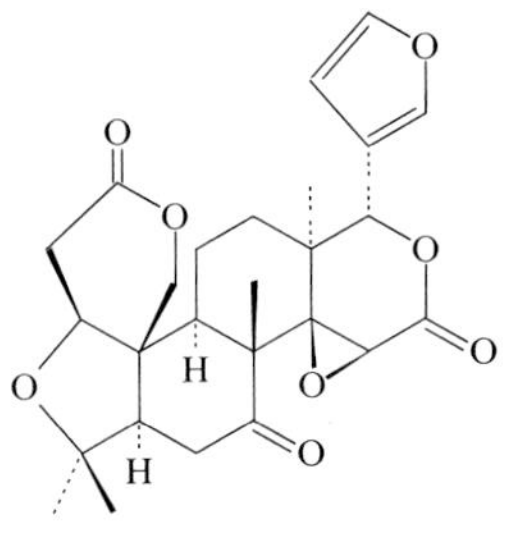

(**180**)

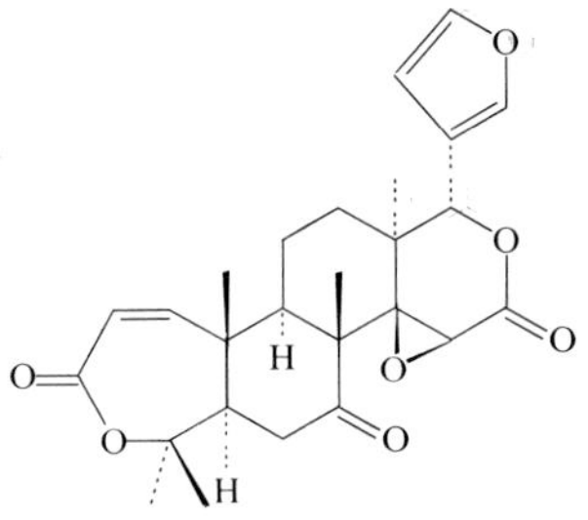

(**181**)

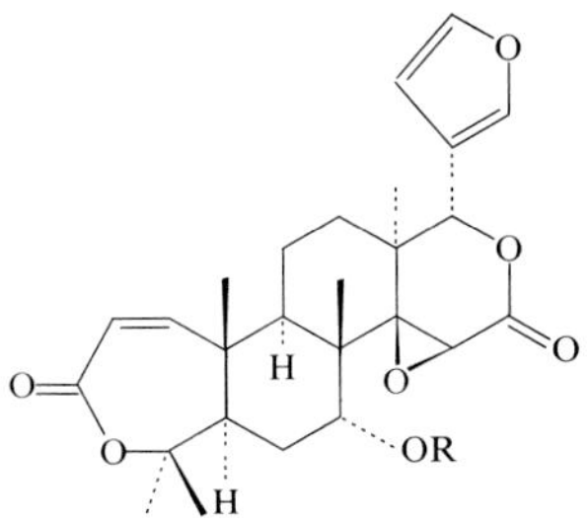

(**182**) R = Ac
(**183**) R = H

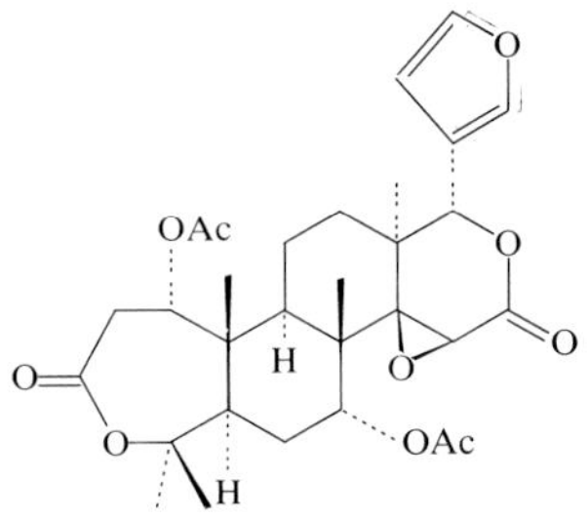

(**184**)

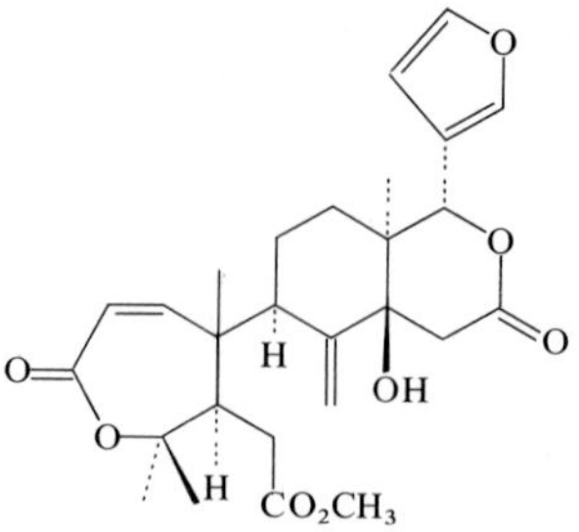

(185)

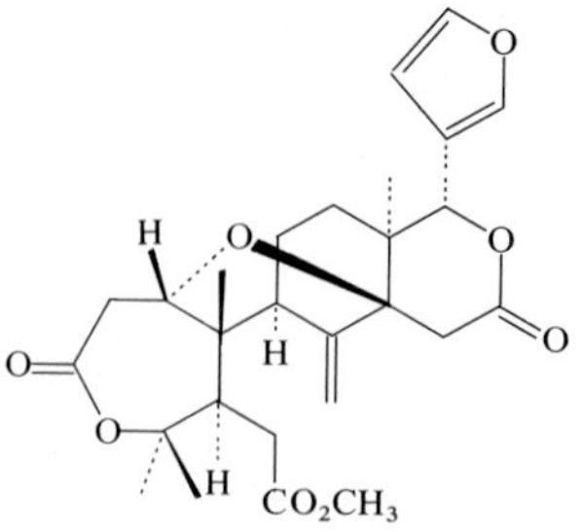

(186)

### 4. 3,4-7,8-16,17-seco-Meliacans (*76*)

The most highly oxidized bitter principle among the tetranortriterpenoids of the Cneoraceae is cneorin-**R** from *Np* with the empirical formula $C_{27}H_{34}O_8$. Spectroscopic data show it to be a tetranortriterpene with four C-methyl singlets, a β-substituted furan ring, and a conjugated ε-lactone as ring A (*98*). Ring D is expanded to a δ-lactone with $\delta_H = 5.65$ s for 17-H and $\delta_C = 78.0$ for C-17. A methyl ester group with $\delta_H = 3.70$ s (3H) and $\delta_C = 173.9$ s/51.8 q and the transformation of one of the original five methyl groups into an exocyclic methylene group with singlets at $\delta_H = 5.22$ and 5.01 and associated triplet with $\delta_C = 118.2$ ppm are indications that ring B is opened to the 7,8-*seco*-meliacolid (*99*).

The molecule also has a tertiary OH group with $\nu_{OH} = 3380\,cm^{-1}$ and $\delta_H = 3.15$ ppm s, exchangeable with $D_2O$, located at one of two possible sites, C-9 or C-14.

Treatment of cneorin-**R** with sodium hydride in glycol dimethyl ether converts it into an internal ether which is identical in all respects with methylivorensate (*100*) (**186**). This transformation proves the position of the tertiary hydroxy group and the constitution and configuration of cneorin-**R**, which are illustrated by formula (**185**).

A DREIDING model of cneorin-**R** shows that the 14β-OH group is directly opposed to the proton on C-1, causing a striking downfield shift of the 1-H signal to $\delta = 6.98$ ppm.

## XII. Protolimonoids

### 1. Tirucallan-(20*S*)-Triterpenoids (*101*)

The discovery of protolimonoids with familiar or novel structures gave support to our previous conjecture (*75*), that the bitter principles of the Cneoraceae are also biogenetically derived from tirucallan-(20*S*)-triterpenoids by oxidative degradation.

*References, pp. 181—187*

The two last decades have seen such a variety of bitter principles isolated from the Rutaceae and Meliaceae described in the literature (*36–38, 57*), that the assumption seems more and more justified, that they are formed by apo-rearrangement (*88, 102*) of derivatives of tirucallol-7-ene or tirucallon-7-ene respectively with highly oxidised side chains.

A typical representative of the protolimonoids is melianone (**187**), which we isolated from the Canary Island species of Cneoraceae under the designation $\mathbf{NP}_{37}$. The compound was first isolated by D. LAVIE *et al.* (*41*) from *Melia azedarach* (Meliaceae), and later by J. POLONSKY *et al.* (*103*) from *Simarouba amara* (Simaroubaceae). The configuration of the melianone side chain was established by C. W. LYONS and D. R. TAYLOR (*104*).

Recent years have also seen confirmation by several partial syntheses (*106, 107*) of the postulate (*105*) that the apo-rearrangement proceeds via 7α,8α-epoxitirucallans to the 7α-hydroxy-$\Delta^{14}$-apotirucallans with migration of a methyl group. To the best of our knowledge, the only natural 7α,8α-epoxitirucallan isolated was found by T. G. HALSALL and T. A. STROKE (*108*) in *Turraeanthus africanus* (Meliaceae).

We then isolated melianone epoxide (**188**) with the empirical formula $C_{30}H_{46}O_5$, calling it cneorin-$\mathbf{NP}_{38}$. The spectra of the epoxide closely resemble those of melianone, but the signals for the trisubstituted double bond are missing. We studied the compound together with V. SINNWELL (*109*), who measured the 400 Mz $^1$H-NMR spectrum which showed (**188**) to be a mixture of two epimers in the ratio 1 : 1. The most significant feature of the spectrum is a characteristic doubling of the signals for the protons in the C-21 region, integration showing each signal to correspond to only half the protons expected (*110*).

The proton at C-7 appears as a broadened singlet at $\delta = 2.95$ ppm with $W_{1/2} = 6$ Hz and a chemical shift characteristic of an epoxide proton. The α-orientation of the oxiran ring across carbon atoms 7 and 8 is deduced from the shift of the 13-$CH_3$ singlet which has moved downfield from $\delta = 0.85$ in melianone to 1.05/1.00 ppm (*110*).

The structure (**188 a** and **b**) assigned to $\mathbf{NP}_{38}$, i. e. 7α,8α-epoximelianone, is confirmed by the key fragment at m/z = 328 for $C_{22}H_{32}O_2$ (**189**) in the mass spectrum, arising from $M^+$ by loss of the side chain and a hydrogen atom.

The following protolimonoids $\mathbf{NP}_{32}$, $\mathbf{NP}_{34}$ and $\mathbf{NP}_{36}$ all have the tetracyclic ring system of melianone, indicated in their mass spectra by the fragments m/z = 297 and 271 with structures (**190**) and (**191**) respectively (*104*) and (*102*). The 7-H multiplets are located at $\delta = 5.30$ ppm in the proton spectra.

Cneorin-$\mathbf{NP}_{36}$, which was isolated only in very small quantities, was studied in cooperation with V. SINWELL (*109*) and G. REMBERG (*111*). The empirical formula $C_{30}H_{46}O_3$ reflects the fact that $\mathbf{NP}_{36}$ has only two oxygen atoms in the side chain instead of the four found in the side chains of the isomers **NP**$_{32}$ and **NP**$_{34}$, $C_{30}H_{46}O_5$.

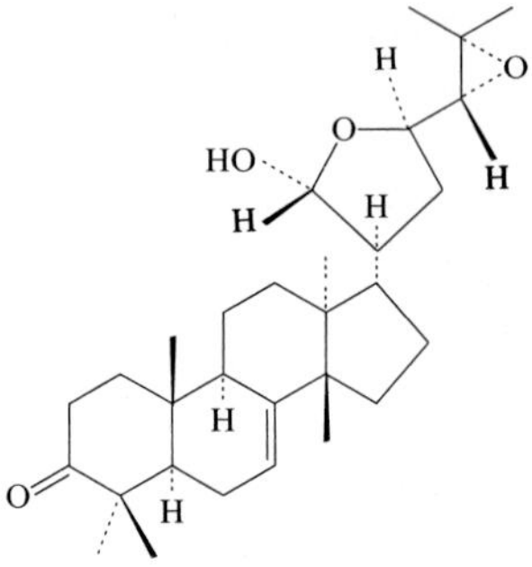

**(187)**

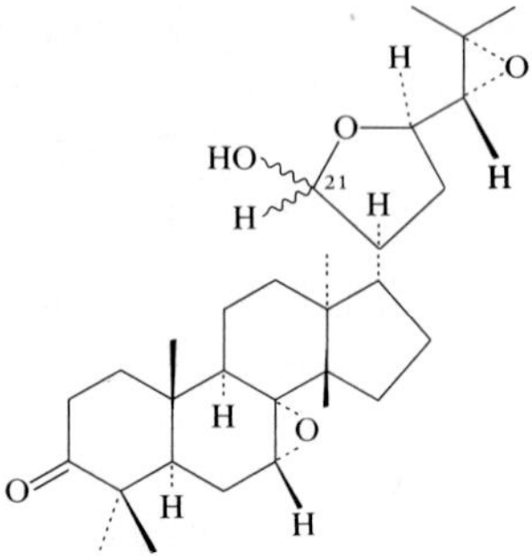

**(188a)** 21α-H
**(188b)** 21β-H

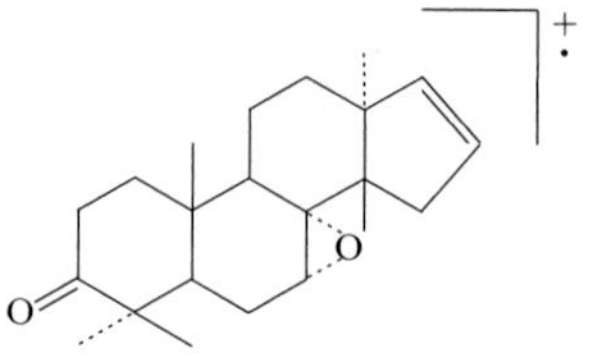

m/z = 328
**(189)**

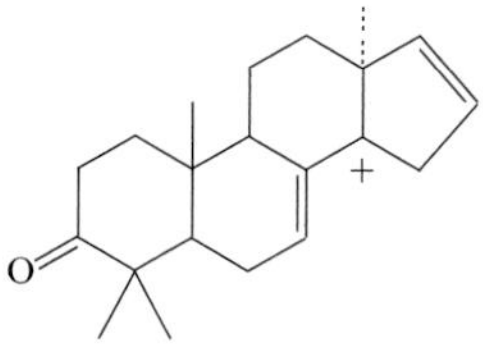

m/z = 297
**(190)**

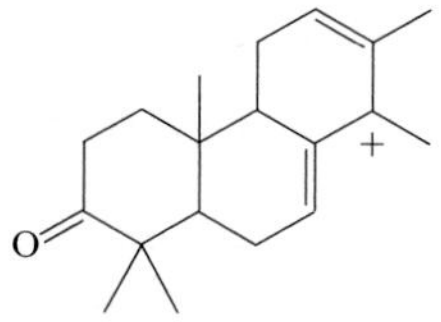

m/z = 271
**(191)**

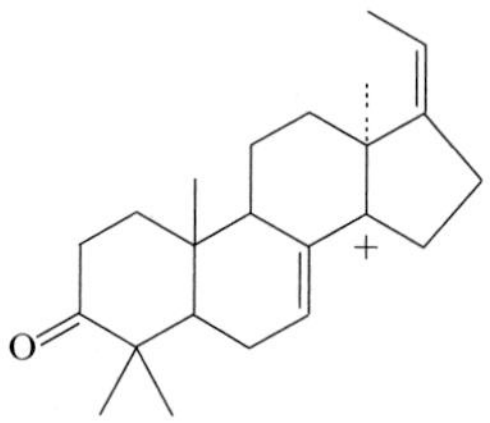

m/z = 325
**(192)**

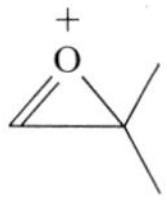

m/z = 71
**(193)**

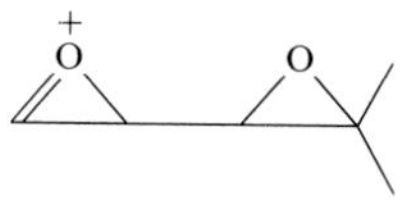

m/z = 113
**(194)**

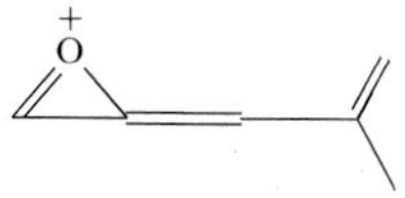

m/z = 95
**(195)**

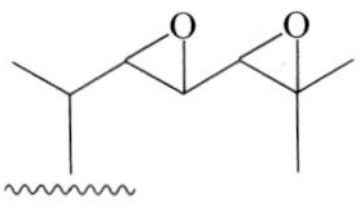

**(196)** 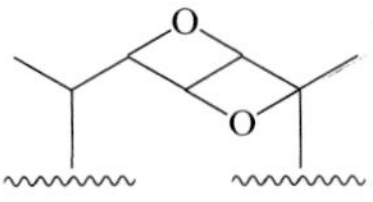 **(197)** 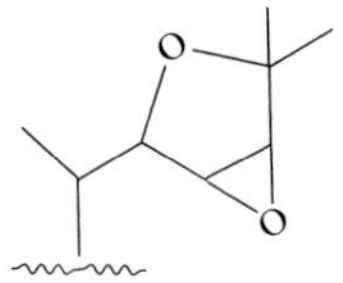 **(198)** 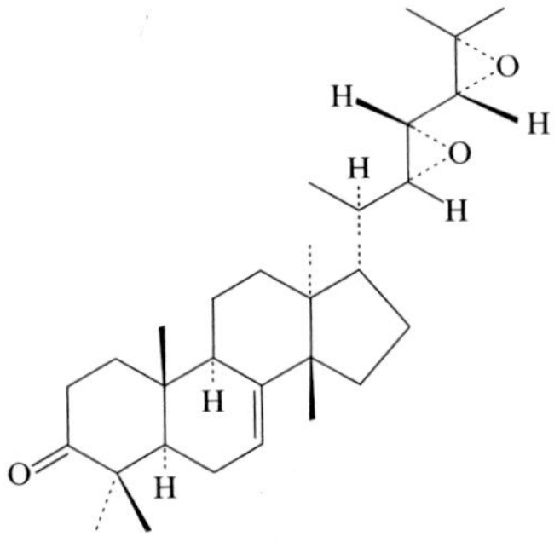 **(199)**

*References, pp. 181—187*

The IR spectrum of $NP_{36}$ has neither a hydroxy nor an additional carbonyl band, but typical epoxide bands are seen at 8, 11 and 12 μ. A terminal epoxide group is revealed by the mass spectrum, with the ion m/z=71 for $C_4H_7O$ (**193**) (*102*). Further key fragments at m/z=113 for $C_6H_9O_2$ (**194**) and m/z=95 for $C_6H_7O$ (**195**) suggest a second epoxide group in the immediate vicinity of the first. This assignment receives further support from the fragment at m/z=325 (**192**), which, with two mass units more, is also found in the mass spectrum of the related hispidols (*112*). In the case of $NP_{36}$ the fragment is formed from $M^+ - CH_3$ by partial loss of the side chain including the two oxygen atoms.

The $^{13}$C-NMR spectrum (100.62 MHz in $CDCl_3$) was measured by the elegant technique of C. LE COCQ and J. Y. LALLEMAND (*113*). The spectrum of $NP_{36}$ has 16 signals for CH and $CH_3$, and 14 for quaternary C and $CH_2$. After deducting the signals for the ring system there remain $7 \times CH/CH_3$ and $1 \times CH/CH_2$ to be assigned. Analysis of the spectrum yields the partial structures (**196**), (**197**) and (**198**), of which only (**196**) is compatible with the $^1$H-NMR spectral data (*114*). $NP_{36}$ has the constitution (**199**) with the diepoxide side chain, a novel feature in protolimonoids. With the aid of INDOR-spectroscopy, the 21-$CH_3$ group and 20-H could be unambiguously identified (*115*). From the biogenetically established (20*S*)-configuration as a starting point, the configuration of the side chain is deduced from the coupling constants and the *trans*-orientation of the epoxide protons 22- and 23-H, as shown in structure (**199**) for $NP_{36}$.

The antibiotic hedamycin, a fungal metabilite from *Streptomyces griseoruber*, has a diepoxide side chain of similar structure (*116*).

The proton spectra of the more highly oxidized protolimonoids $NP_{32}$ and $NP_{34}$ show them each to have a secondary and a tertiary hydroxy group in the side chain. The $Pb(OAc)_4$ test differentiates between the isolated hydroxy groups of $NP_{32}$ and the vicinal hydroxy groups of $NP_{34}$.

The IR carbonyl bands of $NP_{32}$ are consistent with a δ-lactone ring as in (**200**), a structure closely related to bourjotinolan-A (**203**) from *Flindersia bourjotiana* (Rutaceae) (*117*) and sapelin-A (**204**) from *Etandrophragma cylindricum* (Meliaceae) (*118*), both of which compounds have also been isolated from *Trichilia hispida* (Meliaceae) (*119*). The 24-H doublet coupling constant, $J = 8$ Hz, indicates that the protons at C-23 and C-24 are diaxial (*120*), while the proton at C-24 is shifted to lower field in $NP_{32}$ acetate (**201**).Oxidation yields the ketone (**202**), in which the 24-H doublet of $NP_{32}$ at $\delta = 3.88$ is replaced by a singlet at $\delta = 4.52$ ppm.

$NP_{34}$, with the carbonyl band of a γ-lactone, and vicinal hydroxy groups, has formula (**205**) with unknown stereochemistry at carbon atoms 23 and 24. The γ-lactone (*107*) (**207**) of known configuration prepared from melianodiol (*121*) (**206**), and the lactone glycol with the same constitution derived from turraeanthin (*102*, formula XVII) have melting points and specific rotations different from those of **NP**$_{34}$.

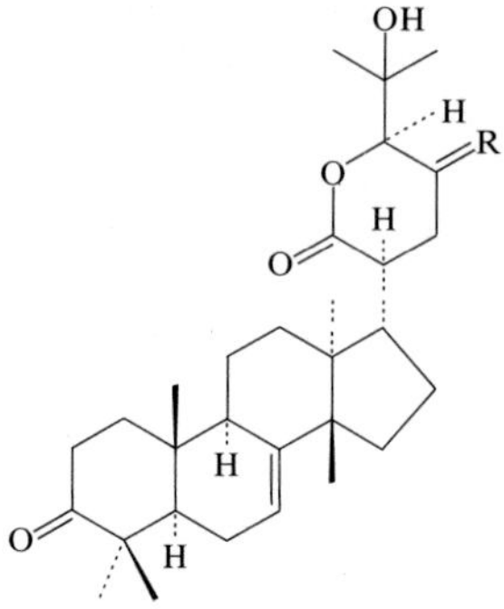

**(200)** R = H/α-OH
**(201)** R = H/α-OAc
**(202)** R = O

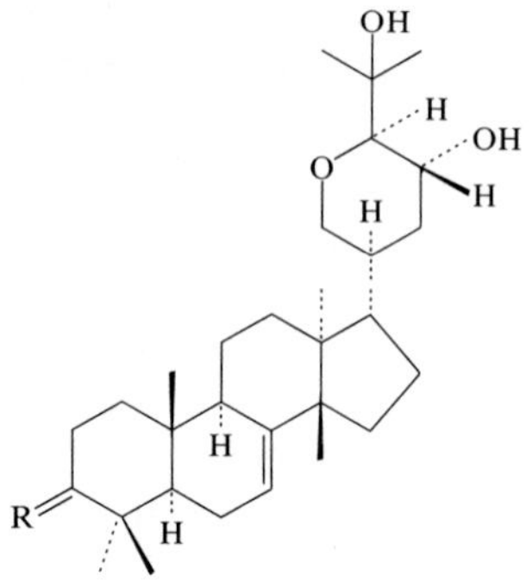

**(203)** R = O
**(204)** R = H/α-OH

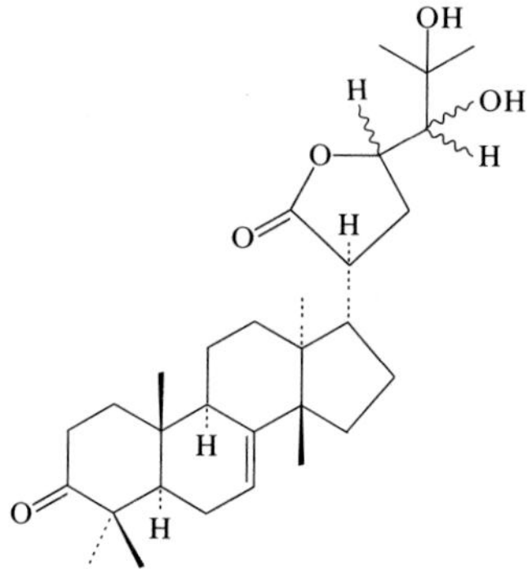

**(205)**

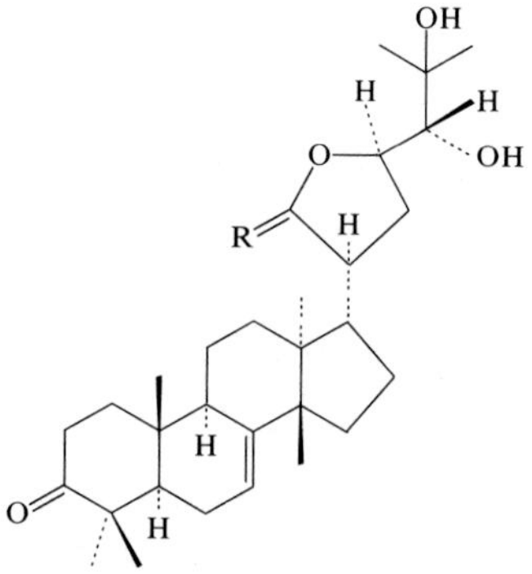

**(206)** R = H/α-OH
**(207)** R = O

## 2. Apotirucallan-(20*S*)-Triterpenoids

Cneorin-**NP**$_{35}$, a compound also found in the Canary Islands species of Cneoraceae, has the empirical formula $C_{32}H_{48}O_6$ and is an acetate with a ring system corresponding to the apotirucallans (**208**). The chemical shift of the olefinic proton 15-H at $\delta = 5.50$ ppm contrasts with that of the olefinic proton 7-H of the $\Delta^7$-tirucallans, which is consistently at $\delta = 5.30$. The mass spectrum also confirms the ring system with the 7α-hydroxy group, having the base peak at m/z = 328 for $C_{20}H_{32}O_2$ (**209**), which is formed by loss of the side chain and an H-atom. Loss of water from the base peak gives the ion at m/z = 310. The nature of the side chain $C_{10}H_{15}O_4$ can be deduced from the loss of acetic acid and the radical $C_4H_7O$ of mass 71 from $M^+$ to yield the fragment m/z = 397 for $C_{26}H_{37}O_4$ (**210**). This proves the position of the acetate group at C-21 and the epoxide group at C-25/26. The side chain of $\mathbf{NP_{35}}$ corresponds to that of turraeanthin acetate (*102*) with unknown configuration at carbon atoms 21, 23 and 24. The structural assignment is confirmed by the $^1$H- and $^{13}$C-NMR spectra.

We found by chance in the literature a compound of the same structure, called "compound A" by J. D. CONNOLLY *et al.*, who isolated it from *Chisochetum paniculatus* (Meliaceae) (*122*). Direct comparison of the two compounds showed them to be identical.

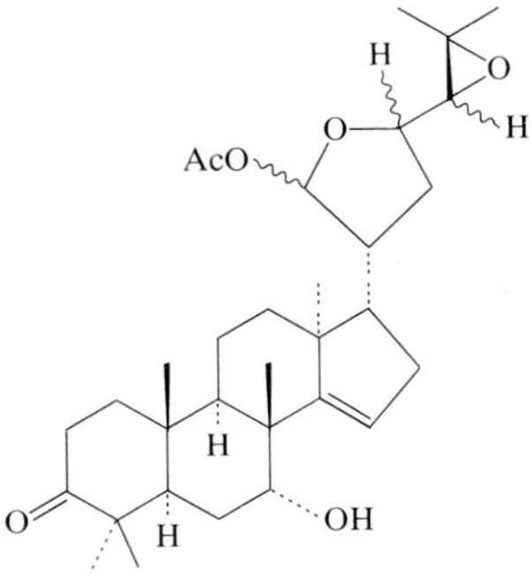

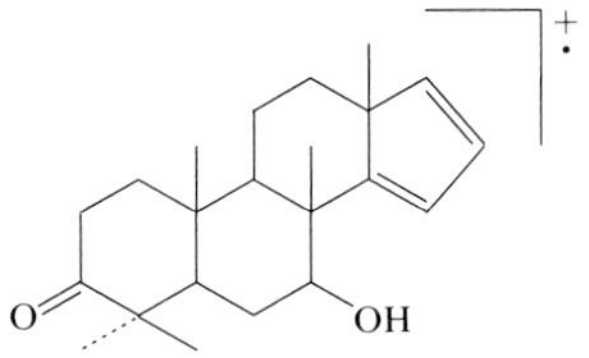

m/z = 328 m/z = 397

(**208**) (**209**) (**210**)

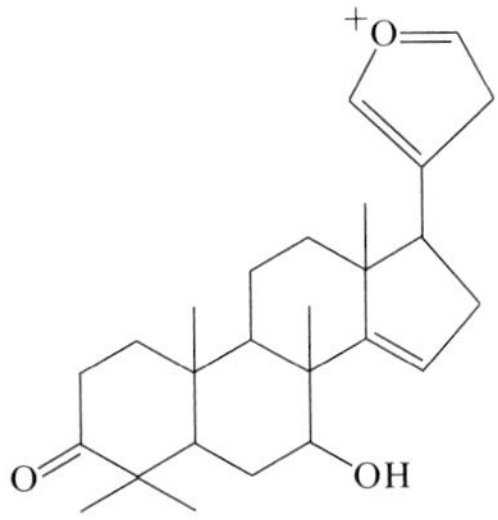

## XIII. Comments on Biosynthesis of Cneorins and Tricoccins

Since the isolation of tetranortriterpenes and triterpenes of the protolimonoid type from *Cneorum tricoccon* and *Neochamaelea pulverulenta* it has become more and more evident that not only the polyterpenoid bitter principles of the Rutaceae, Meliaceae and Simaroubaceae, but also those of the Cneoraceae are formed from 3α- or 3β-hydroxy-tirucalla-7,24-diene (**211**) by oxidative degradation of the side chain and apo-rearrangement. A characteristic feature is the loss of the four terminal C-atoms with formation of the α-oriented β-substituted furan ring, a structural element common to the majority of the limonoids and meliacins. The OH functional groups at carbon atoms 3 and 7 and the $\Delta^{14}$ double bond of the natural precursors are of biogenetic origin. The progressive oxidative degradation of the compounds until obacunone (**181**) is reached, is illustrated by the series of closely related natural products azadirone (*81*) (**212**), azadiradione (*81*) (**213**), and gedunin (*123*) (**214**) in the familiar manner. Simple oxidation to ketones, enones and oxirans, and ring expansions to ε- and δ-lactones predominate. Angular methyl groups are also accessible to oxidative attack, with formation of primary alcohols: the classical example is the biosynthesis of limonin (**180**) from obacunone (**181**) by transformation of ring A into a bicyclic lactone ether (*36, 61*).

Besides obacunone and its derivatives, we have isolated other 3,4-*seco*-meliacans from the Cneoraceae, including $\mathbf{S_{13}}$, $\mathbf{S_{38}}$ and $\mathbf{S_{40}}$. The tricoccins $\mathbf{S_7}$ (**162**) and $\mathbf{S_{43}}$ (**177**) of the 7,8-*seco*-meliacan type are of biogenetic interest. The biosynthesis of such compounds probably proceeds via intermediates such as (**215**), (**216**) and (**217**), followed by rotation about the C-9/C-10 bond and elimination of water to give $\mathbf{S_{43}}$; a subsequent oxidative step at C-16 leads to $\mathbf{S_7}$.

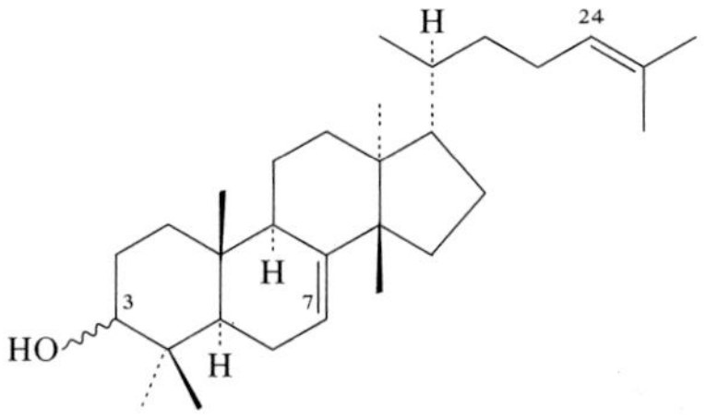

(**211**)

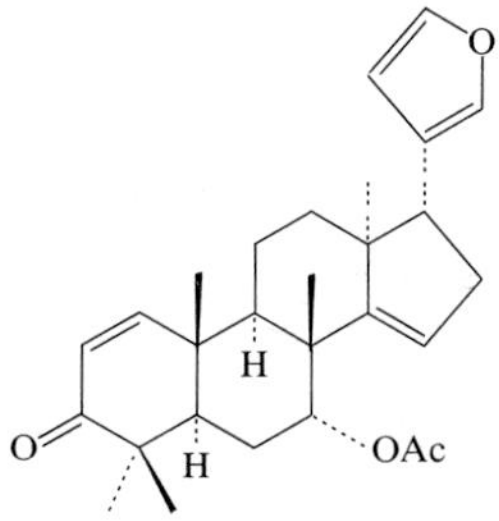

(**212**)

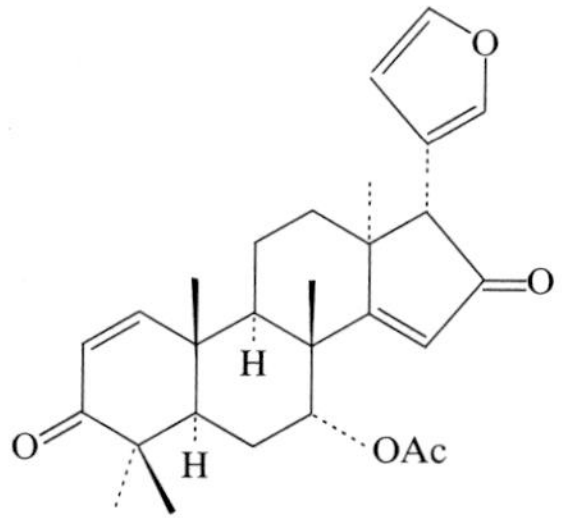

(**213**)

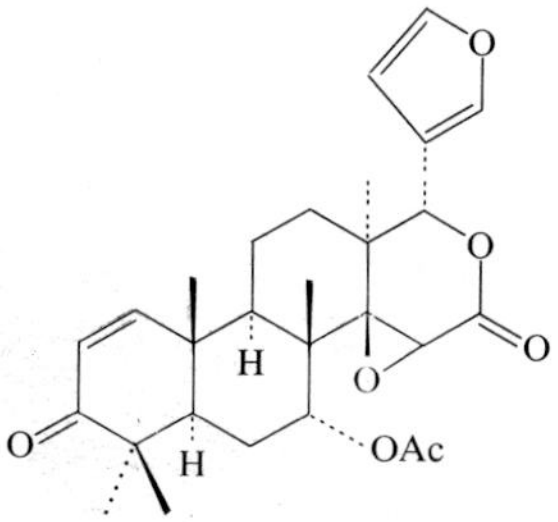

(**214**)

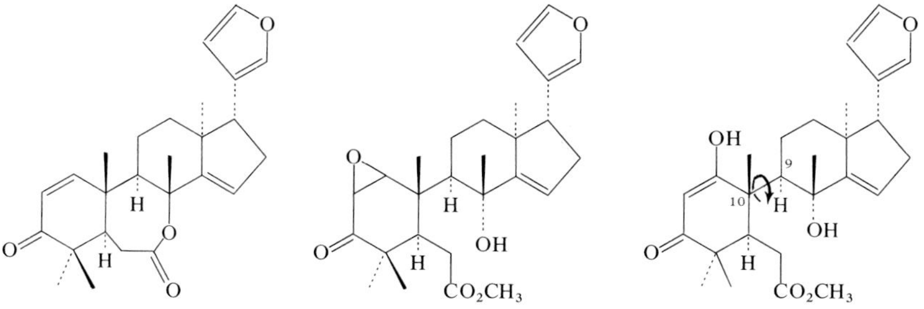

(215) (216) (217)

The cneorins and tricoccins derived from the cneoran skeleton (**20**) play a central role in biosynthetic considerations. They are formed by further oxidation, perhaps from obacunone (**181**) or a precursor with the partial structure (**218**) of a 14β,15β-epoxymeliacolide. The formation of the cyclopropane ring can be initiated by the attack of a proton on the oxygen atom of the oxiran ring with intermediate formation of a cation at C-14. This cation, reacts further with expulsion of a proton from the α-oriented methyl group at C-13 to form the α-hydroxy-lactone (**219**). Decarboxylation follows renewed proton attack, with transient formation of the cation (**220**) with the charge at C-17, resulting in loss of chirality at this centre. This cation, which is sterically virtually unhindered, can react with a water molecule approaching from either side, with expulsion of a proton. Simultaneous dehydrogenation or oxidation of the primary alcohol leads to the 17α- or 17β-hydroxyaldehyde (**221**), from which the γ-lactones (**222**) with (17*R*)- or (17*S*)-configuration are formed directly by oxidation and elimination of water. Alternatively, the C-15 hemiacetals (**223**) can also be formed from (**221**). As only γ-lactols with (17*R*)-configuration have been identified to date, our earlier assumption that these compounds are always precursors of the γ-lactones is unlikely, this route appearing in the light of the scheme as probably a biosynthetic sidetrack.

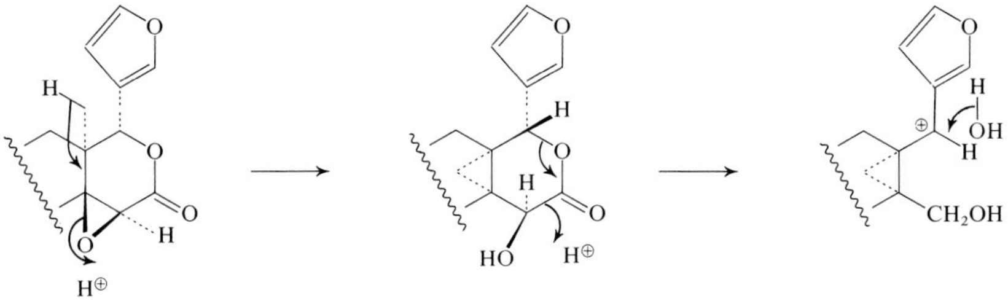

(218) (219) (220)

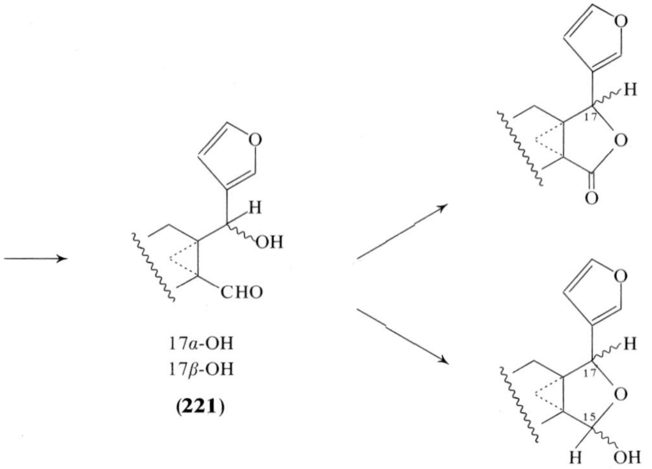

17α-OH
17β-OH
(221)

17α-H
17β-H
(222)

17α-H
17β-H
15α-OH
15β-OH
(223)

Our earlier formulation of the biosynthesis of the Cneoraceae bitter principles (*3*) took as its starting point the dilactone structure of rings A/B shown in formula (**224**). The existence of such structures recently received confirmation through the work of W. KRAUS *et al.* (*124*) who isolated surenolactone from *Toona sureni* (Meliaceae) and identified the dilactone structure (**229**). Hydrolysis of lactone ring B in (**224**) to the carboxylic acid, and elimination of water from the tertiary alcohol to give the *exo*-methylene at C-30 as in partial structure (**225**) is followed by attack by a proton with expulsion of water to yield the seven-membered cyclic ketone (**226**) with a positive charge at C-8. This cation can stabilize itself directly by opening the C-9/C-10 bond and incorporation of an α-oriented hydroxy group at C-10 resulting in a molecule with the partial structure shown in (**227**).

Combination of the partial structures (**222**/**223**) and (**227**) with ring A hydrolysed gives the model precursor (**228**) with defined configuration at the carbon atoms 5, 10 and 13 (14), supplemented by the *R*- and *S*-configuration at carbon atoms 15 and 17. The model (**228**) is either directly, or after transformation of the $\Delta^8$ double bond into the epoxide, diol, C-9 ketone or finally into an allylalcohol with the hydroxy group at C-9 and a $\Delta^{8(30)}$ double bond, the potential precursor of all cneorins and tricoccins with the cneoran skeleton.

Fraxinellone (**233**), found both in *Dictamus albus* (Rutaceae) (*125, 126, 127*) and also in *Melia azedarach* (Meliaceae) (*128*) and the more recently isolated calodendrolite (**231**) from *Calodendrum capense* (*129*) are doubtless elaborations of unknown tetranortriterpenoids. Starting from a possible precursor with the partial structure (**230**), cleavage of the molecule to (**231**) can be formulated as above, by the attack of a proton at the oxygen atom of lactone ring B, leading to an intermediate cation at C-8 and cleavage of the

*References, pp. 181—187*

C-9/C-10 bond. By reduction, catalytic decarboxylation and oxidation of (**231**) the lactol (**234**) is formed, which is oxidized to fraxinellone. Calodendrolite is not necessarily the precursor of fraxinellone, as ring D of (**230**) could in principle be degraded to the γ-lactone before ring B is cleaved to (**232**). This structure is also an intermediate in the synthesis of (±)-fraxinellone described by T. KUBOTA *et al.* (*130*).

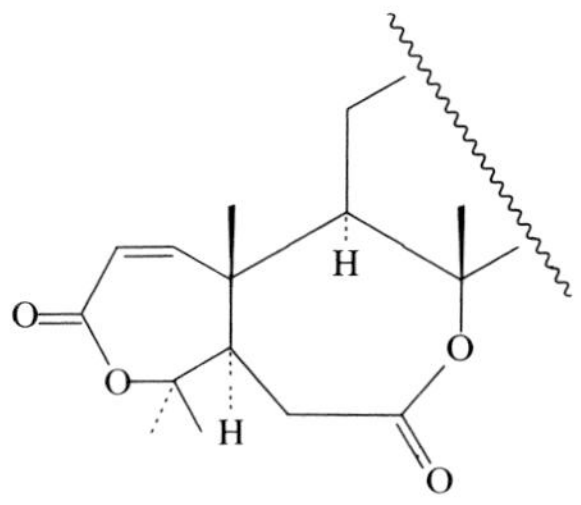

(**224**) 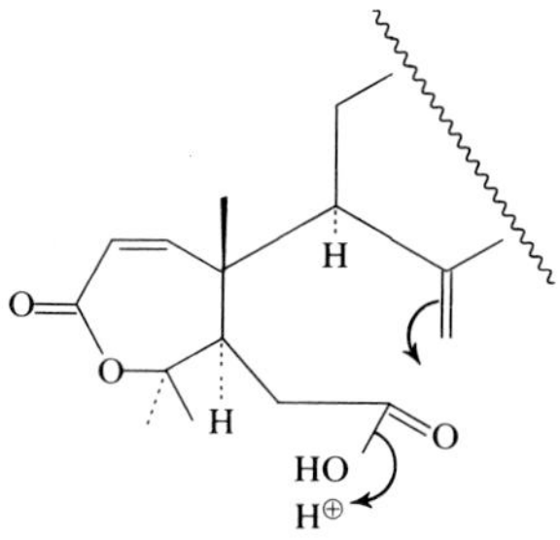 (**225**)

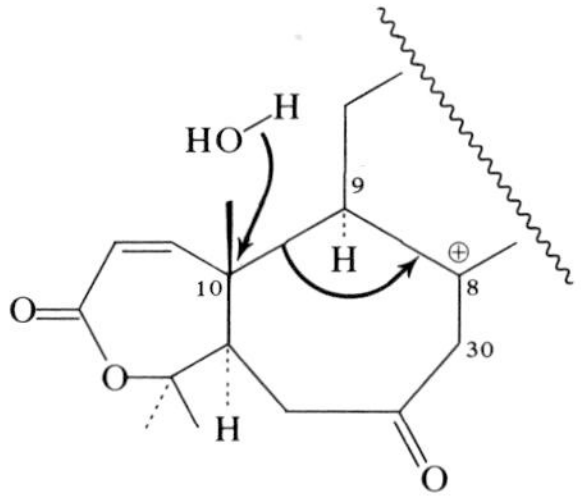

(**226**) 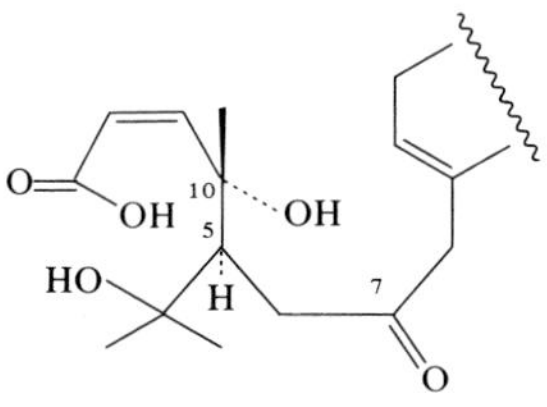 (**227**)

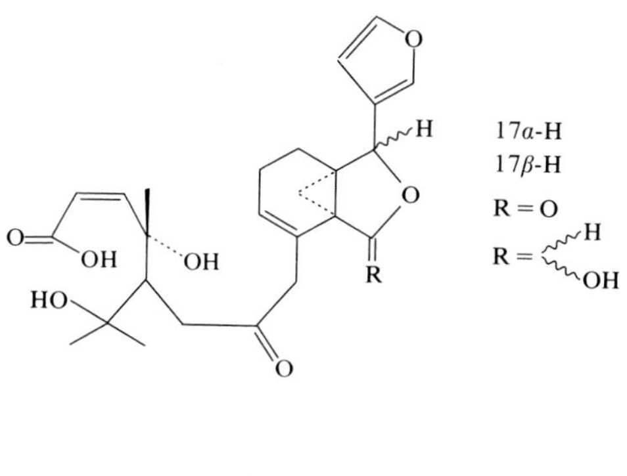

17α-H
17β-H
R = O
R = H, OH

(**228**) 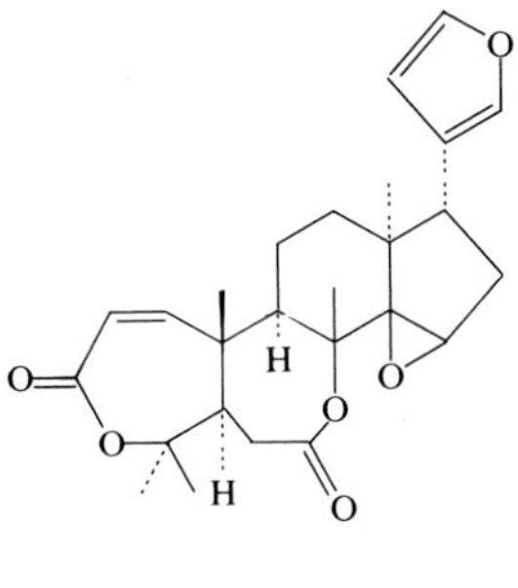 (**229**)

(230)

(231)

(232) (233) (234)

# XIV. Tables of Natural Bitter Principles from Cneoraceae

## 1. Cneorins from Neochamaelea pulverulenta (Vent.) Erdtm. of Known Constitution and Configuration

| Cneorin- | Formula | M. p. (°C) | $[\alpha]_D^{20}$ | Comp. No. | References |
|---|---|---|---|---|---|
| **A** | $C_{26}H_{30}O_8$ | 220–222 | −74,6° | (**45**) | (*3, 23, 27, 42, 45*) |
| **B** | $C_{25}H_{26}O_7$ | 196–198 | +93.0° | (**33**) | (*3, 23, 27, 42*) |
| $\mathbf{B_I}$ | $C_{25}H_{26}O_7$ | 211 | +1.0° | (**34**) | (*3, 23, 45*) |
| $\mathbf{B_{III}}$ | $C_{25}H_{26}O_7$ | 152–154 | +66.4° | (**35**) | (*3, 23, 27, 45*) |
| **C** | $C_{25}H_{26}O_7$ | 200–201 | +66.5° | (**16**) | (*3, 23, 27, 42, 45*) |
| $\mathbf{C_I}$ | $C_{25}H_{26}O_7$ | 185 | −47.3° | (**23**) | (*3, 23, 45*) |
| $\mathbf{C_{III}}$ | $C_{25}H_{26}O_7$ | 187 | +59.8° | (**17**) | (*3, 23, 27, 43, 45*) |
| **D** | $C_{25}H_{28}O_8$ | 214–216 | −67.1° | (**44**) | (*3, 23, 27, 42, 45*) |
| **E**[a] | — | — | — | — | (*24, 45*) |
| **F** | $C_{25}H_{28}O_7$ | 167 | +126.5° | (**62**) | (*3, 24, 45, 64*) |
| **G**[b] | $C_{30}H_{38}O_{10}$ | 258–260 | −40.3° | (**184**) | (*3, 24, 62*) |
| **H** | $C_{25}H_{28}O_8$ | 235 | +1.0° | (**55**) | (*24, 45*) |
| **I**[c] | $C_{28}H_{34}O_8$ | 153 | +53.1° | (**182**) | (*3, 24*) |
| **K** | $C_{25}H_{28}O_7$ | 138 | −2.0° | (**73**) | (*24, 64*) |

*References, pp. 181—187*

| Cneorin- | Formula | M. p. (°C) | $[\alpha]_D^{20}$ | Comp. No. | References |
|---|---|---|---|---|---|
| $\mathbf{K}_I$ | $C_{26}H_{30}O_7$ | 136 | — | **(74)** | *(24, 64)* |
| **M**[d] | — | — | — | — | *(24)* |
| **N** | $C_{25}H_{28}O_8$ | 226 | −5.8° | **(59)** | *(24, 28)* |
| **O** | $C_{25}H_{28}O_8$ | amorph. | +34.5° | **(57)** | *(24, 28)* |
| **P**[e] | — | 175 | — | — | *(8)* |
| **Q** | $C_{25}H_{26}O_8$ | 197 | −104.5° | **(95)** | *(67)* |
| **R** | $C_{27}H_{34}O_8$ | 193 | +78.3° | **(185)** | *(76)* |
| **S**[f] | — | — | — | — | *(8)* |
| $\mathbf{NP}_{27}$ | $C_{25}H_{28}O_{10}$ | 250 | 0.0° | **(61)** | *(8)* |
| $\mathbf{NP}_{29}$ | $C_{25}H_{26}O_8$ | 216 | −28.2° | **(96)** | *(67)* |
| $\mathbf{NP}_{32}$ | $C_{30}H_{46}O_5$ | 225−230 | −66.3° | **(200)** | *(101)* |
| $\mathbf{NP}_{34}$ | $C_{30}H_{46}O_5$ | 230 | −65.8° | **(205)** | *(101)* |
| $\mathbf{NP}_{35}$[g] | $C_{32}H_{48}O_6$ | 212 | +12.8° | **(208)** | *(8)* |
| $\mathbf{NP}_{36}$ | $C_{30}H_{46}O_3$ | 178 | −86.3° | **(199)** | *(8)* |
| $\mathbf{NP}_{37}$[h] | $C_{30}H_{46}O_4$ | 224 | −43.8° | **(187)** | *(101)* |
| $\mathbf{NP}_{38}$ | $C_{30}H_{46}O_5$ | 231 | −67.0° | **(188)** | *(8)* |

[a] Cneorin-$\mathbf{C}_I$
[b] 7α-Acetoxy-dihydronomilin *(96)*
[c] α-Obacunolacetate *(92, 93)*
[d] Tricoccin-$\mathbf{R}_1$
[e] Modification of Cneorin-$\mathbf{B}_I$
[f] Obacunon *(89, 90)*
[g] "Compound A" from *Chisochetum paniculatus* *(122)*
[h] Melianone *(41, 104)*

## 2. Cneorins of Unknown Structure

| Cneorin- | Formula | M. p. (°C) | $[\alpha]_D^{20}$ | Comp. No. | References |
|---|---|---|---|---|---|
| **L** | $C_{25}H_{28}O_8$ | 184 | — | — | *(24)* |
| $\mathbf{NP}_{28}$ | $C_{25}H_{28}O_6$ | 251 | −45.7° | — | *(8)* |

## 3. Tricoccins from Cneorum tricoccon L. of Known Constitution and Configuration

| Tricoccin- | Formula | M. p. (°C) | $[\alpha]_D^{20}$ | Comp. No. | References |
|---|---|---|---|---|---|
| $\mathbf{R}_0$[a] | — | — | — | **(33)** | *(3, 26, 28)* |
| $\mathbf{R}_1$[b] | $C_{26}H_{30}O_8$ | 170 | −62.6° | **(75)** | *(26, 64)* |
| $\mathbf{R}_2$ | $C_{25}H_{26}O_7$ | 204 | −25.3° | **(51)** | *(26, 28)* |
| $\mathbf{R}_3$[c] | — | — | — | **(34)** | *(3, 26, 28)* |
| $\mathbf{R}_5$ | $C_{25}H_{26}O_7$ | 240 | −107.5° | **(53)** | *(26, 28)* |
| $\mathbf{R}_6$ | $C_{25}H_{28}O_8$ | 190 | +116.7° | **(60)** | *(26, 28)* |
| $\mathbf{R}_8$ | $C_{25}H_{28}O_8$ | 250 | −131.0° | **(58)** | *(26)* |
| $\mathbf{R}_9$ | $C_{25}H_{28}O_8$ | 175 | +69.3° | **(69)** | *(26, 64)* |
| *epi*-$\mathbf{R}_9$ | $C_{25}H_{28}O_8$ | amorph. | +62.3° | **(70)** | *(8)* |

| Tricoccin- | Formula | M. p. (°C) | $[\alpha]_D^{20}$ | Comp. No. | References |
|---|---|---|---|---|---|
| $\mathbf{R}_{10}$ | $C_{25}H_{28}O_8$ | amorph. | −11.7° | **(76)** | *(64)* |
| $\mathbf{R}_{12}$ | $C_{25}H_{28}O_8$ | amorph. | — | **(71)** | *(64)* |
| *epi*-$\mathbf{R}_{12}$ | $C_{25}H_{28}O_8$ | amorph. | — | **(72)** | *(64)* |
| $\mathbf{S}_1$ | $C_{25}H_{28}O_8$ | 170 | +1.0° | **(82)** | *(26, 29)* |
| $\mathbf{S}_2$ | $C_{25}H_{28}O_8$ | 201 | −52.2° | **(77)** | *(26, 29)* |
| $\mathbf{S}_3$[d] | $C_{26}H_{30}O_7$ | 225 | −51.0° | **(181)** | *(3, 26)* |
| $\mathbf{S}_4$ | $C_{25}H_{28}O_6$ | 191 | +8.8° | **(113)** | *(26, 65, 73)* |
| $\mathbf{S}_5$ | $C_{25}H_{28}O_8$ | 192 | −16.8° | **(86)** | *(26, 29)* |
| $\mathbf{S}_6$ | $C_{25}H_{30}O_8$ | 205 | −105.9° | **(106)** | *(26, 29, 65)* |
| $\mathbf{S}_7$ | $C_{27}H_{32}O_6$ | 202 | −40.5° | **(162)** | *(26, 62)* |
| $\mathbf{S}_8$ | $C_{27}H_{34}O_7$ | 250 | — | **(173)** | *(26, 62)* |
| $\mathbf{S}_{10}$ | $C_{26}H_{30}O_8$ | 177 | −148.8° | **(78)** | *(26, 29)* |
| $\mathbf{S}_{11}$ | $C_{26}H_{30}O_8$ | 108 | — | **(83)** | *(26)* |
| $\mathbf{S}_{12}$ | $C_{26}H_{30}O_8$ | 170 | −15.2° | **(87)** | *(26)* |
| $\mathbf{S}_{13}$ | $C_{26}H_{36}O_5$ | 238 | −24.6° | **(138)** | *(75)* |
| $\mathbf{S}_{14}$ | $C_{25}H_{28}O_8$ | 152 | +86.3° | **(65)** | *(64)* |
| $\mathbf{S}_{15}$ | $C_{25}H_{28}O_8$ | 195 | −56.0° | **(80)** | *(8)* |
| $\mathbf{S}_{16}$ | $C_{25}H_{30}O_7$ | 180 | +37.3° | **(127)** | *(73)* |
| $\mathbf{S}_{17}$ | $C_{25}H_{28}O_7$ | 220−228 | −39.3° | **(107)** | *(65)* |
| $\mathbf{S}_{18}$ | $C_{25}H_{28}O_7$ | amorph. | +63.0° | **(108)** | *(65)* |
| $\mathbf{S}_{19}$ | $C_{25}H_{34}O_7$ | 259 | +1.0° | **(175)** | *(62)* |
| $\mathbf{S}_{20}$ | $C_{25}H_{28}O_8$ | amorph. | 0.0° | **(84)** | *(8)* |
| $\mathbf{S}_{22}$ | $C_{26}H_{34}O_6$ | 197 | +51.7° | **(154)** | *(76)* |
| $\mathbf{S}_{23}$[e] | $C_{26}H_{32}O_7$ | 135 | +73.6° | **(183)** | *(8)* |
| $\mathbf{S}_{24}$ | $C_{25}H_{30}O_8$ | amorph. | — | **(109)** | *(65)* |
| $\mathbf{S}_{24}$-Ac | $C_{27}H_{32}O_9$ | 168 | +60.0° | — | *(65)* |
| $\mathbf{S}_{26}$ | $C_{25}H_{28}O_8$ | 153 | +21.7° | **(91)** | *(8)* |
| $\mathbf{S}_{27}$ | $C_{25}H_{30}O_8$ | amorph. | — | **(129)** | *(73)* |
| $\mathbf{S}_{27}$-(Ac)$_2$ | $C_{29}H_{30}O_{10}$ | 161 | +64.9° | **(130)** | *(73)* |
| $\mathbf{S}_{28}$ | $C_{25}H_{30}O_7$ | amorph. | — | **(110)** | *(65)* |
| $\mathbf{S}_{28}$-Ac | $C_{27}H_{32}O_8$ | 175 | +107.2° | — | *(65)* |
| $\mathbf{S}_{30}$-Ac | $C_{27}H_{32}O_8$ | amorph. | — | **(136)** | *(8)* |
| $\mathbf{S}_{31}$-Ac | $C_{27}H_{32}O_8$ | amorph. | — | **(137)** | *(8)* |
| $\mathbf{S}_{32}$-Ac | $C_{32}H_{42}O_{10}$ | amorph. | — | **(161)** | *(76)* |
| $\mathbf{S}_{33}$ | $C_{25}H_{28}O_7$ | 237 | −132.6° | **(116)** | *(65)* |
| $\mathbf{S}_{35}$ | $C_{26}H_{32}O_8$ | 123−128 | −53.8° | **(111)** | *(65)* |
| $\mathbf{S}_{36}$ | $C_{25}H_{30}O_8$ | 234 | −93.5° | **(112)** | *(65)* |
| $\mathbf{S}_{38}$ | $C_{26}H_{36}O_6$ | 238 | −40.5° | **(146)** | *(8)* |
| $\mathbf{S}_{39}$ | $C_{25}H_{30}O_8$ | 237 | −14.4° | **(122)** | *(8)* |
| $\mathbf{S}_{40}$ | $C_{26}H_{36}O_6$ | 248 | −17.7° | **(150)** | *(75)* |
| $\mathbf{S}_{41}$[f] | — | — | — | **(182)** | *(8)* |
| $\mathbf{S}_{42}$ | $C_{25}H_{28}O_6$ | amorph. | +47.7° | **(114)** | *(73)* |
| $\mathbf{S}_{43}$ | $C_{27}H_{34}O_5$ | 212 | −69.1° | **(177)** | *(8)* |

[a] Cneorin-**B**
[b] Cneorin-**M**
[c] Cneorin-$\mathbf{B}_I$
[d] Obacunone *(89, 90)*
[e] α-Obacunol *(92, 93)*
[f] α-Obacunolacetate *(92, 93)*

*References, pp. 181—187*

## 4. Tricoccins of Unknown Structure

| Tricoccin- | Formula | M. p. (°C) | $[\alpha]_D^{20}$ | Comp. No. | References |
|---|---|---|---|---|---|
| **R$_4$** | $C_{25}H_{28}O_8$ | 210 | — | — | (*26, 28*) |
| **R$_7$** | $C_{26}H_{28}O_9$ | 211 | — | — | (*26*) |
| **R$_{11}$** | $C_{25}H_{28}O_8$ | 251 | +105.0° | — | (*8*) |
| S$_{21}$ | $C_{25}H_{30}O_8$ | 172 | +51.7° | — | (*8*) |
| S$_{25}$ | $C_{25}H_{28}O_6$ | 220 | — | — | (*8*) |
| S$_{29}$ | $C_{25}H_{28}O_7$ | amorph. | — | — | (*8*) |
| S$_{29}$-Ac | $C_{27}H_{30}O_8$ | 173 | — | — | (*8*) |
| S$_{34}$ | $C_{26}H_{32}O_8$ | 186 – 189 | — | — | (*8*) |
| S$_{37}$ | $C_{25}H_{30}O_8$ | 204 | — | — | (*8*) |

### References


*1.* Takthajan, A: Evolution und Ausbreitung der Blütenpflanzen, Jena 1973.

*2.* Cronquist, A.: An Integrated System of Classification of Flowering Plants. New York: Columbia University Press. 1981.

*3.* Straka, H., F. Albers, and A. Mondon: Die Stellung und Gliederung der Familie *Cneoraceae* (*Rutales*). Beitr. Biol. Pflanzen **52,** 267 (1976).

*4.* Lobreau-Callen, D., S. Nilsson, F. Albers, and H. Straka: Les *Cneoraceae* (*Rutales*): étude taxonomique, palynologique et systématique. Grana **17,** 125 (1978).

*5.* Mondon, A., H. Callsen, and P. Hartmann: Inhaltsstoffe der Cneoraceen, III: Trennverfahren für *Cneorum pulverulentum* und Untersuchung der Wachsfraktion und Phytosterole. Chem. Ber. **108,** 1989 (1975).

*6.* Mondon, A., and U. Schwarzmaier: Inhaltsstoffe der Cneoraceen, I. Untersuchung von *Cneorum tricoccon.* Chem Ber. **108,** 925 (1975).

*7.* Plouvier, C.: Sur la recherche de hétérosides flavoniques dans quelque groupes botaniques. Compt. rend. Acad. Sci., Paris **256,** 4061 (1965).

*8.* Mondon, A., and B. Epe: Unpublished Results.

*9.* Hegnauer, R.: Chemotaxonomie der Pflanzen, Bd. 3, S. 433, Basel und Stuttgart 1964.

*10.* Gibbs, R. D.: Chemotaxonomy of Flowering Plants, Vol. 3, p. 1671. Montreal and London 1974.

*11.* Trautmann, D., B. Epe, U. Oelbermann, and A. Mondon: Konstitution und Konfiguration der Cneorubine. Chem. Ber. **113,** 3848 (1980).

*12.* Mondon, A., H. Callsen, P. Hartmann, G. Cuno, and C. H. Anderson: Inhaltsstoffe der Cneoraceen, II: Studien zur Hydrierung, zum Alkaliabbau und zur Synthese von Chromonen. Chem. Ber. **108,** 934 (1975).

*13.* Mondon, A., and H. Callsen: Inhaltsstoffe der Cneoraceen, IV: Chromone und Cumarine aus *Cneorum pulverulentum.* Chem Ber. **108,** 2005 (1975).

*14.* Trautmann, D., B. Epe, U. Oelbermann, and A. Mondon: Notiz über alte und neue Chromone aus Cneoraceen. Chem. Ber. **109,** 2963 (1976).

*15.* Epe, B., U. Oelbermann, and A. Mondon: Neue Chromone aus Cneoraceen. Chem. Ber. **114,** 757 (1981).

*16.* Gonzalez, A. G., J. P. Castaneda, and B. M. Fraga: Nueva Chromonas de la *Neochamaelea pulverulenta* Erdtm. Anal Quim. **68,** 447 (1972).

*17.* Gonzalez, A. G., B. M. Fraga, and R. Torres: Chromonas del *"Cneorum tricoccum".* Anal. Quim. **70,** 91 (1974).

18. GONZALEZ, A. G., B. M. FRAGA, and O. PINO: Chromenes and Chromones, IV: New Chromone from the Stems of *Cneorum tricoccum*. Phytochemistry **13,** 2305 (1974).
19. — — — Chromenes and Chromones, VI.: Pulverin, a new Chromone from the Fruits of *Neochamaelea pulverulenta*. Phytochemistry **14,** 1656 (1975).
20. — — — Minor Chromones of *Neochamaelea pulverulenta*. Anal. Quim. **73,** 557 (1977).
21. GONZALEZ, A. G., B. M. FRAGA, M. G. HERNANDEZ, O. PINO, and A. G. RAVELO: New Sources of Natural Coumarins, Part 35. New Coumarins from *Cneorum tricoccum*. Rev. Latino Am. Quim. 205 (1978).
22. GONZALEZ, A. G., B. M. FRAGA, O. PINO, J. P. DECLEREQ, G. GERMAIN, and J. FAYOS: X-Ray Structure of Bethancorol, a new Courmarin from *Cneorum tricoccum*. Tetrahedron Letters 1729 (1976).
23. CALLSEN, H.: Über die Inhaltsstoffe der Cneoraceen. Dissertation Universität Kiel 1972.
24. EPE, B.: Konstitution und Konfiguration der Pentanortriterpene aus *Neochamaelea pulverulenta*. Dissertation Universität Kiel 1976.
25. TRAUTMANN, D.: Über Sesterterpene aus *Cneorum tricoccon*. Diplomarbeit Universität Kiel 1974.
26. — Die charakteristischen Inhaltsstoffe von *Cneorum tricoccon L.* Dissertation Universität Kiel 1977.
27. MONDON, A., and H. CALLSEN: Zur Kenntnis der Bitterstoffe aus Cneoraceen, II. Tetrahedron Letters 699 (1975).
28. MONDON, A., D. TRAUTMANN, B. EPE, and U. OELBERMANN: Zur Kenntnis der Bitterstoffe aus Cneoraceen, VI. Tetrahedron Letters 3291 (1976).
29. — — — — Zur Kenntnis der Bitterstoffe aus Cneoraceen, VII. Tetrahedron Letters 3295 (1976).
30. POLONSKY, J.: The Structure of Simarolide, the Bitter Principle of *Simarouba amara*. Proc. Chem. Soc., London 292 (1964); BROWN, W. A. C., and G. A. SIMM: The Constitution and Absolute Stereochemistry of Simarolide, the Bitter Principle of *Simarouba amara*. Proc. Chem. Soc., London, 293 (1964).
31. HIKINO, H., T. OHTA, and T. TAKEMOTO: Stereostructure of Picrasin A, Simaroubolide of *Picrasma quassinoides*. Chem. Pharm. Bull. **18,** 1082 (1970).
32. POLONSKY, J., M. VAN TRI, TH. PRANGÉ, CL. PASCARD, and TH. SEVENET: Isolation and Structure (X-Ray Analysis) of a new $C_{25}$ Quassinoid Soulameolide from *Soulamea tomentosa*. J.C.S. Chem. Comm. 641 (1979).
33. POLONSKY, J., Z. VARON, TH. PRANGÉ, CL. PASCARD, and CH. MORETTI: Structures of Simarinolide and Guanepolide (X-Ray Analysis), New Quassinoids from *Simaba* cf. *orinocensis*. Tetrahedron Letters **22,** 3605 (1981).
34. CHAKRABORTY, D. P., P. BHATTACHARYYA, S. P. BHATTACHARYYA, J. BORDENER, G. L. A. HENNESSEE, and B. WEINSTEIN: Clausenolide: a Novel Pentanortriterpenoid Furanolacton; X-Ray Crystal Structure. J. C. S. Chem. Comm. 246 (1979).
35. KRAUS, W., and R. CRAMER: Pentanortriterpenoide aus *Azadirachta indica* A. Juss (Meliaceae). Chem. Ber. **114,** 2375 (1981).
36. DREYER, D. L.: Limonoid Bitter Principles. Fortschr. Chem. org. Naturst. **26,** 190 (1968).
37. CONNOLLY, J. D., K. H. OVERTON, and J. POLONSKY: The Chemistry and Biochemistry of the Limonoids and Quassinoids. Progress in Phytochemistry (L. REINHOLD and Y. LIEVSCHITZ, eds.), Vol. II, 385. London: Interscience Publishers 1970.
38. CONNOLLY, J. D., and K. H. OVERTON: Chemistry of Terpenes and Terpenoids (ed. A. A. NEWMAN), p. 207. London-New York: Academic Press. 1972.
39. POLONSKY, J.: Quassinoid Bitter Principles. Fortschr. Chem. org. Naturst. **30,** 101 (1973).
40. NAKANISHI, K., T. GOTO, S. ITÔ, S. NATORI, and S. NOZOE: Natural Products Chemistry, Vol. 1, p. 313. Tokyo: Kodensha, and New York-San Francisco-London: Academic Press. 1974.

*41.* LAVIE, D., M. K. JAIN, and I. KIRSON: Terpenoids-V. Melianone from *Melia azedarach L.* Tetrahedron Letters 2049 (1966). — Terpenoids. Part VI. The Complete Structure of Melianone. J. Chem. Soc. (C), 1347 (1967).

*42.* MONDON, A., and H. CALLSEN: Zur Kenntnis der Bitterstoffe aus Cneoraceen, I and II. Tetrahedron Letters 551, 699 (1975).

*43.* MONDON, A., H. CALLSEN, and B. EPE: Zur Kenntnis der Bitterstoffe aus Cneoraceen, III. Tetrahedron Letters 703 (1975).

*44.* HENKEL, G., H. DIERKS, B. EPE, and A. MONDON: Zur Kenntnis der Bitterstoffe aus Cneoraceen, IV. Tetrahedron Letters 3315 (1975).

*45.* MONDON, A., and B. EPE: Zur Kenntnis der Bitterstoffe aus Cneoraceen, V. Tetrahedron Letters 1273 (1976).

*46.* BARTON, D. H. R., and D. ELAD: Colombo Root Bitter Principles. Part I. The Functional Groups of Columbin. J. Chem. Soc. 2085 (1956).

*47.* 90 MHz, $\delta$-values for TMS = O as internal standard. $CDCl_3$ was used as solvent unless stated otherwise.

*48.* In the quoted reference, (*44*) shows the projection and formulation of the antipodes.

*49.* BIJVOET, J. M., A. F. PEERDEMAN, and A. J. van BOMMEL: Determination of the absolute configuration of optically active compounds by means of X-Rays. Nature (London) **168,** 271 (1951).

*50.* "$\alpha$" and "$\beta$" denote the projection of a substituent below or above the plane of the paper respectively.

*51.* OKORIE, D. A., and D. A. H. TAYLOR: Limonoids from the Timber of *Trichilia heudelottii* Planch. ex Oliv. J. Chem. Soc. (C) 1828 (1968).

*52.* FERGUSON, G., P. A. GUNN, W. C. MARSH, R. MCCRINDLE, R. RESTIVO, J. D. CONNOLLY, J. W. B. FULKE, and M. S. HENDERSON: Triterpenoids from *Guarea glabra* (Meliaceae): A New Skeletal Class identified by Chemical Spectroscopic, and X-Ray Evidence. J. C. S. Chem. Comm. 159 (1973); Tetranortriterpenoids and Related Substances. Part XVII. A New Skeletal Class of Triterpenoids from *Guarea glabra* (Meliaceae). J. C. S. Perkin I, 491 (1975).

*53.* The $^{13}$C-NMR spectrum of cneorin-B can not be measured in acetone-$d_6$ owing to insufficient solubility in this solvent.

*54.* HEIDENREICH, H.: Synthesen in der Reihe der Cneorine und Tricoccine. Dissertation, Universität Kiel 1979.

*55.* FUJITA, E., I. UCHIDA, and T. FUJITA: Teucvin, a Novel Furanoid Norditerpene from *Teucrium viscidum* var. *Miquelianum.* J. C. S. Chem. Comm. 793 (1973).

*56.* KRAUS, W., and R. CRAMER: 17-Epiazadiradion and 17$\beta$-Hydroxyazadiradion, zwei neue Inhaltsstoffe aus *Azadirachta indica* A. Juss. Tetrahedron Letters 2395 (1978).

*57.* "Specialist Periodical Reports, Terpenoids and Steroids". The Chemical Society London Vol. 1 – 9 (1971 – 1979); The Royal Society of Chemistry London Vol. 10 (1980).

*58.* cf. PARKER, K. A., and M. R. ADEMCHUK: Intramolecular Diels-Alder Reactions of the Furan Diene. Tetrahedron Letters 1689 (1978); DE CLERCQ, P. J., and L. A. VAN ROYEN: The Intramolecular Diels-Alder Furan Approach in Synthesis: 11-Oxatricyclo-[6.2.1.0$^{1,6}$]undec-9-en-5-one. Synth. Comm. **9,** 771 (1979); STERNBACH, D. D., and D. M. ROSSANA: Intramolecular Diels-Alder Reactions of the Furan Diene: Substituent and Solvent Effects. Tetrahedron Letters **23,** 303 (1982).

*59.* SCOTT, A. I.: Interpretation of the Ultraviolet Spectra of Natural Products. Oxford: Pergamon Press. 1964.

*60.* NARAYANAN, C. R., R. V. PACHAPURKAR, S. K. PRADHAN, V. R. SHAH, and N. S. NARASIMHAN: Structure of Nimbin. Chem. Ind. (London) 322 (1964).

*61.* MELERA, A., K. SCHAFFNER, D. ARIGONI, and O. JEGER: Zur Konstitution des Limonins I. Über den Verlauf der alkalischen Hydrolyse von Limonin und Limonol. Helv. Chim. Acta **40,** 1420 (1957); DREYER, D. L.: Citrus Bitter Principles-II. Application of NMR to

Structural and Stereochemical Problems. Tetrahedron **21,** 75 (1965); RAO, M. M., M. MESHULAM, R. ZELNIK, and D. LAVIE: *Cabralea eichleriana*-II, Structure and Stereochemistry of Limonoids of *Cabralea eichleriana*. Phytochemistry **14,** 1071 (1975); CONNOLLY, J. D., C. LABBÉ, and D. S. RYCROFT: Tetranortriterpenoids and Related Substances. Part 20. New Tetranortriterpenoids from the seeds of *Chukrasia tabularis* (Meliaceae); Simple Esters of Phragmalin and 12α-Acetoxyphragmalin. J. C. S. Perkin I, 285 (1978).

62. MONDON, A., D. TRAUTMANN, B. EPE, U. OELBERMANN, and CH. WOLFF: Zur Kenntnis der Bitterstoffe aus Cneoraceen, VIII. Tetrahedron Letters 3699 (1978).
63. BURKE, B. A., W. R. CHAN, K. E. MAGNUS, and D. R. TAYLOR: Extractives of *Cedrela odorata* L. — III. The Structure of Photogedunin. Tetrahedron **25**, 5007 (1969).
64. MONDON, A., B. EPE, and D. TRAUTMANN: Zur Kenntnis der Bitterstoffe aus Cneoraceen, X. Tetrahedron Letters 4881 (1978).
65. EPE, B., D. TRAUTMANN, and A. MONDON: Zur Kenntnis der Bitterstoffe aus Cneoraceen, XI. Tetrahedron Letters 1365 (1979).
66. The designation "iso-tricoccin-$\mathbf{R}_9$" used in reference (*65*) should be replaced by "tricoccin-$\mathbf{R}_{12}$".
67. EPE, B., U. OELBERMANN, and A. MONDON: Zur Kenntnis der Bitterstoffe aus Cneoraceen, XIII. Tetrahedron Letters 3839 (1979).
68. JOHNSON, R. A., and E. G. NIDY: Superoxide Chemistry. A Convenient Synthesis of Dialkyl Peroxides. J. Org. Chem. **40,** 1680 (1975).
69. TURNER, J. A., and W. HERZ: Fe(II)-Induced Decomposition of Unsaturated Cyclic Peroxides Derived from Butadienes. A Simple Procedure for Synthesis of 3-Alkyl-furans. J. Org. Chem. **42,** 1900 (1977).
70. HORNER, L., and W. JURGELEIT: Die Reduktion organischer Peroxide mit tertiären Phosphinen. Tertiäre Phosphine VI. Liebigs Ann. Chem. **591,** 138 (1955).
71. MATSUMOTO, M., and K. KONDO: Sensitized Photooxygenation of Linear Monoterpenes Bearing Conjugated Double Bonds. J. Org. Chem. **40,** 2259 (1975).
72. COREY, E. J., and J. W. SUGGS: Pyridinium Chlorochromate. An Efficient Reagent for Oxidation of Primary and Secondary Alcohols to Carbonyl Compounds. Tetrahedron Letters 2647 (1975).
73. EPE, B., and A. MONDON: Zur Kenntnis der Bitterstoffe aus Cneoraceen, XIV. Tetrahedron Letters 4045 (1979).
74. The configuration given in reference (*65*) should be corrected.
75. — — Zur Kenntnis der Bitterstoffe aus Cneoraceen, IX. Tetrahedron Letters 3901 (1978).
76. — — Zur Kenntnis der Bitterstoffe aus Cneoraceen, XII. Tetrahedron Letters 2015 (1979).
77. SONDENGAM, B. L., C. S. KAMGA, S. F. KIMBU, and J. D. CONNOLLY: Proceranon, a new Tetranortriterpenoid from *Carapa procera*. Phytochemistry **20,** 173 (1981).
78. AYAFOR, J. F., B. L. SONDENGAM, J. D. CONNOLLY, D. S. RYCROFT, and J. I. OKOGUN: Tetranortriterpenoids and Related Compounds. Part 26. Tecleanin, a Possible Precursor of Limonin, and other New Tetranortriterpenoids from *Teclea grandifolia* Engl. (Rutaceae). J. C. S. Perkin I, 1750 (1981).
79. SONDENGAM, B. L., C. S. KAMGA, and J. D. CONNOLLY: Evodulone, A New Tetranortriterpenoid from *Carapa procera*. Tetrahedron Letters 1357 (1979).
80. KRAUS, W., and K. KYPKE: Surenon and Surenin, two novel Tetranortriterpenoids from *Toona sureni* [Blume] Merrill. Tetrahedron Letters 2715 (1979).
81. LAVIE, D., and M. K. JAIN: Tetranortriterpenoids from *Melia azadirachta* L. Chem. Commun. 278 (1967).
82. CONNOLLY, J. D., I. M. S. THORNTON, and D. A. H. TAYLOR: Partial Synthesis of Mexicanolide from 7-Oxo-deacetoxy-khivorin. Chem. Commun. 17 (1971).

83. BEVAN, C. W. L., J. W. POWELL, and D. A. H. TAYLOR: West African Timbers. Part VI. Petroleum Extracts from Species of the Genera *Khaya, Guarea, Carapa* and *Cedrela.* J. Chem. Soc. 980 (1963); CONNOLLY, J. D., R. MCCRINDLE, and K. H. OVERTON: The Constitution of Mexicanolide. A Novel Cleavage Reaction in a Naturally Occuring Bicyclo[3,3,1]nonane Derivative. Chem. Commun. 162 (1965); Tetranortriterpenoids-IV [Bicyclononanolides II]. The Constitution and Stereochemistry of Mexicanolide. Tetrahedron **24,** 1489 (1968); ADEOYE, S. A., and D. A. BEKOE: The Molecular Structure of *Cedrela odorata* Substance B. Chem. Commun. 301 (1965).
84. cf. CIMINO, G., S. DE STEFANO, A. GUERRIERO, and L. MINALE: Furanosesquiterpenoids in Sponges-I; Pallescensin-1, -2 and -3 from *Disidea pallescens.* Tetrahedron Letters 1417 (1975).
85. The $^{1}$H-NMR data given in reference (*62*) for $S_{19}$ are to be corrected by an upfield shift of 8 Hz.
86. KRAUS, W., W. GRIMMINGER, and G. SAWITZKI: Toonacilin und 6-Acetoxy-toonacilin, zwei neue B-seco-Tetranortriterpenoide mit fraßhemmender Wirkung: Angew. Chem. **90,** 476 (1968); Angew. Chem. Internat. Edn. **17,** 476 (1978).
87. KRAUS, W., and W. GRIMMINGER: Toonafolin, ein neues Tetranortriterpenoid-B-Lacton aus *Toona ciliata.* M. J. Roem. var. *australis* (Meliaceae). Liebigs Ann. Chem. **1981,** 1838.
88. ARIGONI, D., D. H. R. BARTON, E. J. COREY, O. JEGER, L. CAGLIOTI, S. DEV, P. G. FERRINI, E. R. GLAZIER, A. MELERA, S. K. PRADHAN, K. SCHAFFNER, S. STERNHELL, J. F. TEMPLETON, and S. TOBINGA: Constitution of Limonin. Experientia **16,** 41 (1960); ARNOTT, S., A. W. DAVIE, J. M. ROBERTSON, G. A. SIM, and D. G. WATSON: The Structure of Limonin: X-Ray Analysis of Epilimonol Iodoacetate. J. Chem. Soc. 4183 (1961).
89. EMERSON, O. H.: The Bitter Principles of Citrus Fruit. I. Isolation of Nomilin, a New Bitter Principle from the Seeds of Oranges and Lemons. J. Am. Chem. Soc. **70,** 545 (1948).
90. BARTON, D. H. R., S. K. PRADHAN, S. STERNHELL, and J. F. TEMPLETON: Triterpenoids. Part XXV. The Constitution of Limonin and Related Bitter Principles. J. Chem. Soc. 255 (1961).
91. KUBO, J., S. P. TANIS, Y. W. LEE, I. MIURA, K. NAKANISHI, and A. CHAPYA: The Structure of Harrisonin. Heterocycles **5,** 485 (1976).
92. KUBOTA, T., T. MATSUURA, T. TOKOROYAMA, T. KAMIKAWA, and T. MATSUMOTO: Establishment of the Correlation of Obacunone and Limonin. Tetrahedron Letters 325 (1961).
93. KAMIKAWA, T.: Constitution of Obacunone. Nippon Kakaku Zasshi **83,** 625 (1962).
94. DEAN, F. M., and T. A. GEISSMAN: The Functional groups of Nomilin and Obacunone. J. Org. Chem. **23,** 596 (1958).
95. ADESIDA, G. A., and D. A. H. TAYLOR: Isolation of Obacunol from *Lovoa trichiloides.* Phytochemistry **11,** 2641 (1972).
96. AHMED, F. R., A. S. NG, and A. G. FALLIS: 7α-Acetoxydihydronomilin: isolation, spectra and crystal structure. Can. J. Chem. **56,** 1020 (1978); NG, A. S., and A. G. FALLIS: Comment: 7α-Acetoxydihydronomilin and mexicanolide: Limonoids from *Xylocarpus granatum* (Koeng). Can. J. Chem. **57,** 3088 (1979).
97. MARCELLE, G. B., and B. S. MOOTOO: 7α,11β-Diacetoxydihydronomilin, A new Tetranortriterpenoid from *Cedrela mexicana.* Tetrahedron Letters **22,** 505 (1981).
98. IR (KBr) 1670 and 1625 $cm^{-1}$ (conj. ε-lactone); UV ($CH_3OH$) $\lambda_{max}$ 208 nm ($\varepsilon = 13000$); $^{1}$H-NMR $\delta = 6.98$ and 6.00 with $J = 13$ Hz; $^{13}$C-NMR (DMSO-$d_6$) $\delta = 152.2$ d for C-1 and 119.0 d for C-2.
99. For nomenclature cp. OHOCHUKU, N. S., and D. A. H. TAYLOR: Chemical Shift of the Tertiary Methyl Goups in the Nuclear Magnetic Resonance Spectra of some Limonoids. J. Chem. Soc. (C), 864 (1969).

*100.* ADESOGAN, E. K., and D. A. H. TAYLOR: Limonoid Extractives from *Khaya ivorensis*. J. Chem. Soc. (C), 1710 (1970).

*101.* MONDON, A., B. EPE, and U. OELBERMANN: Zur Kenntnis der Bitterstoffe aus Cneoraceen, XV. Tetrahedron Letters 4467 (1981).

*102.* BEVAN, C. W., D. E. U. EKONG, T. G. HALSALL, and P. TOFT: West African Timbers. Part XX. The Structure of Turraeanthin, an oxygenated Tetracyclic Triterpene Monoacetate. J. Chem. Soc. (C), 820 (1967).

*103.* POLONSKY, J., Z. VARON, R. M. RABANAL, and H. JACQUEMIN: 21,20-Anhydromelianon and Melianon from *Simarouba amara* (Simaroubaceae); Carbon-13 NMR Spectral Analysis of $\Delta^7$-Tirucallol-Type Triterpenes. Israel J. Chem. **16,** 16 (1977).

*104.* LYONS, C, W., and D. R. TAYLOR: The Stereochemistry of Melianone and Sapelin F: Correlation with Bourjotinolon A. J. C. S. Chem. Comm. 517 (1975).

*105.* COTTERRELL, G. P., T. G. HALSALL, and M. J. WRIGLESWORTH: A Chemical Model for a Possible Oxidative Rearrangement in the Biosynthesis of Tetranortriterpenes: the Preparation of Methyl 3α-Acetoxy-7-oxoapotirucalla-14,24-dien-21-oate. Chem. Commun. 1121 (1967); The Rearrangement of 7α,8α- and 8,9-Epoxytirucallans. J. Chem. Soc. (C), 1503 (1970);. D. LAVIE and E. C. LEVY: Studies on Epoxides IV. Rearrangements in Triterpenoids. Tetrahedron Letters 2097 (1968).

*106.* BUCHANAN, J. G. ST. C., and T. G. HALSALL: The Conversion of Turraeanthin and Turraeanthin A into Simple Meliacins by a Route Involving an Oxidative Rearrangement of Probable Biogenetic Importance. J. Chem. Soc. (C), 2280 (1970).

*107.* MERRIEN, A., and J. POLONSKY: The Natural Occurence of Melianodiol and its Diacetate in *Samadera madagascariensis* (Simaroubaceae): Model Experiments on Melianodiol directed towards Simarolide. Chem. Commun 261 (1971); MERRIEN, A., B. MEUNIER, CL. PASCARD, and J. POLONSKY: Epoxide Configuration in 3β-Acetoxy-7α-hydroxy-14,15-epoxy-apotirucallane prepared from Tirucalla-7,24-dien-3-one. X-Ray Analysis. Tetrahedron **37,** 2303 (1981).

*108.* HALSALL, T. G., and T. A. STROKE: Abstract Book of 10th JUPAC Symposium of Natural Products, New Zealand, 1976 (cf. ref. (*103*)).

*109.* SINNWELL, V.: $^1$H- and $^{13}$C-NMR spectra measured at 400 and 100 MHz respectively in $CDCl_3$.

*110.* The following signals show the doubling particularly clearly: $\delta = 5.36/5.34$ (t,t, $J = 2.5$ Hz, 1H, 21-H) (Difference $\Delta = 8$ Hz), 2.55/2.53 (d,d, $J = 2.5$ Hz, 1H, exchangeable with $D_2O$, 21-OH) ($\Delta = 8$ Hz), 2.85/2.73 (d,d, $J = 8$ Hz, 1H, 24-H), ($\Delta = 44$ Hz), 1.05/1.00 (s,s, 3H, 13-$CH_3$) ($\Delta = 20$ Hz); in contrast, 23-H appears as a very broad multiplet centred at $\delta = 3.93$ ppm. The chemical shift of the protons 24-H and 13-$CH_3$ of the epimers is influenced most strongly by the α- or β-orientation of the OH group at C-21.

*111.* REMBERG, G.: Mass spectra measured with high resolution.

*112.* JOLAD, S. D., J. J. HOFFMANN, K. H. SCHRAMM, J. R. COLE, M. S. TEMPESTA, and R. B. BATES: Constituents of *Trichilia hispida* (Meliaceae), 4. Hispidols A and B, Two New Tirucallane Triterpenoids. J. Org. Chem. **46,** 4085 (1981).

*113.* LE COÇQ, C., and J.-Y. LALLEMAND: Precise Carbon-13 N.M.R. Multiplicity Determination. J.C.S. Chem. Commun. 150 (1981).

*114.* $^1$H-NMR (400 MHz, $CDCl_3$): $\delta = 2.54$ (d, $J = 6.2$ Hz; 1H, 24-H), 2.82 (dd, $J_1 = 6$, $J_2 = 2.3$ Hz; 1H, 23-H), 2.64 (dd, $J_1 = 2.3$, $J_2 = 8.1$ Hz; 1H, 22-H).

*115.* $^1$H-NMR (400 MHz, $CDCl_3$): Indor-Spectroscopy: $\delta = 1.06$ (d, $J = 7$ Hz; 3H, 21-$CH_3$), 1.30 (quintuplet, $J_1 = 7$, $J_2 = 8.1$, $J_3 = 11$ Hz; 1H, 20-H).

*116.* CERONI, M., and U, SÉQUIN: The Structure of the Antibiotic Hedamycin. IV. Relative Configurations in the Diepoxide Side Chain. Tetrahedron Letters 3707 (1979); ZEHNDER, M., U.SÉQUIN, and H. NADIG: The Structure of the Antibiotic Hedamycin. V. Crystal Structure and Absolute Configuration. Helv. Chim. Acta **62,** 2525 (1979).

*117.* BREEN, G. J. W., E. RITCHIE, W. T. L. SIDWELL, and W. C. TAYLOR: The Chemical

Constituents of Australian *Flindersia* Species. XIX. Triterpenoids from the leaves of *F. bourjotiana*. Austr. J. Chem. **19,** 455 (1966).

*118.* CHAN, W. R., D. R. TAYLOR, and T. YEE: Triterpenoids from *Etandrophragma cylindricum* Sprague. Part I. Structures of Sapelins A, and B. J. Chem. Soc. (C), 311 (1970).

*119.* JOLAD, S. D., J. J. HOFFMANN, J. R. COLE, M. S. TEMPESTA, and R. B. BATES: Constituents of *Trichilia hispida* (Meliaceae). 2. A New Triterpenoid, Hispidone, and Bourjotinolon A. J. Org. Chem. **45,** 3132 (1980).

*120.* OKOGUN, J. I., CH. O. FAKUNLE, D. E. U. EKONG, and J. D. CONNOLLY: Chemistry of the Meliacins (Limonoids): Structure of Melianin A, a new Protomeliacin from *Melia azedarach*. J.C.S. Perkin I, 1352 (1975).

*121.* LAVIE, D., M. K. JAIN, and (Mrs.) S. R. SHPAN-GABRIELITH: A Locust Phagorepellent from Two *Melia* Species. Chem. Commun. 910 (1967).

*122.* CONNOLLY, J. D., C. LABBÉ, D. S. RYCROFT, and D. A. H. TAYLOR: Tetranortriterpenoids and Related Compounds. Part 22. New Apotirucallol Derivatives and Tetranortriterpenoids from the Wood and Seeds of *Chisocheton paniculatus* (Meliaceae). J.C.S. Perkin I, 2959 (1979).

*123.* AKISANYA, A., C. W. L. BEVAN, T. G. HALSALL, J. W. POWELL, and D. A. H. TAYLOR: West African Timbers. Part IV. Some Reactions of Gedunin. J. Chem. Soc. 3705 (1961).

*124.* KRAUS, W., K. KYPKE, M. BOCKEL, W. GRIMMINGER, G. SAWITZKI, and G. SCHWINGER: Surenolacton, ein neues Tetranortriterpenoid-A/B-Dilacton aus *Toona sureni* [Blume] Merrill (Meliaceae). Liebigs Ann. Chem. **1982,** 87.

*125.* THOMS, H.: Die chemischen Inhaltsstoffe der Rutaceen, VII. Über den weißen Diptam, *Dictamus albus* L. Ber. Dtsch. Pharm. Ges. **33,** 68 (1923).

*126.* PAILER, M., G. SCHADEN, G. SPITELLER, and W. FENZL: Die Konstitution des Fraxinellons (aus *Dictamus albus* L.). Monatsh. Chem. **96,** 1324 (1965).

*127.* COGGON, P., A. T. MCPHAIL, R. STORER, and D. W. YOUNG: The Structure and Absolute Configuration of Fraxinellone, a Biogenetically Intriguing Terpenoid from *Dictamus albus* L. Chem. Commun. 828 (1969).

*128.* EKONG, D. E. U., C. O. FAKUNLE, A. K. FASINA, and J. I. OKOGUN: The Meliacins (Limonoids). Nimbolin A and B, Two New Meliacin Cinnamates from *Azadirachta indica* L. and *Melia azedarach* L. Chem. Commun. 1166 (1969).

*129.* CASSADY, J. M., and CH.-SH. LIU: The Structure of Calodendrolide, a Novel Terpenoid from *Calodendrum capense* Thunb. J. C. S. Chem. Comm. 86 (1972).

*130.* FUKUYAMA, Y., T. TOKOROYAMA,and T. KUBOTA: Total synthesis of Fraxinellon. Tetrahedron Letters 3401 (1972); TOKOROYAMA, T., Y. FUKUYAMA, T. KUBOTA, and K. YOKOTANI: Synthetic studies on Terpene Compounds. Part 13. Total Synthesis of Fraxinellone. J. C. S. Perkin I, 1557 (1981).

*(Received May 17, 1982)*

# Chemical and Biological Aspects of Marine Monoterpenes[1]

By S. NAYLOR, F. J. HANKE, L. V. MANES, and P. CREWS
Thimann Laboratories and Center for Coastal Marine Studies
University of California, Santa Cruz, U.S.A.

With 1 Figure

**Contents**

I. Introduction . . . . . . 190

II. Structural Variation . . . . . . 193

III. The Role of Halogens in Biogenesis . . . . . . 195
1. Introduction . . . . . . 195
2. Acyclic Structures . . . . . . 196
3. Monocyclic Structures . . . . . . 199

IV. Relationships Between Taxonomy and Occurrence of Structural Types . . . . . . 200
1. Introduction . . . . . . 200
2. The Plocamiaceae Family . . . . . . 201
3. The Rhizophyllidaceae Family . . . . . . 206
4. The Ceramiaceae Family . . . . . . 207
5. Degraded and Mixed Biogenetic Monoterpenes . . . . . . 209
6. Conclusion . . . . . . 209

V. Metabolite Transfer and Biological Activity . . . . . . 209

VI. Spectroscopic and Chemical Properties . . . . . . 212
1. Introduction . . . . . . 212
2. Halogen Content and Regiochemistry . . . . . . 215
3. Stereochemistry . . . . . . 219
4. Artifacts . . . . . . 221
5. Conclusions . . . . . . 222

---

[1] Dedication — This paper is dedicated to Professor PAUL SCHEUER (University of Hawaii) who is a distinguished pioneer in the field of marine natural products Chemistry.

VII. Physical and Spectroscopic Tables .......... 223
1. Table 13A—H: Summary of Structures and Carbon-13 NMR Chemical Shifts 223
2. Table 14: Physical Properties .......... 233
Acknowledgement .......... 236
References .......... 236

## I. Introduction

The earliest studies on monoterpenes date back to the 1800's. Nevertheless, this area remains of current interest since monoterpenes are abundant in volatile plant and seaweed oils (*1*). As the simplest terpenoids, monoterpenes are formed by the dimerization of isoprene ($C_5$) equivalents, and have carbon skeletons which are subdivided into regular or rearranged isoprene types (*2*). In spite of such a simple genesis, about ten different carbon frameworks had been identified by 1920, and by 1972 they included 30 from a pool of more than 400 compounds (*3*). In recent years the study of monoterpenes has attracted the attention of diverse groups ranging from organic chemists to ecologists (*4—7*). Not surprisingly, the monoterpene literature includes reviews on their biochemistry, biosynthesis, synthesis, or structure determination (*1—4, 8—11*).

An understanding of terrestrial monoterpene chemistry came well before that of marine monoterpenes. In fact, marine derivatives remained unobserved until 1955 when KATAYAMA reported seven common monoterpenes as constituents of the green alga *Ulva pertusa* (*12*). In subsequent studies (*13*), he also observed common monoterpenes from brown, green and red algae as summarized in Table 1. However, these first reports of marine monoterpenes were unexciting and did not stimulate further investigation. The first *unusual* marine monoterpenes were not observed until 1973 when FAULKNER reported isolation of the polyhalogenated metabolites (**1**)* and (**7**) from the sea hare *Aplysia californica* (*14, 15*). In the following year, CREWS and KHO reported another novel metabolite, cartilagineal (**22**), from a red seawed *Plocamium cartilagineum* (*16*). In the decade following these initial discoveries, 101 new marine monoterpenes were reported. These include isoprenoid compounds, degraded monoterpenes or monoterpenes of mixed biogenesis which are summarized in Table 13. The discovery of such a large number of new monoterpenes has greatly expanded the fascinating chemistry of this terpenoid class and has stimulated us to organize the review which follows. Our coverage of monoterpenes jointly considers their important chemical and biological aspects.

* See Table 13 for structures.

Table 1. *Common Monoterpenes from Marine Organisms*

| | 1,8-Cineol | p-Cymene | Linalool | Geraniol | α-Pinene | *d*-Limonene | 2-Terpineol |
|---|---|---|---|---|---|---|---|
| Algae (*13*) | | | | | | | |
| Chlorophyta | | | | | | | |
| *Ulva pertusa* | + | + | + | + | + | + | + |
| *Enteromorpha sp.* | + | + | + | + | − | − | − |
| *Codium fragile* | + | + | + | + | − | − | − |
| Phaeophyta | | | | | | | |
| *Sargassum sp.* | + | + | + | + | − | − | − |
| *Laminaria sp.* | + | + | + | + | + | + | − |
| Rhodophyta | | | | | | | |
| *Porphyra tenera* | + | + | + | + | + | + | − |
| *Digenea simplex* | + | + | + | + | + | + | − |
| Bryozoan (*20*) | | | | | | | |
| *Flustra foliacea* | cis-citral | trans-citral | citro-mellol | + | nerol | | |

Very few marine organisms are known to elaborate monoterpenoids. Only red seaweeds of the families Plocamiaceae and Rhizophyllidaceae or sea hares of the genus *Aplysia* contain oils *rich* in acyclic and cyclic monoterpenes. Some of their chemistry has been summarized in short reviews (*17*, *19*) and an overview of monoterpene structure types vs. species is presented in Table 2.

Table 2. *Sources of Marine Halogenated Monoterpenes*

| *Organism* | Monoterpene Type |
|---|---|
| [Anaspidea, Aplysiidae] | |
| *Aplysia californica, A. limacina* | linear |
| [Gigartinales, Rhizophyllidaceae] | |
| *Chondrococcus hornemanni, C. japonicus* | linear, cyclic |
| *Ochtodes crockeri, O. secundiramea* | cyclic |
| [Gigartinales, Plocamiaceae] | |
| *Plocamium angustum, P. cruciferum, P. oregonum, P. sandvicense* | linear |
| *P. mertensii* | cyclic |
| *P. cartilagineum, P. costatum, P. violaceum* | linear, cyclic |
| [Ceramiales, Ceramiaceae]* | |
| *Microcladia coulteri, M. californica* | cyclic |
| *M. borealis* | linear, cyclic |

* May not be direct sources of monoterpenes (see p. 208).

The structural variations that have been observed for marine monoterpenes are discussed in Section II. It is striking that, as shown by Table 1, examples are rare of monoterpenes which are common to both marine and terrestrial organisms. In addition, only one other compound, myrcene (**37**) found in *Chondrococcus sp.* (*21, 48*) (cf. Table 13D) is among the 400 monoterpenes listed by DEVON and SCOTT (*3*). Nearly all of the marine monoterpenes observed to date are halogenated; this can be contrasted with the fact that no halogenated monoterpenes have, to the best of our knowledge, been found in terrestrial organisms. Actually, utilization of halogens is known in only a few terrestrial microorganisms (*22, 23*). This may not be surprising considering that seawater contains high concentrations of chloride ($1.99 \times 10^7$ µg/L) and bromide ($6.8 \times 10^4$ µg/L) ions (*24*), whereas similar concentrations of halide ions are *not* found in the terrestrial environment.

The biogenesis of halogenated marine monoterpenes is an important topic and is explored in Section III. Carbonium ion induced cyclizations (*3*) are invoked to explain the formation of most non-halogenated monocyclic monoterpenes. By contrast, halonium ion induced cyclizations represent a novel ring-forming strategy employed by marine species. Thus, FENICAL (*25*) has pointed out that $H^+$ initiated cyclizations of geraniol can lead to a non-halogenated ochtodane derivative (**102**), one of several sex pheromones observed from the male boll weevil *Anthonomus grandis*. A parallel cyclization initiated by $Br^+$ can eventually lead to ochtodene (**60**) isolated from the red algae *Ochtodes secundiramea*. An analogous biosynthetic event possibly generates monocyclics of the plocamene series (see compounds in Table 13F) which represents a unique carbon framework without a terrestrial counterpart.

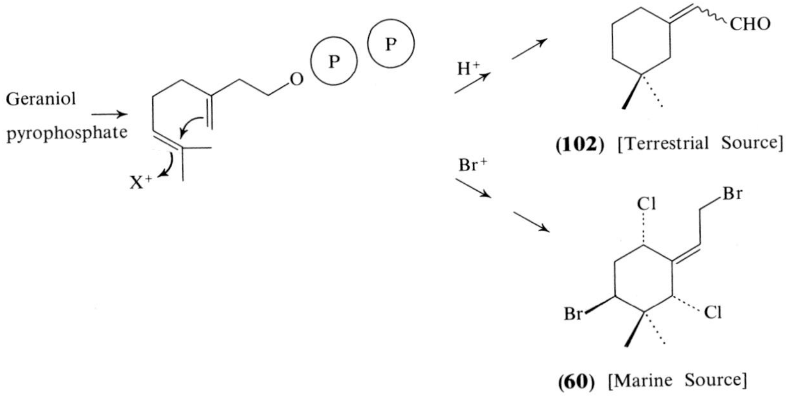

Distinctive odors have occasionally been noted for seaweeds that are rich in monoterpenes. Crude extracts of such seaweeds typically contain complex monoterpene mixtures. Moreover use of these compounds in a chemical approach to taxonomy might seem feasible and is discussed in Section IV.

By analogy to what is known for terrestrial monoterpenes*, marine monoterpenes should and actually do possess potent biological activity. Section V contains examples of marine monoterpenes having bio-activity ranging from fish toxicity (*16*) and fish antifeedant activity (*30, 32*), to various types of pharmacological activity (*31*).

Also considered in Section V are the subjects of metabolite transfer and metabolic fate of marine monoterpenes. The metabolic turnover rate of monoterpenes observed in the peppermint plant *Maytenus piperita* of 5—30 hours (*33*) contradicts earlier ideas that monoterpenes are inert waste products (*34*). A similar conclusion seems in order for marine monoterpenes. Although there is no direct evidence on this point, the statement by STALLARD and FAULKNER that the turnover rate of chemicals (which includes monoterpenes) in the digestive gland of the sea hare *Aplysia californica* "must be less than three months" is certainly relevant (*32*).

Complete structure determination of a monoterpene should be a relatively straightforward task given the well-developed methodologies employed in organic spectroscopy and the availability of $^{13}$C NMR data for model compounds (*35, 36*). However, as discussed in Section VI, polyhalogenated marine monoterpenes sometimes test the limits of both the spectroscopic and the chemical approach to determining halogen regiochemistry or the stereochemistry of other substituents in the molecule (*37*). Other NMR approaches have been utilized and include $^{13}$C (or $^{1}$H) spin-lattice relaxation time ($T_1$) measurements, lanthanide shift studies, and analysis of carbon chemical shifts and coupling constants.

## II. Structural Variation

We have subdivided the marine monoterpenes into two categories. These include a) isoprene dimers (92 compounds) and b) compounds which are degraded or of mixed biogenetic origin (9 compounds). A majority of the isoprene dimers have halogen content in excess of 50% of their total molecular weight, excepting myrcene (**37**), (**69**), (**70**), (**72**), (**73**), and (**74**). It is curious that relatively few (<30%) of the marine monoterpenes are

* It has been well documented that terrestrial plants use monoterpenes to attract seed-dispersing organisms (*27*), to repell predatory insects (*28*) and to inhibit the growth of potential competitors (*5, 29*). Insects also use monoterpenes in defense or communication (*6, 26*).

oxygenated; this contrasts greatly with what is observed for marine sesquiterpenes (*38*) and diterpenes (*39*).

The regular monoterpenoids (isoprene dimers) can be subclassified into four major structural categories. These consist of a linear head-to-tail framework (**A**) and three distinct monocyclic types (**B**, **C**, **D**). The four carbon skeletons along with their numbering system are shown in Figure 1. We have further subdivided the acyclic compounds into four sub-groups: monoenes (Table 13A), dienes (Table 13B), trienes (Table 13C), and myrcenes (Table 13D). Some systematic nomenclature has been used in the literature, but trivial names abound, especially in the case of myrcenes which are usually named as substituted myrcenes, eg. 7-chloromyrcene (**38**) (*21*).

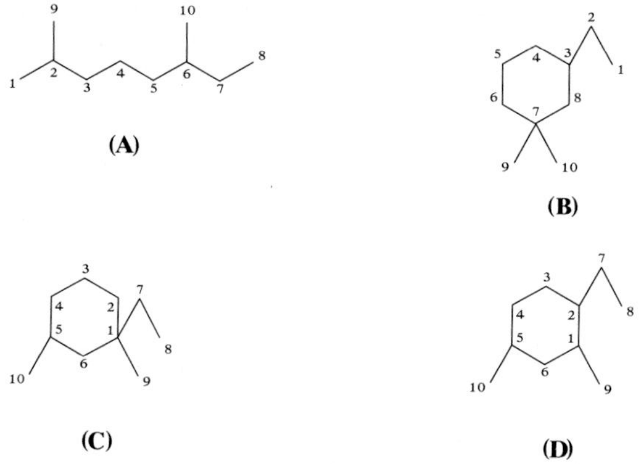

(**A**)

(**B**)

(**C**)

(**D**)

Fig. 1. Structural types of marine monoterpenes

Some interesting generalizations can be made about the various carbon frameworks (**A**—**D**). Even though the head-to-tail acyclic skeleton (**A**) is ubiquitous, only one compound, (**37**) occurs in both marine (Table 2) and terrestrial organisms (*2*). The monocyclic carbon skeleton (**B**) is also well known but is not represented by specific examples of metabolites found in both marine and terrestrial organisms (see pp. 191, 192). Finally, no terrestrial metabolite counterparts are known for compounds with monocyclic carbon skeletons (**C**) and (**D**), both of which have been isolated exclusively from *Plocamium* species. Chemical evidence discussed on p. supports the biogenetic assumption that carbon frame (**C**) rearranges to (**D**) by ethyl migration as opposed to other alternatives such as methyl migration in a 1,5-dimethyl-1-ethylcyclohexane precursor.

The structure determination of a chiral monoterpenoid natural product is not complete until its entire stereochemistry has been defined. The relative stereochemistry has been determined for all of the monocyclic and a majority of the acyclic monoterpenes. The absolute stereochemistry has been assigned in a few instances to crystalline compounds by X-ray analysis; this is indicated in Table 13A—H by brackets. In spite of this, we have preferred *not* to designate absolute stereochemistry for the monoterpenes reviewed in this article. Instead chiral centers are designated in terms of the $R^*$ $S^*$ notations on the following grounds. All of the reports in which absolute stereochemistry was proposed involved comparisons of statistical data for Friedel pairs. Occasionally this method gives ambiguous results (*40*). An important example in the monoterpene area is the absolute stereochemistry of violacene (**76**) (*87*) deduced by X-ray analysis as 1*R*, 2*R*, 3*S*, 4*R* which, when compared with the absolute stereochemistry from X-ray diffraction for (**83**) of 1*S*, 2*S*, 3*R*, 4*S* (*100*) suggests an enantiomer-like relationship between the two which is biogenetically unprecedented.

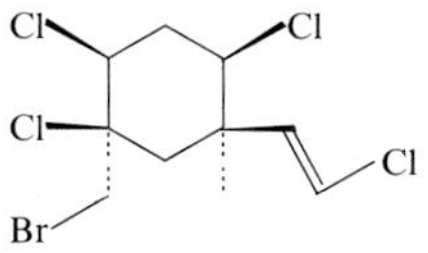

(**76**) 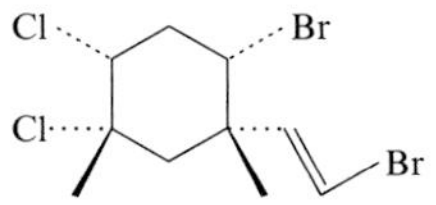 (**83**)

## III. The Role of Halogens in Biogenesis

### 1. Introduction

Most monoterpenes isolated from plants or animals are chiral and arise from enzyme controlled pathways (*41, 42*). Insights into the mechanisms associated with these pathways are the result of extensive biosynthetic experimental work* or are based on analogy to mechanisms deduced for organic reactions (i.e. biogenetic rationalizations). The majority of biogenetic schemes invoke the formation and capture of carbocations. One such example was discussed earlier on p. 192. Biosynthetic studies are so far lacking on how marine monoterpenes are elaborated. Proposed pathways to new marine monoterpene carbon skeletons or compounds with novel functionality are based on either biogenetic considerations** or analogy to biomimetic synthesis.

* Prior to 1977, there were over 800 publications dealing with terrestrial monoterpene biosynthesis (*41*).

** Discussions have appeared on the biogenesis of higher halogenated marine terpenoids (*38, 39, 47*).

The presence of halogen substituents represents a most significant difference between terrestrial and marine terpenoids. Unfortunately, little is known about the pathways of their *in vivo* synthesis. However, it has been shown experimentally that when bromide ion is added to enzymes isolated from red seaweeds various halogenated hydrocarbons and ketones are produced (*43*). Bromine is the predominant halogen in red seaweed sesquiterpenes and diterpenes. However, inspection of halogenated monoterpene structures reveals a new pattern in that many compounds contain either only chlorine or a high chlorine to bromine ratio (except in monoterpenes isolated from the Rhizophyllidaceae family).

It is interesting to note that two different types of haloperoxidase enzymes have been isolated from seaweeds. The chloroperoxidase enzyme is known to oxidize and incorporate Cl, Br, and I into substrate molecules by electrophilic substitution (*44, 45*). The more frequent occurrence of Cl, as compared with Br or I, in a marine natural product could reflect the natural relative halide abundance in seawater (see p. 192), presuming that factors involving selective membrane permeability or halogen-enzyme stability are unimportant (*46*). Alternatively, as purified seaweed bromoperoxidases will not oxidize $Cl^-$ to $Cl^+$ (*46*), it is reasonable to infer that organisms utilizing such enzymes might form halogenated natural products only by $Br^+$ initiated reactions. Such differences in haloperoxidase composition may be important in determining the predominance of a particular array of halogens in algal terpenes.

## 2. Acyclic Structures

On the basis of biogenetic and phyletic considerations we suggest that the acyclic monoterpenes (Tables 13A—D) can be divided into two categories, based upon the myrcene and ocimene skeletons respectively (Scheme I). It is striking that only halogenated myrcene derivatives are isolated from algae of the Rhizophyllidaceae family, while only halogenated ocimene derivatives are found in Plocamiaceae. This is remarkable given that loss of HOPP from geranyl pyrophosphate can directly generate either myrcene or ocimene. It is also noteworthy that the parent substance myrcene has been observed as a minor component of *Chondrococcus* (*48*), while ocimene itself, a known terrestrial monoterpene (*2*), has not yet been reported from red seaweeds.

Enzyme-catalyzed Markovnikov additions of $Br^+$ to a $(CH_3)_2C{=}CHR$ constellation often have been invoked as an important initiating step in the biogenesis of brominated sesquiterpenes (*38*). This process seems to be of similar importance in acyclic monoterpene biogenesis. For example, addition of $Br^+$ to myrcene followed by $Cl^-$ leads to dihalogenated

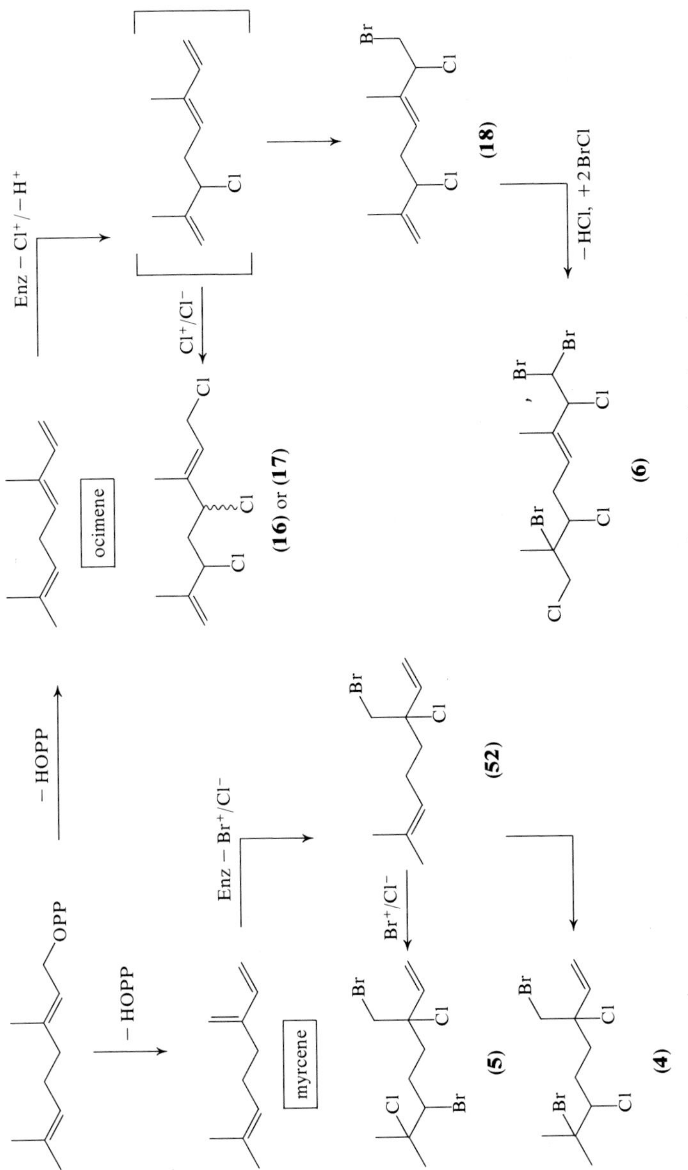


*Scheme I.* Biogenesis of halogenated myrcenes and ocimenes

myrcene (**52**) as shown in Scheme I [note also that loss of $H^+$ from the same intermediate leads to the monobrominated myrcenes (**49**) and (**41**)]. Enzyme-catalyzed addition of $Cl^+$, which does not find analogy in previous algal halide metabolite discussions, seems especially prevalent in Plocamiaceae metabolite biogenesis. Initial addition of $Cl^+$ to ocimene followed by $H^+$ loss, subsequent addition of $Cl^+$ and finally $Cl^-$ capture would lead to (**16**) or (**17**) (preplocamene B and C) as shown in Scheme I. Additional support for the suggestion of a biogenetic $Cl^+$ addition is provided by the observation that twenty-two acyclic compounds have a $C_5(Cl)C_6(Cl)$ constellation possessing either the $5(R^*), 6(S^*)$ or $5(R^*), 6(R^*)$ relative configuration. This can be rationalized by assuming enzyme-catalyzed addition of $Cl^+$ to $C_5{=}C_6$ (in an ocimene derivative) followed by $Cl^-$ capture as shown below. The most important aspects of the biogenesis of acyclic halogenated myrcenes and ocimenes can thus be summarized by the representative examples in Scheme I which includes pathways to compounds (**4**), (**5**), and (**6**) involving addition of either $Cl^+$, or $Br^+$, followed by loss of HX, or addition of $Cl^+/Br^-$, or $Br^+/Cl^-$.

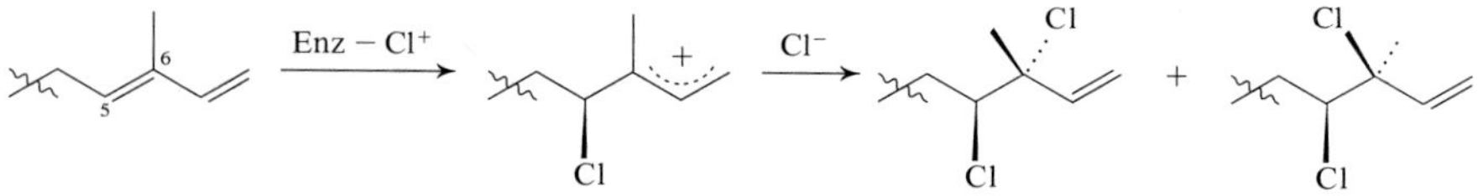

6 5 Enz – Cl⁺ + Cl Cl⁻ Cl Cl + Cl Cl

Results from biomimetic synthesis experiments also provide support for some of the biogenetic postulates shown in Scheme I. YOSHIHARA (*49*) observed 3-bromomyrcene (**103**) as a component of the product mixture which results when myrcene is treated with a reagent that generates $Br^+$ (Scheme II a). FAULKNER (*14*), using N-bromoacetamide as a source of $Br^+$ along wiht excess $Cl^-$, converted 2-acetoxy-6-methyl-2,5-heptadiene (**104**) to trihalogenated ketone (**105**) which was also prepared from compound (**1**) (Scheme II b).

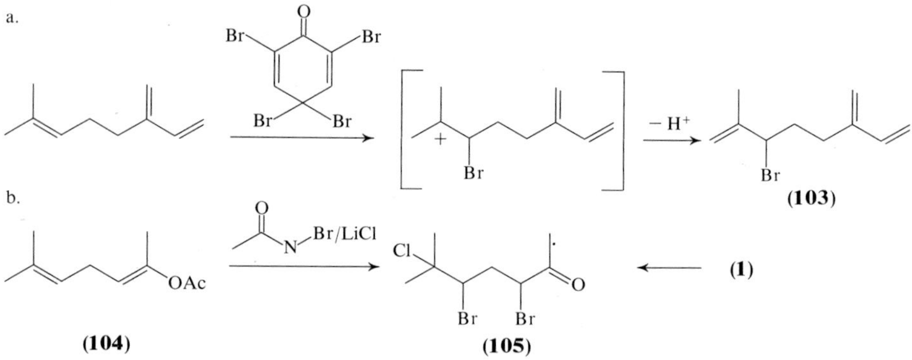

a. O Br Br Br Br + Br – H⁺ Br (**103**)

b. OAc O N Br/LiCl Cl O Br Br (**1**) (**104**) (**105**)

*Scheme II.* Biomimetic synthesis of halogenated acyclics

### 3. Monocyclic Structures

On the basis of biogenetic and phyletic considerations the monocyclic terpenes can also be divided into two categories, one including compounds with ring skeleton (**B**) and one with (**C**) and (**D**) (p. 194).

Compounds containing skeleton (**B**) or the ochtodane framework presumably arise by cyclization of myrcene. These natural products are only available from seaweeds of the Rhizophyllidaceae. Two separate biomimetic syntheses provide support for the biogenetic relationship suggested between myrcene and ring skeleton (**B**). Bromonium ion addition to myrcene yielded 5-bromo-4,4-dimethyl-2-vinyl-1-cyclohexane (**106**), an apparently unpublished ochtodane minor component of *Chondococcus hornemanni* (*49*) (Scheme III a). MASAKI (*50*) used a slightly different approach to convert myrcene to ochtodane (**70**), as illustrated in Scheme III b.

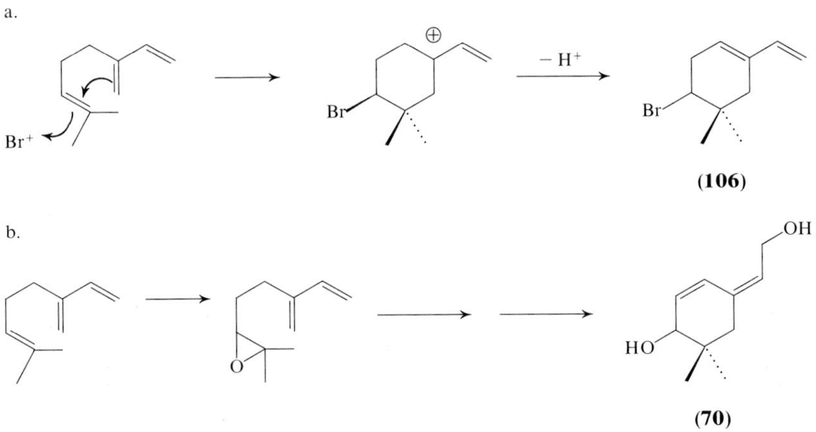

*Scheme III.* Biomimetic syntheses of ochtodanes

A biosynthetic link between the preplocamenes (**15**—**17**) and the monocyclic compounds of type (**C**) (Table 13F) is suggested by the close structural and stereochemical ties between some of their derivatives. The biogenesis of the preplocamenes from ocimene was outlined in Scheme I. Halonium ion addition initiates biogenetic cyclization of (**16**) or (**17**) to (**84**) or (**81**) as shown in Scheme IV. It is significant that the relative stereochemistry shown for (**16**) and (**17**) which are $C_3$ epimers nicely correlates with the relative stereochemistry observed in (**81**) and (**84**) which are epimeric at that same site. That monocyclic terpenes of skeleton (**C**) (Table 13F) might rearrange by ethyl migration to give compounds of type (**D**) (Table 13G) has been previously suggested by FAULKNER (*51*) who subsequently provided experimental support by observing the quantitative conversion of compound (**78**) to (**88**) on treatment with AgOAc/HOAc (*52*).

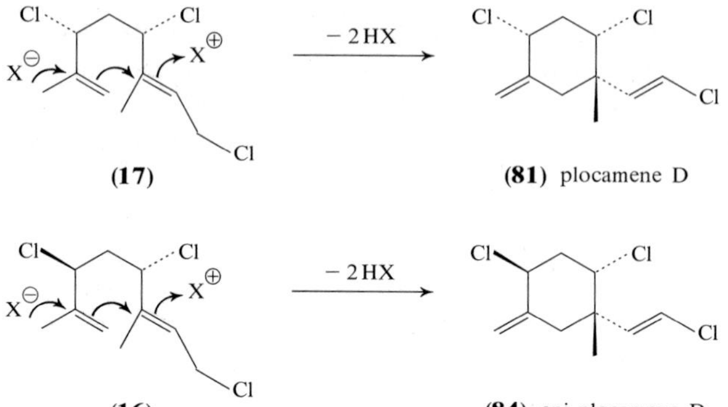

(**17**) $\xrightarrow{-2HX}$ (**81**) plocamene D

(**16**) $\xrightarrow{-2HX}$ (**84**) epi-plocamene D

*Scheme IV.* Biogenesis of 1-ethyl-1,3-dimethylcyclohexanes

## IV. Relationships Between Taxonomy and Occurrence of Structural Types

### 1. Introduction

Correlations exist between the taxonomic classifications of marine organisms and the structures of their terpenoid products. For example, it has been found that halogenated sesquiterpenes occur primarily in red seaweeds of the Rhodomelaceae family or their herbivorous predators (*38*). Similarly, as noted earlier (see p. 191), only a few marine organisms, such as certain species of red algae or sea hares, are sources of monoterpenes (Table 2). Furthermore, marine monoterpenes of mixed biogenesis, the cymopols (**95**—**101**) (Table 13H), are found in one species of green alga in the family Dasycladaceae (*55*).

The distribution of marine terpenoids raises the question: could variations within a single terpenoid group (e. g. monoterpenoids) still reflect phyletic differences? Several important prerequisites must be satisfied before this question can be considered. Firstly, a large array of monoterpenes must exist; this is satisfied by the 92 regularly constituted monoterpenes listed in Table 13. Secondly, structures assigned to the various metabolites must be secure since errors complicate attempts to correlate biogenetic capability with taxonomy. Next, the composition of the monoterpene mixtures isolated from the various species should not depend on habitat, i.e. there should be no ecotypes. Finally, the composition of the mixtures should not vary among individuals of a single species, i.e. there should be no chemotypes.

In at least one instance the composition of halogenated terpene mixtures from marine organisms has been shown to be constant in spite of ecological variations. KHO-WISEMAN (*56*) observed that specimens of *Plocamium cartilagineum* collected in the intertidal zone and specimens grown in culture from spores obtained from the same field site gave mixtures of monocyclic halogenated monoterpenes with similar composition*. However, the observation that different morphologically indistinguishable specimens of *P. violaceum* and *P. cartilagineum* collected in the same locale produce halogenated monoterpene mixtures of different composition (*58, 59*) violates one of the requirements described above**.

## 2. The Plocamiaceae Family

Seaweeds of the family Plocamiaceae, which are distributed world-wide, have been extensively investigated. Each of eight different *Plocamium* species contain halogenated monoterpenes of structure types (**A**), (**C**) or (**D**). At least three different monoterpenes occur in more than one species, a phenomenon which suggests a common biosynthetic link between them, and are listed in Table 3. The largest variety of halogenated monoterpenes has been isolated from *Plocamium cartilagineum,* the most cosmopolitan *Plocamium* species, with collections ranging from such diverse locations as the Pacific coasts of North America and Australia, the Mediterranean, the Isle of Wight and the inhospitable Janus Island of Antarctica.

The taxonomy of *Plocamium cartilagineum* has been difficult in that two distinct morphological forms seem to exist (*61*). One form is coarse, whereas the other is smaller and more delicate (*62*). But in 1967, DIXON combined the two previously recognized varietal forms into a single species. Unfortunately, the variations in monoterpene mixtures are too complex to permit a correlation with taxonomy. For example, *P. cartilagineum* from La Jolla, Ca (*64*) elaborates only acyclic monoterpenes of type (**A**), whereas

---

* HOWARD and co-workers (*57*) demonstrated that mixtures of halogenated sesquiterpenes and diterpenes from *Laurencia* species are constant even when the organism is grown under laboratory culture conditions where temperature and photo period are varied.

** The work of FENICAL and NORRIS (*60*) also provides an *indirect* observation of this kind. They found that specimens of *"Laurencia pacifica"* collected from the same vicinity could be arranged into three groups based upon differences in their mixtures of halogenated terpenes and $C_{15}$ fatty acids.

a collection from the United Kingdom (*52*) afforded compounds of type (**C**) and (**D**) and collections from different locations in central California (*17, 58*) contained only (**A**) type acyclic or (**C**) and (**D**) type monocyclic terpenoids as summarized in Table 4. An even more dramatic example of this phenomenon is the chemical variation observed among individuals. MYNDERSE and FAULKNER (*59*) observed three different chemical patterns in *P. cartilagineum* individuals from a single location whose extracts varied in their content of acyclic 1,3,7-trienes or 1,7-dienes of type (**A**) differing in *E* or *Z* stereochemistry about $C_1{=}C_2$.

Table 3. *Monoterpenes Common to more than One* Plocamium *Species*

| Compound | Algae | Location | Reference |
|---|---|---|---|
| 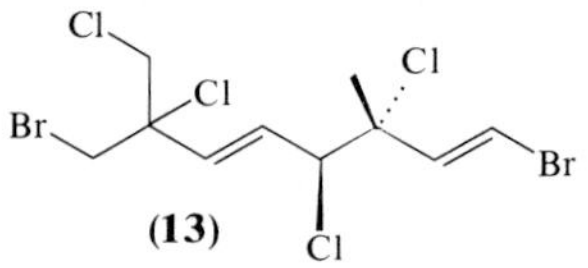 **(13)** | *Plocamium sp.* | Antarctica | (*92*) |
| | *P. oregonum* | Whidby Island, Washington | (*97*) |
| 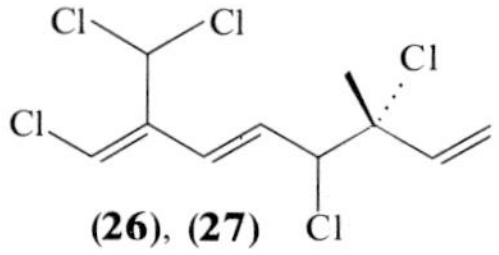 **(26)**, **(27)** | *P. cartilagineum* | San Diego, Ca | (*64*) |
| | *P. sandvicense* | Maui, Hawaii | (*17*) |
| 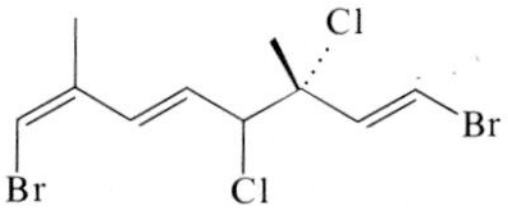 **(29)**, **(30)** | *P. cartilagineum* | San Diego, Ca | (*64*) |
| | | South Wales, U.K. | (*52*) |
| | *P. oregonum* | Whidby Island, Washington | (*97*) |

In 1977 it was discovered that certain central California collections of *P. violaceum* (*65*) contained only type (**A**) acyclic rather than the normal (**C**) or (**D**) monocyclic terpenes. This prompted a detailed investigation (*17, 58*) of the chemical patterns of *P. violaceum* over its habitat range from southern California to the Oregon coast. Examination of specific life history forms of this plant from different collection sites showed that its monoterpene content was constant with location and that the monoterpene content of plants from a specific location did not vary seasonally. At least three distinct chemical forms of *P. violaceum* were identified (*56, 58*), the most unusual form being one which elaborated only acyclic terpenoids and was found in only two localized, yet separate regions (*56*). These patterns can be seen from the data in Table 5. While the factors responsible for these intraspecific variations are unknown, the natural products chemistry of many other marine plants and invertebrates exhibits apparent chemical type differences (*66—67*).

Table 4. *Variation of* P. cartilagineum *Metabolites*

| Location | Major Components | Reference |
|---|---|---|
| 1. Bird Rock, La Jolla, Ca. | **(33)** **(35)** | (*64*) |
| 2. Isle of Whight or Overton, S. Wales, U.K. | **(78)** **(89)** | (*52*) |
| 3. Four Mile Beach, Santa Cruz, Ca. | **(76)** **(87)** | (*17*) |
| 4. Point Joe, Monterey, Ca. | **(15)** | (*58*) |

Intraspecific variations of halogenated monoterpenes from *P. cartilagineum* or *P. violaceum* provide an unexpected windfall for the chemist engaged in isolation studies. More specifically, within a given structural family, one isomer may be the major component at one collection site while a different isomer dominates at another location (Table 6). Consequently, before isolation is begun, it is advantageous to survey (e.g. by GC/MS) oils obtained from specimens collected at different sites to determine which samples will most easily provide specific compounds. As an example, the isolation of two $C_{10}H_{13}Cl_3$ isomers, plocamene D (**81**) and E (**92**) from *P. violaceum* (*68*) was conveniently accomplished by isolating the former from a Four Mile Beach collection and the latter from a Pigeon Point collection (Table 6). As might be expected, broadening the geographical range of collection sites for a specific species usually affords new metabolites. This is the case for compounds from *P. cartilagineum* (Table 6) and also for *Chondrococcus hornemanni* (see below and Table 7). It also follows that the monoterpenes isolated from predators of the Plocamiaceae such as *Aplysia californica* or *Aplysia limacina* may vary as the latter travel to different algal stands.

Table 5. *Geographical Survey*

| Chemical Type | Location | 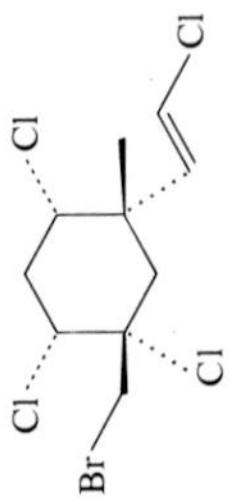 (76) | 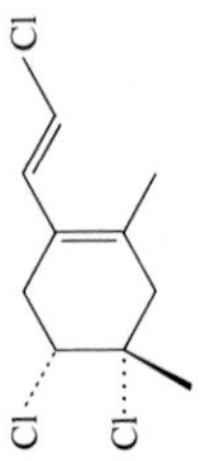 (86) | 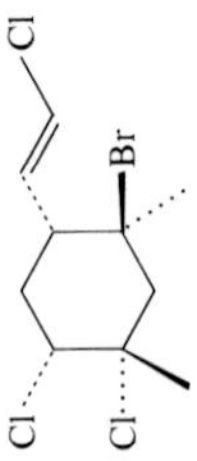 (87) | 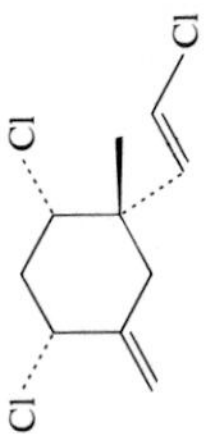 (81) |
|---|---|---|---|---|---|
| | **Southern Oregon** | | | | |
| α | Cape Arago, North | 94.8 | 1.3 | — | 0.3 |
| α | Cape Arago, South | 66.4 | 23.3 | — | 2.3 |
| α | Simpson's Reef | 69.4 | 22.5 | — | 2.8 |
| α | Harris Beach | 84.7 | 10.7 | — | — |
| | **Northern California** | | | | |
| | *Humbolt County* | | | | |
| α | Patrick's Point | — | — | — | 47.4 |
| | *Mendocino County* | | | | |
| δ | Todd's Point | 4.5 | 4.9 | 3.6 | — |
| β | Sea Rock Motel | — | — | — | — |
| β | Russian Gulch | — | — | — | — |
| | **Central California** | | | | |
| | *San Mateo County* | | | | |
| α | Montara Lighthouse | 32.3 | 11.4 | 5.6 | — |
| α | Moss Beach | 4.7 | 80.5 | 4.8 | — |
| α | Pescadero Beach | 96.6 | 3.4 | — | — |
| α | Bean Hollow | 32.5 | 33.8 | — | 33.7 |
| α | Waddell Creak | 85.2 | 6.5 | — | 8.3 |
| | *Santa Cruz County* | | | | |
| α | Davenport Landing | 26.3 | 42.1 | 17.6 | 8.7 |
| α | Bonny Doon | 34.1 | 41.3 | 15.4 | 6.7 |
| α | Four Mile Beach | 49.3 | 23.6 | 5.1 | 20.9 |
| α | Pigeon Point (I) | 18.7 | 33.4 | 16.1 | 3.0 |
| α | Pigeon Point (II) | 7.1 | 31.5 | 17.1 | — |
| | *Monterey County* | | | | |
| δ | Asilomar Beach | 4.5 | — | — | — |
| δ | Middle Reef Moss Beach | 70.4 | 9.1 | 6.1 | — |
| δ | Fanshell Beach | 52.9 | 13.5 | — | 24.5 |
| β | Point Joe | — | — | — | — |
| β | Pesdadero Point | — | — | — | — |
| | *San Luis Obispo Co.* | | | | |
| δ | San Simeon | 38.0 | 18.1 | — | — |
| δ | Leffingwell Creek | 77.4 | 19.3 | 2.3 | — |
| δ | Montana Del Oro | 38.3 | 30.7 | 18.9 | — |

*of* P. violaceum (*56*)

| **(82)** | **(84)** | **(92)** | **(17)** | **(16)** | **(15)** | (no No.) a, b x = Cl/Br |
|---|---|---|---|---|---|---|
| — | — | — | — | — | — | — |
| — | — | 6.0 | — | — | — | — |
| — | — | 5.5 | — | — | — | — |
| — | — | 4.6 | — | — | — | — |
| 52.6 | — | — | — | — | — | — |
| — | — | 46.1 | 31.8 | 9.1 | — | — |
| — | — | — | — | — | 92.5 | 7.5 |
| — | — | — | 37.0 | 8.4 | — | 54.6 |
| — | — | 45.7 | — | — | — | — |
| — | — | 9.8 | — | — | — | — |
| — | — | — | — | — | — | — |
| — | — | — | — | — | — | — |
| — | — | — | — | — | — | — |
| — | — | 5.3 | — | — | — | — |
| — | — | 2.5 | — | — | — | — |
| — | — | 1.1 | — | — | — | — |
| — | — | 28.8 | — | — | — | — |
| — | — | 44.3 | — | — | — | — |
| — | 45.9 | — | 10.5 | 13.0 | 11.1 | 15.0 |
| — | — | — | 1.3 | 4.9 | — | 8.0 |
| — | — | — | 2.0 | 4.3 | — | 2.8 |
| — | — | — | 3.2 | — | 71.4 | 12.4/12.0 |
| — | — | — | 24.5 | 29.5 | 3.6 | 30.9/11.5 |
| — | — | — | 16.3 | 7.7 | — | 19.9 |
| — | — | — | — | — | — | — |
| — | — | — | 3.5 | 3.1 | — | 6.4 |

Table 6. *Variation of Seaweed Monoterpene Isomers with Collection Locations*

| Species<br>Location | Relative % Composition | | |
|---|---|---|---|
| A. *P. violaceum* (*65, 68*) | | 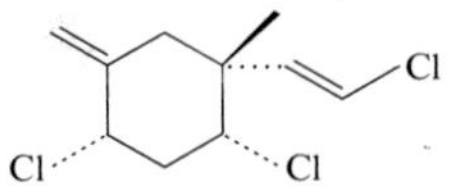 [structure] | 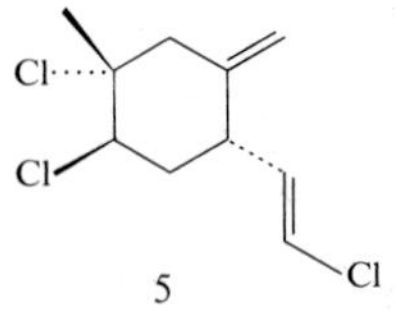 [structure] |
| Four Mile Beach, Ca. | | 95 | 5 |
| Davenport Landing, Ca. | | 62 | 38 |
| Pigeon Point, Ca. | | 9 | 91 |
| | 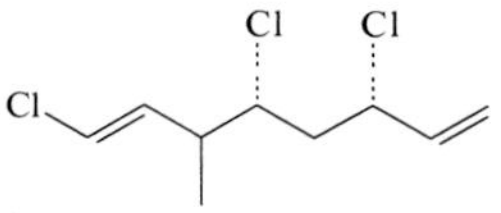 [structure] | 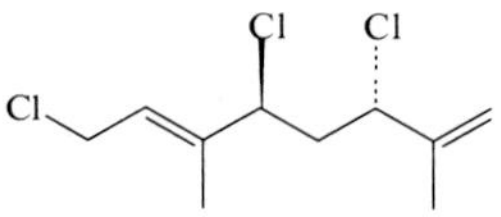 [structure] | 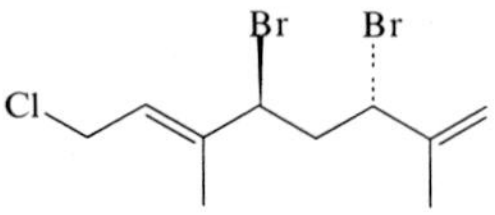 [structure] |
| Point Joe, Ca. | 4 | 0 | 96 |
| Pescadero Point, Ca. | 43 | 51 | 6 |
| Asilomar Beach, Ca. | 30 | 38 | 32 |
| B. *Plocamium cartilagineum* (*17*) | | 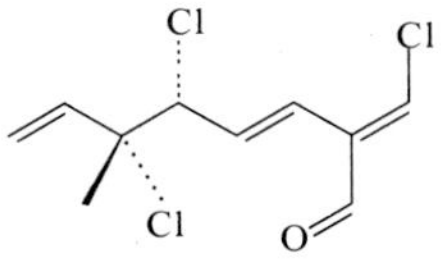 [structure] | 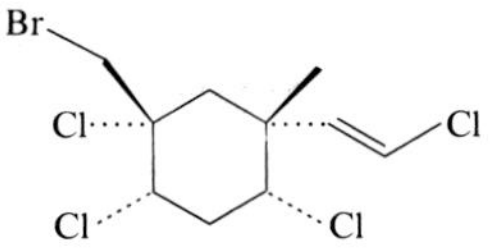 [structure] |
| Four Mile Beach, Ca. | | 0 | 100 |
| Davenport Landing, Ca. | | 100 | 0 |

## 3. The Rhizophyllidaceae Family

Halogenated monoterpenes have been isolated from the genus *Chondrococcus* and *Ochtodes* of the Rhizophyllidaceae family. The variation in the composition of monoterpenes from this family mirrors that observed for the Plocamiaceae. The best example is provided by *Chondrococcus hornemanni,* the most widely distributed species, whose major components vary from collection sites in Hawaii, Japan and Sri Lanka as shown in Table 7. Variations have also been noted for collections from locations in Hawaii which are only six miles apart.

Eighteen monoterpenes [all type (**B**) monocyclics] have been isolated from the two *Ochotodes* species investigated which are *O. secundiramea* (*69*) from the Caribbean and *O. crockeri* (*30*) from the Pacific. Interestingly, only two metabolites, chondrocole A (**55**) and chondrocole C (**58**) were common constituents of *Ochotodes* and *Chondrococcus* species (*70*).

Table 7. *Variation of* C. hornemanni *Metabolites*

| Location | Major metabolites, %'s | Reference |
|---|---|---|
| 1. Black Point, Hawaii | 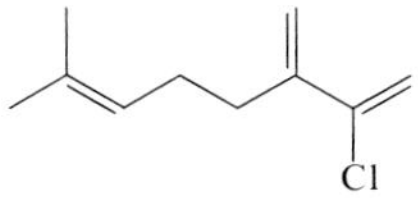 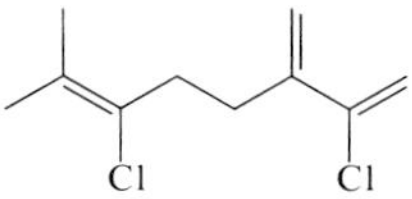 60%; 20% | (*95*) |
| 2. Halona blowhole, Hawaii | 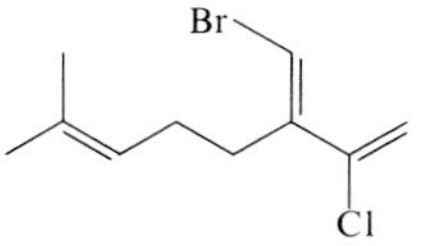 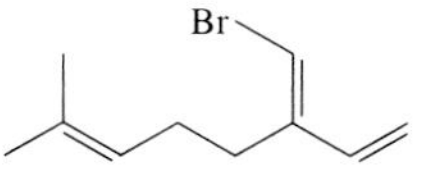 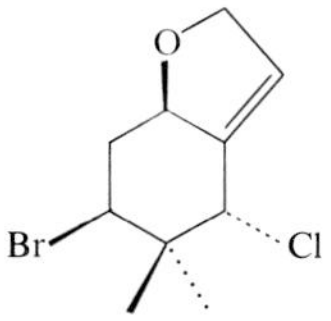 50%; 20%; 15% | (*95*) |
| 3. Sri Lanka | 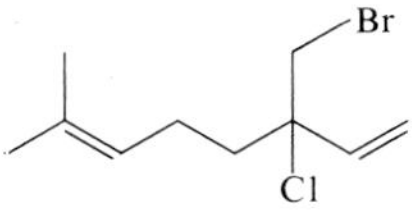 | (*101*) |
| 4. Amami Island, Japan | 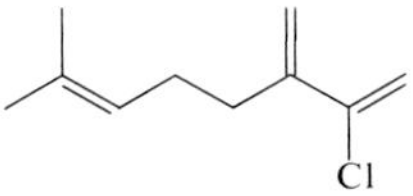 | (*48*) |

## 4. The Ceramiaceae Family

Proximity associations between *Plocamium* and *Microcladia* are common in the lower intertidal communities of central California (*72*). These alga are morphologically similar, yet, as noted below, they are taxonomically far apart. Common associated pairs found in central California are *Plocamium cartilagineum*/*Microcladia coulteria, P. cartilagineum*/*M. californica, P. violaceum*/*M. borealis,* and, in the Pacific northwest, *P. oregonium*/*M. borealis*. Hypotheses to explain these specific plant/plant associations include the requirement for a similar environment or the ability of both plants to resist predation by herbivores or competition by epiphytes, endophytes or parasites. The chemistry of *Microcladia* was originally investigated because it was intriguing to consider that in view of the association similar toxic compounds might exist in *Microcladia* and *Plocamium*. Suspicions of this chemical similarity were confirmed by

GC/MS examination of the mixtures of halogenated monoterpenes published in two separate reports (*17, 71*).

Violacene (*76*) has been observed by GC/MS (*71*) in semi-purified extracts of all *Microcladia* species *(M. californica, M. borealis,* or *M. coulteria)* available in central California. This observation is rather surprising because *Microcladia* (Ceramiales) is taxonomically quite distinct from seaweeds of the Plocamiaceae or Rhizophyllidaceae (both in Gigartinales) (*61*) which are the source for all the regular monoterpenes summarized in Tables 13A—G. This peculiarity led to the conjecture that *Microcladia* species are not responsible for the direct synthesis of halogenated monoterpenes found in them (*17*).

Similar mixtures of type (**C**) and (**D**) monocyclic terpenes are found in *M. borealis* and in *P. violaceum* which occur next to one another in central California. Similar mixtures of type (**A**) acyclic terpenes come from *P. oregonium* and from *M. borealis* which occur next to one another at Whidby Island, Washington. The yields of halogenated monoterpenes from *Microcladia* are uniformly much lower than those from *Plocamium,* but their concentrations are high enough to permit isolation and structural verification by $^{1}$H NMR analysis (*71*). The two most obvious deductions that might be drawn from these observations are (a) although taxonomically dissimilar, *Microcladia* and *Plocamium* can produce identical halogenated compounds or (b) *Microcladia* species are able to "concentrate"* halogenated metabolites produced by adjacent *Plocamium* plants. We favor the latter hypothesis because it is difficult to imagine that these two unrelated seaweeds have simultaneously evolved the complex, enzymatically controlled, multistep, biosynthetic capability discussed on pp. 196 to 200. Additional support for this view is provided by the low levels of monoterpenes observed in *Microcladia.* In an analogous observation. FENICAL (*73*) (U.C., San Diego, Scripps Institute) indicated to us some time ago that his examination of a *Microcladia* species failed to reveal halogenated monoterpene metabolites in specimens collected from the sea of Cortez where no *Plocamium* occurs**.

---

* No specific mechanism can be implied. However, it is conceivable that such a process could range from one involving lipophylic attraction between a plant's outer surface and nonpolar natural products dispersed in seawater, to one involving storage of exogenous metabolites in special locations within a seaweed.

** Several coralline algae from Japan (eg. *Marginiosporum aberanns, Amphiroa zonata* and *Coralline pilulifera)* were found to contain numerous halogenated sesquiterpenes in minute yields, all of which have been observed in *Laurencia* sp. (*74*). In a survey of halogenated compounds from algae in the Sea of Cortez, using GC-MS, RINEHART *et al.* (*75*) observed halogenated sesquiterpenes common to *Laurencia* seaweeds from a variety of other seaweeds including *Plocamium cartilagineum* and also in other organisms including a sponge, brittle star and bryozoan!

### 5. Degraded and Mixed Biogenetic Monoterpenes

An unusual bisnor-monoterpene alcohol (**93**) was isolated (*76*) from *Plocamium cruciferum* together with a regular monoterpene (**6**). The carbon skeleton of the former can arise from degradation of a type (**A**) compound by loss of either $C_8$ and $C_9$ or loss of $C_7$ and $C_8$. A GC/MS study of the algal extract indicated numerous $C_{10}$ compounds, which were not isolated or identified; this observation favors the possibility that (**93**) is of isoprenoid origin.

Seven prenylated bromohydroquinones, the cymopols (**95—101**) were isolated from the lipid extract of the calcareous alga *Cymopolia barbata* (*55*). The cymopols are of mixed biogenesis and probably arise by condensation of a geranyl derivative with a shikimate derived quinol.

### 6. Conclusion

It appears that no simple correlation exists between occurrence patterns of halogenated monoterpenes and taxonomy. MABRY (*77*) has suggested that variations in secondary metabolites are the result of varying ecosystem influences upon a species which produces a "physiologically different organism". The response of the organism might then be to produce a variation in secondary metabolites, with structural modifications including different carbon skeletons, ring systems, or different oxygenation and halogenation patterns. The extent to which this is true will destroy attempts to correlate secondary metabolites and organism taxonomy.

## V. Metabolite Transfer and Biological Activity

It seems remarkable that herbivorous opistobranch molluscs, such as *Aplysia sp.* actually feed on seaweeds which contain copious quantities of halogenated terpenes. Even more unusual is the observation that these sea hares appear to concentrate metabolites in their digestive glands (*78*)*. Not unexpectedly, the digestive glands of *Alysia californica* and *A. limacina,* which graze upon *Plocamium,* contains large quantities of halogenated monoterpenes. For example, IMPERATO and co-workers (*81*) found that 24% (1.2 g.) of the extract oil from the digestive glands of five adult *A. limacina* consisted of the pentachloro monoterpene (**27**).

* The other major instances of secondary metabolite transfer between marine invertebrates and their prey are the numerous chemical studies on sponge-nudibranch associations. [See (*79, 80*) and references therein.]

Unfortunately, there is no direct evidence for dietary transfer of monoterpenes. To date, only acyclic type (**A**) compounds have been observed in *Aplysia;* three of these metabolites are also common to *Plocamium cartilagineum* collected from the same vicinity (see Table 8). Another relevant observation is that *A. californica* collected at Isla Carman, Mexico, a location devoid of marine algae, does not contain any halogenated terpenes (*78*).

Table 8. *Monoterpenes Common to Sea Hares and* Plocamium

| Compound | Organism | Location | References |
|---|---|---|---|
| Cl, Cl, Cl, Cl, Cl (**27**) | *A. limacina*<br>*P. cartilagineum* | Bay of Naples<br>1. Bay of Naples<br>2. San Diego | (*81*)<br>(*81*)<br>(*64*) |
| Cl, Cl, Br, Br, Br, Cl (**7**) | *A. californica*<br>*P. cartilagineum* | San Diego<br>San Diego | (*15*)<br>(*64*) |
| Cl, OH, Br, Br, Br (**1**) | *A. californica*<br>*P. cartilagineum* | San Diego<br>San Diego | (*14*)<br>(*32*) |

The presence of novel sea hare monoterpenes, which are closely related to algal metabolites may indicate the capability of the animal to perform certain biosynthetic transformations* (see Table 9). Thus, the acetate (**23**), isolated exclusively from *A. californica (83)*, may be a transformation product of either cartilagineal (**22**) or the trienes (**24**) or (**28**). The isolation from *A. californica* of monoterpene (**9**) (*83*) which was not detected in over two hundred samples of *Plocamium* collected in the vicinity of the sea hare supports this hypothesis**.

* It has been shown from labelling studies that *A. californica* can carry out *in vivo* synthetic modifications on halogenated sesquiterpenes (*82*).

** However, FAULKNER (*83*) pointed out that the algal source of the metabolites probably has not been discovered.

Table 9. *Monoterpenes Unique to Sea Hares*

| Compound | Organism | Reference |
|---|---|---|
| 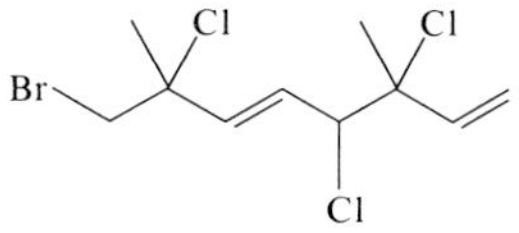 **(8)** | *A. californica* | (*83*) |
| 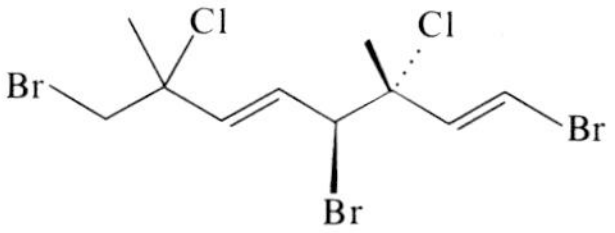 **(9)** | *A. californica* | (*83, 92*) |
| 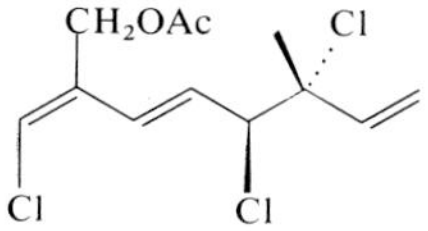 **(23)** | *A. californica* | (*83*) |
| 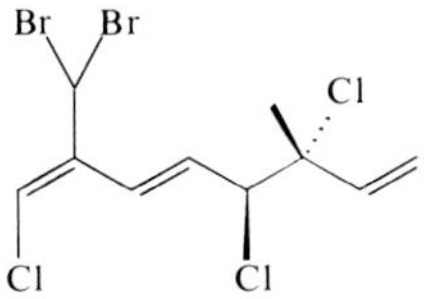 **(24)** | *A. limacina* | (*91*) |

The relatively high yield of halogenated terpenes from *Aplysia* digestive glands indicates that sea hares can concentrate and store algal metabolites. It has been suggested that these compounds are chemical deterrents against sea hare predators (*32*). This view is supported by the varied biological activities of halogenated monoterpenes which include ichthotoxicity (*68*), eg. (**76**), fish antifeedant activity (*30*), eg. (**66**), antimicrobial activity (*32*), eg.* (**1**), antifungal activity (*31*), eg. (**86**), antiinsect activity (*68*), eg. (**86**), and pharmacological activity in model assays including anti-cell division in sea urchin eggs (*31*), eg. (**36**).

* A direct correlation has been observed between the presence of halogenated metabolites and antimicrobial activity in approximately 1200 marine organisms (*84*).

# VI. Spectroscopic and Chemical Properties

## 1. Introduction

Two techniques, nuclear magnetic resonance and mass spectrometry, have been used routinely in the characterization of a majority of the marine monoterpenes. In some instances, ambiguities from spectroscopic analysis were clarified by X-ray analysis (*85*) or by chemical interconversion and degradation studies. While X-ray crystallography has been carried out on only a small number of compounds* (*85*), it includes representatives from almost every structural sub-category.

A wide range of structural and stereochemical features have been established by analysis of spectral parameters such as NMR coupling constants ($J$'s), NMR chemical shifts ($\delta$'s), mass spectral parent ion cluster shapes ($M^+$) or mass spectral fragmentation peaks. The practice of comparing spectral data of a halogenated marine monoterpene with those in reference collections has been rarely employed because the combinations of halogen substituents differed greatly from those found in standard compounds.

Nuclear magnetic resonance data, mainly $^1$H and sometimes $^{13}$C, have provided crucial structural information for essentially every marine monoterpene reported so far. Fortunately, many marine monoterpenoids give relatively simple NMR spectra**. Generally, chemical shifts and coupling constants of individual protons could be entirely assigned, especially in spectra taken at high magnetic fields. Unfortunately, $^1$H chemical shifts were not useful in the difficult problem of differentiating between CHBr and CHCl, while $^1$H $J$'s were extremely useful in deducing substituent stereochemistry. Assignment of stereochemistry in the $-CH(X)-CH_2-$ subunit found in ring types (**B**), (**C**), and (**D**) was commonly accomplished by straight-forward application of the Karplus rule for evaluating $^3J$'s. References to the publications containing $^1$H NMR chemical shifts and coupling constants for the monoterpenes reviewed here are listed in Table 14.

The availability of $^{13}$C NMR data has proved to be invaluable in solving difficult monoterpene regiochemical and stereochemical problems and in providing partial structures. For example, $^{13}$C NMR spectroscopy is especially useful in solving such difficult problems as determining the location of tetrasubstituted double bonds [as in (**86**)] or the stereochemistry about a trisubstituted double bond [as in (**15**)—(**17**)], or defining the structure near oxygen especially in ethers [as in (**55**)]. Inconsistencies noted

* As pointed out on p. 195 no absolute stereochemistry is inferred in this review.

** See references (*71, 86*) for examples of actual spectra.

between the $^{13}C$ NMR spectra of a compound and its published structure have stimulated reinvestigation of the initially incorrect structures for such seminal monoterpenes as violacene (**76**) (*87*), and chondrocole A (**55**) (*88*).

When $^{13}C$ spectra are available for appropriate model compounds, deductions about the structure of a newly isolated monoterpene natural product are facilitated. Fairly extensive reference $^{13}C$ NMR parameters are available for non-halogenated monoterpenes (*35*), but the information about halogenated model compounds or halogenated monoterpenes is skimpy (*36*). Consequently we have summarized in Table 13 $^{13}C$ NMR chemical shifts which are known for more than 60% of the regular monoterpene structures.

Mass spectrometry has been of great utility in studying halogenated monoterpenes because both Cl and Br have two naturally abundant isotopes. Comparing observed multiple clusters for both $M^+$ and fragment ion clusters with those expected for different halide atom combinations [see Figure 128 in (*89*)] has been a standard and extremely useful practice. Unfortunately, overinterpretation of fragment ion clusters has led to errors in halogen regiochemical assignments. Another difficulty has been that many highly halogenated compounds fail to exhibit a clear $M^+$. However, a careful evaluation of odd m/z fragment ion peaks generates, by construction of a fragmentation map, fairly reliable projections of the unknown X (Cl vs. Br) and the non-detectable $M^+$ mass. An example is outlined in Scheme V for oregonene-A (**13**).

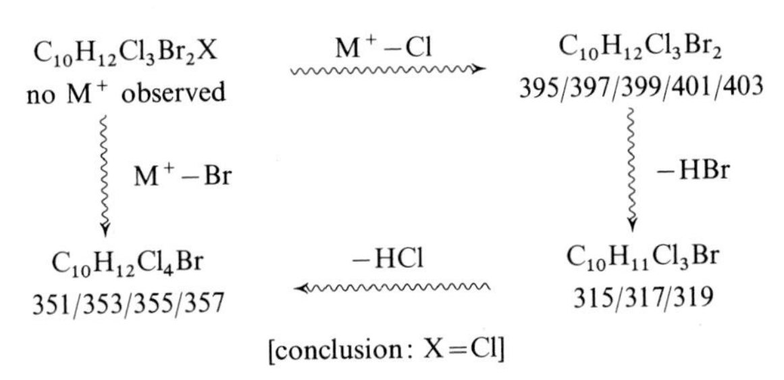


*Scheme V.* Determination of halogen content in oregonene A (**13**) when no $M^+$ is observed

Ultraviolet and infrared spectra have not been of any special value in structure elucidation of marine monoterpenes. Yet $\lambda_{max}$ values of conjugated dienes can yield insights about the conformation of the diene chromophore (*90*) present in a number of monoterpenes, such as myrcene (**37**)*.

---

* The diene moiety of myrcene was assigned the *s-trans* configuration based upon detailed $^1H$ NMR coupling constant analysis (*91*).

Consideration of the combined $^{1}H$, $^{13}C$ and mass spectroscopic properties for the various regular monoterpenes in Table 13 reveals a simple procedure by which a compound can be placed into structural category (**A**), (**B**), (**C**), or (**D**). Actually, a minimum of information is needed, as summarized in Scheme VI, which includes the following key facts: the molecular formula (from an m. s. $M^{+}$, from $^{13}C$ NMR, or from combustion analysis) which is also used to calculate an unsaturation number, a count of $sp^2$ carbons (eg. the double bond count from $^{13}C$ NMR data), a count of isolated methyls (eg. $-\overset{|}{\underset{|}{C}}-CH_3$ from $^{13}C$ and $^{1}H$ NMR data), and information about the presence of a $-CH-C=CHCl$ or $-C(CH_3)=C-C=CHCl$ group. This strategy is also useful for rapidly determining whether a newly isolated halogenated marine monoterpene falls into an existing structural class or represents a new one.

| Structure Type | Key Sub-Units | Important Spectroscopic Properties |
|---|---|---|
| (**A**) | | Unsaturation number equal to the number of C=C ($^{13}C$ NMR) |
| (**B**) |  | $^{13}C$ ($CH_3$'s δ 20—30)<br>$^{1}H$ ($CH_3$'s δ 0.8—1.4) |
| (**C**) | 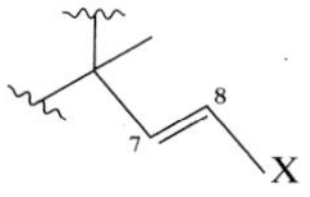 7, 8, X | $^{13}C$ ($CH_3$ δ 26—30; $C_7$ δ 134—140)<br>$^{1}H$ ($CH_3$ δ 1.2—1.3; $H_7$ δ 5.9—6.6) |
| (**D**) | 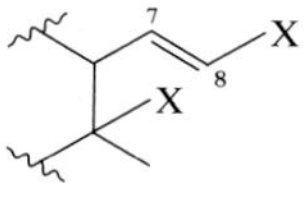 7, 8, X, X | $^{13}C$ ($CH_3$ δ 28—32; $C_7$ δ 130—133)<br>$^{1}H$ ($CH_3$ δ 1.9; $H_7$ δ 5.9—6.1, dd)<br>m/z = 131 (base peak) |
| | 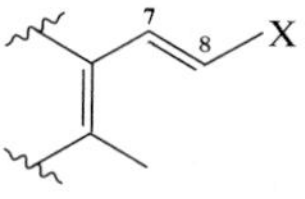 7, 8, X | $^{13}C$ ($CH_3$ δ 18—19; $C_7$ δ 6.7—6.8)<br>$^{1}H$ ($CH_3$ δ 1.7; $H_7$ δ 5.7—5.9, d)<br>m/z = 131 (base peak) |

*Scheme VI.* Correlation of structure types (**A**), (**B**), (**C**), and (**D**) with diagnostic spectral features

Fortunately structures of only seven monoterpenes have needed revision due to errors in interpreting spectroscopic data or the results of chemical interconversions. The original and corrected structures are

*References, pp. 236—241*

summarized in Table 10. It should be emphasized that corrections were not needed for any *carbon framework,* but rather involved reassignment of stereochemical features, halogen regiochemistry or halogen content.

## 2. Halogen Content and Regiochemistry

Accurate determination of the total number of halogens and their exact location represents the most difficult problem associated with structure elucidation of marine monoterpenes. It is usually advantageous to cross-check a molecular formula obtained from mass spectral data with one which can be determined from $^{13}C$ NMR data (see p. 214). An accurate count of $-CHX$- or $-CH_2X$ groups can easily be made from NMR data, especially from $^1J_{CH}$ values. By contrast, differentiation between Cl and Br in a mixed halogenated metabolite is not a trivial problem, yet is relevant for more than 70% of the known monoterpenes. Some of the spectroscopic and chemical approaches that have been used for these purposes will now be discussed.

The process used to arrive at the correct structure of (**9**) illustrates the difficulties in establishing the total halogen count. In the original report on the acyclic diene (**9**) it was noted that an $M^+$ peak was not observed and that fragment ion peaks supported the formula $C_{10}H_{13}Cl_2Br_2X$ (*83*). Similarities in both the $H_5$ NMR signals and the mass spectral fragmentations between (**7**) and (**9**) led to assignment of X=Cl to $C_5$ and thus a halogen count of $Cl_3Br_2$ for (**9**). When $^{13}C$ NMR spectral information became available (*92*) it was clear, on comparing the chemical shifts of $C_5-X$ in (**9**) (60.1) with (**10**) (67.6) whose structure had been established by X-ray analysis, that revision was in order, and a new halogen count of $Cl_2Br_3$ was proposed.

Analysis of intense mass spectral fragment ion peak clusters has been of special value for assignment of Br and Cl in type (**A**) compounds. Illustrative examples are observation of the fragment m/z at 89/91 (*64*) for compounds (**23**)—(**28**) and observation of the fragment m/z at 123/125 (*92*) for compounds (**11**)—(**12**) which indicate attachment of Cl's to, respectively, $C_5$ and $C_5$ and $C_6$.

Carbon-13 NMR spectroscopy has been very useful in determining halogen regiochemistry and appears to be far superior to $^1H$ NMR spectroscopy for this purpose. Crews and co-workers (*37*) have generated a complete set of substituent constants for $C-X$ in both acyclic and cyclic systems. Surprisingly, the difference in chemical shift between $C-Br$ and $C-Cl$ varies considerably with carbon type and also depends on whether carbon is part of an acyclic or cyclic array. These trends are summarized in Table 11 from which it can be seen that the two can be distinguished most reliably at a primary site in an acyclic molecule and least reliably at a tertiary site in a cyclic molecule.

Table 10. *Summary of Structures That Have Been Revised*

| Original | Revised |
| --- | --- |
| 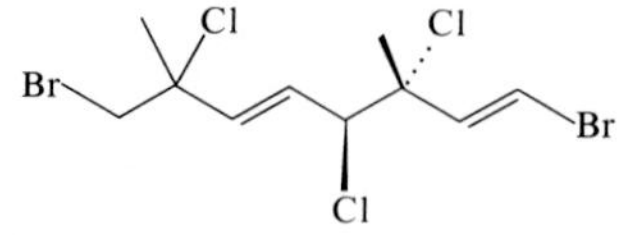 | 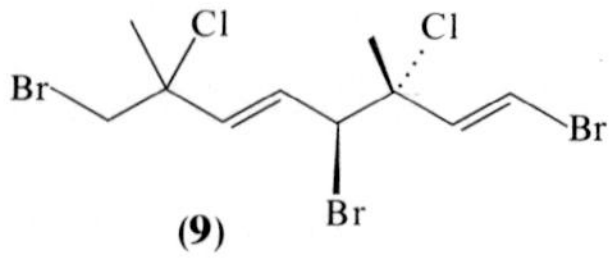 (**9**) |
| 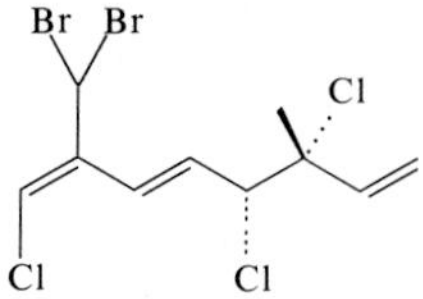 | 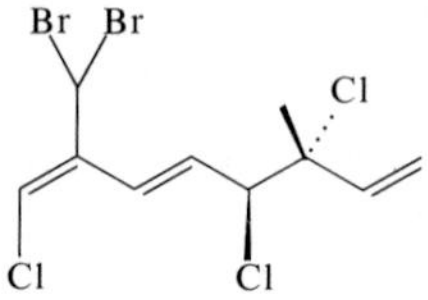 (**24**) |
| 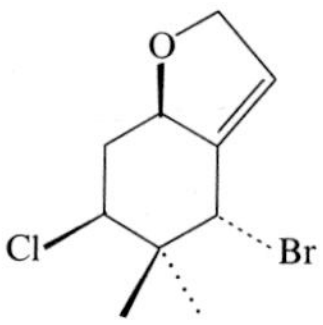 | 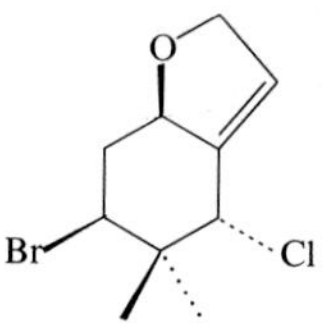 (**55**) Chondrocole A |
| 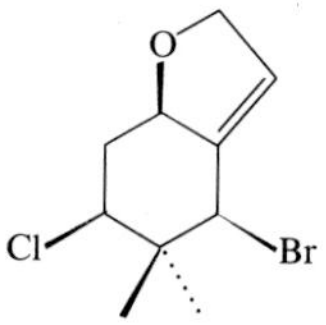 | 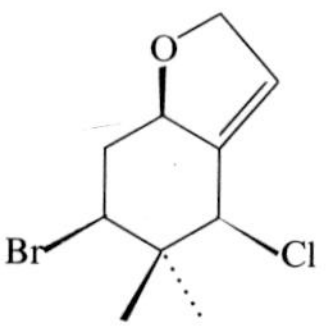 (**56**) Chondrocole B |
| 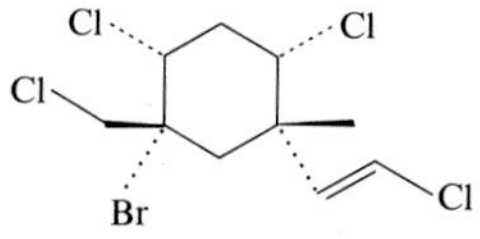 | 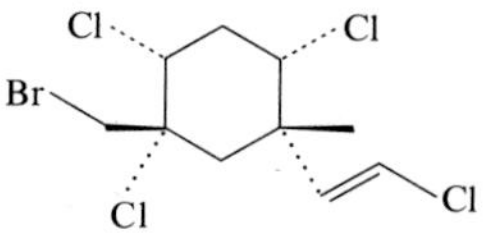 [**76**] Violacene |
| 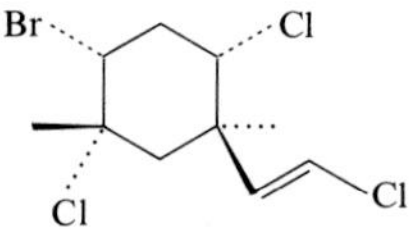 | 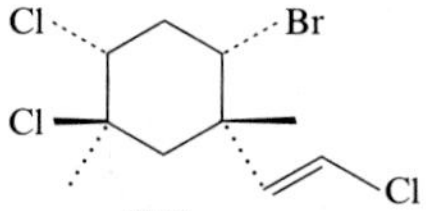 (**80**) |
| 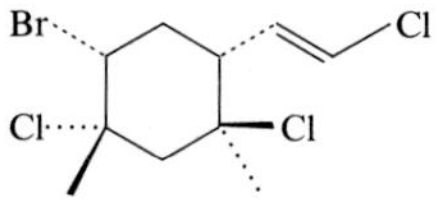 | 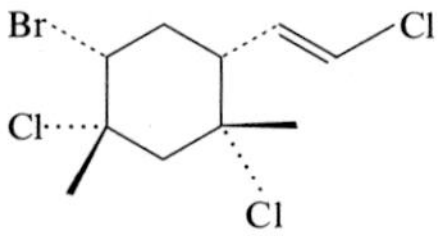 (**91**) |

*References, pp. 236—241*

Table 11. *$^{13}$C NMR Chemical Shift Differences for Br vs. Cl at Cα: $\Delta\delta C\alpha$ (eg. $\delta C_{Br} - \delta C_{Cl}$)*

| Type | Acyclic | Cyclic |
|---|---|---|
| $RCH_2X$ | 12 | — |
| $R_2CHX$ | 8 | 4 |
| $R_3CX$ | 3 | 1 |

$^{13}$C NMR chemical shifts suggested that the $-CH_2Cl$ moiety originally proposed (*93*) for violacene (**76**) due to an apparent $M^+ - CH_2Cl$ mass spectral fragment peak was questionable. The appropriate reference shifts are summarized below; the revised structure (**76**) based on X-ray analysis (*87*) is consistent with the $^{13}$C NMR shifts in (**13**). A procedure utilizing $T_1$'s was the basis for locating the halogens in the structure originally proposed for mertensene (**80**) (*94*). However, a revised structure has been recently suggested by CREWS (*37*) based upon an analysis of substituent $^{13}$C NMR additivity parameters. Moreover, direct comparison of the $-CHX$ $^{13}$C shifts of (**80**) to (**83**), whose structure was established by X-ray crystallography, indicates that the revised structure (**80**) provides a better fit of the observed shifts (cf. Table 10).

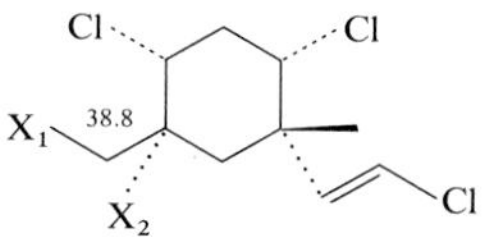

$X_1 = Cl$, $X_2 = Br$ (**76a**)
$X_1 = Br$, $X_2 = Cl$ (**76**)

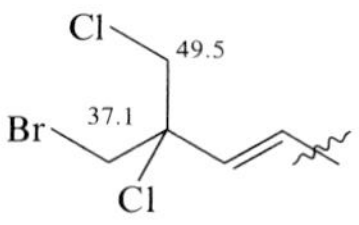

(**13**)

Difficulties were also encountered in assigning Cl and Br to $C_6$ or $C_8$ in chondrocole A (**55**) and chondrocole B (**56**) (cf. Table 10). These compounds were shown to be $C_8$ epimers from $^1$H NMR analysis. The original placements of the halogens were based upon the appearance of an intense $M^+ - Br$ fragment ion, which was interpreted as involving the loss of an allylic bromine, and were consistent with the biogenetic relationship envisioned between the acyclic compounds (**40**) & (**41**) and (**55**) & (**56**), respectively. The structure of chondrocole A was subsequently revised as the result of its oxidation to the known monoterpene chondrocolactone (**57**), whose structure had been determined by X-ray analysis (*88*). Using available spectroscopic data, such as the $^1$H NMR spectra of (**55**), (**56**), and (**57**) of Table 14, and the epimeric relationship between (**55**) and (**56**) it seems obvious that the structure of chondrocole B should be (**56**) (see Table 10). Analogously, for a similar metabolite (**59**) two structures have been proposed, differing in the attachment of Cl and Br to $C_6$ and $C_8$ (*36*, *96*). CREWS (*37*) has suggested that even after analysis of $^{13}C$ NMR shifts, halogen regiochemistry in (**59**) can not be decided.

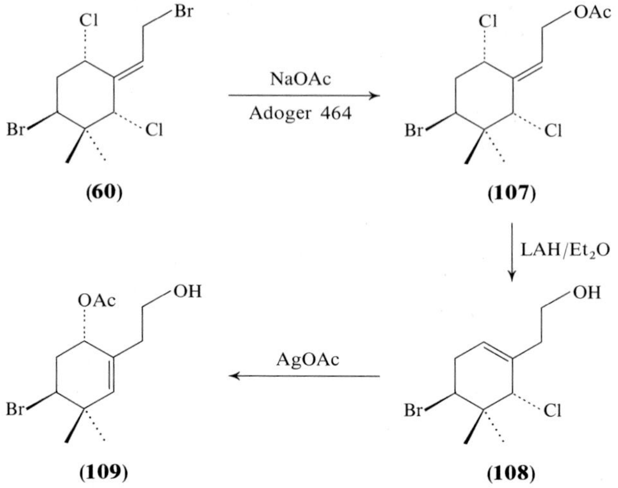

NaOAc / Adoger 464: (**60**) → (**107**)

$LAH/Et_2O$: (**107**) → (**108**)

AgOAc: (**108**) → (**109**)

Some workers have preferred to rely on chemical degradation schemes to locate Br vs. Cl. A representative example is provided by the approach used by FENICAL (*69*) in unambiguously determining the structure of ochtodene (**60**). The key reactions involved the conversion of (**60**) to (**109**). However, the interesting observation that violacene (**76**) when treated with $Li_2CO_3$/DMF gave a mixture of (**110**) and (**111**) (*87*) suggests that some degradations may give ambiguous information. Furthermore, the observed transformation (*94*) of mertensene (**80**) to (**112**) with $CrClO_4$/DMF is not consistent with the structural revision proposed for (**80**) based upon $^{13}C$ NMR arguments (*37*).

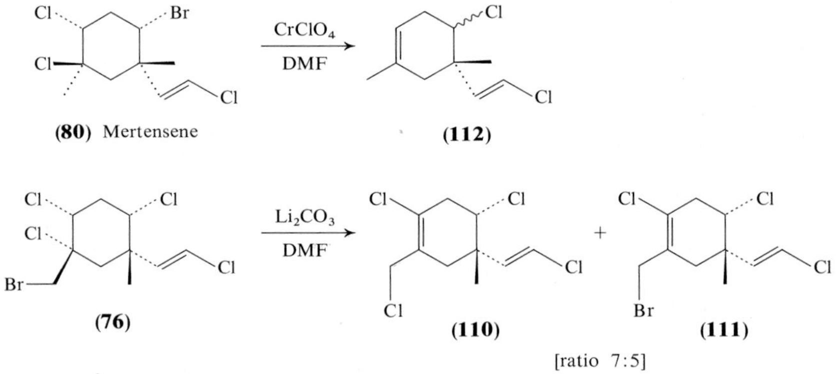

$CrClO_4$ / DMF: (**80**) Mertensene → (**112**)

$Li_2CO_3$ / DMF: (**76**) → (**110**) + (**111**) [ratio 7:5]

*References, pp. 236—241*

## 3. Stereochemistry

NMR parameters have provided useful insights into the stereochemistry of acyclic monoterpenes. A rigid conjugated planar diene chromophore can be characterized by long range $^{1}H$ NMR coupling constants of 1—2 Hz. Thus, the observation of a $^{4}J$ of 2 Hz between $H_3$ and $H_9$ in cartilagineal (**22**) is indicative of a conformationally fixed enal chromophore (*16*). Determination of the relative stereochemistry at $C_2$, $C_5$ and $C_6$ represents a more important problem because it is common to twenty-four of the Type (**A**) acyclic compounds. Proton and carbon NMR strategies have been proposed to solve the $C_5-C_6$ stereochemistry, but to date only X-ray analysis has provided information about the relative stereochemistry at $C_2$. FAULKNER (*64*) first observed that $^{1}H$ NMR chemical shifts could be diagnostic of $C_5-C_6$ stereochemistry in the acyclic monoterpenes. The $C_6-Me$ shift of δ 1.79 was characteristic of 5 (*R**) 6 (*R**) whereas δ 1.73 was characteristic of 5 (*R**) 6 (*S**), but compounds such as (**11**) with shifts of δ 1.76 cannot be assigned to one of the two categories. CREWS (*97*) observed a much larger $^{13}C$ NMR $C_6-Me$ shift difference of δ 28 for 5 (*R**) 6 (*R**) and δ 25 for 5 (*R**) 6 (*S**). These trends are summarized in Table 12.

It has been suggested that the larger shift difference observed in the $^{13}C$ NMR spectra makes this the method of choice when assigning such an element of stereochemistry and that the original stereochemistry of compound (**24**) [5 (*R**) 6 (*R**)] should be revised to 5 (*R**) 6 (*S**) (Table 10) in view of the more consistent $C_6$-methyl shifts in the $^{13}C$ NMR spectrum as shown in Table 12.

Table 12. *Relative Stereochemical Assignment at $C_5$ & $C_6$ in Acyclic Monoterpenes Using $C_6-Me$ Chemical Shifts*

| Monoterpene | Configuration | | Methyl Shift (ppm) | |
|---|---|---|---|---|
| | $C_5$ | $C_6$ | $^{1}H$ | $^{13}C$ |
| 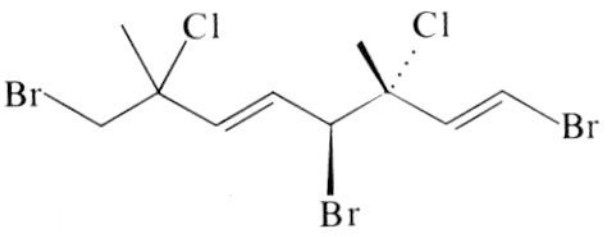 (**9**) | *R** | *S** | 1.78 | 26.3 |
| 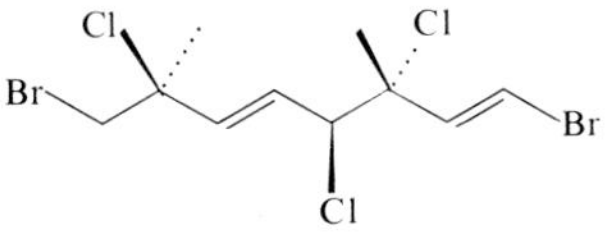 [**10**] | *R** | *S** | 1.75 | 25.5 |

Table 12. *(continued)*

| Monoterpene | Configuration | | Methyl Shift (ppm) | |
|---|---|---|---|---|
| | $C_5$ | $C_6$ | $^1H$ | $^{13}C$ |
| 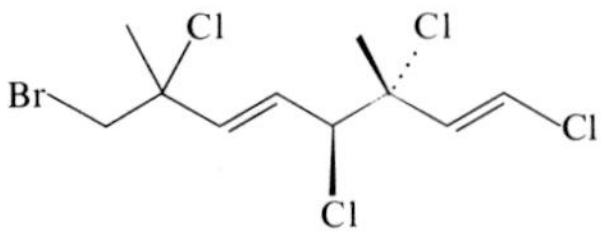 **(11)** | *R** | *S** | 1.76 | 25.7 |
| 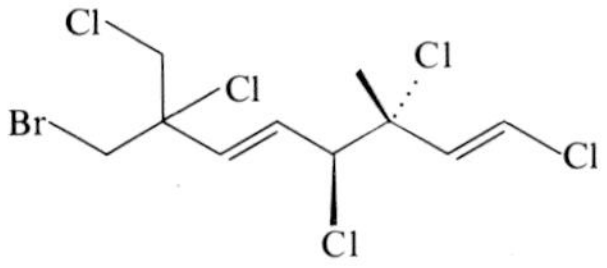 **(12)** | *R** | *S** | 1.75 | 25.4 |
| 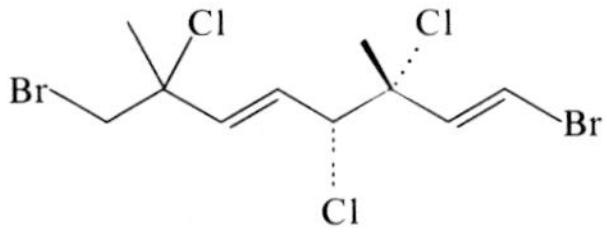 **(14)** | *R** | *R** | 1.81 | 28.0 |
| 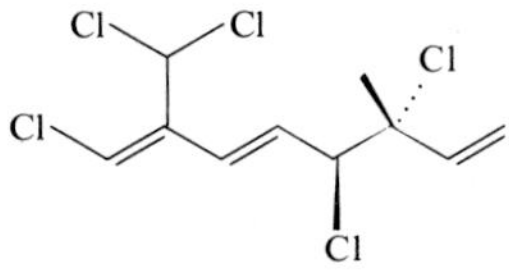 **(27)** | *R** | *S** | 1.73 | 25.1 |
| 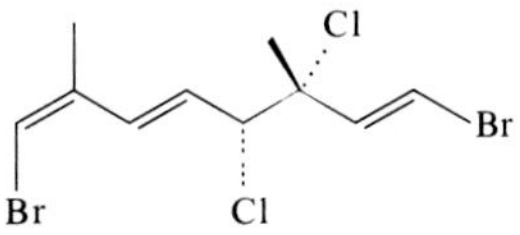 **(29)** | *R** | *R** | 1.79 | 28.4 |

The angular dependency of $^3J$ and $^4J$ magnitudes has often been utilized to assign substituent stereochemistry in six-membered ring monoterpenes. The magnitude of certain $^3J$'s also indicates that almost all of the monoterpene monocyclic rings are conformationally honogeneous. A good example is **(61)** in which the values of $J_{5,6}$ (11.2 & 4.8 Hz) clearly indicate that the $C_6-Br$ is a pseudo-axial in a non-inverting half-chair (*69*). Valuable stereochemical information has also come from $^4J$'s; for example, stereochemical assignments made at $C_6$ in **(55)**, **(56)**, **(58)**, and **(75)** were based upon differences in couplings involving $H_8$ and $H_2$ (*37, 95*). Assignment of stereochemistry to $C_3$ and $C_5$ of the epimers (**16**) and (**17**) was based on considering the symmetry environment of the protons on $C_4$. As these are either homotopic-like or diastereotopic-like, a decision between (**16**) and (**17**) was reached by matching the observed signals with the pattern expected for the $H_4$ multiplets (*65*).

Analysis of carbon chemical shifts has been found to be effective in deducing the stereochemistry of six-membered ring side chains (*98*). It is well known that in the absence of ring hetero atoms signals of axial methyls occur 6—8 ppm upfield from those of equatorial ones and that β-effects do not alter this trend [eg. (*37*)]. Correlations originally proposed by CREWS and KHO (*99*) were effectively used to assign stereochemistry to methyl groups attached to quaternary carbons in ring Types (**C**) and (**D**), such as in (**81**), (**82**), (**84**), (**86**), (**92**), (**69**) and (**70**), or in the Cl in (**75**). Recently, corrections of six-membered ring substituent stereochemistries in mertensene (**80**) and in (**89**) were proposed (*37*) based on a similar analysis of $^{13}$C-methyl shifts. Inconsistencies in shifts of carbon atoms also raised doubts about applicability of a procedure which used $^1$H or $^{13}$C NMR $T_1$'s to solve six-membered ring side chain stereochemistry (*94*).

A minor stereochemical revision seems needed for monocyclic diol (**73**) which is diastereomeric with (**72**) (*30*). Using NMR data, the two hydroxyl groups of (**72**) were deduced to be *trans* and pseudoequatorial whereas in (**73**) $C_6-OH$ was deduced to be pseudoaxial although the $C_5-OH$ stereochemistry could not be defined. Now a *trans* bis-pseudoaxial arrangement of the two hydroxyl groups would make (**73**) formally enantiomeric to (**72**) as a *trans* diaxial conformer should invert to the more stable bis-pseudoequatorial conformer; hence it seems logical that (**73**) possesses *cis*-diol stereochemistry.

## 4. Artifacts

Most of the halogenated regular monoterpenes are reported to be stable in spite of possessing such labile features as tertiary allylic halogens. This is fortunate since the presence of labile halogen substituents could greatly complicate structural analysis. Two examples which illustrate possible complications due to formation of artifacts have been reported. SIMS (*53*) observed that (**82**) rearranges to (**113**) at room temperature. On the basis of this observation we suggest that (**85**) might actually be a rearrangement product of (**84**). Also, as might be expected, protic solvents should be used with caution for extraction and chromatography. While FENICAL (*30, 69*) was able to isolate compounds (**60**)—(**74**) using isopropyl alcohol as extraction solvent, ICHIKAWA (*70*) found that extraction of *Chondrococcus*

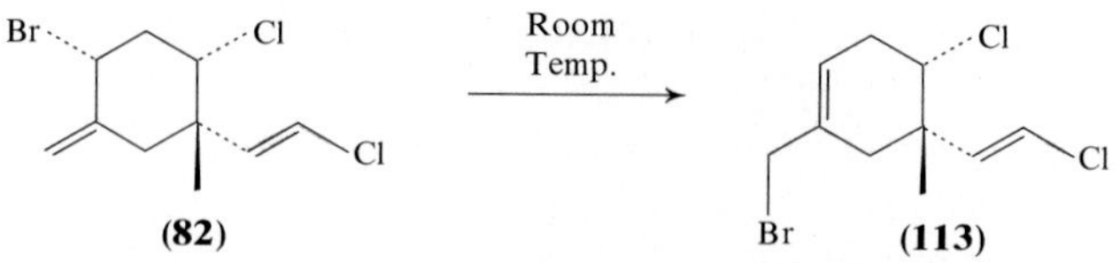

(82) → Room Temp. → (113)

*japonicus* with acetone followed by hexane yielded (**53**) and (**54**) but that when methanol was the extraction solvent only compounds (**114**)—(**117**) were obtained. Treatment of these same compounds with refluxing methanol also gave (**114**)—(**117**) (*70*). Reasonable extraction and isolation procedures were used to isolate the other compounds in Table 13 so the probability of any of them being artifacts is low.

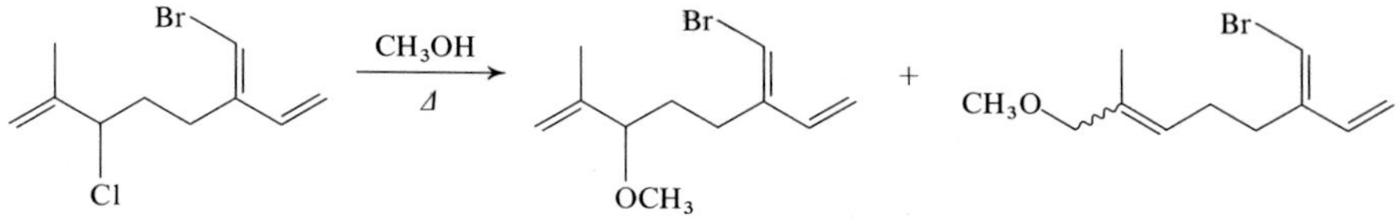

(53) → $CH_3OH$, Δ → (114) + (115)

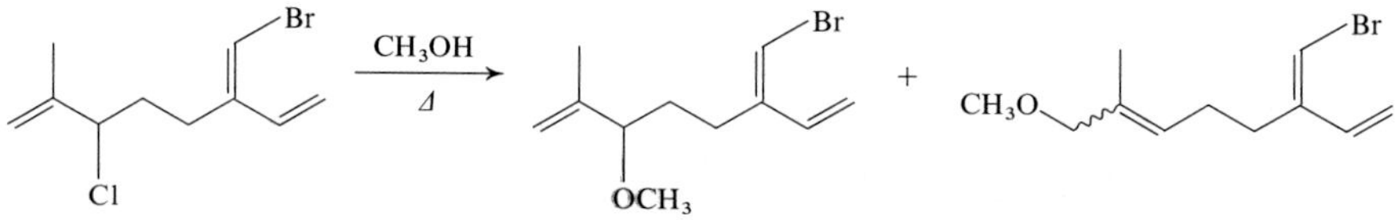

(54) → $CH_3OH$, Δ → (116) + (117)

As an interesting contrast to the above, treatment of costatone (**20**) with strong base yielded the degraded monoterpene costatolide (**94**) by loss of the relatively stable $-CHBr_2$ fragment (*54*).

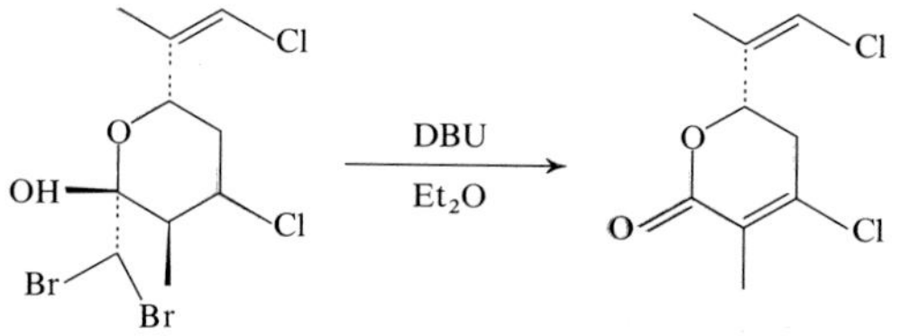

(20) → DBU, $Et_2O$ → (94)

## 5. Conclusions

Comparison of the similarities and differences observable in the $^1H$ and $^{13}C$ NMR data in Tables 13 and 14 indicates that $^{13}C$ is the most reliable way to distinguish between closely related compounds. (This might be especially valuable when one sample is unavailable for direct comparison.) Shift differences among halogenated monoterpenes which are epimeric at secondary carbons are easily observed; examples include the pairs (**16**) and (**17**) or (**81**) and (**84**). Also, perceptible carbon shift differences are observable among monocyclic compounds containing different tertiary halogens as exemplified by (**89**) and (**91**). A further example of the superiority of $^{13}$C NMR for making comparisons is that separate samples of (**87**) gave vastly different melting points (78.5—79; 43.5—44.5) and optical rotations ($[\alpha]_D = -84, -73.6$) (*68, 51*), yet their carbon NMR spectra were identical. These and all other physical properties of marine monoterpenes appear in Table 14.

## VII. Physical and Spectroscopic Tables

Table 13. *Summary of Structures and Carbon-13 NMR Chemical Shifts*

A. 2,6-Dimethyloctenes

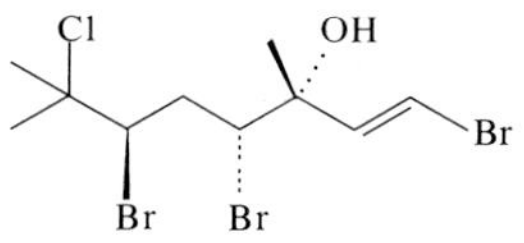

(**1**) 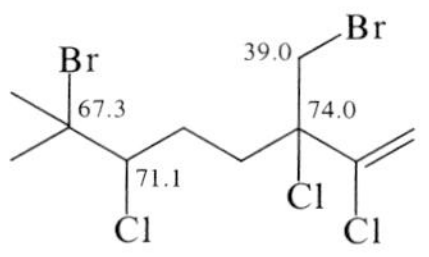 (**2**)

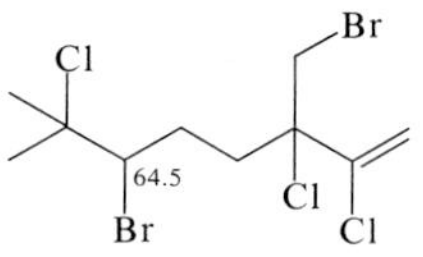

(**3**) 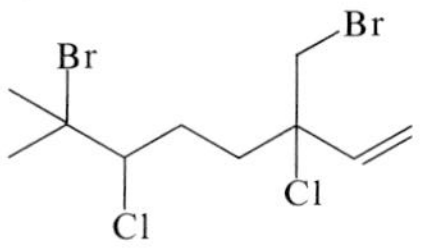 (**4**)

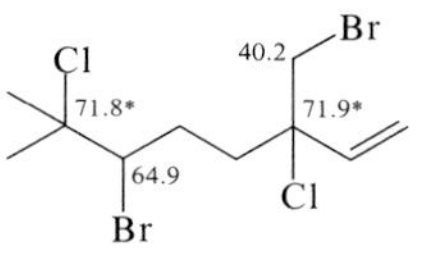

(**5**) 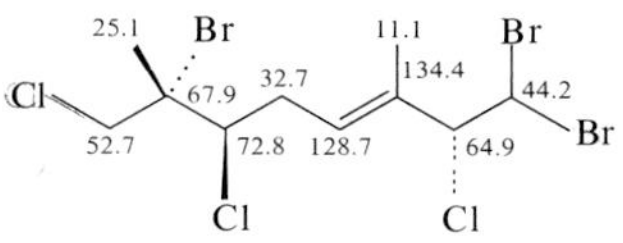 [**6**]

[] – Denotes structure determination by X-ray analysis.
* – Carbon shift assignments may be interchangeable.

Table 13 *(continued)*

B. 2,6-Dimethyloctadienes

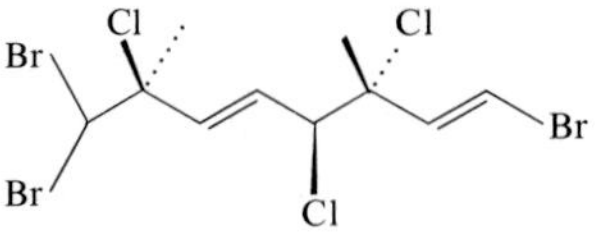

[7]

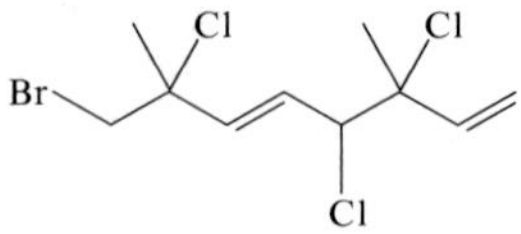

(8)

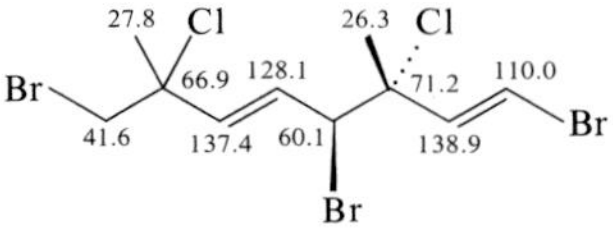

(9)

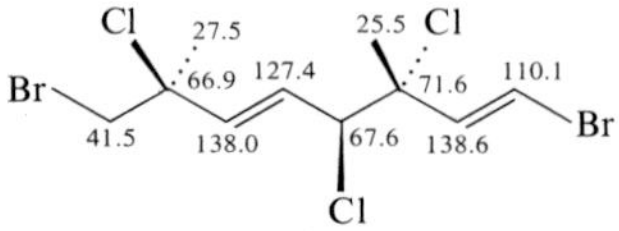

[10]

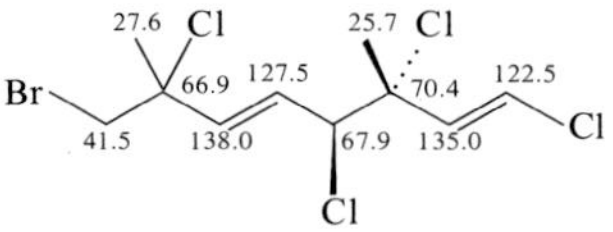

(11)

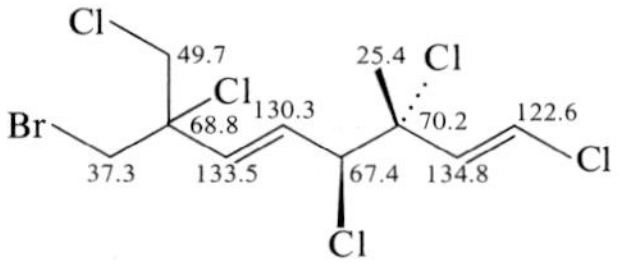

(12)

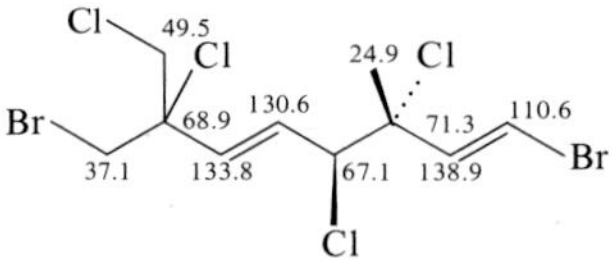

(13) Oregonene A

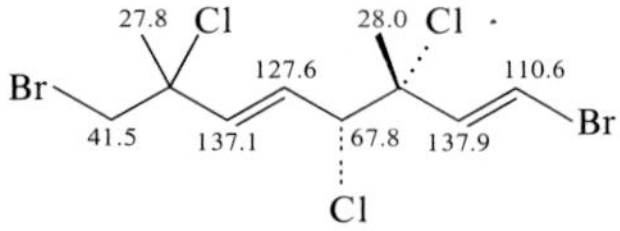

(14)

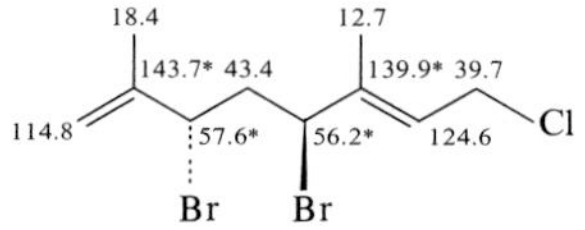

(15) Preplocamene A

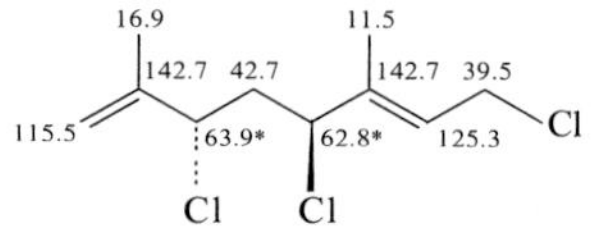

(16) Preplocamene B

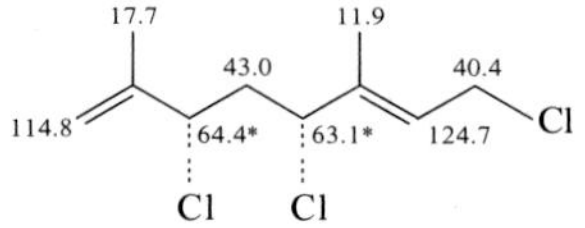

(17) Preplocamene C

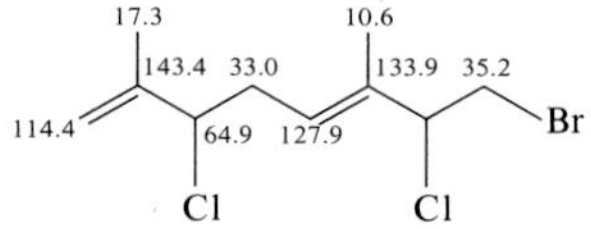

(18)

[] – Denotes structure determination by X-ray analysis.

* – Carbon shift assignments may be interchangeable.

*References, pp. 236—241*

Table 13 *(continued)*

B. 2,6-Dimethyloctadienes (continued)

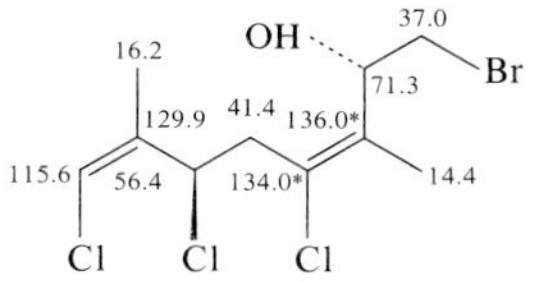

[**19**] Costatol

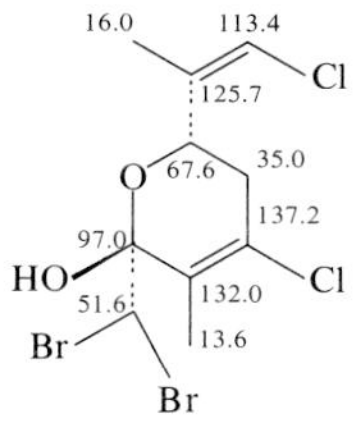

[**20**] Costatone

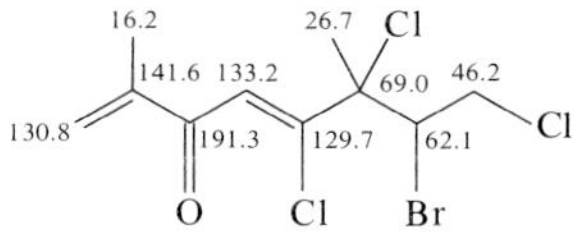

(**21**)

C. 2,6-Dimethyloctatrienes

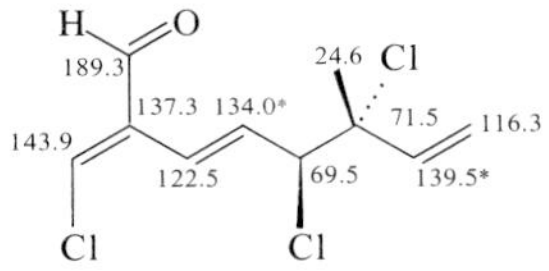

(**22**) Cartilagineal

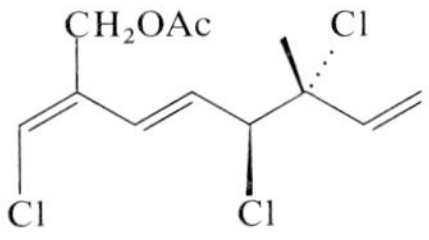

(**23**)

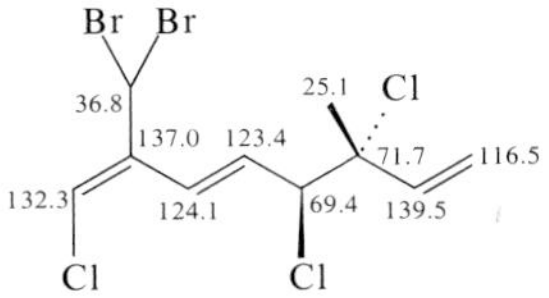

(**24**)

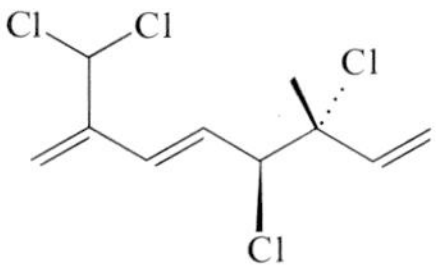

(**25**)

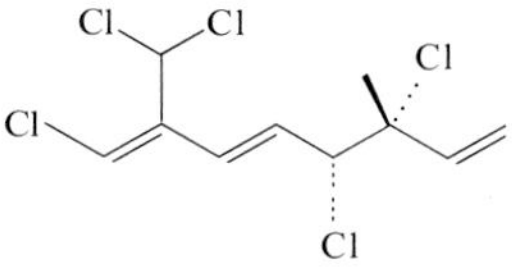

(**26**)

[ ] – Denotes structure determination by X-ray analysis.
* – Carbon shift assignments may be interchangeable.

Table 13 *(continued)*

C. 2,6-Dimethyloctatrienes (continued)

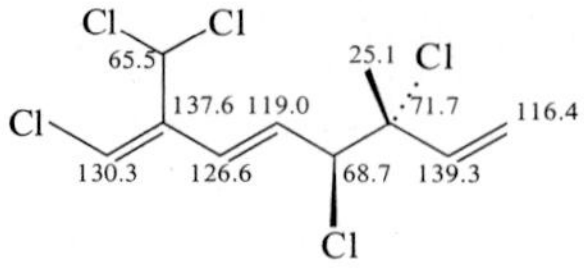

Cl Cl 65.5 25.1 Cl 137.6 119.0 71.7 116.4 Cl 130.3 126.6 68.7 139.3 Cl

**(27)**

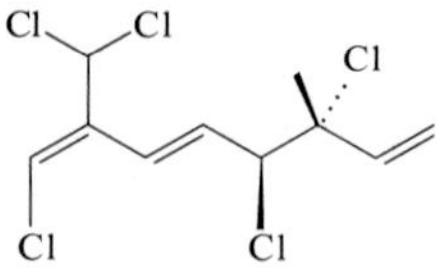

Cl Cl Cl Cl Cl

**(28)**

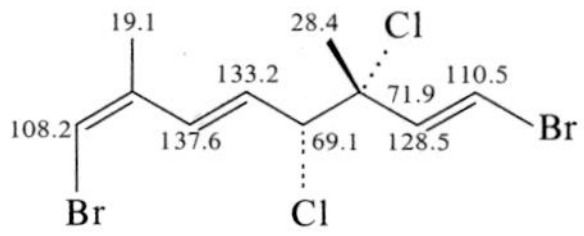

19.1 28.4 Cl 133.2 110.5 71.9 108.2 137.6 69.1 128.5 Br Br Cl

**(29)**

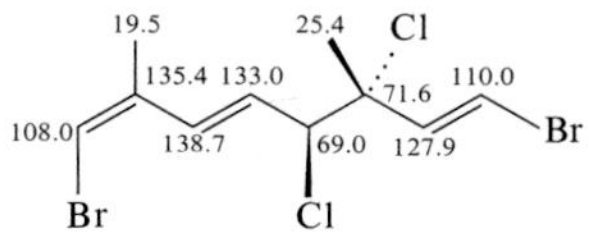

19.5 25.4 Cl 135.4 133.0 110.0 71.6 108.0 138.7 69.0 127.9 Br Br Cl

**(30)**

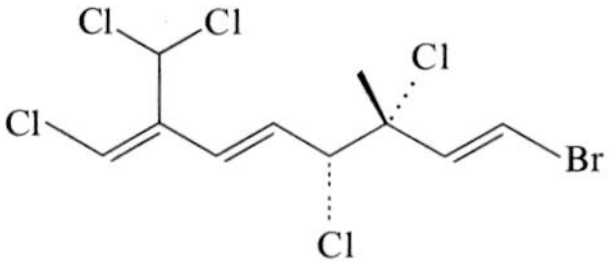

Cl Cl Cl Cl Br Cl

**(31)**

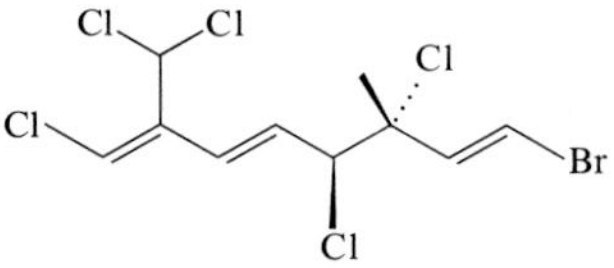

Cl Cl Cl Cl Br Cl

**(32)**

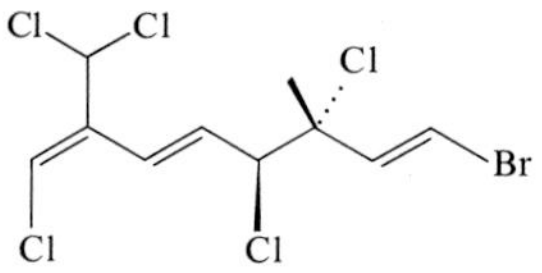

Cl Cl Cl Br Cl Cl

**(33)**

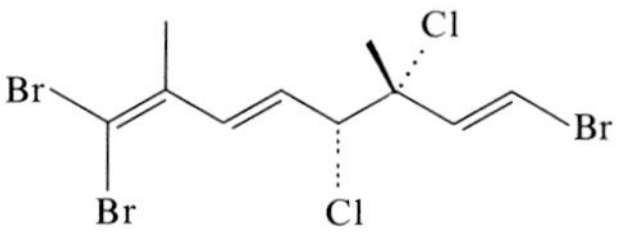

Cl Br Br Br Cl

**(34)**

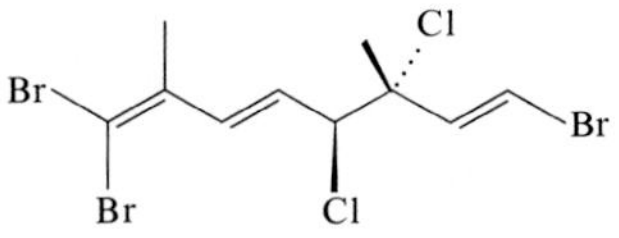

Cl Br Br Br Cl

**(35)**

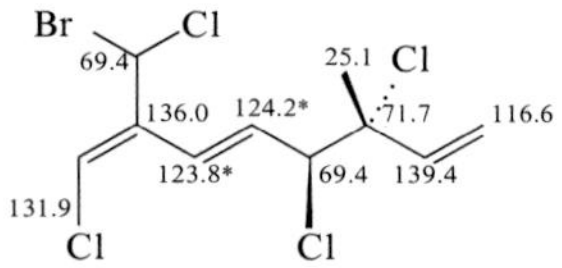

Br Cl 69.4 25.1 Cl 136.0 124.2* 71.7 116.6 123.8* 69.4 139.4 131.9 Cl Cl

**(36)**

[] – Denotes structure determination by X-ray analysis.
* – Carbon shift assignments may be interchangeable.

*References, pp. 236—241*

Table 13 *(continued)*

D. Myrcenes

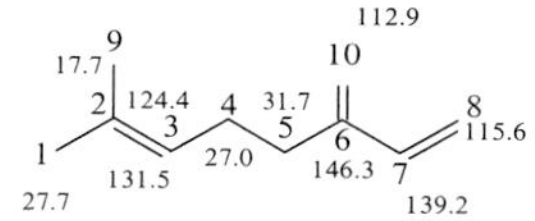

9
17.7
2 124.4 4 31.7
3 5
1 27.0 6 146.3 7
27.7 131.5
112.9
10
8
115.6
139.2

**(37)** Myrcene

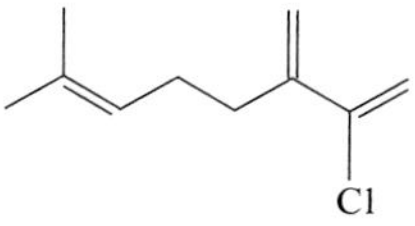

Cl

**(38)**

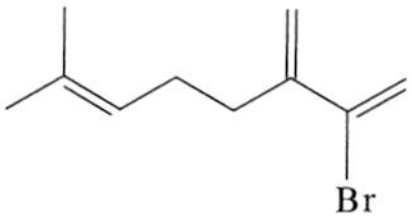

Br

**(39)**

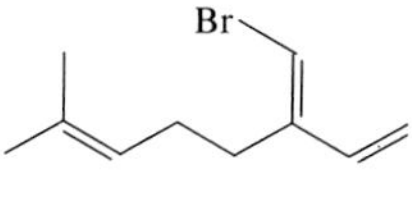

Br

**(40)**

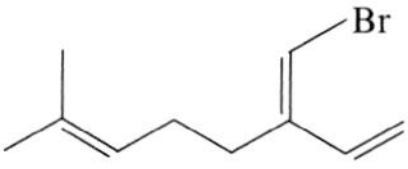

Br

**(41)**

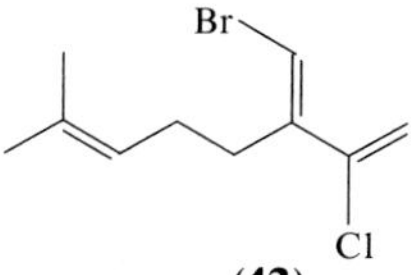

Br
Cl

**(42)**

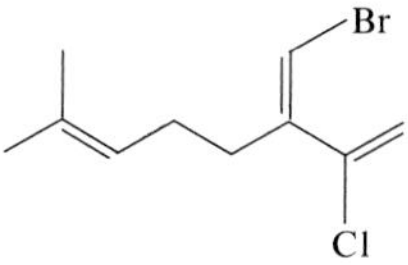

Br
Cl

**(43)**

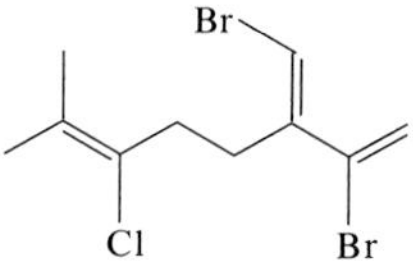

Br
Cl Br

**(44)**

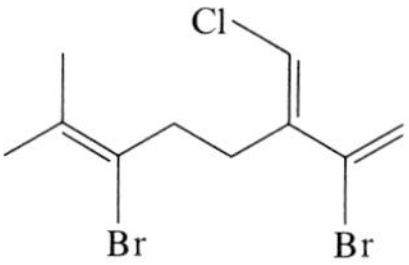

Cl
Br Br

**(45)**

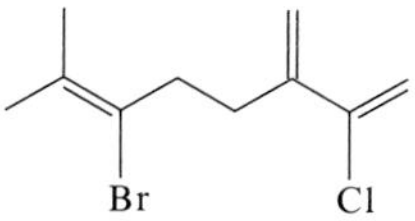

Br Cl

**(46)**

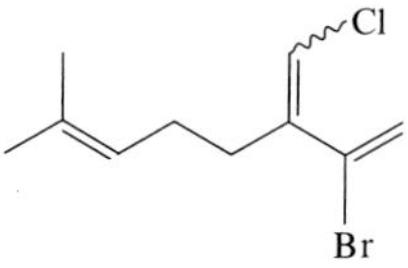

Cl
Br

**(47)**

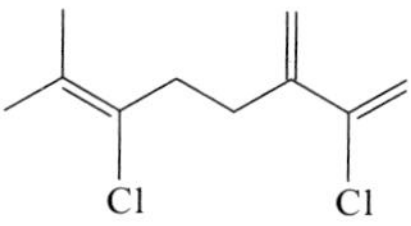

Cl Cl

**(48)**

[] – Denotes structure determination by X-ray analysis.
* – Carbon shift assignments may be interchangeable.

Table 13 *(continued)*

D. Myrcenes (continued)

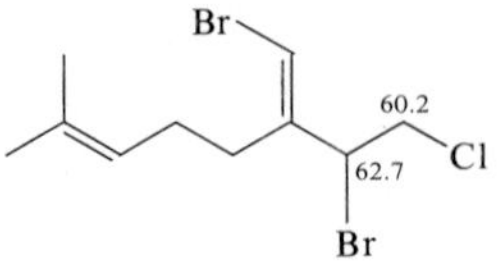

**(49)** 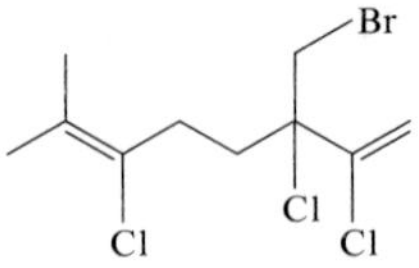 **(50)**

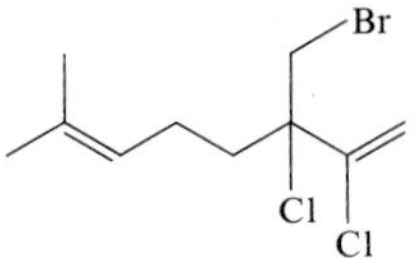

**(51)** 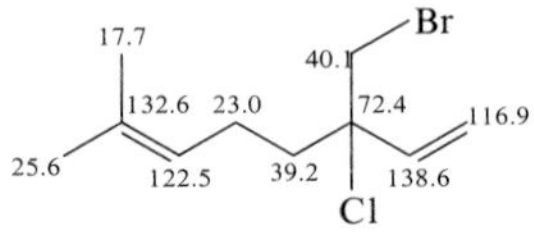 **(52)**

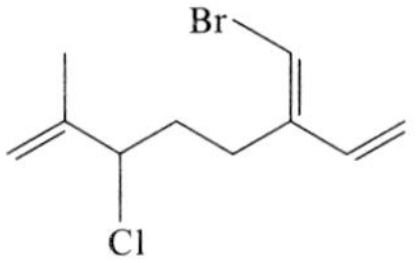

**(53)** 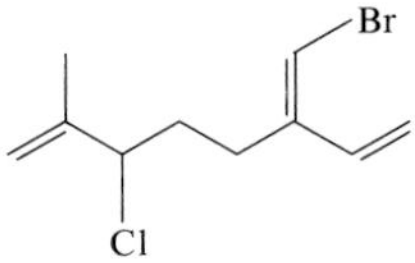 **(54)**

E. 1-Ethyl-3,3-dimethylcyclohexanes (Ochtodanes)

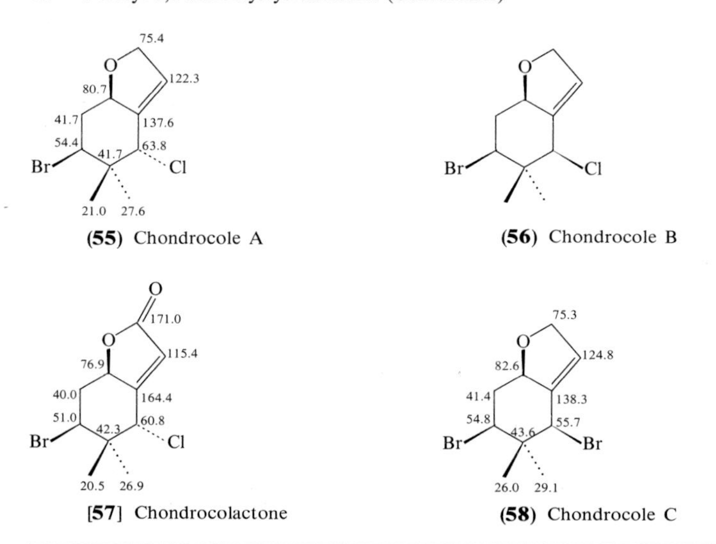


**(55)** Chondrocole A

**(56)** Chondrocole B

**[57]** Chondrocolactone

**(58)** Chondrocole C

[] – Denotes structure determination by X-ray analysis.
* – Carbon shift assignments may be interchangeable.

*References, pp. 236—241*

Table 13 *(continued)*

E. 1-Ethyl-3,3-dimethylcyclohexanes (Ochtodanes) (continued)

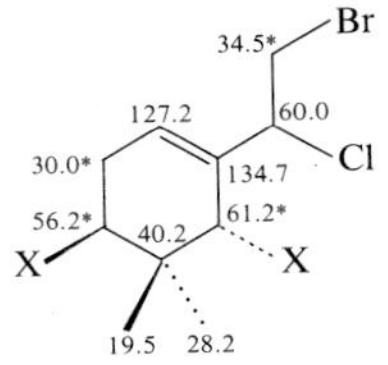

X = Br, Cl **(59)**

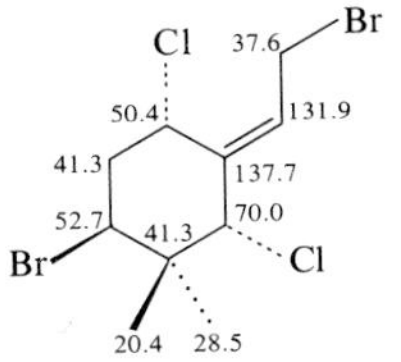

**(60)** Ochtodene

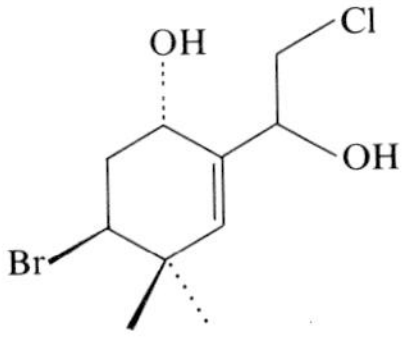

**(61)** Ochtodiol

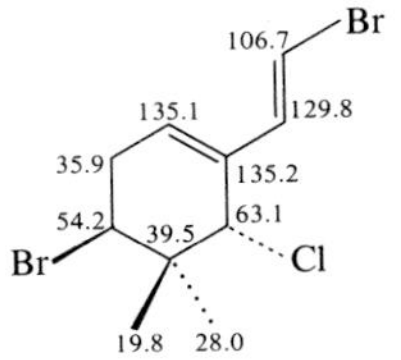

**(62)**

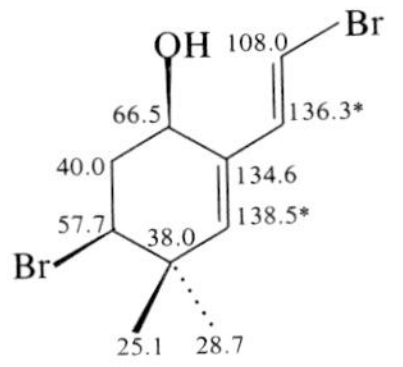

**(63)**

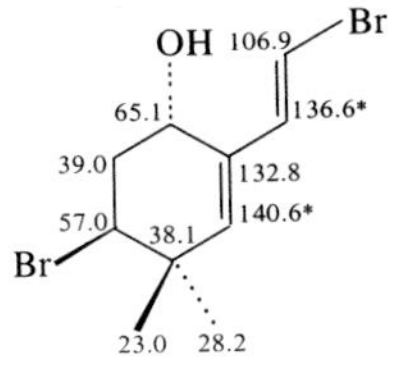

**(64)**

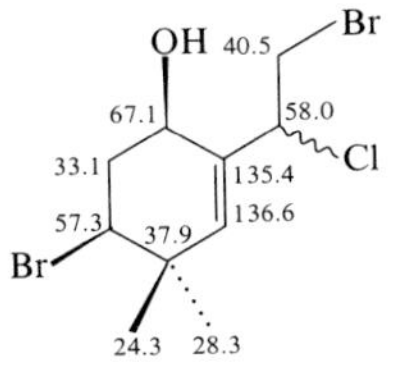

**(65)**

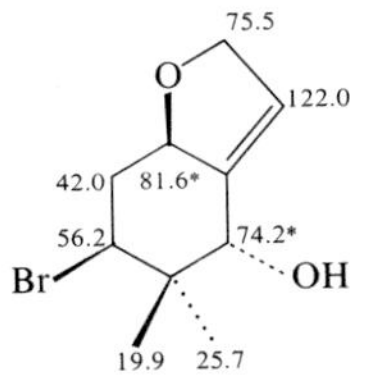

**(66)**

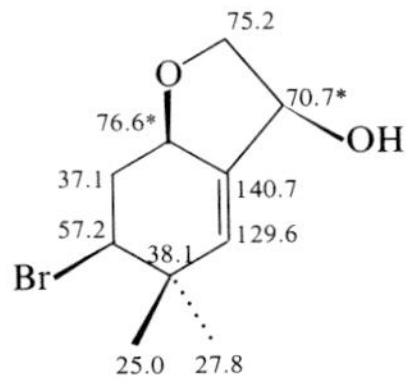

**(67)**

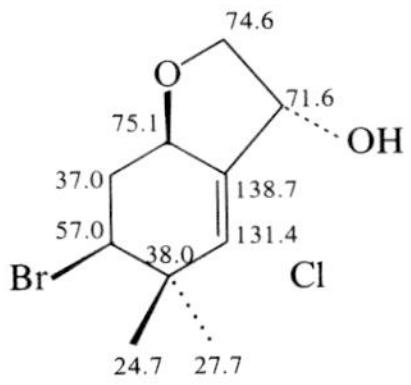

**(68)**

[] – Denotes structure determination by X-ray analysis.

* – Carbon shift assignments may be interchangeable.

Table 13 *(continued)*

E. 1-Ethyl-3,3-dimethylcyclohexanes (Ochtodanes) (continued)

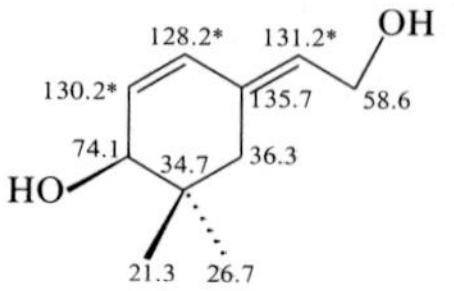

**(69)**

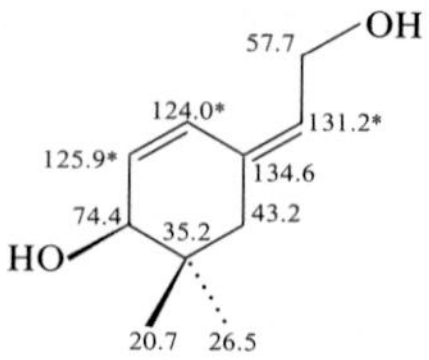

**(70)**

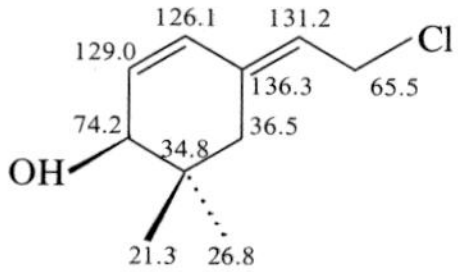

**(71)**

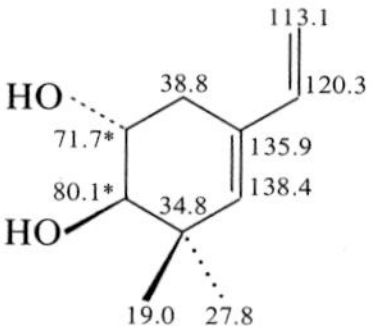

**(72)**

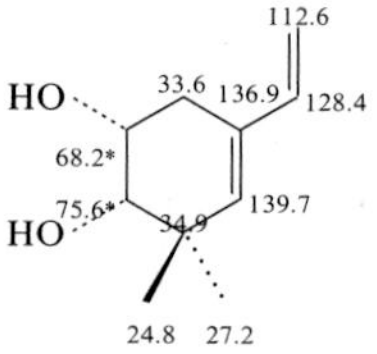

**(73)**

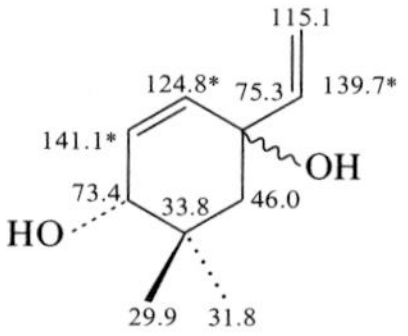

**(74)**

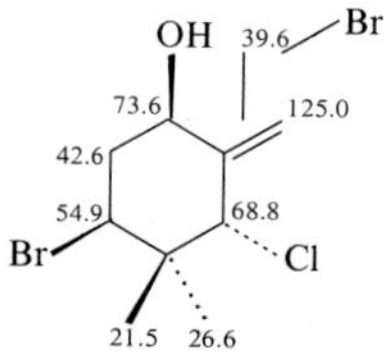

**(75)**

[] – Denotes structure determination by X-ray analysis.
* – Carbon shift assignments may be interchangeable.

*References, pp. 236—241*

Table 13 *(continued)*

F. 1-Ethyl-1,3-dimethylcyclohexanes

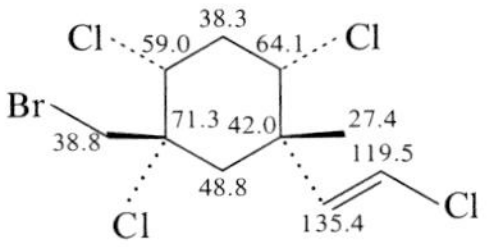

[**76**] Violacene

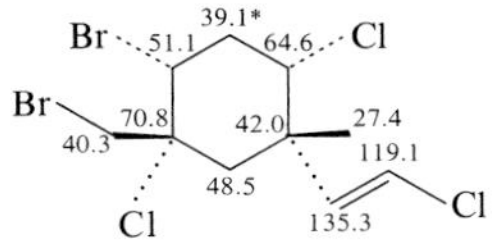

(**77**)

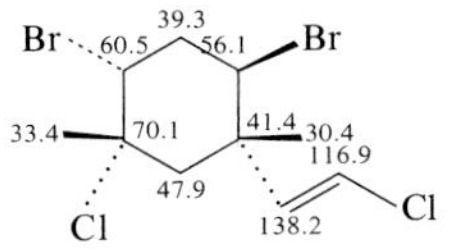

(**78**)

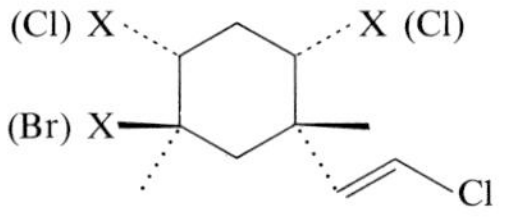

(**79**)

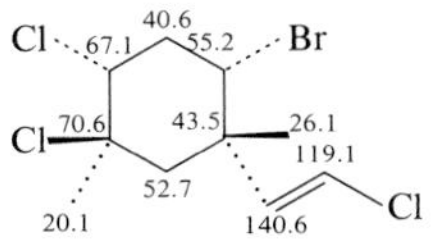

(**80**) Mertensene

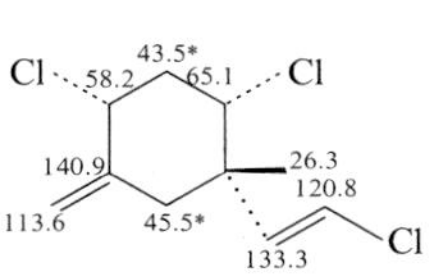

(**81**) Plocamene D

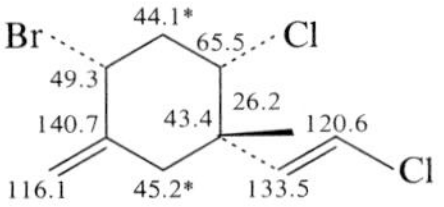

(**82**) Plocamene-D′

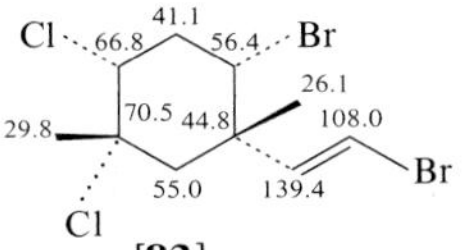

[**83**]

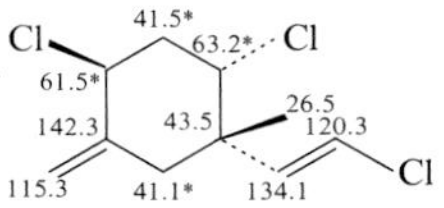

(**84**) Epi-plocamene-D

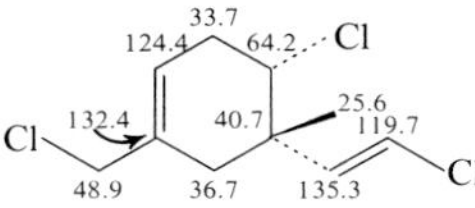

(**85**)

G. 1-Ethyl-2,4-dimethylcyclohexanes

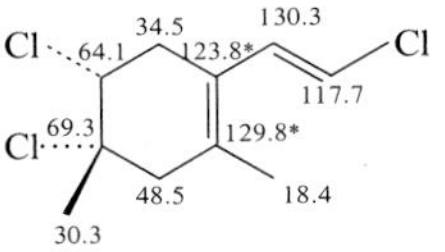

(**86**) Plocamene-B

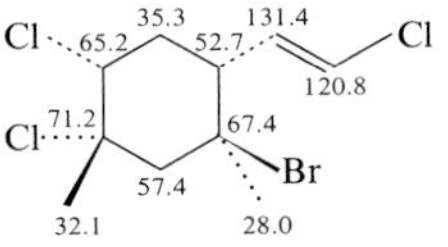

[**87**] Plocamene-C or Violacene-2

[ ] – Denotes structure determination by X-ray analysis.

* – Carbon shift assignments may be interchangeable.

Table 13 *(continued)*

G. 1-Ethyl-2,4-dimethylcyclohexanes (continued)

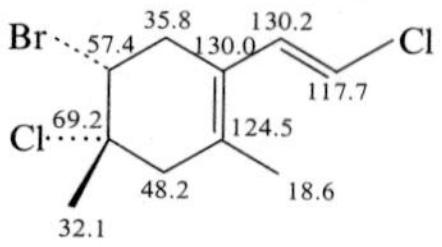

**(88)**

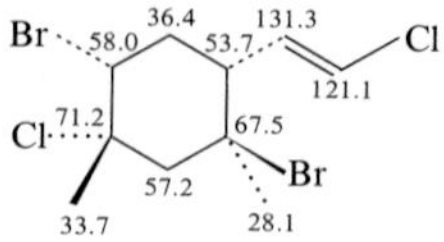

**[89]**

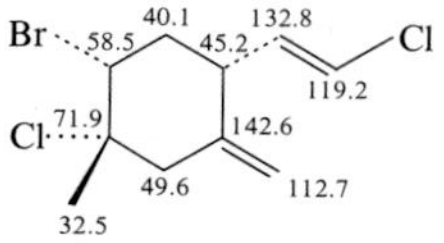

**(90)**

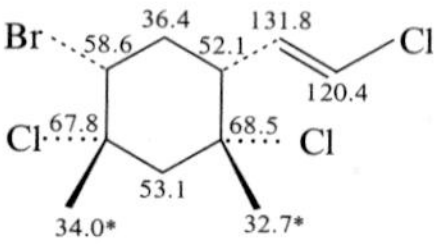

**(91)**

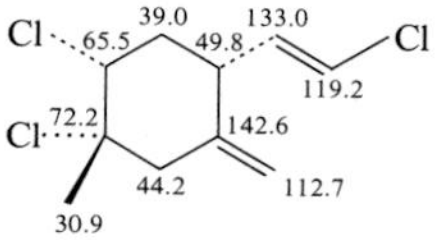

**(92)** Plocamene-E

H. Degraded or Mixed Biogenesis Monoterpenes

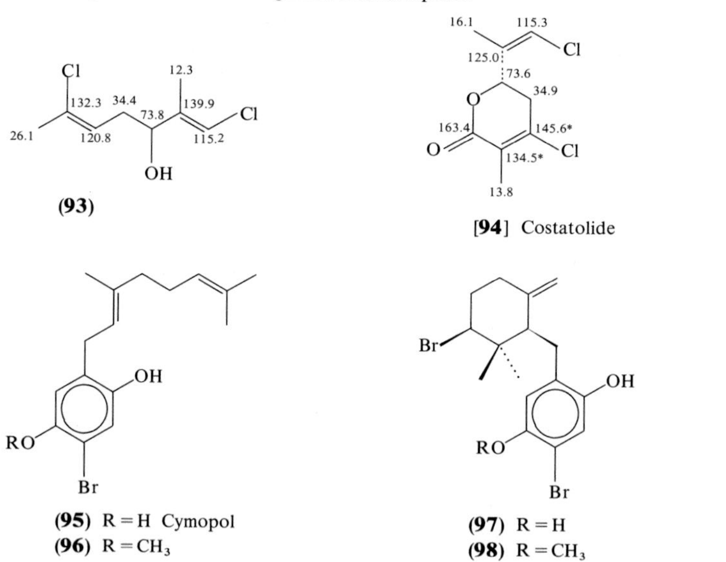


**(93)**

**[94]** Costatolide

**(95)** R = H Cymopol
**(96)** R = $CH_3$

**(97)** R = H
**(98)** R = $CH_3$

[] – Denotes structure determination by X-ray analysis.
* – Carbon shift assignments may be interchangeable.

*References, pp. 236—241*

Table 13 *(continued)*

H. Degraded or Mixed Biogenesis Monoterpenes (continued)

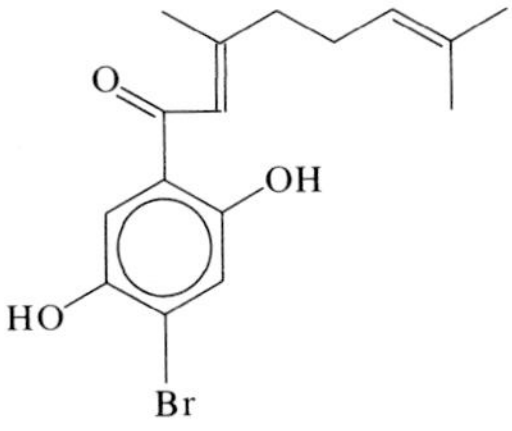

**(99)**

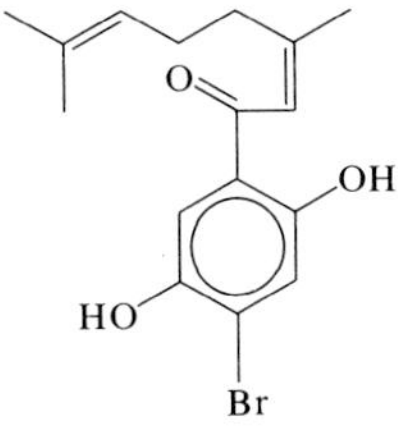

**(100)**

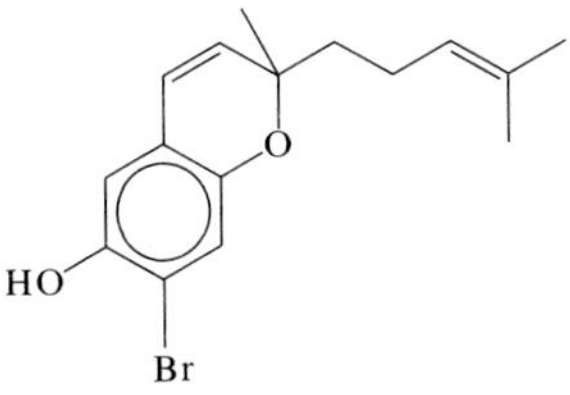

**(101)**

[ ] – Denotes structure determination by X-ray analysis.
* – Carbon shift assignments may be interchangeable.

Table 14. *Physical Properties*

| Compound | α | MP °C | $^{1}H$ | $^{13}C$ | Reference |
|---|---|---|---|---|---|
| **(1)** | −64 | oil | √ | — | *(14, 32)* |
| | — | — | — | — | *(103)* |
| **(2)** | — | — | √ | √ | *(96)* |
| **(3)** | — | — | √ | √ | *(96)* |
| **(4)** | — | — | √ | — | *(96)* |
| **(5)** | — | — | √ | √ | *(96)* |
| **(6)** | +99 | 60.5—61.5 | √ | √ | *(76)* |
| **(7)** | −50.2 | 54 | — | — | *(15)* |
| | — | — | √ | — | *(64)* |
| **(8)** | — | — | √ | — | *(83)* |
| **(9)** | — | 20 | √ | — | *(83)* |
| | — | — | √ | √ | *(92)*[a] |
| **(10)** | −46.3 | 48.5—49.0 | √ | √ | *(92)* |
| **(11)** | −48.3 | — | √ | √ | *(92)* |
| | — | — | — | — | *(53)* |

[a] Reference reporting structure revision.

Table 14 *(continued)*

| Compound | α | MP °C | $^{1}$H | $^{13}$C | Reference |
|---|---|---|---|---|---|
| **(12)** | −20.2 | — | √ | √ | *(92)* |
| | — | — | — | — | *(53)* |
| **(13)** | — | oil | √ | √ | *(97)* |
| | −19.4 | — | √ | √ | *(92)* |
| **(14)** | — | oil | √ | √ | *(97)* |
| **(15)** | — | oil | √ | √ | *(65)* |
| **(16)** | — | oil | √ | √ | *(65)* |
| **(17)** | — | oil | √ | √ | *(65)* |
| **(18)** | +19 | oil | √ | √ | *(104)* |
| **(19)** | −103 | 88—89 | √ | √ | *(105)* |
| **(20)** | −52.4 | 70 | √ | √ | *(105)* |
| | −54.0 | 70—70.5 | √ | √ | *(54)* |
| **(21)** | −14.3 | oil | √ | √ | *(104)* |
| **(22)** | — | oil | √ | √ | *(16)* |
| | — | — | — | — | *(65)* |
| **(23)** | — | — | √ | — | *(83)* |
| **(24)** | −9.7 | — | √ | √ | *(81, 37)*[a] |
| **(25)** | −5.7 | — | √ | — | *(64)* |
| **(26)** | +34.6 | — | √ | — | *(64)* |
| **(27)** | +5.1 | — | √ | — | *(64)* |
| | +4.8 | — | √ | √ | *(81)* |
| **(28)** | −39.3 | — | √ | — | *(64)* |
| **(29)** | +62.8 | — | √ | — | *(64)* |
| **(30)** | — | oil | — | √ | *(97)* |
| | −46.0 | — | √ | — | *(64)* |
| **(31)** | +4.7 | — | √ | — | *(64)* |
| **(32)** | −4.4 | — | √ | — | *(64)* |
| **(33)** | −22.9 | — | √ | — | *(64)* |
| **(34)** | +44.1 | — | √ | — | *(64)* |
| **(35)** | −34.7 | — | √ | — | *(64)* |
| **(36)** | — | — | √ | √ | *(37)* |
| **(37)** | — | — | √ | — | *(48)* |
| | — | — | — | √ | *(35)* |
| **(38)** | — | — | √ | — | *(48)* |
| **(39)** | — | — | √ | — | *(48)* |
| **(40)** | — | — | √ | — | *(48)* |
| **(41)** | — | — | √ | — | *(48)* |
| **(42)** | — | — | √ | — | *(48)* |
| **(43)** | — | — | √ | — | *(48)* |
| **(44)** | — | — | √ | — | *(48)* |
| **(45)** | — | — | √ | — | *(48)* |
| **(46)** | — | — | — | — | *(48)* |
| | — | — | √ | — | *(21)* |
| **(47)** | — | — | — | — | *(48)* |
| **(48)** | — | — | √ | — | *(21)* |
| | — | — | — | — | *(95)* |
| **(49)** | — | — | √ | √ | *(96)* |

[a] Reference reporting structure revision.

*References, pp. 236—241*

Table 14 *(continued)*

| Compound | α | MP °C | $^1$H | $^{13}$C | Reference |
|---|---|---|---|---|---|
| **(50)** | — | — | √ | — | *(96)* |
| **(51)** | — | — | √ | — | *(96)* |
| **(52)** | −3.7 | oil | √ | √ | *(101)* |
| **(53)** | +5.0 | — | √ | — | *(70)* |
| **(54)** | −9.3 | — | √ | — | *(70)* |
| **(55)** | −16.0 | — | √ | √ | *(95)* |
| | — | — | — | √ | *(88)*[a] |
| **(56)** | — | — | √ | — | *(95)*[b] |
| **(57)** | −48.0 | 107—108 | √ | √ | *(88)* |
| **(58)** | — | — | — | — | *(96)* |
| | — | — | √ | √ | *(30)* |
| **(59)** | — | — | √ | — | *(96, 37)*[c] |
| | — | — | — | √ | *(36)* |
| **(60)** | — | — | √ | √ | *(69)* |
| **(61)** | — | — | √ | — | *(69)* |
| **(62)** | +16.7 | — | √ | √ | *(30)* |
| **(63)** | −45.4 | — | √ | √ | *(30)* |
| **(64)** | −71.2 | — | √ | √ | *(30)* |
| **(65)** | −26.8 | — | √ | √ | *(30)* |
| **(66)** | +30.0 | — | √ | √ | *(30)* |
| **(67)** | −64.6 | — | √ | √ | *(30)* |
| **(68)** | −55.0 | — | √ | √ | *(30)* |
| **(69)** | +4.3 | — | √ | √ | *(30)* |
| **(70)** | 0.0 | — | √ | √ | *(30)* |
| **(71)** | +5.7 | — | √ | √ | *(30)* |
| **(72)** | 0.0 | — | √ | √ | *(30)* |
| **(73)** | 0.0 | — | √ | √ | *(30)* |
| **(74)** | −14.8 | — | √ | √ | *(30)* |
| **(75)** | — | oil | √ | √ | *(37)* |
| **(76)** | — | 71.0—71.5 | √ | √ | *(93, 87)*[a] |
| | — | — | √ | √ | *(52)* |
| **(77)** | −43.8 | 74—75 | √ | √ | *(52)* |
| | −67.8 | — | √ | √ | *(53)* |
| **(78)** | +32.3 | oil | √ | √ | *(52)* |
| **(79)** | — | oil | √ | — | *(52)* |
| **(80)** | — | crystals | √ | √ | *(94, 37)*[a] |
| **(81)** | −4.1 | oil | √ | √ | *(68)* |
| **(82)** | — | oil | √ | √ | *(68)* |
| **(83)** | −18 | 72—73 | √ | √ | *(100)* |
| **(84)** | −63.3 | — | √ | √ | *(53)* |
| **(85)** | −110.0 | — | √ | √ | *(53)* |
| **(86)** | −48 | 100—101 | √ | √ | *(86)* |
| **(87)** | — | 43.5—44.5 | √ | — | *(51)* |
| | −84 | 78.5—79 | √ | √ | *(68)* |
| **(88)** | −13.2 | 104—104.5 | √ | √ | *(52)* |

[a] Reference reporting structure revision.
[b] Revision suggested in this review.
[c] Reference reporting ambiguous structure.

Table 14 *(continued)*

| Compound | α | MP °C | $^{1}$H | $^{13}$C | Reference |
|---|---|---|---|---|---|
| **(89)** | −35.7 | 86—87 | √ | √ | *(52)* |
| | −88 | 121—122 | √ | √ | *(100)* |
| | — | — | √ | √ | *(94)* |
| **(90)** | −70.5 | — | √ | √ | *(52)* |
| **(91)** | — | — | √ | √ | *(94, 37)*[a] |
| **(92)** | −105 | oil | √ | √ | *(68)* |
| **(93)** | −9.8 | oil | √ | √ | *(102)* |
| **(94)** | −152 | oil | √ | √ | *(54)* |
| **(95)** | — | 59—61 | √ | — | *(55)* |
| **(96)** | — | oil | — | — | *(55)* |
| **(97)** | — | — | — | — | *(55)* |
| **(98)** | — | glass | √ | — | *(55)* |
| **(99)** | — | 81—81.5 | √ | — | *(55)* |
| **(100)** | — | 55—56 | √ | — | *(55)* |
| **(101)** | — | oil | √ | — | *(55)* |

[a] Reference reporting structure revision.

## Acknowledgement

Support for the research reviewed here from the laboratory of Prof. PHILIP CREWS was under a grant from the California Sea Grant Program Department of Commerce. We are indebted to Ms. MELISSA QUEEN and Ms. JUDITH TILLSON who skillfully typed this manuscript and patiently did all of the lay-out.

## References


*1.* MAYO, P. DE: Mono- and Sesquiterpenoids. In: The Chemistry of Natural Products, Vol. II, pp. 1—179 (K. W. BRATLEY, ed.). New York: Interscience. 1959.

*2.* RUZICKA, L.: History of the Isoprene Rule. Proc. Chem. Soc. 341 (1959).

*3.* DEVON, T. K., and A. I. SCOTT: Handbook of Naturally Occurring Compounds. Vol. II, p. 576. New York: Academic Press. 1972.

*4.* CAVILL, G. W. K.: Insect Terpenoids and Nepetalactone. In: Cyclopentanoid Terpene Derivatives, pp. 203—235 (W. I. TAYLOR and A. R. BATTERSBY, eds.). New York: Marcel Dekker. 1969.

*5.* HALLIGAN, J. P.: Toxic Terpenes from *Artemisia californica*. Ecology **56,** 999 (1975).

*6.* BOWERS, W. S.: Insect Juvenile Hormones and Pheromones of Isopentenoid Biogenesis. Lipids **13,** 736 (1978).

*7.* EPSTEIN, W. W., and L. A. GAUDIOSO: A Non-head-to-tail Monoterpenoid with a New Skeletal System from *Artemisia tridentata rothrockii*. J. Org. Chem. **47,** 175 (1982).

*8.* FRANCIS, M. J. O.: Monoterpene Biosynthesis. In: Aspects of Terpenoid Chemistry and Biochemistry, pp. 29—48 (T. W. GOODWIN, ed.). New York: Academic Press. 1971.

*9.* POULTER, C. D., and H. C. RILLING: The Prenyl Transfer Reaction. Enzymatic and Mechanism Studies of the 1—4 Coupling Reaction in the Terpene Biosynthetic Pathway. Accounts Chem. Res. **11,** 307 (1978).

*10.* LOOMIS, W. D.: Biosynthesis and Metabolism of Monoterpenes. In: Terpenoids in Plants, pp. 59—80 (J. B. PRIDHAM, ed.). New York: Academic Press. 1967, and references therein.
*11.* WHITTAKER, D.: The Monoterpenes. In: Chemistry of Terpenes and Terpenoids, pp. 11—82 (A. A. NEWMAN, ed.). New York: Academic Press. 1972.
*12.* KATAYAMA, T.: The Volatile Constituents of Seaweed. III. The Terpenes of Volatile Constituents of *Ulva pertusa*. Bull. Japan Soc. Sci. Fisheries **21,** 412 (1955).
*13.* — Volatile Constituents. In: Physiology and Biochemistry of Algae, pp. 467—473 (R. A. LEWIS, ed.). New York: Academic Press. 1962.
*14.* FAULKNER, D. J., and M. O. STALLARD: 7-Chloro-3,7-dimethyl-1,4,6-tri-bromo-1-octen-3-ol, A Novel Monoterpene Alcohol from *Aplysia californica*. Tetrahedron Letters 1171 (1973).
*15.* FAULKNER, D. J., M. O. STALLARD, J. FAYOS, and J. CLARDY: (3R, 4S, 7S)-*trans,trans*-3,7-dimethyl-1,8,8-tribromo-3,4,7-trichloro-1,5-octadiene, a Novel Monoterpene from the Sea Hare, *Aplysia californica*. J. Amer. Chem. Soc. **95,** 3413 (1973).
*16.* CREWS, P., and E. KHO: Cartilagineal. An Unusual Monoterpene Aldehyde from Marine Alga. J. Org. Chem. **39,** 3303 (1974).
*17.* CREWS, P.: The Breadth of Monoterpene Synthesis by Marine Red Algae: Potential Difficulties in their Application as Taxonomic Markers. In: Marine Natural Products Chemistry (D. J. FAULKNER and W. FENICAL, eds.), IV: 1, pp. 211—223. New York: Plenum Press. 1977.
*18.* FAULKNER, D. J.: Interesting Aspects of Marine Natural Products Chemistry. Tetrahedron **33,** 1421 (1977); Natural Organohalogen Compounds. In: The Handbook of Environmental Chemistry (D. HUTZINGER, ed.). Berlin-Heidelberg-New York: Springer. 1980.
*19.* MOORE, R. E.: Marine Aliphatic Natural Products. In: Aliphatic and Related Natural Product Chemistry. Vol. I, pp. 20—67 (F. D. GUNSTONE, ed.). The Chemical Society. 1979.
*20.* CHRISTOPHERSEN, C., and J. S. CARLE: Chemical Signals from a Marine Bryozoan. Naturwiss. **65,** 440 (1978).
*21.* ICHIKAWA, N., Y. NAYA, and S. ENOMOTO: Halogenated Monoterpene Derivatives from *Desmia (Chondrococcus) japonicus*. Proc. Japan Acad. **51,** 562 (1975).
*22.* SIUDA, J. F., and J. F. DEBARNARDIS: Naturally Occurring Halogenated Organic Compounds. Lloydia **36,** 107 (1973).
*23.* ANDREWS, A. G., C. L. JENKINS, M. P. STARR, J. SHEPHERD, and H. HOPE: Structure of Xanthomonadin I. A Novel Dibrominated Aryl-polyene Pigment Produced by the Bacteium *Xanthomonas juglandis*. Tetrahedron Letters 4023 (1976).
*24.* RILEY, J. P., and R. CHESTER: Introduction to Marine Chemistry, p. 465. New York: Academic Press. 1971.
*25.* FENICAL, W.: Natural Products Chemistry in the Marine Environment. Science **215,** 923 (1982).
*26.* WEATHERSTON, J.: The Chemistry of Anthropod Defense Substances. Quart. Rev. **21,** 287 (1967).
*27.* DODSON, C. H., R. L. DRESSLER, H. G. HILLS, R. M. ADAMS, and N. H. WILLIAMS: Biologically Active Compounds in Orchid Fragrances. Science **164,** 1243 (1969).
*28.* HANOVER, J. W.: Physiology of Tree Resistance to Insects. An. Rev. Entomol. **20,** 75 (1975).
*29.* MULLER, C. H., and C. H. CHOU: Phytotoxins: An Ecological Phase of Phytochemistry. In: Phytochemical Ecology, pp. 201—215 (J. B. HARBORNE, ed.). New York: Academic Press. 1972.
*30.* PAUL, J. V., O. J. MCCONNELL, and W. FENICAL: Cyclic Monoterpenoid Feeding Deterrents from the Red Marine Alga *Ochtodes crockeri*. J. Org. Chem. **45,** 3401 (1980).

31. JACOBS, R., and P. CREWS: Unpublished results.
32. STALLARD, M. O., and D. J. FAULKNER: Chemical Constituents of the Digestive Gland of the Seahare *Aplysia californica* — I. Importance of Diet. Comp. Biochem. Physiol. **49 B,** 25 (1974).
33. BURBOTT, A. J., and W. D. LOOMIS: Evidence for Metabolic Turnover of Monoterpenes in Peppermint. Plant Physiol. **44,** 173 (1969).
34. SANDERMANN, W.: Terpenoids: Structure and Distribution. In: Comparative Biochemistry, Vol. 3, pp. 503—590 (M. FLORKIN and H. S. MASON, eds.). New York: Academic Press. 1972.
35. WEHRLI, F. W., and T. NISHIDA: The Use of Carbon-13 Nuclear Magnetic Resonance Spectroscopy in Natural Products Chemistry. Fortschr. Chem. Org. Naturstoffe **36,** 1—229 (1979).
36. SIMS, J. J., A. F. ROSE, and R. R. IZAC: Application of $^{13}$C-NMR to Marine Natural Products. In: Marine Natural Products. Chemical and Biological Perspectives, Vol. II, pp. 297—378 (P. J. SCHEUER, ed.). New York: Academic Press. 1978.
37. CREWS, P., F. J. HANKE, S. NAYLOR, E. R. HOGUE, E. KHO, and R. BRASLAU: Halogen Regiochemistry and Substituent Stereochemistry Determination in Marine Monoterpenes by $^{13}$C NMR. J. Org. Chem. (In press.)
38. MARTIN, J. D., and J. DARIAS: Algal Sesquiterpenoids. In: Marine Natural Products. Chemical and Biological Perspectives, Vol. I, pp. 125—171 (P. J. SCHEUER, ed.). New York: Academic Press. 1978.
39. FENICAL, W.: Deterpenoids. In: Marine Natural Products. Chemical and Biological Perspectives, Vol. II, pp. 173—245 (P. J. SCHEUER, ed.). New York: Academic Press. 1978.
40. See Footnote 13 in: SELOVER, S. J., and P. CREWS: Kylinone, a New Sesquiterpene Skeleton from the Marine Alga *Laurencia pacifica*. J. Org. Chem. **45,** 69 (1980).
41. CHARLWOOD, B. V., and D. V. BANTHORPE: The Biosynthesis of Monoterpenes. Progr. Phytochem. **5,** 65 (1978).
42. CROTEAU, R.: Biosynthesis of Monoterpenes. In: Biosynthesis of Isoprenoid Compounds, Vol. I, pp. 225—282 (J. W. PORTER and S. L. SPURGEON, eds.). New York: Wiley and Sons. 1981.
43. THEILER, R. F., J. C. COOK, and L. P. HAGAR: Halohydrocarbon Synthesis by Bromoperoxidase. Science **202,** 1094 (1978).
44. HAGER, L. P., D. R. MORRIS, F. S. BROWN, and H. EBERWEIN: Chloroperoxidase II. Utilization of Halogen Anions. J. Biol. Chem. **241,** 1769 (1966).
45. PEDERSEN, M.: A Brominating and Hydroxylating Peroxidase from the Red Alga *Cystoclonium purpureum*. Physiol. Plant **37,** 6 (1976).
46. THEILER, R. F., J. S. SIUDER, and L. P. HAGER: Bromoperoxidase from the Red Algae *Bonnemasisonia hamifera*. In: Food and Drugs from the Sea, pp. 153—169 (P. W. KAUL and C. J. SINDEMANN, eds.). Oklahoma: University of Oklahoma Press. 1978.
47. FENICAL, W.: Molecular Aspects of Halogen-based Biosynthesis of Marine Natural Products. Recent Adv. Phytochem. **13,** 214 (1979).
48. ICHIKAWA, N., Y. NAYA, and S. ENOMOTO: New Halogenated Monoterpenes from *Desmia (Chondrococcus) hornemanni*. Chem. Lett. (Japan) 1333 (1974).
49. YOSHIHARA, K., and Y. HIROSE: The Derivation of a Brominated Algal Component from Myrcene. Bull. Chem. Soc. (Japan) **51,** 653 (1978).
50. MASAKI, Y., K. HASHIMOTO, K. SAKUMA, and K. KAJI: New Ochtodane Syntheses from Myrcene. Tetrahedron Letters 1481 (1982).
51. MYNDERSE, J. S., D. J. FAULKNER, J. FINER, and J. CLARDY: (1R, 2S, 4S, 5R)-1-Bromo-*trans*-2-chlorovinyl-4,5-dichloro-1,5-dimethyl cyclohexane, a New Monoterpene Skeletal Type from the Red Alga *Plocamium violaceum*. Tetrahedron Letters 2175 (1975).

52. HIGGS, M. D., D. J. VANDERAH, and D. J. FAULKNER: Polyhalogenated Monoterpenes from *Plocamium cartilagineum* from the British Coast. Tetrahedron **33,** 2775 (1977).
53. STIERLE, D. B., and J. J. SIMS: Marine Natural Products — XV. Polyhalogenated Cyclic Monoterpenes from the Red Alga *Plocamium cartilagineum* of Antarctica. Tetrahedron **35,** 1261 (1969).
54. STIERLE, D. B., R. M. WING, and J. J. SIMS: Marine Natural Products XI, Costatone and Costatolide, New Polyhalogenated Monoterpenes from the Red Seaweed *Plocamium costatum*. Tetrahedron Letters 4455 (1976).
55. HOGBERG, H. E., R. H. THOMPSON, and T. J. KING: The Cymopols, A Group of Prenylated Bromohydroquinones from the Green Calcareous Alga *Cymopolia barbata*. J. Chem. Soc. Perkin I, 1696 (1976).
56. KHO-WISEMAN, E.: Natural Products Chemistry of the Red Marine Algae *Plocamium violaceum* (Farlow) and *Plocamium cartilagineum* (Dixon). Ph. D. Thesis, University of California, Santa Cruz (1978).
57. HOWARD, B. M., A. M. NONOMURA, and W. FENICAL: Chemotaxonomy in Marine Algae. Secondary Metabolite Synthesis by *Laurencia* in Unialgal Culture. Biochem. Syst. Ecol. **8,** 329 (1980).
58. CREWS, P., L. CAMPBELL, and E. HERON: Different Chemical Types of *Plocamium violaceum* (Rhodophyta) from the Monterey Bay Region, California. J. Phycol. **13,** 297 (1977).
59. MYNDERSE, J. S., and D. J. FAULKNER: Variations in the Halogenated Monoterpene Metabolites of *Plocamium cartilagineum* and *P. violaceum*. Phytochem. **17,** 237 (1978).
60. FENICAL, W., and J. N. NORRIS: Chemotaxonomy in Marine Algae: Chemical Separat ion of Some *Laurencia* Species (Rhodophyta) from the Gulf of California. J. Phycol. **11,** 104 (1975).
61. ABBOTT, I. A., and G. J. HOLLENBERG: Marine Algae of California, p. 827. Stanford, Ca.: Stanford University Press. 1973.
62. DIXON, P. S.: Biology of the Rhodophyta, p. 285. Edinburgh: Oliver and Boyd. 1973.
63. — The Typification of *Fucus cartilagineus* and *F. corneus* Huds. Blumea **15,** 55 (1967).
64. MYNDERSE, J. S., and D. J. FAULKNER: Polyhalogenated Monoterpenes from the Red Alga *Plocamium cartilagineum*. Tetrahedron **31,** 1963 (1975).
65. CREWS, P., and E. KHO-WISEMAN: Acyclic Polyhalogenated Monoterpenes from the Red Alga *Plocamium violaceum*. J. Org. Chem. **42,** 2812 (1977).
66. FENICAL, W.: "Chemical Variation in *Laurencia pacifica* (Kylin) Sesquiterpenes", unpublished results.
67. TURSCH, B.: Some Recent Developments in the Chemistry of Alcyonaceans. Pure Appl. Chem. **48,** 1 (1976).
68. CREWS, P., E. KHO-WISEMAN, and P. MONTANA: Halogenated Alicyclic Monoterpenes from the Red Alga *Plocamium*. J. Org. Chem. **43,** 116 (1978).
69. MCCONNELL, O. J., and W. FENICAL: Ochtodene and Ochtodiol: Novel Polyhalogenated Cyclic Monoterpenes from the Red Seaweed *Ochtodes secundiramea*. J. Org. Chem. **43,** 4238 (1978).
70. NAYA, Y., Y. HIROSE, and N. ICHIKAWA: Labile Halogenated Monoterpenes from *Desmia (Chondrococcus) japonicus* (Harvey). Chem. Letters (Japan) 839 (1976).
71. CREWS, P., P. NG., E. KHO-WISEMAN, and C. PACE: Halogenated Monoterpenes from the Red Alga *Microcladia*. Phytochem. **15,** 1707 (1976).
72. SMITH, G. M.: Marine Algae of the Monterey Peninsula, California, p. 752. Stanford, Ca.: Stanford University Press. 1969.
73. FENICAL, W.: Personal communication.
74. OHTA, K., and M. TAKAGI: Halogenated Sesquiterpenes from Marine Red Alga *Marginisporum aberrans*. Phytochem. **16,** 1062 (1977).
75. RINEHART, K. L., JR., R. D. JOHNSON, J. F. SIUDA, G. E. KREJCAREK, P. D. SHAW, J. A.

MCMILLAN, and I. C. PAUL: Structures of Halogenated and Antimicrobial Organic Compounds from Marine Sources. In: The Nature of Seawater, pp. 651—666 (E. D. GOLDBERG, ed.). Berlin: Abakon Verlagsgesellschaft. 1975.

76. BATES, P., J. W. BLUNT, M. P. HARTSHORN, A. J. JONES, M. H. G. MUNRO, W. T. ROBINSON, and S. C. YORKE: Halogenated Metabolites of the Red Alga *Plocamium cruciferum*. Aust. J. Chem. **32,** 2545 (1979).
77. MABRY, T. J.: The Chemistry of Geographical Races. Pure Appl. Chem. **34,** 377 (1973).
78. FAULKNER, D. J., and C. IRELAND: The Chemistry of Some Opisthobranch Molluscs. In: Marine Natural Products Chemistry, Vol. IV: 1, pp. 23—24 (D. J. FAULKNER and W. FENICAL, eds.). New York: Plenum Press. 1977.
79. SCHULTE, G. R., and P. J. SCHEUER: Defense Allonomes of Some Marine Molluscs. Tetrahedron **38,** 1857 (1982).
80. THOMPSON, J. E., R. P. WALKER, S. J. WRATTEN, and D. J. FAULKNER: A Chemical Defense Mechanism for the Nudibranch *Cadlina luteomarginata*. Tetrahedron **38,** 1865 (1982).
81. IMPERATO, F., L. MINALE, and R. RICCIO: Constituents of the Digestive Gland of Molluscs of the Genus Aplysia. II. Halogenated Monoterpenes from *Aplysia limacina*. Experientia **33,** 1273 (1977).
82. STALLARD, M. O., and D. J. FAULKNER: Chemical Constituents of the Digestive Gland of the Sea Hare *Aplysia californica* — II. Chemical Transformations. Comp. Biochem. Physiol. **49B,** 37 (1974).
83. IRELAND, C., M. O. STALLARD, D. J. FAULKNER, J. FINER, and J. CLARDY: Some Chemical Constituents of the Digestive Gland of the Sea Hare *Aplysia californica*. J. Org. Chem. **41,** 2461 (1976).
84. HAGAR, L. P.: Mother Nature Likes Some Halogenated Compounds. Basic Life Sci. **19,** 415 (1982).
85. FINER, J., K. HIROTSU, and J. CLARDY: X-ray Diffraction and the Structure of Marine Natural Products. In: Marine Natural Products Chemistry, Vol. IV, pp. 147—158 (D. J. FAULKNER and W. FENICAL, eds.). New York: Plenum Press. 1977.
86. CREWS, P., and E. KHO: Plocamene B, a New Cyclic Monoterpene Skeleton from a Red Marine Alga. J. Org. Chem. **40,** 2568 (1975).
87. VAN ENGEN, D., J. CLARDY, E. KHO-WISEMAN, P. CREWS, M. D. HIGGS, and D. J. FAULKNER: Violacene: A Reassignment of Structure. Tetrahedron Letters 29 (1978).
88. WOOLARD, F. X., R. E. MOORE, D. VAN ENGEN, and J. CLARDY: The Structure and Absolute Configuration of Chondrocolactone, a Halogenated Monoterpene from the Red Alga *Chondrococcus hornemanni*, and a Revised Structure for Chondrocole A. Tetrahedron Letters 2367 (1978).
89. BENYON, J. H.: Mass Spectrometry and Its Applications to Organic Chemistry. New York: Elsevier Publishing Co. 1960.
90. SQUILLACOTE, M. E., R. S. SHERIDAN, O. L. CHAPMAN, and F. A. L. ANET: Planar *s-cis*-1,3-butadiene. J. Amer. Chem. Soc. **101,** 3657 (1979).
91. HARRIS, R. K., and A. V. CUNLIFFE: Nuclear Magnetic Resonance Studies of 1,3-butadienes. IX — The $^1H$ Spectra of Isoprene and Related Compounds. Org. Mag. Resonance **9,** 483 (1977).
92. STIERLE, D. B., R. M. WING, and J. J. SIMS: Marine Natural Products — XVI. Polyhalogenated Acyclic Monoterpenes from the Red Alga *Plocamium* of Antarctica. Tetrahedron **35,** 2855 (1979).
93. MYNDERSE, J. S., and D. J. FAULKNER: Violacene, a Polyhalogenated Monocyclic Monoterpene from the Red Alga *Plocamium violaceum*. J. Amer. Chem. Soc. **96,** 6771 (1974).
94. NORTON, R. S., R. G. WARREN, and R. J. WELLS: Three New Polyhalogenated Monoterpenes from *Plocamium* species. Tetrahedron Letters 3905 (1977).

*95.* BURRESON, B. J., F. X. WOOLARD, and R. E. MOORE: Chondrocole A and B, Two Halogenated Dimethylhexahydrobenzofurans from the Red Alga *Chondrococcus hornemanni* (Mertens) Schmitz. Tetrahedron Letters 2155 (1975).
*96.* — — — Evidence for the Biogenesis of Halogenated Myrcenes from the Red Alga *Chondrococcus hornemanni.* Chem. Letters (Japan) 1111 (1975).
*97.* CREWS, P.: Monoterpene Halogenated by the Red Alga *Plocamium oregonum.* J. Org. Chem. **42,** 2634 (1977).
*98.* LEVY, G. C., R. L. LICHTER, and G. L. NELSON: Carbon-13 Nuclear Magnetic Resonance Spectroscopy. 2nd Edition, p. 338. Wiley-Interscience. 1980.
*99.* CREWS, P., and E. KHO-WISEMAN: Stereochemical Assignments in Marine Natural Products by $^{13}$C NMR Gamma Effects. Tetrahedron Letters 2483 (1978).
*100.* GONZALEZ, A. G., J. M. ARTEAGA, J. D. MARTIN, M. L. RODRIGUEZ, J. FAYOS, and M. MARTINES-RIPOLLS: Two New Polyhalogenated Monoterpenes from the Red Alga *Plocamium cartilagenium.* Phytochem. **17,** 947 (1978).
*101.* WOOLARD, F. X., R. E. MOORE, M. MAHINDRAN, and A. SIVAPALAN: (−)-3-Bromomethyl-3-chloro-7-methyl-1,6-octa-Diene from Sri Lankan *Chondrococcus hornemanni.* Phytochem. **15,** 1069 (1976).
*102.* BLUNT, J. W., M. P. HARTSHORN, M. H. G. MUNRO, and J. C. YORKE: A novel, $C_8$ Dichlorodienol Metabolite of the Red Alga *Plocamium cruciferum.* Tetrahedron Letters 4417 (1978).
*103.* WILCOTT, M. R., R. E. DAVIS, D. J. FAULKNER, and M. O. STALLARD: The Configuration and Conformation of 7-Chloro-1,6-dibromo-3,7-dimethyl-3,4-epoxy-1-octene. Tetrahedron Letters 3967 (1973).
*104.* DUNLOP, R. W., P. T. MURPHY, and R. J. WELLS: A New Polyhalogenated Monoterpene from the Red Alga *Plocamium angustum.* Aust. J. Chem. **32,** 2735 (1979).
*105.* KAZLAUSKAS, R., P. T. MURPHY, R. J. QUINN, and R. J. WELLS: Two Polyhalogenated Monoterpenes from the Red Alga *Plocamium costatum.* Tetrahedron Letters 4451 (1976).

*(Received December 10, 1982)*

# The C-Nucleoside Antibiotics

By J. G. Buchanan, Department of Chemistry,
Heriot-Watt University, Riccarton, Currie, Edinburgh, U. K.

**Contents**

I. Introduction ... 243
II. General Aspects of *C*-Nucleosides ... 246
III. Showdomycin ... 250
IV. The Formycins ... 264
V. Pyrazofurin (Pyrazomycin) ... 270
VI. Oxazinomycin (Minimycin) ... 276
VII. The Ezomycins ... 279
VIII. Biosynthesis of *C*-Nucleoside Antibiotics ... 284
References ... 290

## I. Introduction

Nucleosides, in which a carbohydrate unit is linked to a heterocyclic base, are fundamental components of nucleic acids and nucleotide coenzymes, where they occur in phosphorylated form. In most nucleosides the sugar, either D-ribose or 2-deoxy-D-*erythro* pentose ("2-deoxyribose"), is linked to a nitrogen atom of a pyrimidine or purine, as in the ribonucleosides uridine (**1**) and adenosine (**2**).

With the introduction of improved analytical methods in the post-war era a new uracil-containing nucleotide was discovered in hydrolysates of ribonucleic acids (RNA) (*1, 2*). The parent nucleoside was obtained by enzymic dephosphorylation and shown by Cohn (*3, 4*), Allen (*2, 5*) and their coworkers to be 5-β-D-ribofuranosyluracil (**3**) containing a *C*-ribosyl linkage (*6*). It was given the name pseudouridine (*3*). Pseudouridine itself has been isolated from urine (*7*), from culture filtrates of *Strepto-verticillicum ladakanus* (*8*) and from *Streptomyces platensis* var. *clarensis* together with its 1-methyl derivative (**4**) (*9*).

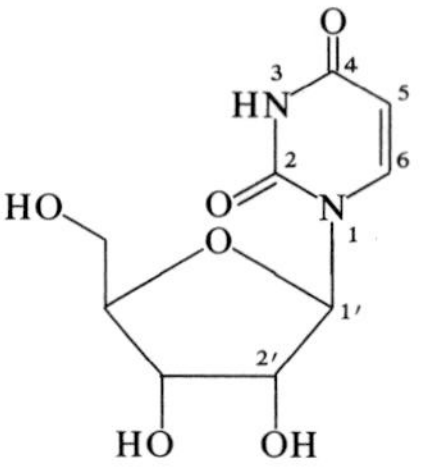

**(1)** 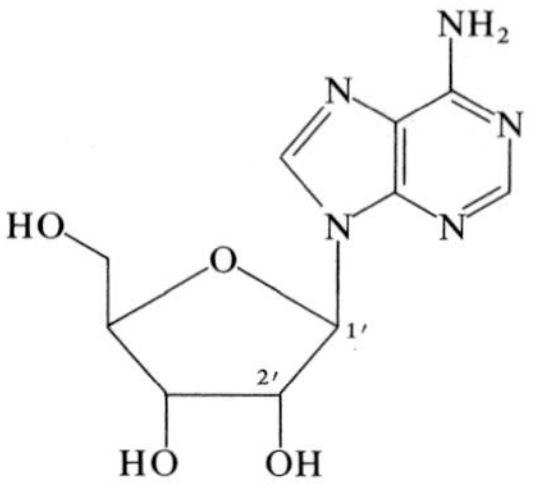 **(2)** 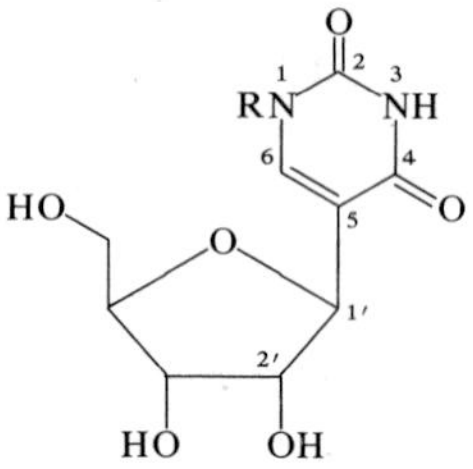 **(3)** R = H
**(4)** R = Me

In the past thirty years a number of nucleosides have been discovered in microorganisms and shown to have antibiotic properties. The chemistry of these "nucleoside antibiotics" has been described in several reviews (*10—14*). They differ from the common nucleosides by modifications to the sugar or heterocyclic base component. They owe their biological activity to the way in which they, or their corresponding nucleotides, inhibit important enzymic processes such as nucleotide and nucleic acid metabolism, protein biosynthesis, biological methylation and chitin biosynthesis (*12, 15*).

A number of the nucleoside antibiotics contain a β-D-ribofuranosyl unit linked to a carbon atom in a heterocyclic ring and are termed *C*-glycosyl nucleosides or simply *C*-nucleosides. This Review concerns the chemistry of the *C*-nucleoside antibiotics of natural occurrence, some six in number (**5—10**). Although neither pseudouridine (**3**) nor its 1-methyl derivative (**4**) show antibiotic activity their chemistry will be dealt with insofar as it impinges on the general theme. There is a vast literature on the synthesis of *C*-nucleoside analogues (*16—18*), some of them close analogues of the natural compounds, others barely recognisable as such; these aspects will not be considered.

In the next section we shall discuss briefly the features common to *C*-nucleosides which enable them to be distinguished from the more common *N*-nucleosides. The following sections then deal with the antibiotics in turn, including reference to biological activity. The final section describes the biosynthesis of *C*-nucleosides, as far as it is known.

*References, pp. 290—299*

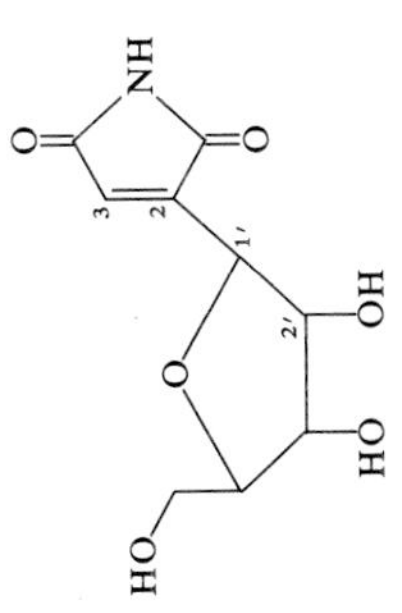

**(5)** showdomycin

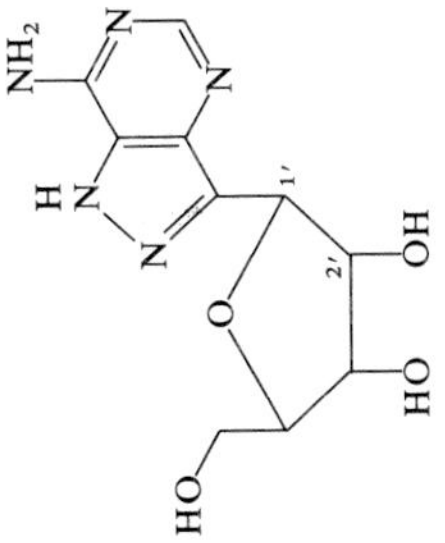

**(6)** formycin

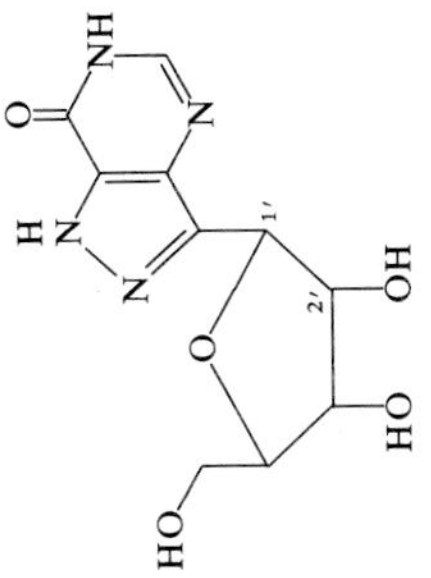

**(7)** formycin B

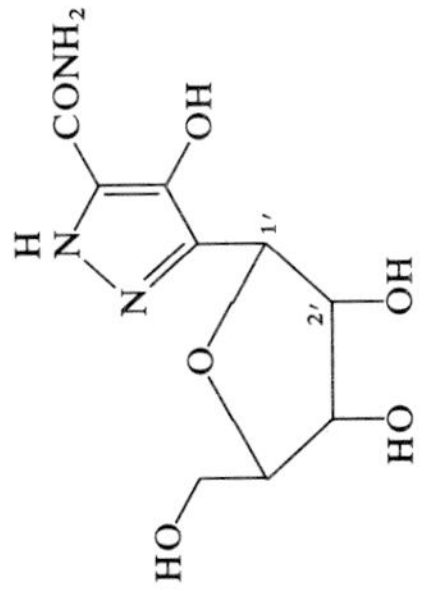

**(8)** pyrazofurin (pyrazomycin)

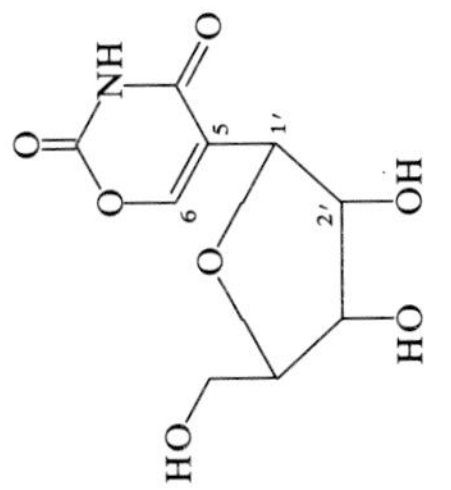

**(9)** oxazinomycin (minimycin)

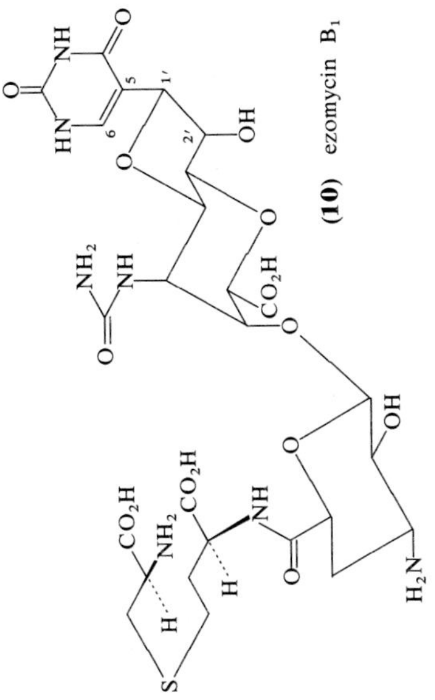

**(10)** ezomycin $B_1$

## II. General Aspects of C-Nucleosides

### 1. Behaviour Towards Chemical Reagents

#### *a) Aqueous Acid*

The *N*-ribofuranosyl derivatives of purines and pyrimidines undergo hydrolysis in hot dilute mineral acid. In the purine nucleosides, hydrolysis is sufficiently rapid so that both ribose and purine may be isolated. More vigorous conditions are required to hydrolyse pyrimidine nucleosides, resulting in decomposition of the ribose produced (*19*). As would be expected, the ribosyl-carbon linkage in *C*-nucleosides is stable under acidic conditions and, it may be noted in passing, towards enzymic cleavage.

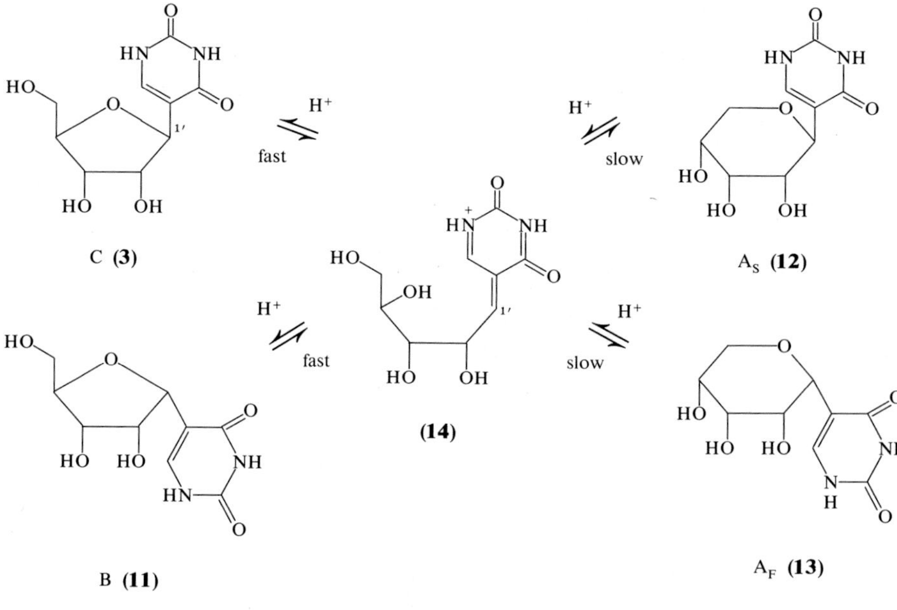

C (**3**) $\underset{\text{fast}}{\overset{H^+}{\rightleftharpoons}}$ (**14**) $\underset{\text{slow}}{\overset{H^+}{\rightleftharpoons}}$ $A_S$ (**12**)

B (**11**) $\underset{\text{fast}}{\overset{H^+}{\rightleftharpoons}}$ (**14**) $\underset{\text{slow}}{\overset{H^+}{\rightleftharpoons}}$ $A_F$ (**13**)

*Scheme 1*

One of the most interesting discoveries in the chemistry of pseudouridine (**3**) is that the ribose moiety undergoes isomerisation in aqueous acid (*3, 6*). When pseudouridine is treated with M-hydrochloric acid at 100° three other isomers are formed, together with pseudouridine itself, and can be separated by borate ion exchange chromatography (*3*) (Scheme 1). In the early stages of the reaction pseudouridine (**3**), termed pseudouridine C, is in equilibrium with its α-furanose isomer, pseudouridine B (**11**). Later, appreciable amounts of the pyranose isomers $A_S$ (**12**) and $A_F$ (**13**) are produced and preponderate at equilibrium (*3*). Prolonged contact with acid affords other compounds which have not been characterised and at no stage is free ribose formed. The kinetics of the reaction resemble the Fischer glycosidation of a sugar with acidic methanol where the first-formed furanosides are gradually superseded by the thermodynamically more stable pyranosides (*20*). The behaviour of pseudouridine is due to the ability of the pyrimidine ring to stabilise a cationic centre at C-1′ of the ribose unit, as in (**14**) (*6*), i.e. C-1′ has benzylic character. In the synthesis of pseudouridine (*21—23*) the behaviour of the intermediate acyclic polyol derivatives (**15**) and (**16**) towards aqueous acid has been studied (Scheme 2) (*22*). Under mild reaction conditions the D-*altro*-pentitol (**15**) affords mainly pseudouridine (**3**) and the D-*allo*-pentitol (**16**) mainly the α-anomer* (**11**) before equilibration ensues. The initial ring closure therefore involves predominant inversion of configuration at C-1′ and is a stereospecific $S_N2$-type process (*22, 23*).

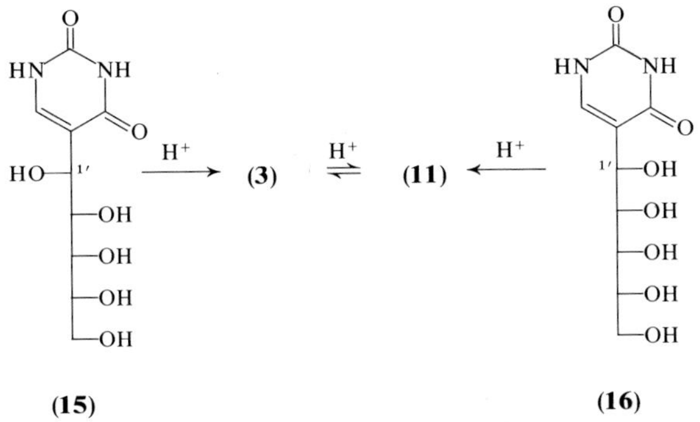

(**15**) $\xrightarrow{H^+}$ (**3**) $\underset{}{\overset{H^+}{\rightleftharpoons}}$ (**11**) $\xleftarrow{H^+}$ (**16**)

*Scheme 2*

Pseudouridine also rearranges under alkaline conditions (Scheme 3). A mechanism involving the unsaturated intermediate (**17**), analogous to (**14**), has been invoked (*6*).

Isomerisation of the sugar portions in pyrazofurin (**8**) and ezomycin $B_1$ (**10**) occurs and will be discussed in the appropriate sections (V and VII, respectively).

* Strictly speaking, the terms anomer and anomeric refer only to glycosides and should not be applied to *C*-glycosyl compounds. In company with other authors, we have used them in this Review for reasons of simplicity.

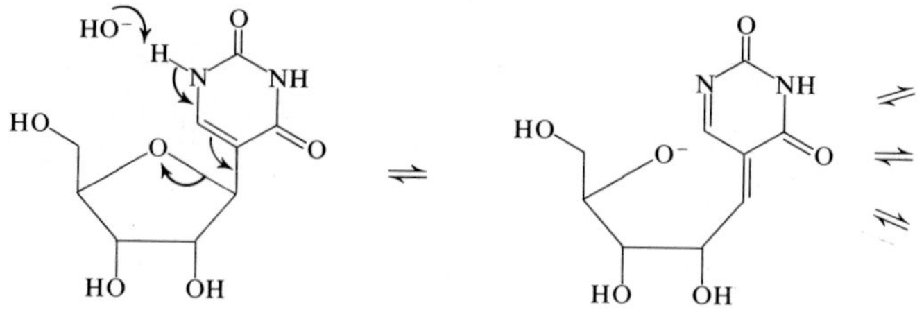

Scheme 3

### *b) Aqueous Hydrazine*

In a successful attempt to demonstrate the presence of ribose in pseudouridine (**3**) DAVIS and ALLEN (*2*) treated it with hydrazine, followed by cleavage of the sugar hydrazone with benzaldehyde; ribose was detected chromatographically. The mechanism of the reaction has not been determined but must be more complex than the hydrazinolysis of uridine (**1**) which also yields ribose under these conditions (*24*). Ribose has been obtained from showdomycin (**5**) (*25*) and oxazinomycin (minimycin) (**9**) (*26*) by this degradation.

## 2. Spectroscopic Properties

The spectroscopic methods giving the most reliable guide to the presence of a glycosyl carbon bond are nmr and mass spectrometry (*27, 28*).

### *a) Nuclear Magnetic Resonance*

In the $^{1}H$ nmr spectra of *N*-ribonucleosides the signal due to H-1′ of the ribose portion, the "anomeric" proton, is easily recognised by being downfield of the other ribose methine and methylene protons. This is due to the deshielding effect of the attachment of C-1′ to both oxygen and nitrogen atoms. In the spectra of *C*-nucleosides the corresponding signal, as to be expected, is at higher field, but is still deshielded relative to the other ribose protons. Table 1 illustrates the difference between the H-1′ chemical shift in *N* and *C*-nucleosides. $^{1}H$-Nmr spectroscopy was used in the final stages of the structure determination of pseudouridine and its isomers (*3*) and played a vital part in the case of the other *C*-nucleosides.

Table 1*. *$^{1}$H-Nmr and $^{13}$C-Nmr Signals (p.p.m. from TMS or DSS) at the Anomeric Position in Nucleosides and C-Nucleosides*

| Compound | H-1′ | | C-1′ | |
|---|---|---|---|---|
| | $D_2O$ | $d_6$-DMSO | $D_2O$ | $d_6$-DMSO |
| Uridine (**1**) | 6.0 (*25*)<br>5.90 (*29*) | 5.79 (*29*) | 89.5 (*36*) | |
| Adenosine (**2**) | | 6.15 (*30*) | | 88.4 (*36*) |
| Pseudouridine (**3**) | 4.72 (*25*)<br>4.67 (*29*) | 5.08 (*30*)<br>4.47 (*29*) | 78.6 (*36*) | 79.13 (*37*) |
| Showdomycin (**5**) | 4.82 (*25*) | 4.63 (*31*) | | 77.52 (*37*) |
| Formycin (**6**) | 5.37 (*25*)<br>5.24 (*32*) | 5.14 (*25, 30*) | | 78.18 (*38*) |
| Formycin B (**7**) | 5.24 (*32*) | 5.07 (*30*) | | 77.53 (*38*) |
| Pyrazofurin (**8**) | 5.42 (*33*) | | 75.4 (*36*) | |
| Oxazinomycin (**9**) | 4.65 (*26*) | 4.5 (*34*) | 78.8 (*39*) | |
| Ezomycin $B_1$ (**10**) | 5.01 (*35*) | | | |

* Data were obtained under a variety of experimental conditions. The original papers should be consulted before detailed comparisons are made.

In the $^{1}$H-nmr spectra of all β-ribonucleosides H-1′ is coupled to H-2′, the value of $J_{1',2'}$ varying from 2—7 Hz. For all of the *N*-ribonucleosides and the pyrazole-containing *C*-nucleosides the H-1′ signal appears as a doublet. In the spectrum of showdomycin (**5**), however, H-1′ shows allylic coupling to the maleimide ring proton, H-3 ($J_{1',3}$ 1.5 Hz) (*25, 31*). Similarly in the spectra of pseudouridine (**3**) and oxazinomycin (**9**) H-1′ appears as a double doublet due to allylic coupling between H-1′ and H-6 [0.8 Hz (*40*) and 1.2 Hz (*34*), respectively].

The $^{13}$C nmr spectra of ribonucleosides are particularly informative (*41*). The low-field region contains the signals due to the heterocyclic portion of the molecule while the ribose carbon atoms lie in the range 60—90 p.p.m. from TMS (*36*). In the *N*-nucleosides, typefied by uridine (**1**) and adenosine (**2**), C-1′ has the lowest-field ribose signal (Table 1). In *C*-nucleosides C-1′ is ca. 10 p.p.m. upfield, and the remaining ribose signals are essentially unchanged; C-4′ is then the lowest-field signal.

The nmr spectra of ezomycin $B_1$ (**10**) and its derivatives have special features which are dealt with in section VII.

### *b) Mass Spectrometry*

The mass spectra of *N*-nucleosides show intense peaks due to the parent heterocyclic base plus one or two hydrogen atoms. These arise by cleavage of the C-1′-N bond (*42*). In *C*-nucleosides the base peak corresponds to the heterocycle +30 (*28*). Although an ion of this mass is also produced from *N*-nucleosides its intensity is much less. The ion has a protonated formyl group attached to the base as shown in structure (**18**) (Scheme 4) for showdomycin (**5**) (*43*). The "B + 30" ion is formed by all *C*-nucleosides studied by this technique (*28, 43, 44*). Its formation is due in part to the benzylic character of C-1′ in *C*-nucleosides [see section II 1 (a)].

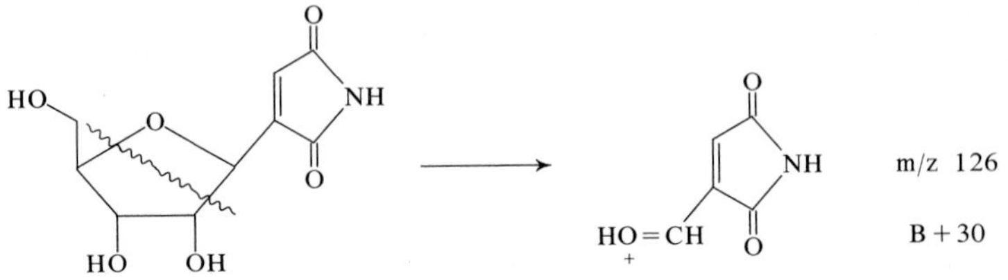

**(5)** → **(18)** m/z 126 B + 30

*Scheme 4*

### *c) Other Physical Methods*

The heterocyclic portions of *C*-nucleosides give characteristic ultraviolet spectra which have proved valuable in structure determinations. Measurements of optical rotatory dispersion and circular dichroism provide evidence as to the anomeric configuration and the conformation of nucleosides (*45*). Nucleoside conformations have also been extensively studied by nmr methods (*46*).

The structures of most of the *C*-nucleoside antibiotics have been confirmed by X-ray crystallography.

## III. Showdomycin

### 1. Isolation and Structure Determination

In 1964 NISHIMURA *et al.*, of the Shionogi Research Laboratory isolated showdomycin from *Streptomyces showdoensis* (*47*). It had antibiotic activity *in vitro,* particularly towards *Streptococcus hemolyticus* and inhibited EHRLICH ascites tumour in mice. Showdomycin was also isolated in the Merck, Sharp and Dohme Research Laboratories, as antibiotic MSD 125 A. The structure was shown to be 2-β-D-ribofuranosylmaleimide (**5**) by R. K. ROBINS and his coworkers (*25*) and by a Shionogi group (*31*), working independently.

ROBINS (*25*) was able to degrade showdomycin to ribose by hydrazinolysis [section II 1 (b)]. The uv absorption spectrum was similar to that of maleimide and the $^{1}H$ nmr spectrum strongly resembled that of pseudouridine (**3**). Other data confirmed the structure as (**5**).

NAKAGAWA *et al.* (*31*) prepared a number of derivatives, including the acetonide (**19**) (Scheme 5). Reaction of (**19**) with aqueous potassium carbonate followed by methylation afforded the *spiro* derivative (**20**) via an internal Michael addition of O-5′ to the maleimide double bond. Removal of the isopropylidene group followed by acetylation gave the diacetate (**21**) whose $^{1}H$ nmr spectrum was well resolved. An X-ray study of the hydrobromide salt (**22**) of the derived pyrrolidine completely settled the structure (*31, 48*).

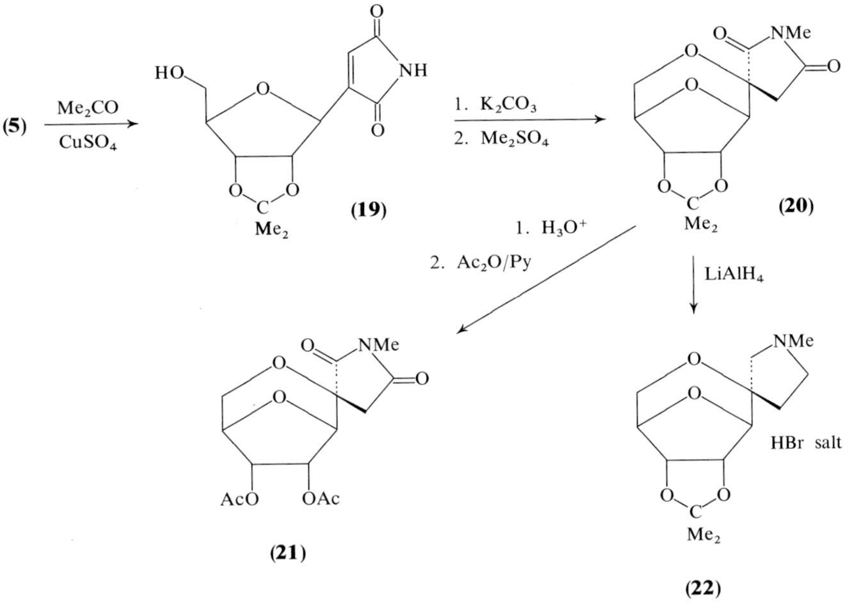

*Scheme 5*

## 2. Synthesis

The synthesis of showdomycin, a popular synthetic target, has been tackled by two distinct routes. The first uses D-ribose as a starting material, leading to optically active showdomycin. In the second the ribose unit is derived from furan by cycloaddition, followed by stereoselective hydroxylation. The earlier work, particularly relating to the preparation of precursors, has been reviewed (*17, 18*).

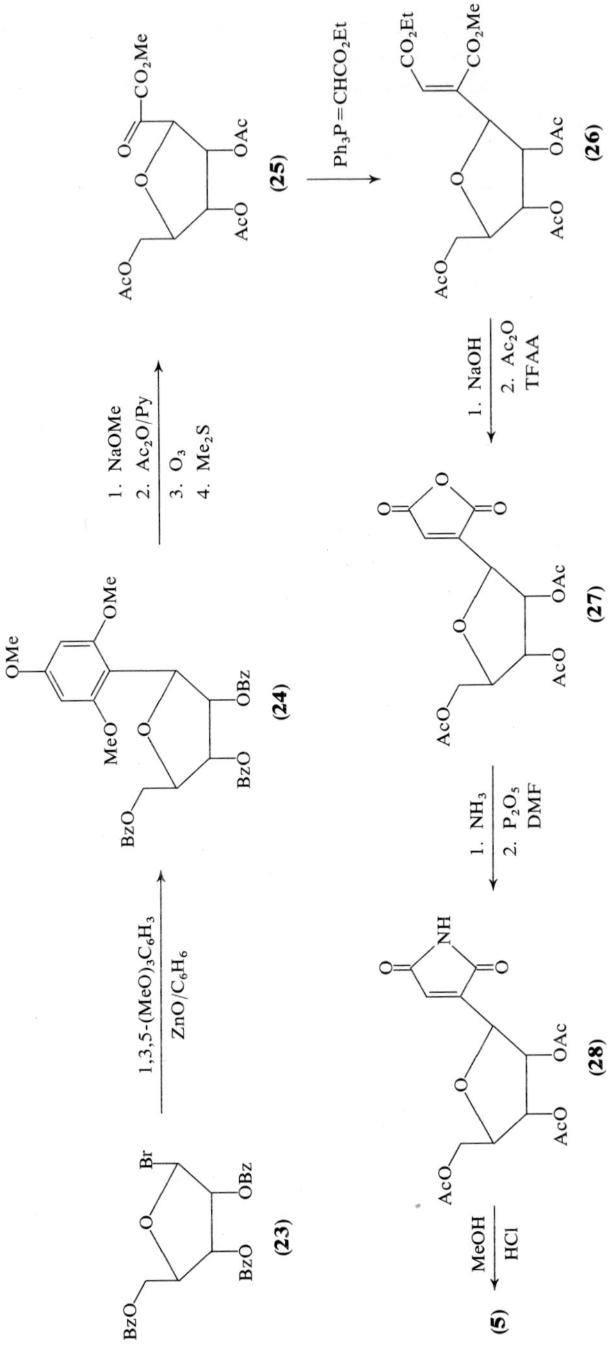

(23) → 1,3,5-$(MeO)_3C_6H_3$, $ZnO/C_6H_6$ → (24) → 1. NaOMe 2. $Ac_2O/Py$ 3. $O_3$ 4. $Me_2S$ → (25) → $Ph_3P{=}CHCO_2Et$ → (26) → 1. NaOH 2. $Ac_2O$ TFAA → (27) → 1. $NH_3$ 2. $P_2O_5$ DMF → (28) → MeOH HCl → (5)

*Scheme 6*

In syntheses from D-ribose a key reaction is the formation of the C-glycosyl bond. The first synthesis, by KALVODA, FARKAŠ and ŠORM used a reaction of the Friedel-Crafts type to link ribose, as the bromide (**23**), to the activated aromatic ring of 1,3,5-trimethoxybenzene (Scheme 6). The symmetry of the product (**24**) allows ozonolysis to the ketoester (**25**) after change of protecting groups. Reaction with the phosphorane afforded mainly the maleate isomer (**26**) which was converted into the anhydride (**27**). The maleimide (**28**) was obtained by ring closure of the maleamic acid and deprotected by acid methanolysis to give showdomycin (**5**). KALVODA (*50*) later devised a second synthesis, from the nitrile (**29**) (*51*) (Scheme 7). Hydrolysis of the cyano group followed by exchange of protecting groups gave the 2,5-anhydro-D-allonic acid (**30**). The acid chloride was treated with HCN in the presence of an excess of phosphorane to yield an E, Z mixture [58% overall from the nitrile (**29**)] of which the *E* isomer (**32**) constituted 72%. The mixture was converted into showdomycin by acetolysis followed by methanolysis [40% yield from (**32**)]. Similar intermediates can be used in a synthesis of formycin (**6**) (*52*) (section IV).

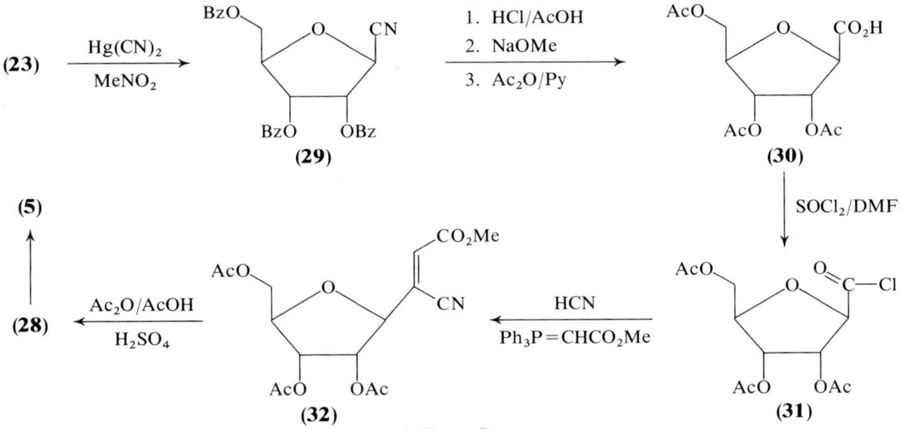

(23) —$Hg(CN)_2$ / $MeNO_2$→ (29) [BzO, CN, BzO, OBz] —1. HCl/AcOH 2. NaOMe 3. $Ac_2O$/Py→ (30) [AcO, $CO_2H$, AcO, OAc] —$SOCl_2$/DMF→ (31) [AcO, O=C—Cl, AcO, OAc] —HCN / $Ph_3P{=}CHCO_2Me$→ (32) [AcO, $CO_2Me$, CN, AcO, OAc] —$Ac_2O$/AcOH / $H_2SO_4$→ (28) → (5)

*Scheme 7*

MOFFATT's approach to showdomycin (*53*) used the 2,5-anhydro-D-allose derivative (**35**) (*54*) derived from the nitrile (**29**) (Scheme 8). Reductive hydrolysis of the nitrile (**29**) in the presence of *N,N'*-diphenylethylenediamine afforded the imidazolidine (**33**) whose benzoyl protecting groups were exchanged for benzyl under basic conditions to give the tribenzyl ether (**34**). Careful acid hydrolysis gave the aldehyde (**35**) which was converted into the ketoester (**36**) via the epimeric hydroxyamides and hydroxyesters. Reaction with the amide Wittig reagent was accompanied by cyclization to give directly the maleimide (**37**) which was deprotected using boron trichloride.

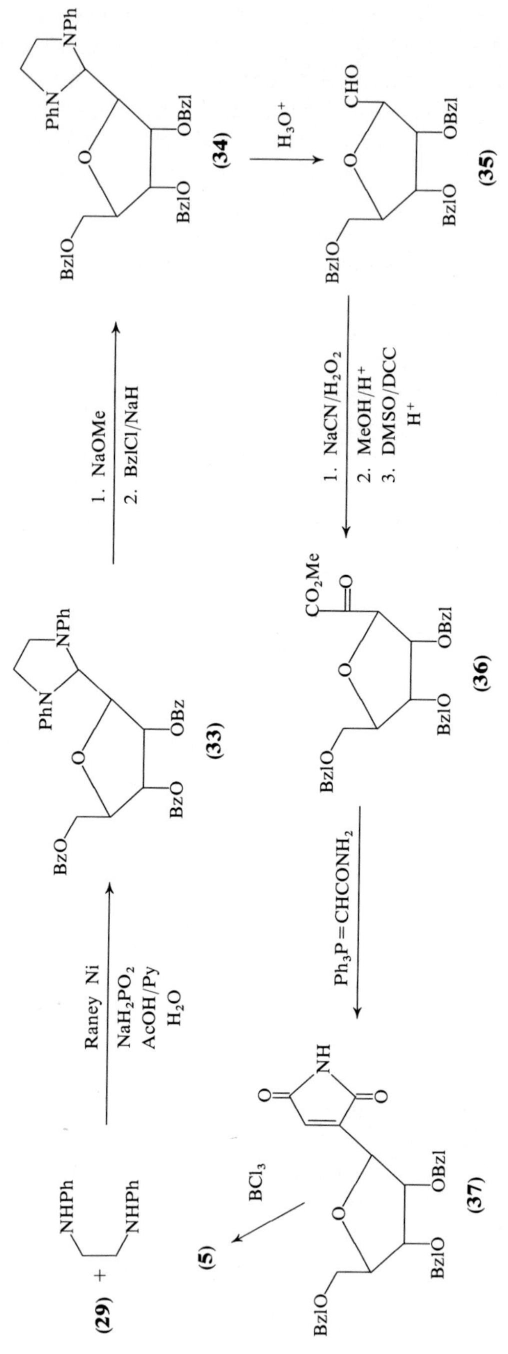

*Scheme 8*

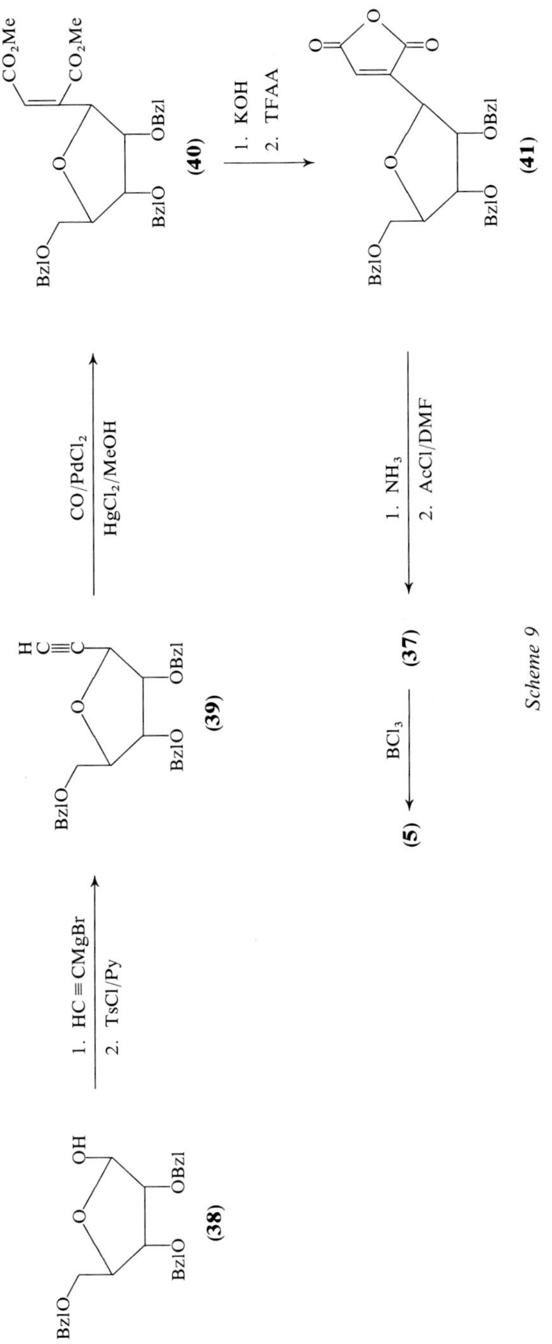

BzlO, BzlO, OBzl, OH
**(38)**

1. $HC \equiv CMgBr$
2. TsCl/Py

BzlO, BzlO, OBzl, $C \equiv CH$
**(39)**

$CO/PdCl_2$
$HgCl_2/MeOH$

BzlO, BzlO, OBzl, $CO_2Me$, $CO_2Me$
**(40)**

1. KOH
2. TFAA

BzlO, BzlO, OBzl, O, O, O
**(41)**

1. $NH_3$
2. AcCl/DMF

**(5)** ← $BCl_3$ — **(37)**

*Scheme 9*

The same intermediate (**37**) was obtained by a sequence involving acetylenic precursors (Scheme 9) (*55, 56*). 2,3,5-Tri-*O*-benzyl-D-ribofuranose (**38**) (*57*) reacted with ethynylmagnesium bromide to form a mixture of epimeric diols. Cyclization with toluene-*p*-sulphonyl chloride in pyridine, followed by chromatography, gave the crystalline ethyne (**39**) in 52% overall yield (*55*). The ethyne (*39*) was converted into the maleic diester (**40**) by dicarbomethoxylation (*58*) and thence into the anhydride (**41**) and imide (**37**) by methods similar to those used by KALVODA *et al.* (*49*).

A recent synthesis of showdomycin from D-ribose, where the four carbon atoms of the maleimide are added as a unit, is shown in Scheme 10 (*59*). The acetate (**42**) reacts under Lewis acid catalysis with the bis enol ether (**43**) to give the cyclobutanone (**44**) in 92% yield. Conversion of (**44**) into its TMS enol ether followed by nitrosation gave the oxime (**45**) which rearranged spontaneously into the malic acid nitrile (**46**). Ammonolysis and ring closure then gave showdomycin (**5**).

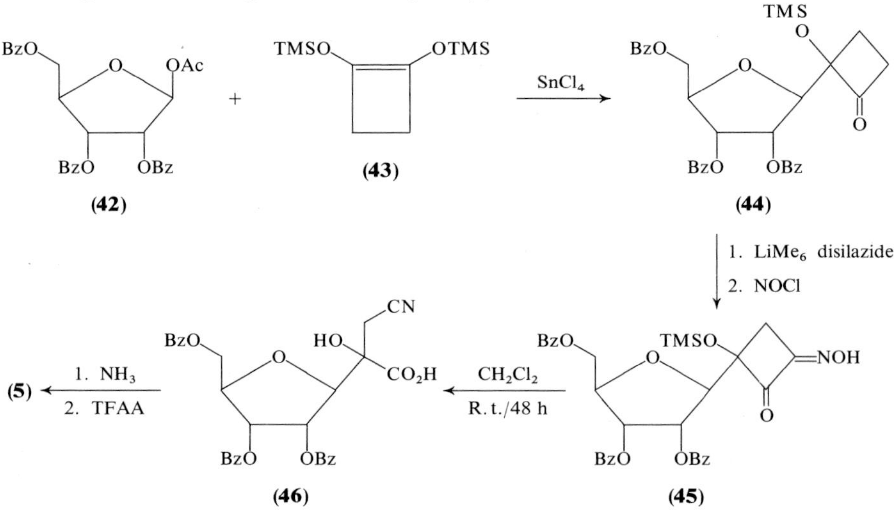

*Scheme 10*

The first approach to *C*-nucleoside synthesis via cycloadditions was by JUST and his colleagues at McGill University; the synthesis of (±) showdomycin is shown in Scheme 11 (*60, 61*). Diels-Alder addition of *E*-methyl nitroacrylate (**47**) to furan was followed by *cis*-hydroxylation and conversion into the acetonides (**48**) and (**49**). Elimination of nitrous acid gave the unsaturated ester (**50**) which was subjected to ozonolysis and reduction to give the epimeric hydroxy esters (**51**). Protection of the primary hydroxyl group, oxidation to the ketoester, and formation of the maleimide ring by the Wittig procedure (*53*) gave the protected (±)-showdomycin (**52**), which was easily hydrolysed to (±)-showdomycin (**5**).

*References, pp. 290—299*

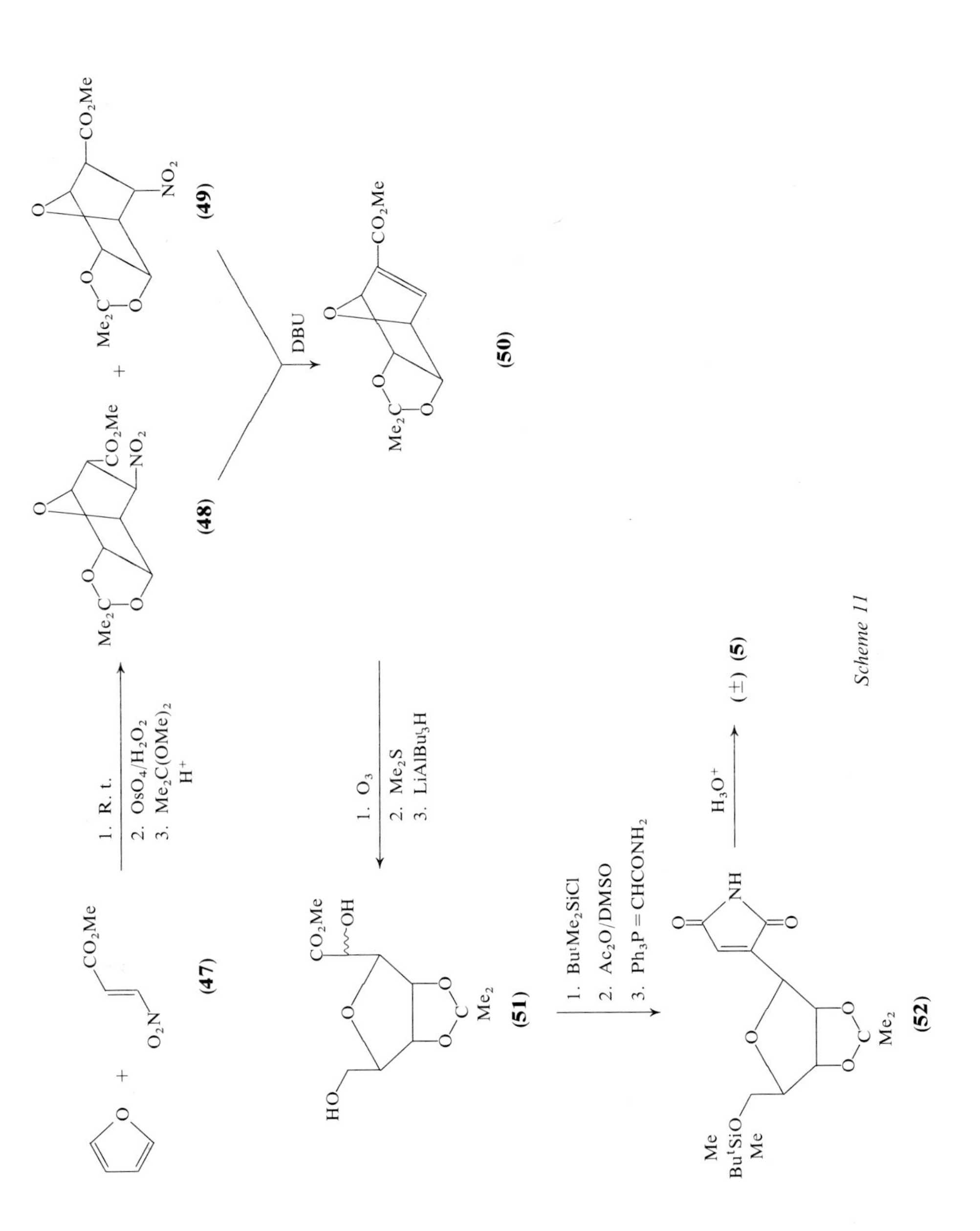
(47)
(48)
(49)
(50)
(51)
(52)
1. R. t.
2. $OsO_4/H_2O_2$
3. $Me_2C(OMe)_2$
$H^+$
DBU
1. $O_3$
2. $Me_2S$
3. $LiAlBu^t_3H$
1. $Bu^tMe_2SiCl$
2. $Ac_2O/DMSO$
3. $Ph_3P=CHCONH_2$
$H_3O^+$
(±) (5)

*Scheme 11*

Optically active showdomycin has been prepared by the Diels-Alder approach using an enzymic resolution (*62*) (Scheme 12). The adduct (**53**) from furan and dimethyl acetylenedicarboxylate (*63, 64*) was converted into the acetonide (**54**) in two stages. Pig liver esterase catalysed the hydrolysis of one ester function with complete stereoselectivity to give the resolved monoester monocarboxylic acid (**55**). Reaction with isobutene in the presence of acid afforded the t-butyl ester whose methyl ester group could be selectively saponified to give the monoester (**56**). It is interesting that such selective saponification is possible, in contrast to the behaviour of other maleic esters where a maleic anhydride intermediate is less strained (*65*). The monoester (**56**) was ozonised and decarboxylated to give the ketoester (**57**) which was converted, in three steps, into the maleic derivative (**58**). Cyclization and deprotection gave showdomycin (**5**).

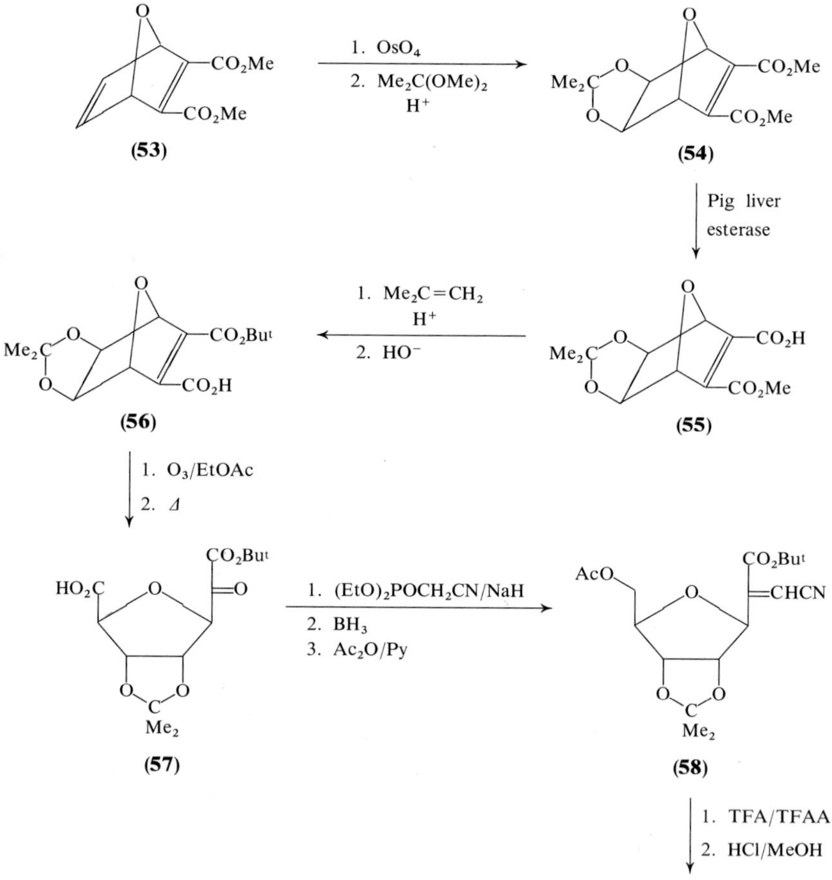

(**53**) → 1. $OsO_4$; 2. $Me_2C(OMe)_2$, $H^+$ → (**54**)

(**54**) → Pig liver esterase → (**55**)

(**55**) → 1. $Me_2C{=}CH_2$, $H^+$; 2. $HO^-$ → (**56**)

(**56**) → 1. $O_3$/EtOAc; 2. $\Delta$ → (**57**)

(**57**) → 1. $(EtO)_2POCH_2CN$/NaH; 2. $BH_3$; 3. $Ac_2O$/Py → (**58**)

(**58**) → 1. TFA/TFAA; 2. HCl/MeOH → (**5**)

*Scheme 12*

*References, pp. 290—299*

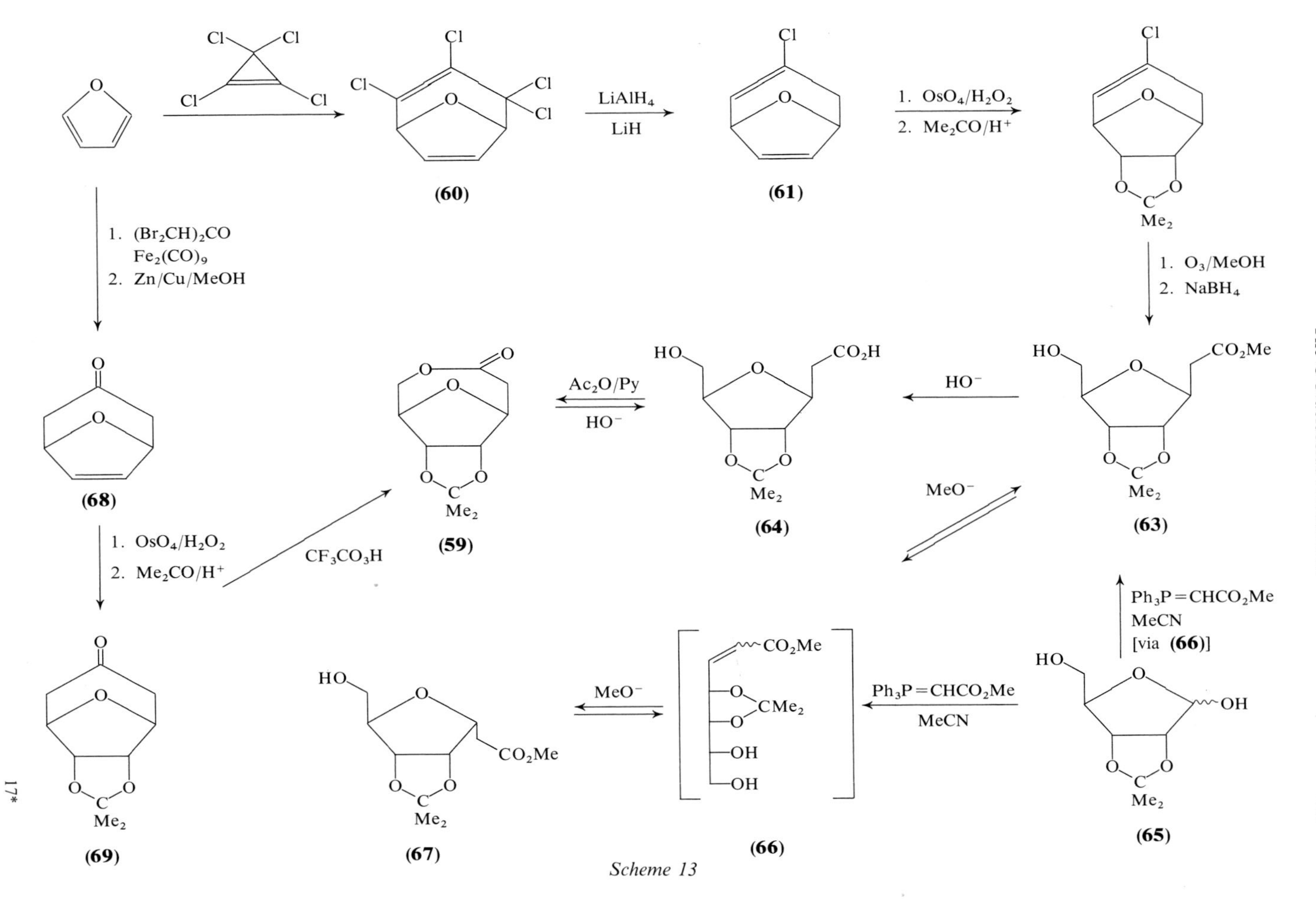

LiAlH4
LiH
1. OsO4/H2O2
2. Me2CO/H+
(60)
(61)
1. (Br2CH)2CO
Fe2(CO)9
2. Zn/Cu/MeOH
1. O3/MeOH
2. NaBH4
Ac2O/Py
HO−
HO−
(68)
(59)
(64)
(63)
MeO−
1. OsO4/H2O2
2. Me2CO/H+
CF3CO3H
Ph3P=CHCO2Me
MeCN
[via (66)]
MeO−
Ph3P=CHCO2Me
MeCN
(69)
(67)
(66)
(65)


*Scheme 13*

Two syntheses of showdomycin have stemmed from the lactone (**59**) whose preparation by several routes is shown in Scheme 13. Cycloaddition of furan and tetrachlorocyclopropene forms the tetrachlorobicyclic compound (**60**) which gives the monochloro compound (**61**) on reduction (*66*). *cis*-Hydroxylation followed by reaction with acetone affords, stereoselectively, the isopropylidene derivative (**62**) which was converted into the methyl ester (**63**) by ozonolysis and borohydride reduction. Alkaline hydrolysis of the ester (**63**) gave the acid (**64**) which was lactonised using acetic anhydride in pyridine. The D-enantiomer of the ester (**64**) was prepared from 2,3-*O*-isopropylidene-D-ribose (**65**) by MOFFATT (*67*). Under the conditions of the Wittig reaction (Scheme 13) the intermediate alkene (**66**) cyclises to give mainly the β-anomer (**63**). A thorough study of the interconversion of the esters (**63**) and (**67**) under basic conditions was undertaken, with the unexpected finding that the α-anomer (**67**) is preferred at equilibrium (*67, 68*).

The bicyclic ketone (**68**) is readily available (*69*) and has been converted into the acetonide (**69**). Baeyer-Villiger oxidation gave the lactone (**59**) directly (Scheme 13). Resolution of the lactone was effected by hydrolysis to the acid (**64**), crystallisation of the cinchonidine salt, and relactonisation (*69*).

The lactone (**59**) has been converted into showdomycin by two routes as shown in Scheme 14 (*69, 70*). There is an advantage in operating from the rigid lactone (**59**) instead of the ester (**63**) because of the tendency of the ester to epimerise under basic conditions. The lactone was converted into the dimethylaminomethylene lactone (**70**) which is a precursor of pseudouridine (**3**) and its derivatives (*69, 71*). Ozonolysis of (**70**) gave the unstable ketolactone (**71**) which formed the maleimide on treatment with the amide Wittig reagent; deprotection gave showdomycin (**5**). Alternatively, and in improved overall yield, the lactone afforded the aldol adduct (**72**) with furfuraldehyde. Dehydration and methanolysis gave the alkene (**73**), protection and ozonolysis of which gave the ketoester (**74**). Maleimide ring formation and deprotection yielded showdomycin.

The most recent synthesis of showdomycin by the Diels-Alder approach is by KOZIKOWSKY (*72*) who used 1,3-dicarbomethoxyallene and furan (Scheme 15). The adduct was converted into the acetonide (**75**) in two steps, as usual, and the key ketoester intermediate (**76**) produced by ozonolysis. Alkylation with iodoacetonitrile to give ketoester (**77**) was followed by retro-Dieckmann cleavage to give the acid (**78**). Selective reduction of the acid group with diborane, and protection of the resulting primary alcohol gave the silyl ether (**79**). When the nitrile group in (**79**) was converted into an amide under mild alkaline conditions it ring-closed to the succinimide (**80**), as expected. Dehydrogenation by the selenenyl-selenoxide method gave the maleimide which was deprotected with aqueous acid to give showdomycin (**5**).

*References, pp. 290—299*

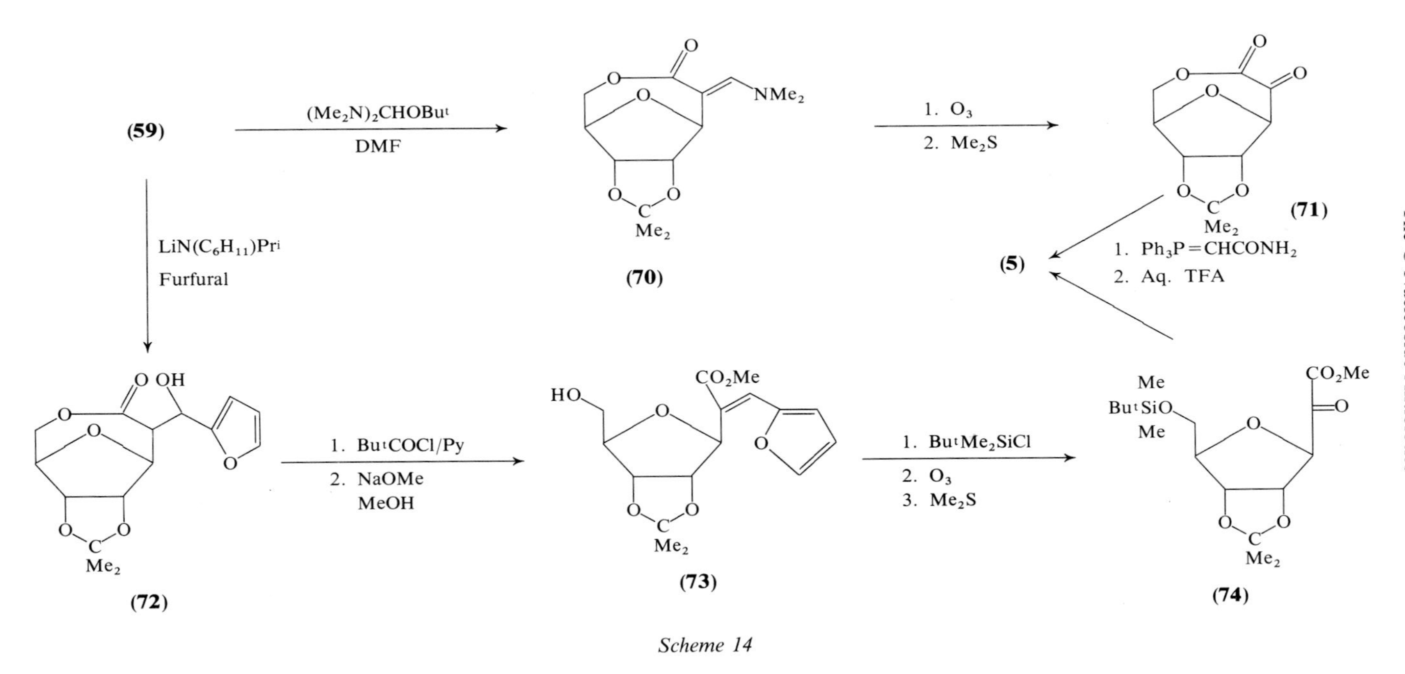
(59)
(Me2N)2CHOBut
DMF
(70)
1. O3
2. Me2S
(71)
1. Ph3P=CHCONH2
2. Aq. TFA
(5)
LiN(C6H11)Pri
Furfural
(72)
1. ButCOCl/Py
2. NaOMe
MeOH
(73)
1. ButMe2SiCl
2. O3
3. Me2S
(74)

Scheme 14

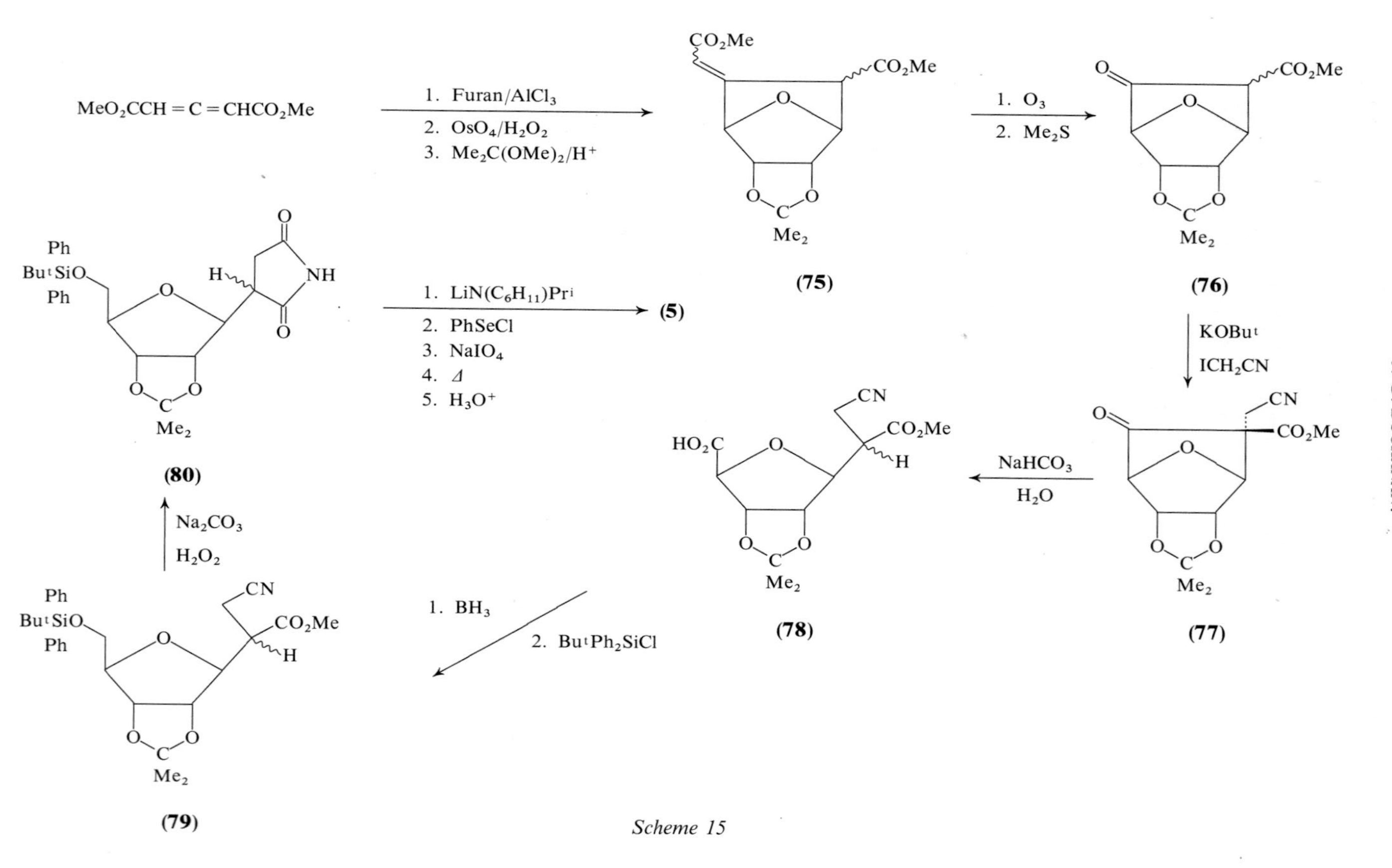
MeO2CCH = C = CHCO2Me
1. Furan/AlCl3
2. OsO4/H2O2
3. Me2C(OMe)2/H+
(75)
1. O3
2. Me2S
(76)
KOBut
ICH2CN
(77)
NaHCO3
H2O
(78)
1. BH3
2. ButPh2SiCl
(79)
Na2CO3
H2O2
(80)
1. LiN(C6H11)Pri
2. PhSeCl
3. NaIO4
4. Δ
5. H3O+
(5)

*Scheme 15*

## 3. Biological Activity

Showdomycin (**5**) is active against gram-positive and gram-negative bacteria (*47*) and against tumours (*47, 73*). Its structural resemblance to uridine (**1**) and pseudouridine (**3**) might suggest that showdomycin could interfere with metabolic pathways involving these compounds. It appears, however, that showdomycin owes its activity to reaction of the maleimide ring with thiol groups in particular enzymes. The nucleoside structure of showdomycin enables it to penetrate cells (*74*) and it does not appear to be converted into the nucleotide form. The cytotoxicity of showdomycin may be due to its more general reaction with thiol groups within the cell. The biological activity of showdomycin has been extensively reviewed (*11—13, 15, 18, 75*).

## 4. Isoshowdomycin

Showdomycin is transformed by *Streptomyces* sp. No. 383 into a crystalline isomer, isoshowdomycin. Spectroscopic studies on isoshowdomycin and its crystalline triacetate have led to the structure (**81**) [or (**82**)] (Scheme 16) (*76*). Isoshowdomycin is also produced by the action of methanolic triethylamine on showdomycin (*76, 77*).

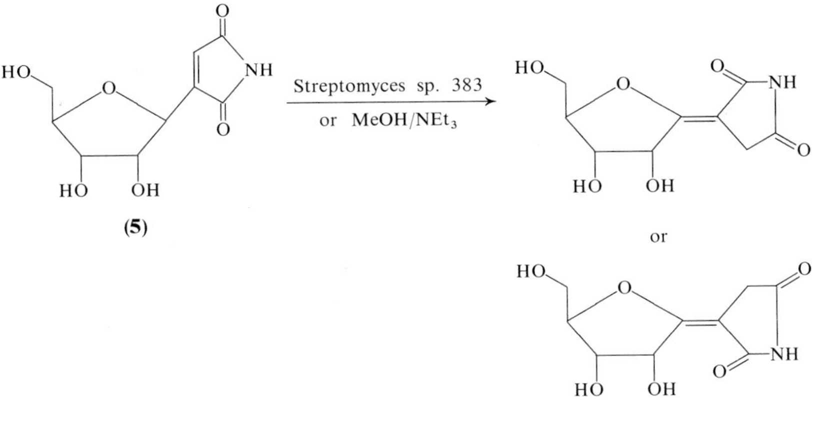

(**5**) —Streptomyces sp. 383 or MeOH/$NEt_3$→ (**81**) or (**82**)

*Scheme 16*

## IV. The Formycins

### 1. Isolation and Structure Determination

Formycin (**6**) was first isolated in the Institute of Microbial Chemistry, Tokyo, from a strain of *Nocardia interforma* during a screening programme for antitumour activity (*78*). Formycin B (**7**) was obtained from the same organism (*79*) and both formycin and formycin B (also called laurusin) were isolated from *Streptomyces lavendulae* (*80*). Formycin (*78*) and especially formycin B (*79, 80*) are active against *Xanthomonas oryzae*, the organism responsible for a rice plant disease, but only formycin is an antitumour agent. Formycin can be deaminated to give formycin B by nitrous acid (*81, 82*), acid hydrolysis (*82*) and enzymically by adenosine deaminase (*81*—*84*). Oxoformycin B (**83**) is a metabolic product of formycin and formycin B in animals (*85*). It also arises by oxidation of formycin and formycin B in microorganisms (*86*), but is inactive as an antibiotic.

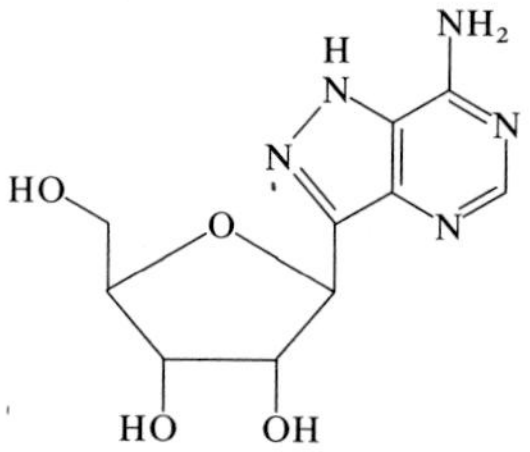

HO, H, N, N, $NH_2$, N, N, O, HO, OH

(**6**)

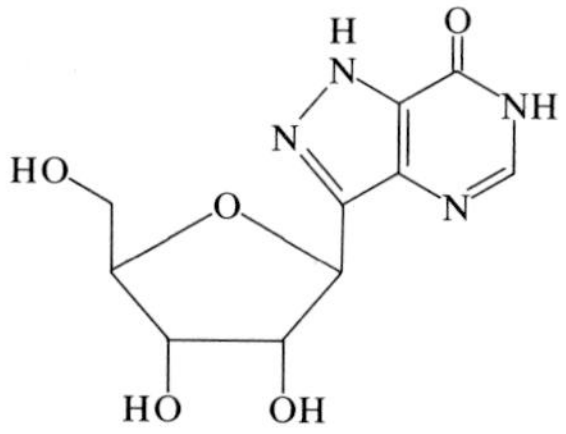

HO, H, N, N, O, NH, N, O, HO, OH

(**7**)

HO, H, N, N, O, NH, N, H, O, O, HO, OH

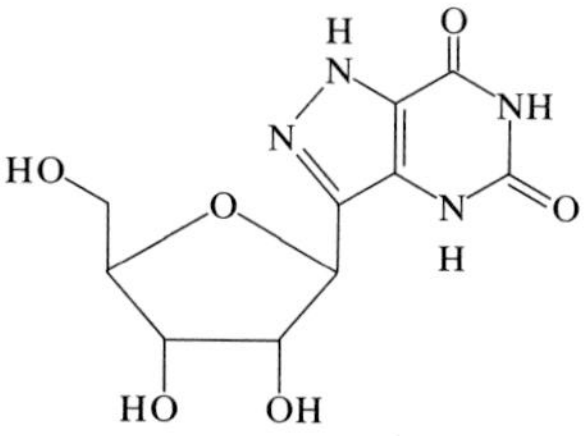

(**83**)

The structures of formycin and formycin B (laurusin) were determined in Japan (*32, 87*) and the U.S.A. (*30*). The chemical degradations, shown in Scheme 17, were much assisted by the earlier syntheses of pyrazolo [4,3-*d*] pyrimidines by ROBINS and his coworkers (*88, 89*). Periodate oxidation of formycin (**6**) afforded the aldehyde (**84**), via over-oxidation (*6*). Aerial oxidation of (**84**) under alkaline conditions gave the acid (**85**) which was decarboxylated to the known pyrazolopyrimidine (**86**) (*87*); a similar sequence was carried out on formycin B. The uv absorption spectra of the *C*-methyl derivative (**87**) (*89*) strongly resembled those of formycin B (**7**) and both were converted by chromic acid oxidation to the acid (**88**). The β-D-ribofuranosyl portion was demonstrated by comparison of the $^{1}H$ nmr spectra of formycin (**6**) and adenosine (**2**). The structure of formycin was established decisively by X-ray crystallography (*32, 90, 91*) which shows that formycin exists in the *syn* conformation in the crystal. $^{1}H$-Nmr data show this to be true for solutions as well (*46*). Formycin B and oxoformycin B have also been studied by X-ray crystallography (*92*).

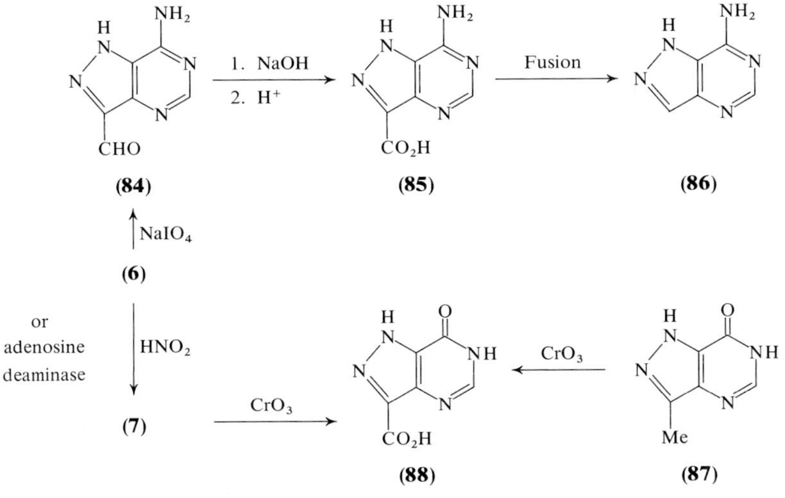

*Scheme 17*

The mass spectra of formycin and formycin B show the expected "B + 30" peaks (*43*). The $^{13}C$ nmr spectra of formycin and its derivatives are temperature dependent because of tautomeric equilibria (*38*). The $^{13}C$ signals of the pyrimidine ring carbons in formycin have recently been reassigned (*93*).

## 2. Synthesis

Oxoformycin (**83**) was the first of the formycins to be synthesized by FARKAŠ and ŠORM (*94*) (Scheme 18). The versatile nitrile (**29**) was converted into the primary amine, with loss of benzoyl groups, by lithium aluminium hydride. Reaction with nitrourea was followed by *O*-benzylation to give the ureido compound (**89**). Nitrosation and then base treatment gave the diazoalkane (**90**) which underwent cycloaddition with dimethyl acetylenedicarboxylate, affording the pyrazole (**91**). Formation of the amide hydrazide (**92**) was followed by conversion to the acid azide which underwent Curtius rearrangement and ring closure to the pyrimidine. Finally, deprotection gave oxoformycin (**83**). The intermediate (**92**) was later exploited by ACTON *et al.* (*95*) to give formycin B (**7**). Careful nitrosation afforded the anhydride (**93**) which was ring opened with decarboxylation and the resulting amino acid converted into the ester (**94**). Reaction with formamide at 218° and deprotection gave formycin B. The hydrazide (**92**) has also been obtained from D-glucose in 19 stages (*96*). Formycin (**6**) has been prepared from formycin B (**7**) (*97*) via the chloro derivative (Scheme 18).

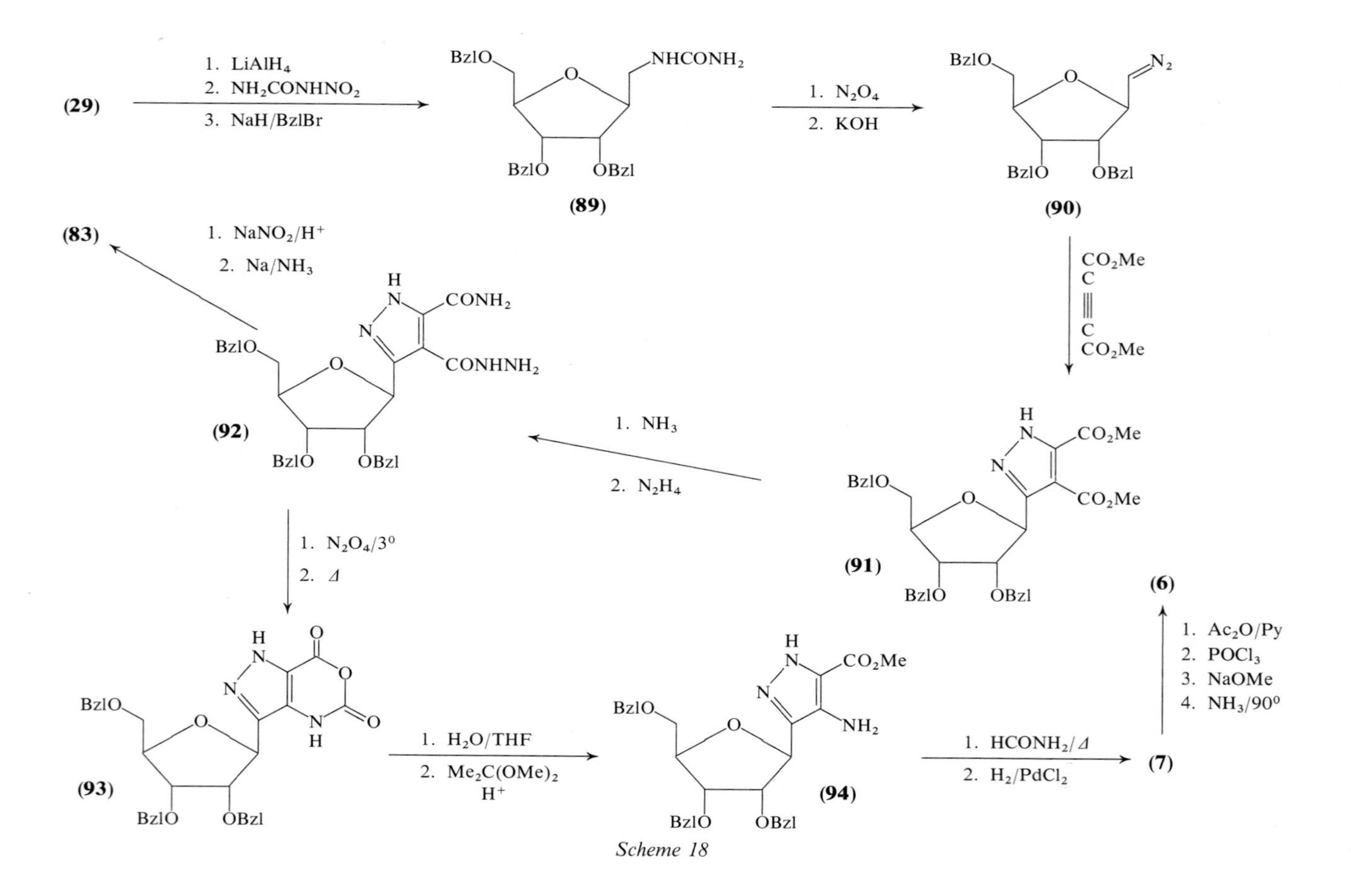
(29)
1. LiAlH$_4$
2. NH$_2$CONHNO$_2$
3. NaH/BzlBr
(89)
1. N$_2$O$_4$
2. KOH
(90)
(91)
1. NH$_3$
2. N$_2$H$_4$
(92)
1. NaNO$_2$/H$^+$
2. Na/NH$_3$
(83)
1. N$_2$O$_4$/3°
2. Δ
(93)
1. H$_2$O/THF
2. Me$_2$C(OMe)$_2$
H$^+$
(94)
1. HCONH$_2$/Δ
2. H$_2$/PdCl$_2$
(7)
1. Ac$_2$O/Py
2. POCl$_3$
3. NaOMe
4. NH$_3$/90°
(6)

Scheme 18

Another synthesis of formycin from the nitrile (**29**) has been described, requiring fewer steps (*98*). Controlled acid hydrolysis to the anhydro-D-allonic acid, conversion to the acid chloride and reaction with HCN in the presence of a phosphorane gave the unsaturated ester (**95**) (Scheme 19) [cf. (**31**) → (**32**) (Scheme 7)]. In the key reaction (**95**) underwent cycloaddition with diazoacetonitrile to give the pyrazole (**96**). The ester (**96**) was converted into the acid and subjected to Curtius degradation in the presence of trichloroethanol affording the carbamate (**97**) which was converted by standard procedures into formycin (**6**). Formycin B was obtained in a parallel sequence via the diester (**98**).

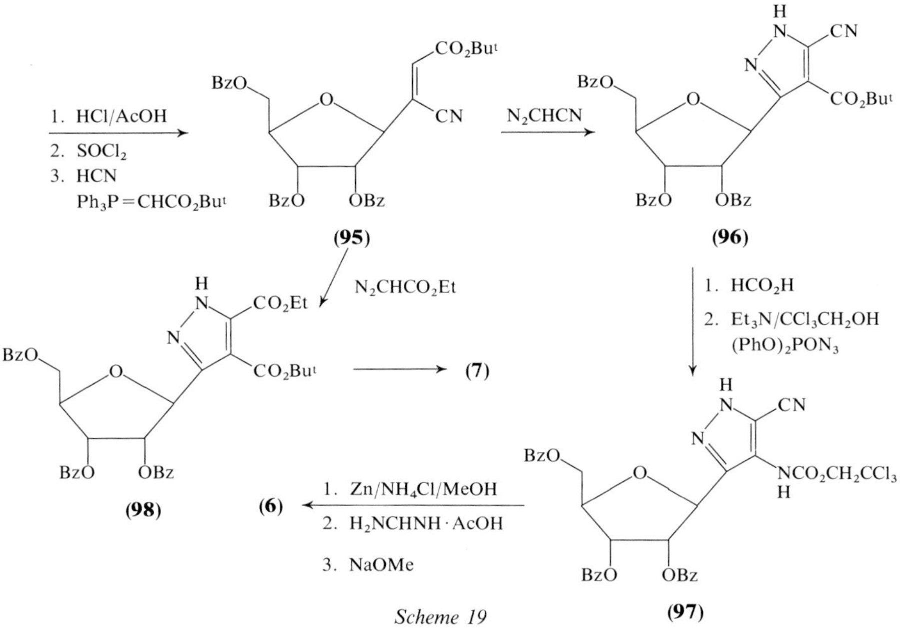

1. HCl/AcOH
2. $SOCl_2$
3. HCN
$Ph_3P{=}CHCO_2Bu^t$

(**95**) → $N_2CHCN$ → (**96**)

(**95**) → $N_2CHCO_2Et$ → (**98**) → (**7**)

(**96**) → 1. $HCO_2H$; 2. $Et_3N/CCl_3CH_2OH$, $(PhO)_2PON_3$ → (**97**)

(**97**) → 1. $Zn/NH_4Cl/MeOH$; 2. $H_2NCHNH \cdot AcOH$; 3. NaOMe → (**6**)

*Scheme 19*

Formycin has been synthesized from an acetylenic precursor, as shown in Scheme 20 (*99, 100*). The acetal (**99**) was prepared from tri-*O*-benzyl-D-ribose (**38**) by an extension of the earlier route (Scheme 9) (*55*). Acid hydrolysis in the presence of hydrazine formed the pyrazole ring which was protected as the 2,4-dinitrophenyl derivative (**100**). Nitration was followed by exchange of protecting groups to give the triacetate (**101**). The base-labile dinitrophenyl group was removed and the resulting pyrazole (**102**) converted into the 1,4-dinitro compound (**103**). A key reaction was the *cine*-substitution (*101*) of the *N*-nitro group with cyanide ion to give the nitrile (**104**) which was reduced to the amine (**105**). Formation of the pyrimidine ring was followed by deprotection to give formycin (**6**). The amine (**105**) is also a precursor of pyrazofurin (**8**) (*102*).

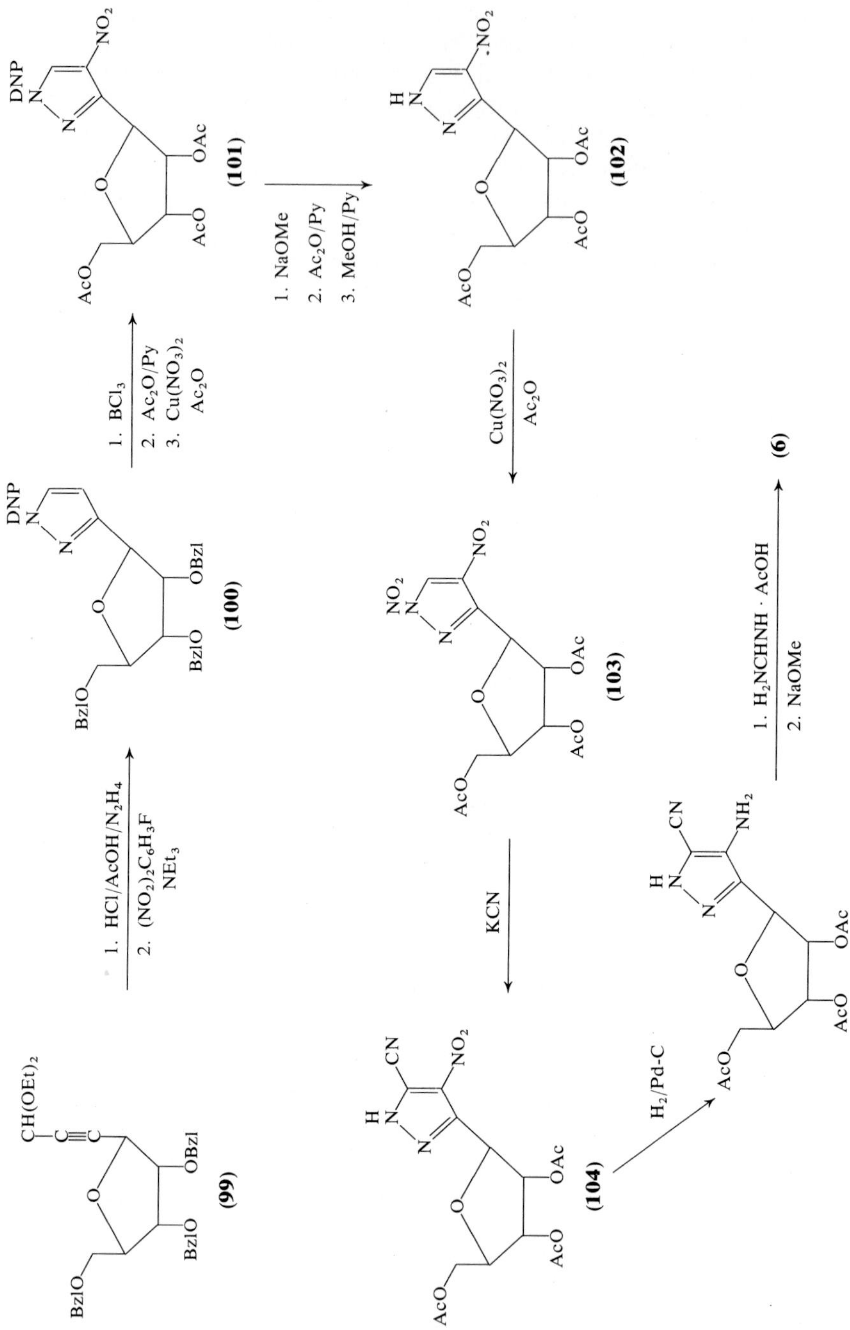

(99) → 1. HCl/AcOH/$N_2H_4$; 2. $(NO_2)_2C_6H_3F$, $NEt_3$ → (100) → 1. $BCl_3$; 2. $Ac_2O$/Py; 3. $Cu(NO_3)_2$, $Ac_2O$ → (101) → 1. NaOMe; 2. $Ac_2O$/Py; 3. MeOH/Py → (102) → $Cu(NO_3)_2$, $Ac_2O$ → (103) → KCN → (104) → $H_2$/Pd-C → (105) → 1. $H_2NCHNH \cdot AcOH$; 2. NaOMe → (6)

*Scheme 20*

## 3. Biological Activity

Formycin (**6**) shows antitumour properties towards a number of experimental tumours (*103*) and mouse leukaemia L. 1210 (*104*). Formycin B (**7**) has little antitumour activity (**81**) except towards L. 5178Y in mice (*105*), but both formycin and formycin B inhibit the rice disease bacterium *Xanthomonas oryzae* (*78, 80, 106*) and the influenza A1 virus (*107*). Formycin is also active against other viruses (*108*).

Formycin, by virtue of its close similarity to adenosine (**2**), is able to participate in a number of enzymic reactions which normally use adenosine as substrate. Thus, it can be phosphorylated at the 5′ position to give mono-, di-, and triphosphates (FMP, FDP, and FTP) (*109, 110*). Once in the nucleotide form other transformations are possible (*110*). FDP is a substrate for polynucleotide phosphorylase in *E. coli* (*110*) and FTP replaces ATP as a substrate in the DNA-dependent RNA polymerase (*110, 111*). FTP can be used to introduce a formycin residue in place of adenosine at the 3′-terminus of transfer RNAs, which are directly involved in protein biosynthesis (*110, 112*).

These modified RNAs show interesting physical and biochemical properties due, it is believed, to the tendency of formycin derivatives to exist mainly in the *syn* conformation* with the pyrimidine portion directed towards the ribose unit (*90—92, 113, 114*). Because of the characteristic uv absorption and fluorescence of the pyrazolopyrimidine ring system, molecules in which formycin replaces adenosine are valuable as experimental probes for the study of enzymic mechanisms (*115—117*).

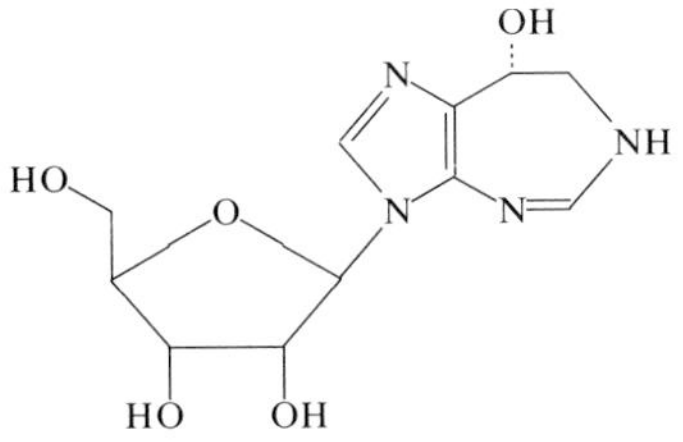

OH
N
NH
HO
O
N
N=
HO OH

**(106)**

Formycin is deaminated to formycin B by adenosine deaminase (*83, 84*). This is prevented by coformycin (**106**), an inhibitor of adenosine deaminase,

* The formulae (**2**) and (**6**) are drawn in the *anti* conformation which for adenosine (**2**) is the preferred conformation.

which is produced together with formycin by *Nocardia interforma* (*14, 118—120*). In the presence of coformycin the effect of formycin on Ehrlich carcinoma is enhanced and its antibacterial spectrum extended (*109, 118*).

The mode of action of formycin against tumours involves conversion into nucleotides, but the precise location of its inhibitory effect remains in doubt. It is probable that it interferes with nucleotide metabolism and HENDERSON et al. (*121*) have reported that FMP inhibits the synthesis of 5-phosphoribosylpyrophosphate, an important enzyme in the *de novo* biosynthesis of purine and pyrimidine nucleotides.

Formycin B does not undergo phosphorylation to the 5′-nucleotide in any of the systems studied. It owes its activity against *Xanthomonas oryzae* to the prevention of the uptake of nucleosides from the medium (*106*). The mechanism of action of formycin on this organism probably differs from that of formycin B.

Further discussion of the biological activity of the formycins is beyond the scope of this article. Attention is directed to several reviews (*11—13, 15, 122*).

## V. Pyrazofurin

### 1. Isolation and Structure Determination

As a result of screening fermentation broths for antiviral activity (*122*) pyrazofurin (**8**) (originally called pyrazomycin) was isolated from *Streptomyces candidus* by workers at Lilly Research Laboratories (*123—126*). The structure was determined by spectroscopy and chemical degradation (*123, 124*) and confirmed by X-ray crystallography (*127*). The enolic function was inferred by titration in conjunction with uv spectroscopy and the amide group was detected by ir spectroscopy. In the $^1$H nmr spectrum the ribose unit showed a signal for H-1′ at about 5 p.p.m., characteristic of a *C*-nucleoside [Section II 2(a)]. Pyrazofurin was converted by diazomethane into the *O, N*-dimethyl derivative (**107**) which was oxidized by permanganate to the carboxylic acid (**108**), identified by synthesis.

The α-anomer of pyrazofurin, pyrazofurin B (**109**) was also isolated from *Streptomyces candidus* and its structure proved by spectroscopy and X-ray crystallography (*33, 36*). It is now believed (*125*) that pyrazofurin B (**109**) is an artifact arising from pyrazofurin itself by acid-catalysed isomerisation. By analogy with the behaviour of pseudouridine (Scheme 1) an intermediate of structure (**110**) is probably involved (Scheme 21).

*References, pp. 290—299*

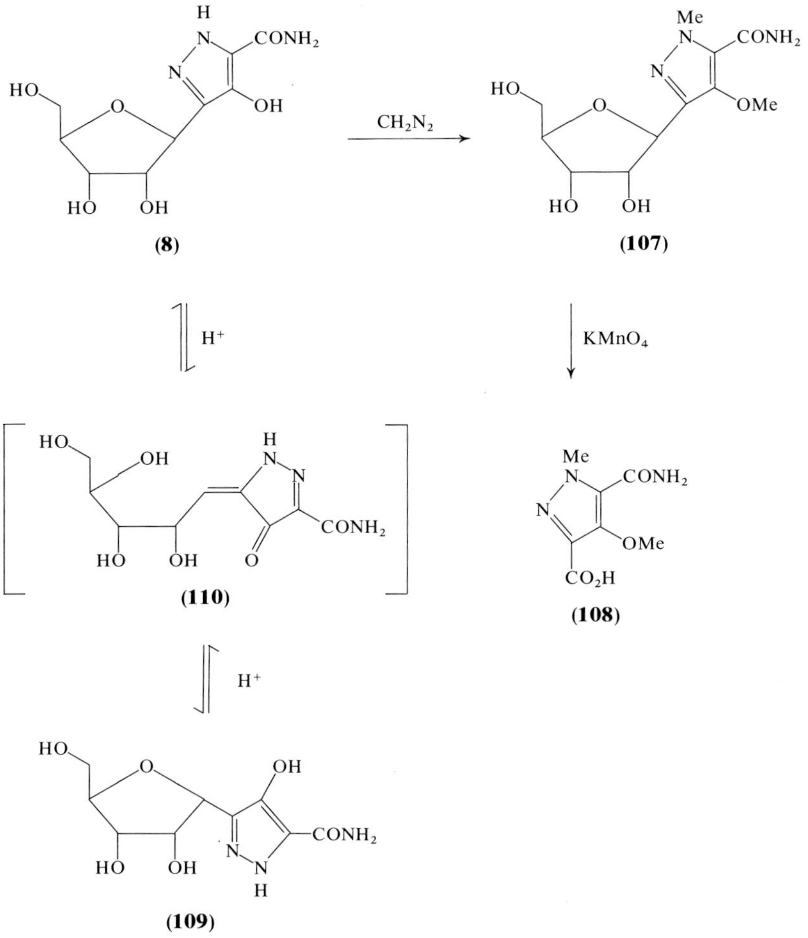

(8) $\xrightarrow{CH_2N_2}$ (107) $\xrightarrow{KMnO_4}$ (108)

(8) $\rightleftharpoons$ ($H^+$) [(110)] $\rightleftharpoons$ ($H^+$) (109)

*Scheme 21*

## 2. Synthesis

Farkaš and Šorm (*44*) (Scheme 22) synthesized pyrazofurin (**8**) from the ketoester (**25**) already used for their synthesis of showdomycin (Scheme 6). Reaction with the protected hydrazine (**111**) gave the hydrazone (**112**) which underwent ring closure in the presence of acetic anhydride and sodium acetate. Formation of the methyl ester (**113**) was followed by amide formation and debenzylation to give pyrazofurin (**8**).

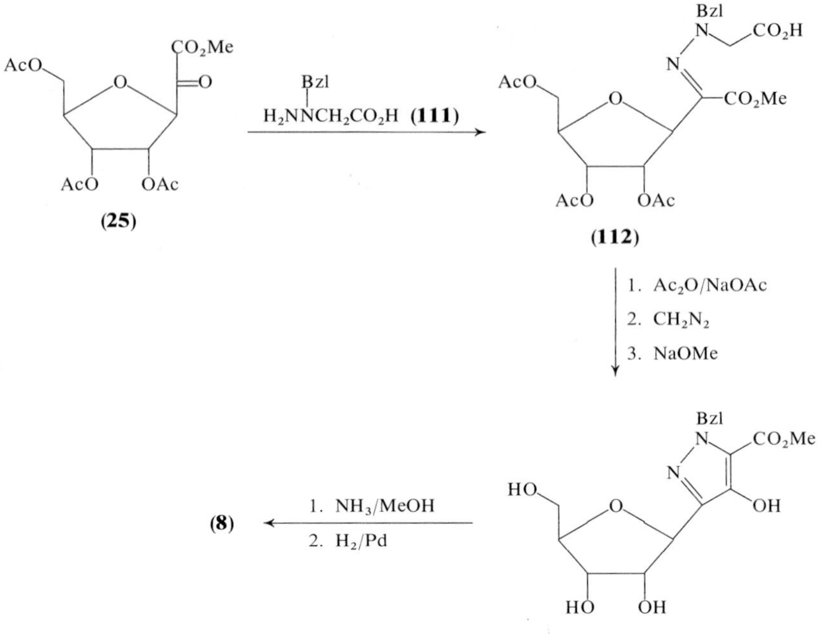

AcO, CO₂Me, O, O, AcO, OAc (25) — Bzl, $H_2NNCH_2CO_2H$ (111) → (112) AcO, Bzl, N, CO₂H, N, CO₂Me, AcO, OAc

1. $Ac_2O$/NaOAc
2. $CH_2N_2$
3. NaOMe

(8) ← 1. $NH_3$/MeOH 2. $H_2$/Pd — (113) HO, Bzl, N, $CO_2Me$, N, OH, HO, OH

*Scheme 22*

A second synthesis, by DE BERNARDO and WEIGELE (*128*), used 2,3-*O*-isopropylidene-D-ribose (**65**) as starting material (Scheme 23). The crystalline bromide (**114**) reacted with the potassium salt of diethyl acetonedicarboxylate in the presence of 18-crown-6 to give the α-anomer (**115**) of the diester in 43% yield. Azo group transfer from toluene-*p*-sulphonyl azide gave directly the pyrazolinone (**116**). Selective ethanolysis of the quaternary ethoxycarbonyl group afforded the monoester (**117**), which was treated with methanolic ammonia at 95° for 3 hr to give the derivative (**118**) of pyrazofurin B as the major product. When the reaction with ammonia was prolonged (12 hr) anomerization took place and the major product was 2,3-*O*-isopropylidene pyrazofurin (**119**). The parent *C*-nucleosides (**109**) and (**8**) were obtained by acid hydrolysis. Presumably the mechanism of conversion of the α-anomer (**118**) into the β-anomer (**119**) resembles that for the isomerization of pseudouridine under basic conditions (Scheme 3). It is interesting that the β-anomer is favoured over the α-anomer at equilibrium in contrast to a number of epimerizable *C*-glycosyl compounds containing a 2,3-*O*-isopropylidene group studied by OHRUI *et al.* (*67, 68*), e.g. (**67**) and (**63**) in Scheme 13. It may be related to the finding that the stable sodium salt of 5-(2,3-*O*-isopropylidene-5-*O*-trityl-D-ribofuranosyl)-barbituric acid is the β-form (**120**) (*67, 129*).

*References, pp. 290—299*

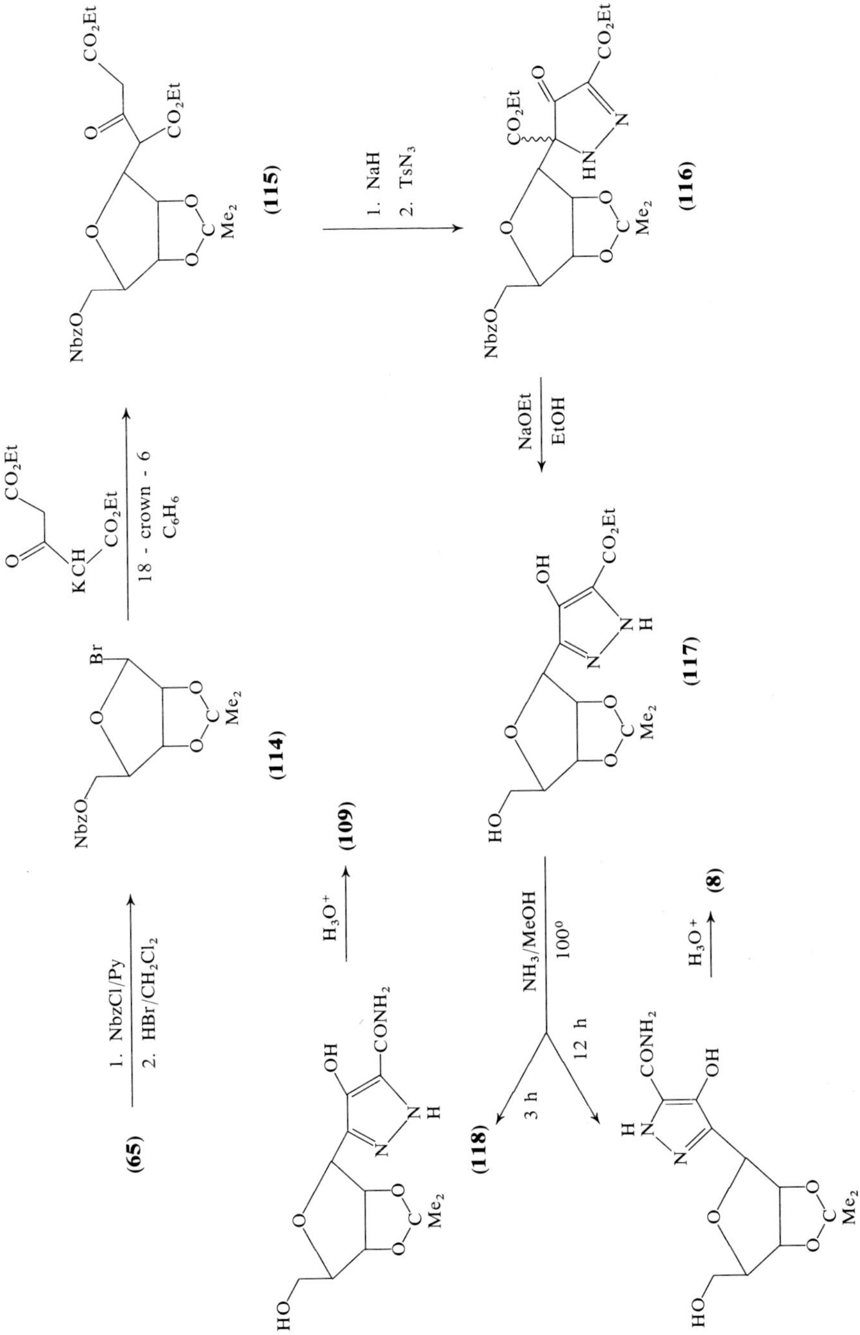

*Scheme 23*

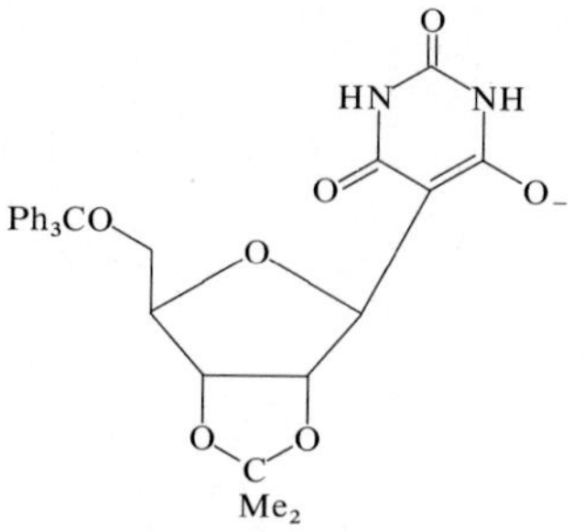

(120)

A very recent synthesis (*130*) uses similar chemistry in the later stages (Scheme 24). 2,3-*O*-Isopropylidene-5-*O*-trityl-D-ribofuranose (**121**) (*129*) reacted with the phosphorane to give the αβ mixture (**123**) in 95% yield. Reaction with toluene-*p*-sulphonyl azide and triethylamine afforded the diazo compound (**124**) which cyclized when treated with sodium hydride to give the hydroxypyrazole (**125**) (42%) and its α-anomer (**126**) (21%). Ammonolysis and acid hydrolysis then gave pyrazofurin (**8**) and pyrazofurin B (**109**) respectively.

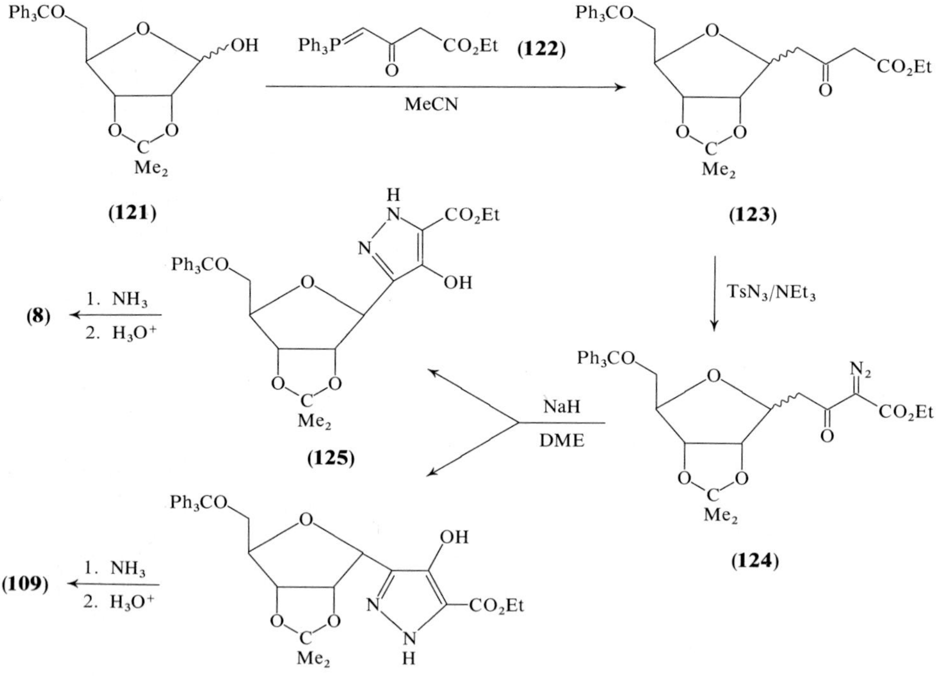

*Scheme 24*

*References, pp. 290—299*

An earlier synthesis of formycin from acetylenic precursors (Scheme 20) has been extended to pyrazofurin (**8**) (Scheme 25) (*102*). The aminonitrile (**105**) was converted into the diazo compound (**127**) which was remarkably stable to acid treatment, but was converted by photolysis to the hydroxynitrile (**128**). Hydrolysis of the nitrile was effected in the presence of nickel(II) acetate to give the amide (**129**) which afforded pyrazofurin (**8**) on deprotection.

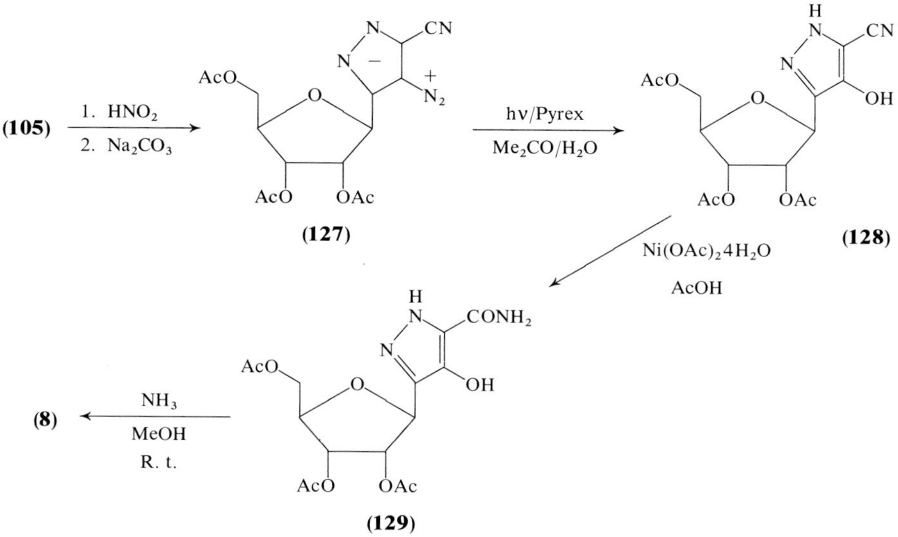

*Scheme 25*

## 3. Biological Activity

Pyrazofurin shows marked antiviral activity (*122, 125, 131*) and is also active against certain tumours (*125, 132, 133*). It has undergone clinical trials as an anticancer agent (*125, 133*), but there are problems of toxicity (*131, 133, 134*).

Early in the work on pyrazofurin it was discovered that its antiviral effect was reversed by uridine and uridine-5′-phosphate [UMP (**130**)], but not by cytidine (*122*). This can be explained by the finding that when pyrazofurin (**8**) is converted into its 5′-phosphate (**131**) (*122, 125, 133*) it inhibits OMP decarboxylase, the enzyme responsible for the decarboxylation of orotidine-5′-phosphate (**132**) to UMP (Scheme 26). This enzyme is essential for the *de novo* biosynthesis of pyrimidine nucleotides. It is interesting that the barbituric acid nucleotide (**133**), which has a structural resemblance to the pyrazofurin nucleotide (**131**), is a very powerful inhibitor of OMP decarboxylase (*135*)*.

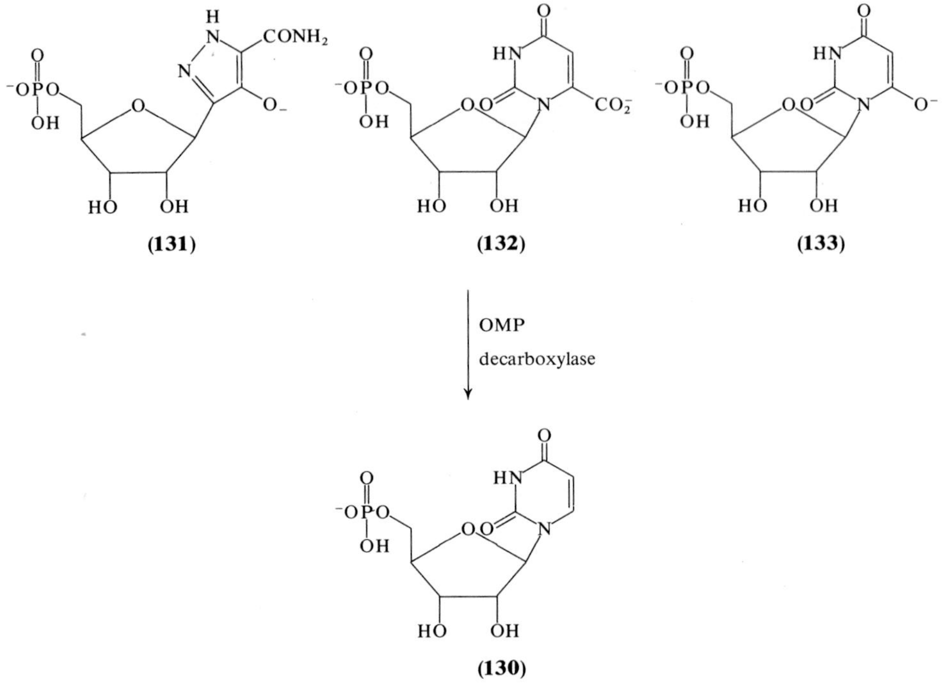

(**131**) (**132**) (**133**)

OMP decarboxylase

(**130**)

*Scheme 26*

## VI. Oxazinomycin (Minimycin)

### 1. Isolation and Structure Determination

Oxazinomycin (**9**) was isolated from *Streptomyces tanesashinensis* by a group from the Sankyo Company in Japan (*34*) and is the same as minimycin isolated from *Streptomyces hygroscopicus* by the Kaken Chemical Company (*26, 136, 137*). It may also occur, along with

* We thank Dr. M. L. SINNOTT for drawing our attention to this paper.

pyrazofurin (**8**), in cultures of *Streptomyces candidus* (*125*). Oxazinomycin is active against bacteria and tumours (*34, 137*), but shows no activity against yeast or fungi (*137*). The structure was determined mainly using physical measurements, particularly $^{1}H$ nmr where the spectrum strongly resembles those of pseudouridine (**3**) and showdomycin (**5**) (*26, 34*). Oxazinomycin (**9**) is relatively stable under acidic conditions, but unstable in base. Ribose was detected only after hydrazinolysis [Section III b] and ammonia converted it into pseudouridine (**3**) in low yield (*26*) (Scheme 27). The structure was determined by X-ray crystallography (*34*).

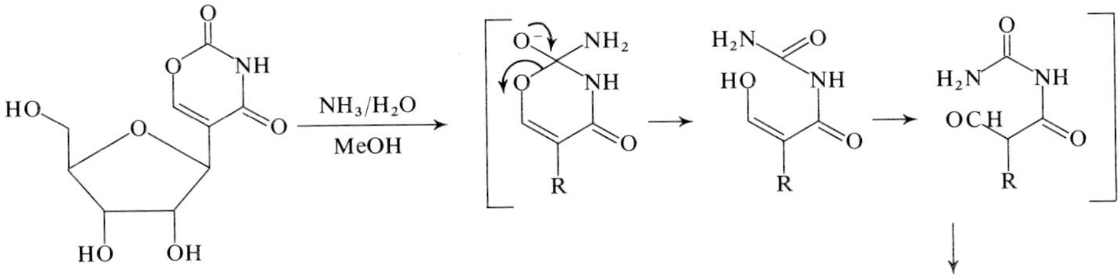

(**9**) $\xrightarrow[\text{MeOH}]{NH_3/H_2O}$ ... (**3**)

*Scheme 27*

## 2. Synthesis

Several research groups have prepared potential intermediates for the synthesis of oxazinomycin (**9**) but only one successful synthesis has been reported (*138*) (Scheme 28). The acetonide (**121**) (*129*) reacted with sodium diethyl cyanomethylphosphonate to give the anomeric mixture of ribofuranosylacetonitriles (**134**). Formylation with bis(dimethylamino)-*tert*-butyloxymethane afforded the enamines (**135**).

Reaction with hydroxylamine gave a mixture of isoxazoles (**136**) which were converted into the enamines (**137**) by hydrogenolysis. The aldehydes, represented as their enols (**138**), were obtained by very careful acid hydrolysis and converted into the oxazines using carbonyldiimidazole. The pure β-anomer (**139**) was isolated and converted into oxazinomycin (**9**) by deprotection. The α-anomer of oxazinomycin was also prepared. The intermediates (**134**), (**135**) and (**138**) all undergo anomeric equilibration via reversible ring opening (cf. Schemes 1 and 3). The key to the synthesis lies in the mildness of the conditions used to prepare the very labile formyl derivatives (**138**) and (**139**).

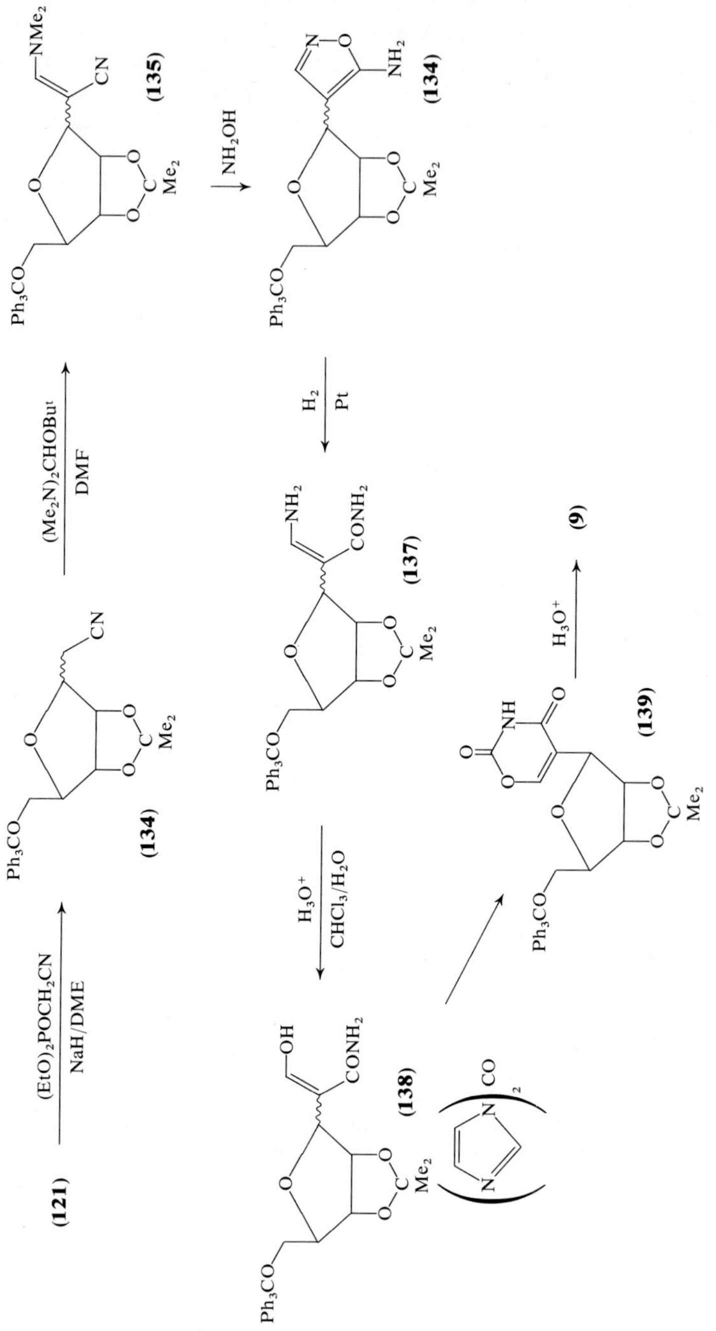

*Scheme 28*

*References, pp. 290—299*

### 3. Biological Activity

Little has been reported since the original isolation work described in Section VI 1 (*34, 137*). It has been suggested (*122*) that oxazinomycin owes its antibiotic activity to the inhibition of nucleotide reductase since the activity is reversed by 2′-deoxycytidine.

## VII. The Ezomycins

The ezomycins are a group of antifungal antibiotics first discovered in the laboratories of the Sankyo Company, Japan (*139*). They are produced by a *Streptomyces* and are active against *Sclerotinia sclerotiorum* de Bery which causes stem rot of French bean (*Phaseolus vulgaris* L.). SAKATA and his colleagues have separated the components of the ezomycin complex and established their structures (*35, 140—149*). The important components are ezomycins $A_1$ (**140**) and $B_1$ (**10**). The others are probably artifacts, derived from them by partial hydrolysis and rearrangement. Ezomycin $A_1$ (**140**) is a 1-glycosyl derivative of cytosine, containing a glycosyl-nitrogen bond. Ezomycin $B_1$ (**10**) is a 5-glycosyluracil and therefore related to pseudouridine (**3**); this is reflected in its chemistry, particularly under acidic conditions.

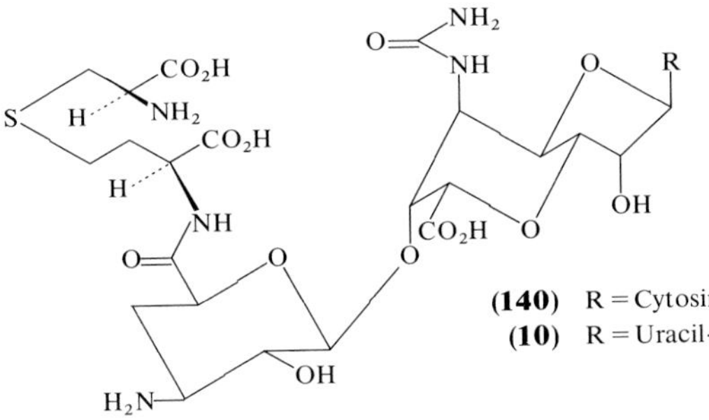

(**140**) R = Cytosine-1-yl
(**10**) R = Uracil-5-yl

Although ezomycin $A_1$ is an *N*-glycosyl compound we shall discuss its chemistry because it is closely related to the other members of the group. Acidic hydrolysis of ezomycin $A_1$ (**140**) using Dowex 50 ($H^+$) resin gives ezomycin $A_2$ (**141**) and L-cystathionine (**142**), identified by comparison with an authentic sample (*140, 144*) (Scheme 29). More vigorous acid hydrolysis of ezomycin $A_2$ (**141**) gave cytosine and ezoaminuroic acid (**143**) (*141, 144*) whose structure was determined mainly by physical methods. The molecular weight was obtained by mass spectrometry of the TMS ether. A study of the benzoates, (**144**) and (**145**), by $^1$H nmr and CD spectroscopy conclusively proved their structures and these have been confirmed by the synthesis of the β-anomer (**145**) (*150*) (Scheme 30). The ČERNÝ epoxide (**146**) (*151*) underwent regiospecific ring opening with azide ion to give the diaxial azidoalcohol which was converted into the benzoate (**147**). Reaction with antimony pentachloride followed by methanol gave the β-glycoside (**148**) selectively which was oxidized and esterified to give the methyl ester (**149**). Hydrogenolysis of the azide group and *N*-benzoylation afforded the derivative (**145**) of ezoaminuroic acid. ZAMOJSKI (*152*) has described a synthesis of the racemic α- and β-glycosides (**150**), also by an epoxide route.

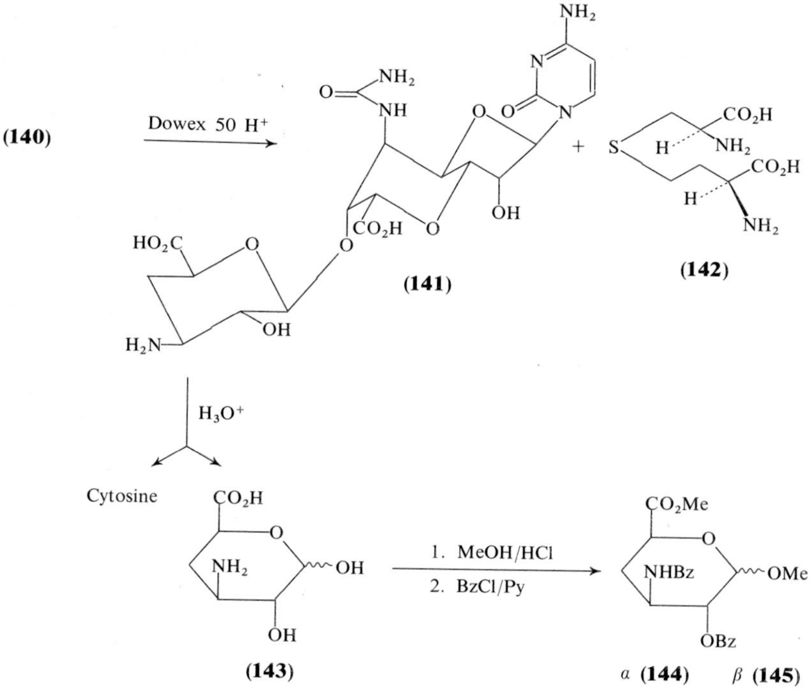

(**140**) Dowex 50 H$^+$ → (**141**) + (**142**)

(**140**) $H_3O^+$ → Cytosine + (**143**)

(**143**) 1. MeOH/HCl 2. BzCl/Py → α (**144**) β (**145**)

*Scheme 29*

Some other degradations of ezomycin $A_1$ (**140**) are shown in Scheme 31 (*143, 146*). Alkaline hydrolysis caused elimination of the ezoaminuroic acid fragment to give the unsaturated nucleoside (**151**). Periodate oxidation of ezomycin $A_1$ (**140**) followed by mild acid hydrolysis gave the parent nucleoside (**152**). The structures of the nucleosides (**151**) and (**152**) were determined, mainly by $^1$H nmr and CD. Other chemical degradations (*143,*

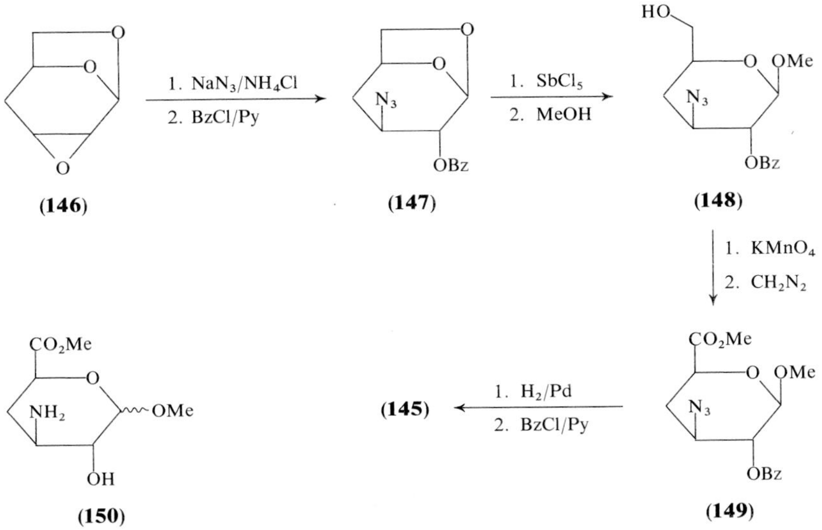

O O O (146) 1. $NaN_3/NH_4Cl$ 2. BzCl/Py O O $N_3$ OBz (147) 1. $SbCl_5$ 2. MeOH HO O OMe $N_3$ OBz (148)

1. $KMnO_4$ 2. $CH_2N_2$

$CO_2Me$ O $NH_2$ OMe OH (150) (145) 1. $H_2/Pd$ 2. BzCl/Py $CO_2Me$ O OMe $N_3$ OBz (149)

*Scheme 30*

*146*) enabled the total structures of ezomycins $A_1$ (**140**) and $A_2$ (**141**) to be established. A notable feature of these compounds is the anhydrooctose unit in place of the usual ribose in nucleosides. A related anhydrooctose is present in the octosyl acids, e.g. octosyl acid A (**153**), isolated from *Streptomyces cacaoi* var. *asoensis* (*14, 153*).

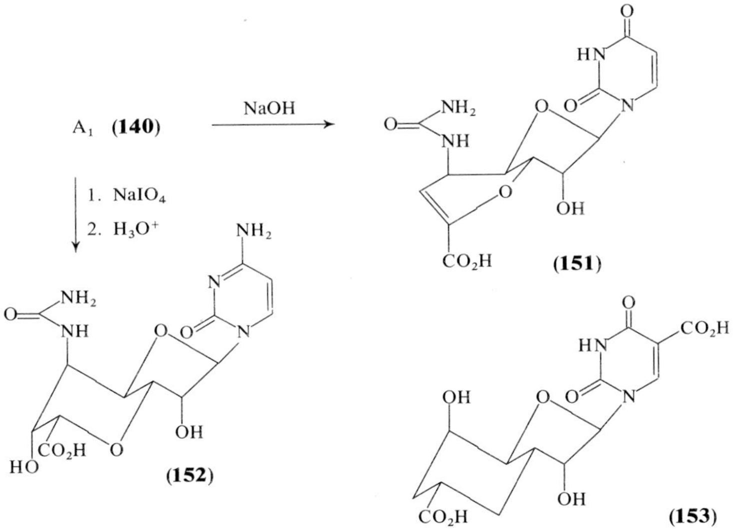

$A_1$ (**140**) NaOH O HN O $NH_2$ O NH O O N O OH $CO_2H$ (151)

1. $NaIO_4$ 2. $H_3O^+$

$NH_2$ N $NH_2$ O NH O O N OH $CO_2H$ O HO (152)

O HN $CO_2H$ OH O O N OH $CO_2H$ (153)

*Scheme 31*

Ezomycin $B_1$ (**10**) was isolated by careful fractionation of the ezomycin components (*145, 148*). It was shown to liberate L-cystathionine (**142**) (*35, 140, 145*) and ezoaminuroic acid (**143**) (*35, 145*) on acid hydrolysis. The uv spectrum suggested that it was a 5-substituted uracil. Ezomycin $B_2$ (**154**), another member of the group, differs from $B_1$ in the absence of the L-cystathionine residue (cf. $A_1$ and $A_2$). Periodate oxidation of $B_2$, followed by reaction with phenylhydrazine, gave the nucleoside (**155**) (Scheme 32). Comparison of (**155**) with the nucleoside (**152**) by mass spectrometry of the TMS derivatives and by $^1H$ nmr showed that the sugar portions were identical. The uv spectrum of (**155**), the chemical shift (δ 4.96) of H-1′, and the absence of a signal due to H-5 proved that (**155**) was a *C*-nucleoside related to pseudouridine (**3**) (*35*) [see Section II 2(a)]. The presence of the strained anhydrooctose ring and other aspects were confirmed by $^{13}C$ nmr measurements (*147, 149*).

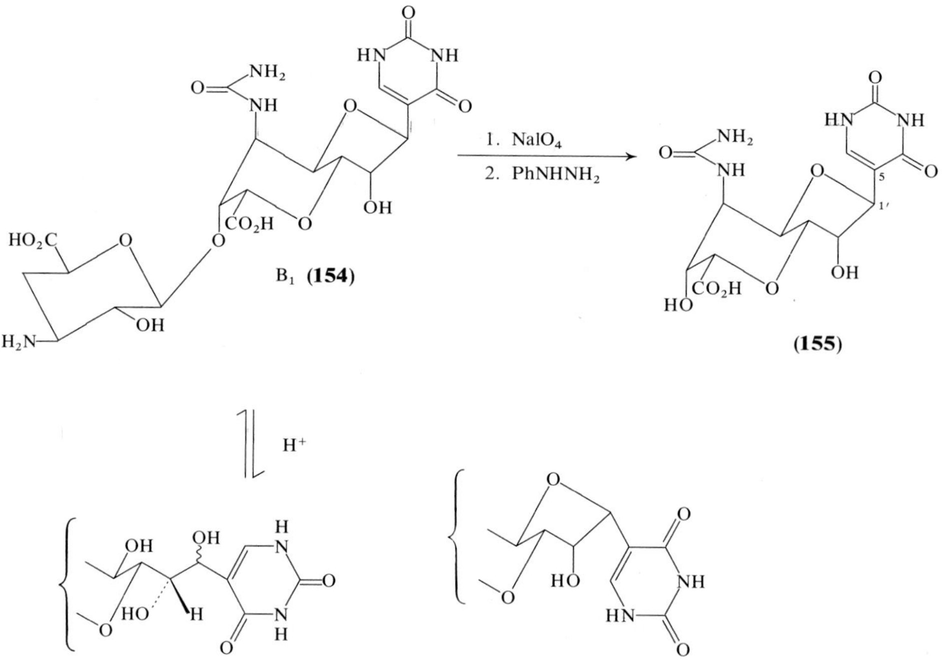

1. $NaIO_4$
2. $PhNHNH_2$

$B_1$ (**154**)

(**155**)

$H^+$

$D_2$ (**157**)

$C_2$ (**156**)

*Scheme 32*

Ezomycin $B_1$ (**10**) and $B_2$ (**154**) undergo isomerisation under acidic conditions (*35*). $B_1$ is converted into a mixture with $C_1$, its α-anomer, and similarly $B_2$ is partially converted into $C_2$ (**156**) (Scheme 32). This behaviour is to be expected from a knowledge of the chemistry of pseudouridine (**3**) [Section II 1 (a)]. Of particular interest is the presence in the equilibrium mixtures of acyclic derivatives, $D_1$ and $D_2$ (**157**) (Scheme 32), to the extent of about 25% (*35*). This is a consequence of the strain imposed by the *trans*-fused rings in the anhydrooctose system. Related effects are observed in the chemistry of the octosyl acids (*14, 153*).

Table 2 summarizes the relationships between the ezomycins, all of which are present in the crude mixture isolated from the *Streptomyces*.

Table 2. *The Ezomycins*

| Ezomycin | Heterocycle. | Anomeric configuration | L-Cystathionine (**142**) | Biological activity |
|---|---|---|---|---|
| $A_1$ (**140**) | cytosine-1-yl | β | + | + + |
| $A_2$ (**141**) | cytosine-1-yl | β | − | − |
| $B_1$ (**10**) | uracil-5-yl | β | + | + + |
| $B_2$ (**154**) | uracil-5-yl | β | − | − |
| $C_1$ | uracil-5-yl | α | + | + |
| $C_2$ (**156**) | uracil-5-yl | α | − | − |
| $D_1$ | uracil-5-yl | acyclic | + | − |
| $D_2$ (**157**) | uracil-5-yl | acyclic | − | − |

## 2. Synthesis

The complete synthesis of an ezomycin *C*-nucleoside, even one lacking the cystathionine residue as in $B_2$, poses several difficult problems. ANZAI (*154, 155*) and SZAREK (*156, 157*) have synthesized anhydrooctose systems similar to those present in the ezomycins and octosyl acids.

## 3. Biological Activity

The ezomycins are inhibitory towards a limited number of fungi, particularly species of *Sclerotinia, Botrytis* and *Gleosporium* (*142, 148*). Bacteria and yeasts are unaffected. Only ezomycins $A_1$, $B_1$, and to a lesser extent $C_1$, show antifungal activity. All three of these contain a cystathionine residue and the two most active have the anhydrooctose unit in the β-configuration. SAKATA has shown that no isomerization of the ezomycin *C*-nucleosides takes place under the test conditions (*35*). It is believed that $C_1$, $C_2$, $D_1$, and $D_2$ are artifacts, arising by rearrangement of the β-anomers (**145**). The mechanism of action of the ezomycins is unknown at present (*35*).

## VIII. Biosynthesis of C-Nucleoside Antibiotics

The *de novo* biosynthesis of the purine and pyrimidine *N*-ribonucleosides has been thoroughly investigated (*158, 159*). Some of the intermediates in the purine pathway are shown in Scheme 33. Formation of the ribosylnitrogen bond occurs as the first step, in which 5-phospho-α-D-ribofuranosyl pyrophosphate (PRPP) (**158**) reacts with L-glutamine to give the ribofuranosylamine (**159**). Glycine contributes two carbon atoms and a nitrogen atom, giving the amide derivative (**160**) which is converted, in several steps, into the important amino amide (AICAR) (**161**) whose amide carbon (C-6) is derived from carbon dioxide. Inosinic acid (IMP) (**162**) is the first purine nucleotide to be formed and this is converted into adenylosuccinic acid (succinyl AMP) (**163**) and finally AMP (**164**). In pyrimidine nucleoside biosynthesis L-aspartic acid and carbon dioxide provide the carbon atoms of the heterocyclic ring and the ribose is derived from PRPP (**158**).

In view of the structural relationship between *C*-nucleosides and the common ribonucleosides it was of great interest to find out whether the biosynthetic pathways had any features in common. The conclusion is that the ribose portion of *C*-nucleosides is derived from ribose, presumably via PRPP (**158**), but that the biosynthesis of the heterocyclic portion is different.

The most thoroughly investigated biosynthesis is that of showdomycin (**5**) by SUHADOLNIK and his colleagues (*160—162*). The first experiments used [5-$^{14}$C]- and [1-$^{14}$C]-labelled α-ketoglutarate, [5-$^{14}$C]- and [2-$^{14}$C]-DL-glutamate, and [1-$^{14}$C]-labelled D-ribose. Labelling patterns, determined by chemical degradation, indicated that carbons 2—5 of α-ketoglutarate or glutamate were incorporated into the maleimide ring and that ribose was the source of the ribofuranosyl unit (*160*). More detailed degradations and the discovery that no labelling resulted from [4-$^{3}$H]-glutamate required that carbons 2, 3, 4, and 5 of α-ketoglutarate (**165**) or glutamate (**166**) become carbons 4, 3, 2, and 1, respectively, in showdomycin (**5**) (Scheme 34) (*161*). A study of the incorporation of [1-$^{13}$C]- and [2-$^{13}$C]-acetate confirmed these conclusions (*162*). It has been shown recently, by $^{13}$C nmr, that the label in [5-$^{13}$C]-DL-glutamate is incorporated into C-1 of showdomycin (*163*). In addition, $^{2}$H nmr spectroscopy has shown that the deuterium in [5-$^{2}$H]-D-ribose is specifically incorporated into H-5′ of the ribose moiety, proving that ribose is introduced as an intact unit (*163*).

In the biosynthesis of formycin (**6**) the later stages were the first to be investigated (*86*). During a study of the conversion of formycin B (**7**) into oxoformycin B (**83**) in *Nocardia interforma* it was discovered that $^{3}$H-formycin B was also converted into $^{3}$H-formycin (**6**) by a transformation at the nucleotide level. Formycin B 5′-phosphate (**167**), and FMP (**168**) were

*References, pp. 290—299*

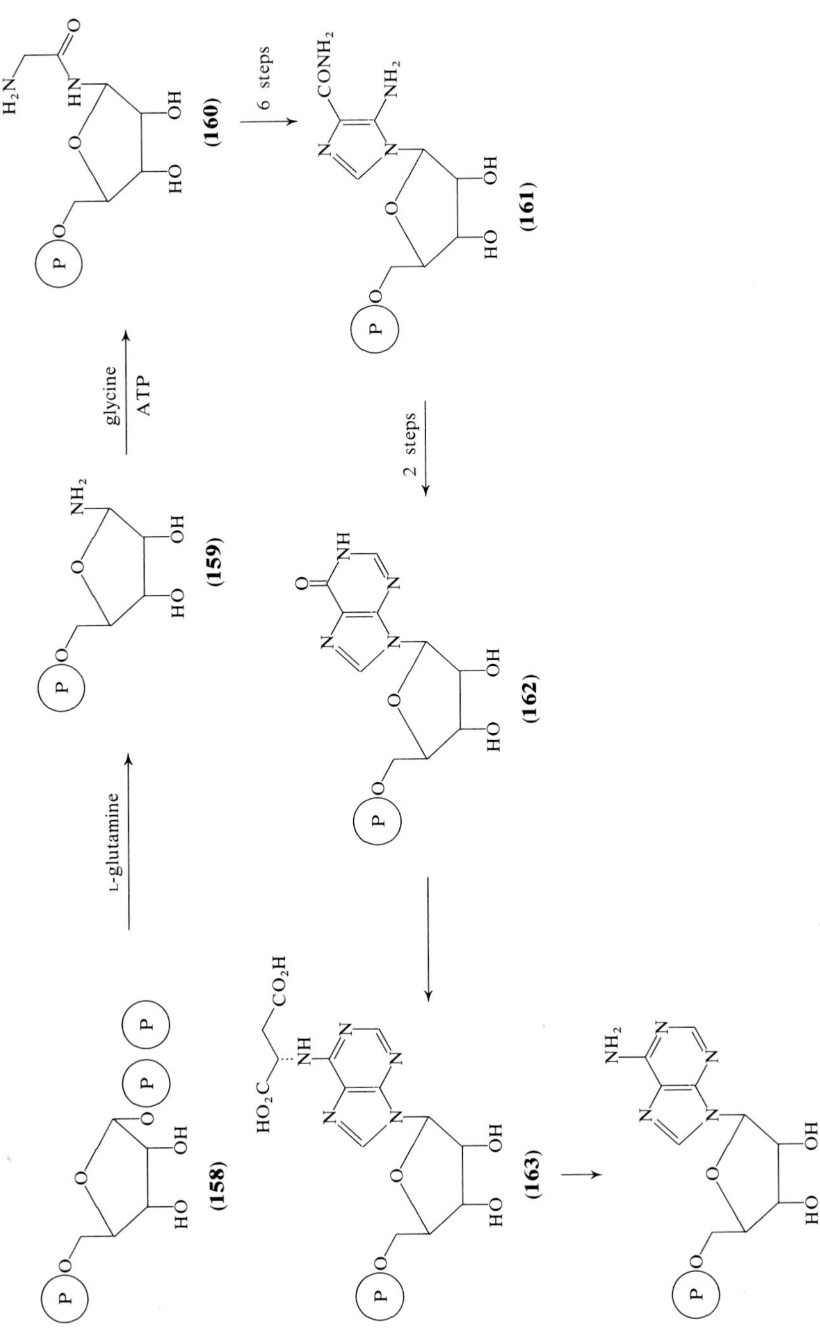

(158) L-glutamine (159) glycine / ATP (160) 6 steps (161) 2 steps (162) (163) (164)

Scheme 33

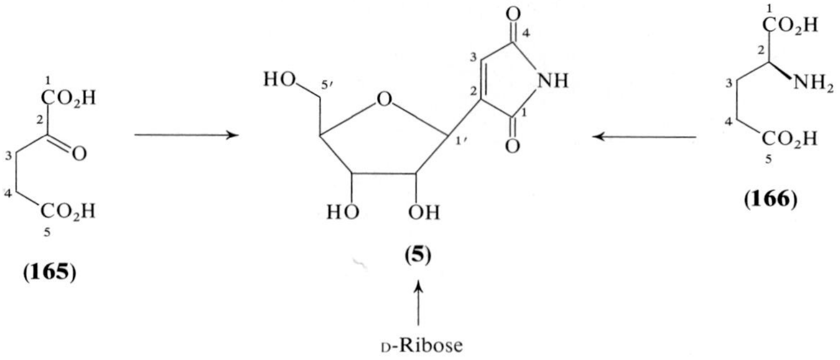

1 $CO_2H$, 2, 3, =O, 4, $CO_2H$, 5

**(165)**

→

HO, 5′, O, 1′, 2, 3, 4, O, NH, 1, O, HO, OH

**(5)**

↑

D-Ribose

←

1 $CO_2H$, 2, $NH_2$, 3, 4, $CO_2H$, 5

**(166)**

*Scheme 34*

identified by comparison with authentic compounds (*164*) and the biosynthetic sequence shown in Scheme 35 was proposed (*82, 86*). The sequence **(167)** → **(168)** is very similar to the purine nucleoside counterpart, **(162)** → **(164)** (Scheme 33). It was proposed that the amination of **(167)** involved an intermediate analogous to **(163)** because the reaction was inhibited by hadacidin, a known inhibitor of adenylosuccinate synthetase (*86*). OCHI et al. (*165*) have studied the later stages of formycin biosynthesis in *Streptomyces* sp. MA 406-A-1. Using a replacement culture technique they found that formycin B (**7**) is converted into formycin (**6**) without the intermediate formation of 5′-nucleotides. Aspartate was again the amino-group donor and the enzyme was inhibited by the formycin chromophore moiety (**86**).

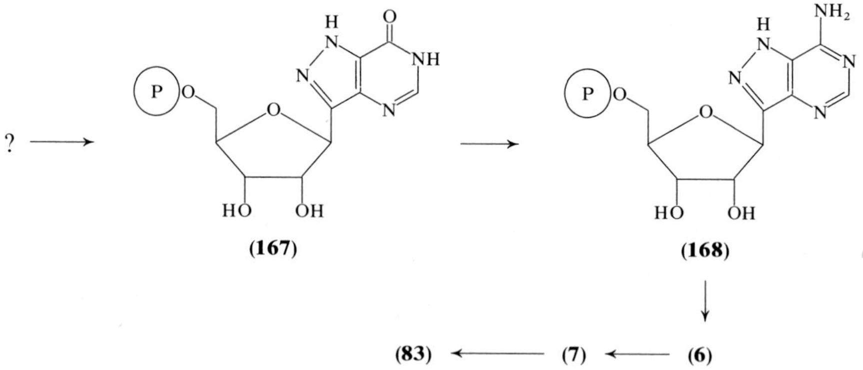

? →

P O, O, H N N, O, NH, N, HO, OH

**(167)**

→

P O, O, H N N, $NH_2$, N, N, HO, OH

**(168)**

↓

**(83)** ← **(7)** ← **(6)**

*Scheme 35*

*References, pp. 290—299*

KUNIMOTO *et al.* (*166*) first studied the earlier stages of the biosynthesis in *Nocardia interforma* and found that [U-$^{14}$C]-D-ribose was incorporated into the ribose portion. A similar conclusion was reached by OCHI *et al.* (*167, 168*) for *Streptomyces* sp. MA 406-A-1.

KUNIMOTO *et al.* (*166*) concluded that the biosynthesis of the heterocyclic portion differed from *de novo* purine biosynthesis (Scheme 33) and that adenine or other purine bases were not intermediates. OCHI has examined the incorporation of various amino acids under his conditions (*167—169*). L-Lysine and L-glutamate appeared to be precursors of the heterocyclic portion of formycin. In the most recent paper (*168*) the incorporation of L-lysine and L-glutamate was studied using the $^{13}$C nmr technique, and the surprising conclusion was reached that [U-$^{13}$C]-L-glutamate (*166*) was incorporated into C-9, C-4, C-5, and C-2 of formycin (**6**) (Scheme 36), with C-6 being labelled after feeding of [U-$^{13}$C]-L-lysine (**169**). The $^{13}$C nmr spectrum of the pyrazolopyrimidine moiety of formycin is difficult to analyse because of tautomerism (*38, 170*). By protonation of the heterocyclic moiety OCHI was able to resolve the signals due to C-2 and C-6. Unfortunately these were incorrectly assigned (*168*) and it was later shown by off-resonance decoupling that C-6, and not C-2, was the lowest field signal (*93*). This leads to the conclusion that glutamate is incorporated into four contiguous positions, C-9, C-4, C-5, and C-6. Since OCHI also reported that [1-$^{14}$C]-glutamate was incorporated, these carbon atoms in formycin must arise from C-4 to C-1 of glutamate (Scheme 36). This reassignment implies that lysine provides C-2 of formycin, as well as being an important source of nitrogen atoms (*168*); the mechanism of lysine incorporation is not at all clear.

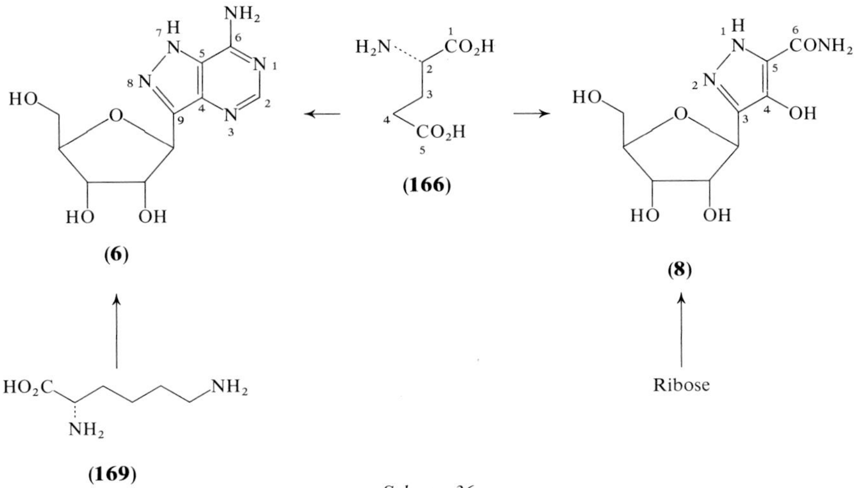

*Scheme 36*

The biosynthesis of the other pyrazole-containing *C*-nucleoside antibiotic pyrazofurin (**8**) has been investigated (*12, 93, 171*). It was found that [1-$^{14}$C]- and [5-$^{2}$H]-D-ribose are incorporated into pyrazofurin by *S. candidus*. $^{2}$H Nmr studies of the labelled pyrazofurin showed that the deuterium was confined to the H-5′ position, proving that an intact molecule of ribose had been transferred into pyrazofurin (*93*).

Glycine and bicarbonate, established precursors of the purine ring system (Scheme 33), were not incorporated as units into pyrazofurin (*93*). [U-$^{14}$C]- and [1-$^{14}$C]-L-glutamate were incorporated, presumably into the pyrazole ring, and it was suggested (*93*) that C-1, C-2, C-3, and C-4 of glutamate (**166**) became C-6, C-5, C-4, and C-3 of pyrazofurin (**8**) (Scheme 36). Experiments with [2-$^{14}$C]-acetate in which C-3, C-4, and C-5 of pyrazofurin became labelled are in agreement with this idea.

The biosynthesis of oxazinomycin (minimycin) (**9**) in *S. hygroscopicus* has been investigated by ISONO and SUHADOLNIK (*172, 173*) and ISONO and UZAWA (*39*). Experiments with [1-$^{14}$C]-D-ribose showed that it gives rise to the ribofuranosyl portion of oxazinomycin (*172, 173*) (Scheme 37). The feeding of [5-$^{14}$C]-DL-glutamate, [U-$^{14}$C]-L-glutamate and [3-$^{3}$H]-DL-glutamate led to the conclusion that C-3, C-4, and C-5 of glutamate became C-6, C-5, and C-4, respectively in oxazinomycin (*173*) (Scheme 37). The use of [1,2-$^{13}$C]-acetate and a study of the satellites due to $^{13}$C-$^{13}$C coupling demonstrated the transfer of an intact acetate unit into C-4 and C-5 of oxazinomycin via C-5 and C-4 of glutamate (*39, 173*) (Scheme 37). The origin of C-2 of oxazinomycin was shown to be carbon dioxide (*172*).

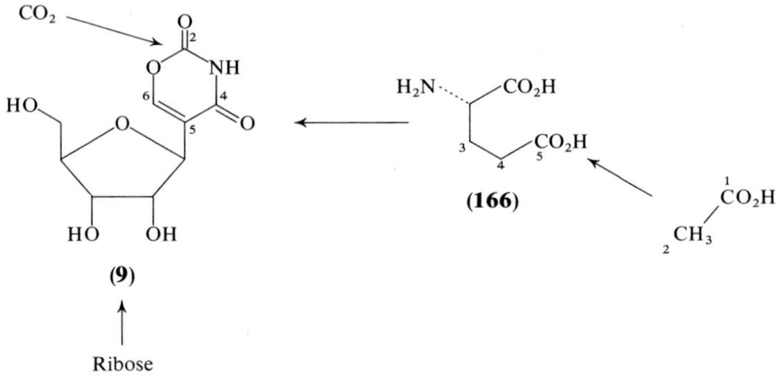

*Scheme 37*

The heterocyclic portion of the *C*-nucleoside antibiotics showdomycin (**5**), formycin (**6**), pyrazofurin (**8**) and oxazinomycin (**9**) is derived biosynthetically from L-glutamate, and attention has been drawn to its possible significance (*12, 93, 168, 171, 173*). The ribofuranosyl unit is derived from D-ribose and the present evidence suggests that the carbon atom of the heterocycle which is linked to C-1 of the ribose corresponds to C-4 of L-glutamate. This has led to the suggestion that the ribosylglutamate derivative (**170**) or a related compound might be a biosynthetic precursor common to these four antibiotics (*93*) (Scheme 38). It should be pointed out that the asymmetric incorporation of glutamate has been demonstrated satisfactorily only for showdomycin (*161*) and oxazinomycin (*39, 173*).

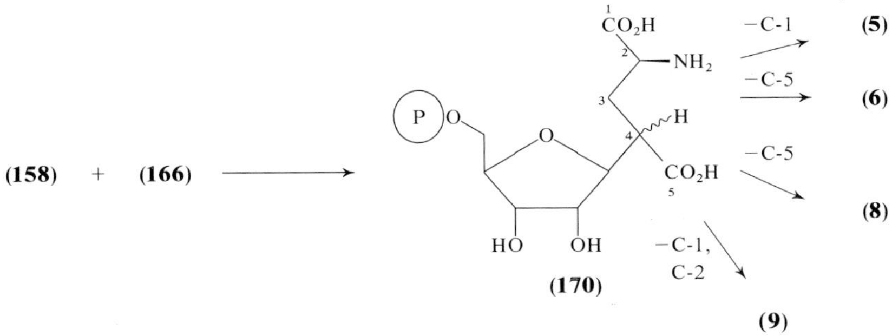

(**158**) + (**166**) ⟶ (**170**)

−C-1 → (**5**); −C-5 → (**6**); −C-5 → (**8**); −C-1, C-2 → (**9**)

*Scheme 38*

The biosynthesis of ezomycin $A_1$ (**140**) and $B_1$ (**10**) has not been investigated, but there are two interesting features. Isono and his colleagues have shown that the anhydrooctosyl unit in the octosyl acids [e.g. (**153**), Scheme 31] is derived by the addition of phosphoenolpyruvate to the 5′-aldehyde of uridine (**2**) (*174, 175*). Ezomycin $B_1$ (**10**) is related to pseudouridine (**3**) and it has been suggested that there may be similarities in their biosynthesis (*145*). In particular it was proposed that ezomycin $A_1$ was converted into the uracil derivative which underwent intramolecular rearrangement to the *C*-nucleoside. Although it is known that pseudouridine residues in transfer RNAs arise by rearrangement of uridine phosphate units within the macromolecule, Uematsu and Suhadolnik (*8, 176*) found that the pseudouridine in culture filtrates of *Streptoverticillicum ladakanus* did not arise through this pathway nor by rearrangement of uridine itself.

## Abbreviations

| | |
|---|---|
| AICAR | Aminoimidazolecarboxamide ribotide |
| AMP | Adenosine-5′ monophosphate |
| ATP | Adenosine-5′ triphosphate |
| Bzl | Benzyl |
| CD | Circular dichroism |
| DCC | Dicyclohexylcarbodiimide |

| | |
|---|---|
| DME | 1,2-Dimethoxyethane |
| DMF | Dimethyl formamide |
| DMSO | Dimethyl sulphoxide |
| DNA | Deoxyribonucleic acid |
| DSS | Sodium 2,2-dimethyl-2-silapentane-5-sulphonate |
| FDP | Formycin-5′ diphosphate |
| FMP | Formycin-5′ monophosphate |
| FTP | Formycin-5′ triphosphate |
| IMP | Inosine-5′ monophosphate (inosinic acid) |
| nmr | Nuclear magnetic resonance |
| OMP | Orotidine-5′ phosphate |
| ORD | Optical rotatory dispersion |
| PRPP | 5′-Phosphoribosyl pyrophosphate |
| Py | Pyridine |
| RNA | Ribonucleic acid |
| TFA | Trifluoroacetic acid |
| TFAA | Trifluoroacetic anhydride |
| TMS | Tetramethylsilane *or* trimethylsilyl |
| UMP | Uridine-5′ monophosphate |
| uv | Ultraviolet |

## References

*1.* COHN, W. E., and E. VOLKIN: Nucleoside-5′-Phosphates from Ribonucleic Acid. Nature **167,** 483 (1951).

*2.* DAVIS, F. F., and F. W. ALLEN: Ribonucleic Acids from Yeast which Contain a Fifth Nucleotide. J. Biol. Chem. **227,** 907 (1957).

*3.* COHN, W. E.: Pseudouridine, a Carbon-Carbon Linked Ribonucleoside in Ribonucleic Acids: Isolation, Structure, and Chemical Characteristics. J. Biol. Chem. **235,** 1488 (1960).

*4.* MICHELSON, A. M., and W. E. COHN: Cyclo-pseudouridine and the Configuration of Pseudouridine. Biochemistry **1,** 490 (1962).

*5.* YU, C.-T., and F. W. ALLEN: Studies on an Isomer of Uridine Isolated from Ribonucleic Acids. Biochim. Biophys. Acta **32,** 393 (1959).

*6.* CHAMBERS, R. W.: The Chemistry of Pseudouridine. Progr. Nucleic Acid Res. Mol. Biol. **5,** 349 (1966).

*7.* ADLER, M., and A. B. GUTMAN: Uridine Isomer (5-Ribosyluracil) in Human Urine. Science **130,** 862 (1959).

*8.* UEMATSU, T., and R. J. SUHADOLNIK: Pseudouridine, Isolation and Biosynthesis of the Nucleoside Isolated from the Culture Filtrates of *Streptoverticillicum Ladakanus.* Biochemistry **11,** 4669 (1972).

*9.* ARGOUDELIS, A. D., and S. A. MIZSAK: 1-Methylpseudouridine, a Metabolite of *Streptomyces Platensis.* J. Antibiotics **29,** 818 (1976).

*10.* FOX, J. J., K. A. WATANABE, and A. BLOCH: Nucleoside Antibiotics. Progr. Nucleic Acid Res. Mol. Biol. **5,** 251 (1966).

*11.* SUHADOLNIK, R. J.: Nucleoside Antibiotics. New York: J. Wiley. 1970.

*12.* SUHADOLNIK, R. J.: Nucleosides as Biological Probes. New York: J. Wiley. 1979.

*13.* SUHADOLNIK, R. J.: Naturally Occurring Nucleoside and Nucleotide Antibiotics. Progr. Nucleic Acid Res. Mol. Biol. **22,** 193 (1979).

14. BUCHANAN, J. G., and R. H. WIGHTMAN: Chemistry of Nucleoside Antibiotics. In: Topics in Antibiotic Chemistry, **6** (P. G. SAMMES, ed.), p. 229. Chichester: E. Horwood. 1982.
15. GOODCHILD, J.: The Biochemistry of Nucleoside Antibiotics. In: Topics in Antibiotic Chemistry, **6** (P. G. SAMMES, ed.), p. 99. Chichester: E. Horwood. 1982.
16. TRONCHET, J. M. J.: Synthèse de Nouveaux Types de *C*-Nucléosides. Biol. Med. **4,** 83 (1975).
17. HANESSIAN, S., and A. G. PERNET: Synthesis of Naturally Occurring *C*-Nucleosides, their Analogs, and Functionalised *C*-Glycosyl Precursors. Adv. Carbohydrate Chem. Biochem. **33,** 111 (1976).
18. DAVES, G. D., JR., and C. C. CHENG: The Chemistry and Biochemistry of *C*-Nucleosides. Progr. Medicin. Chem. **13,** 303 (1976).
19. MICHELSON, A. M.: The Chemistry of Nucleosides and Nucleotides. New York: Academic Press. 1963.
20. LEMIEUX, R. U.: Rearrangements and Isomerizations in Carbohydrate Chemistry. In: Molecular Rearrangements, Part 2 (P. DE MAYO, ed.), pp. 710—719. New York: J. Wiley. 1964.
21. BROWN, D. M., M. G. BURDON, and R. P. SLATCHER: A Synthesis of Pseudouridine and 5-β-D-Ribofuranosyluridine. J. Chem. Soc. (C) **1968,** 1051.
22. LERCH, U., M. G. BURDON, and J. G. MOFFATT: *C*-Glycosyl Nucleosides. I. Studies on the Synthesis of Pseudouridine and Related Compounds. J. Organ. Chem. (USA) **36,** 1507 (1971).
23. BROWN, D. M., and R. C. OGDEN: A Synthesis of Pseudouridine. J. Chem. Soc. Perkin I **1981,** 723.
24. BARON, F., and D. M. BROWN: Nucleotides. Part XXXIII. The Structure of Cytidylic Acids a and b. J. Chem. Soc. (London) **1955,** 2855.
25. DARNALL, K. R., L. B. TOWNSEND, and R. K. ROBINS: The Structure of Showdomycin, a Novel Carbon-Linked Nucleoside Antibiotic Related to Uridine. Proc. Nat. Acad. Sci. (USA) **57,** 548 (1967).
26. SASAKI, K., Y. KUSAKABE, and S. ESUMI: The Structure of Minimycin, a Novel Carbon-linked Nucleoside Antibiotic Related to β-Pseudouridine. J. Antibiotics **25,** 151 (1972).
27. TOWNSEND, L. B.: Nuclear Magnetic Resonance Spectroscopy in the Study of Nucleic Acid Components and Certain Related Derivatives. In: Synthetic Procedures in Nucleic Acid Chemistry, Vol. 2 (W. W. ZORBACH and R. S. TIPSON, eds.), p. 267. New York: J. Wiley. 1973.
28. RICE, J. M., and G. O. DUDEK: Mass Spectra of Uridine and Pseudouridine: Fragmentation Patterns Characteristic of a Carbon-Carbon Nucleoside Bond. Biochem. Biophys. Res. Comm. **35,** 383 (1969).
29. DESLAURIERS, R., and I. C. P. SMITH: A Comparison of the Conformations of Uridine, β-Pseudouridine, and Dihydrouridine in Dimethyl Sulfoxide and Water. A $^1$H Nuclear Magnetic Resonance Study. Canad. J. Chem. **51,** 833 (1973).
30. ROBINS, R. K., L. B. TOWNSEND, F. CASSIDY, J. F. GERSTER, A. F. LEWIS, and R. L. MILLER: Structure of the Nucleoside Antibiotics Formycin, Formycin B and Laurusin. J. Hetero. Chem. **3,** 110 (1966).
31. NAKAGAWA, Y., H. KANŌ, Y. TSUKUDA, and H. KOYAMA: Structure of a New Class of C-Nucleoside Antibiotic, Showdomycin. Tetrahedron Letters **1967,** 4105.
32. KOYAMA, G., K. MAEDA, H. UMEZAWA, and Y. IITAKA: The Structural Studies of Formycin and Formycin B. Tetrahedron Letters **1966,** 597.
33. GUTOWSKI, G. E., M. O. CHANEY, N. D. JONES, R. L. HAMILL, F. A. DAVIES, and R. D. MILLER: Pyrazomycin B: Isolation and Characterization of an α-*C*-Nucleoside Antibiotic Related to Pyrazomycin. Biochem. Biophys. Res. Comm. **51,** 312 (1973).
34. HANEISHI, T., T. OKAZAKI, T. HATA, C. TAMURA, M. NOMURA, A. NAITO, I. SEKI, and M.

ARAI: Oxazinomycin, a New Carbon-linked Nucleoside Antibiotic. J. Antibiotics **24,** 797 (1971).

*35.* SAKATA, K., A. SAKURAI, and S. TAMURA: Studies on Ezomycins, Antifungal Antibiotics, Part X. Structures of Ezomycins $B_1$, $B_2$, $C_1$, $C_2$, $D_1$ and $D_2$. Agr. Biol. Chem. **41,** 2033 (1977).

*36.* WENKERT, E., E. W. HAGAMAN, and G. E. GUTOWSKI: Carbon-13 Nuclear Magnetic Resonance Spectral Analysis of *C*-Nucleosides. The Structure of Pyrazomycin B. Biochem. Biophys. Res. Comm. **51,** 318 (1973).

*37.* CHENON, M.-T., R. J. PUGMIRE, D. M. GRANT, R. P. PANZICKA, and L. B. TOWNSEND: Carbon-13 NMR Spectra of *C*-Nucleosides. Showdomycin and β-Pseudouridine. J. Hetero. Chem. **10,** 427 (1973).

*38.* CHENON, M.-T., R. P. PANZICKA, J. C. SMITH, R. J. PUGMIRE, D. M. GRANT, and L. B. TOWNSEND: Carbon-13 Magnetic Resonance Spectra of *C*-Nucleosides. 3. Tautomerism in Formycin and Formycin B and Certain Pyrazolo[4,3-*d*]pyrimidines. J. Amer. Chem. Soc. **98,** 4736 (1976).

*39.* ISONO, K., and J. UZAWA: $^{13}$C-Nmr Evidence for the Biosynthetic Incorporation of Acetate into Minimycin and Compounds Related to Krebs Cycle. FEBS Letters **80,** 53 (1977).

*40.* HRUSKA, F. E., A. A. GREY, and I. C. P. SMITH: A Nuclear Magnetic Resonance Study of the Molecular Conformation of β-Pseudouridine in Aqueous Solution. J. Amer. Chem. Soc. **92,** 4088 (1970).

*41.* JONES, A. J., D. M. GRANT, M. W. WINKLEY, and R. K. ROBINS: Pyrimidine and Purine Nucleosides. J. Amer. Chem. Soc. **92,** 4079 (1970).

*42.* DEJONGH, D. C.: Mass Spectrometry of Nucleic Acid Components. In: Synthetic Procedures in Nucleic Acid Chemistry, Vol. 2 (W. W. ZORBACH and R. S. TIPSON, eds.), p. 145. New York: J. Wiley. 1973.

*43.* TOWNSEND, L. B., and R. K. ROBINS: The Mass Spectra of Formycin, Formycin B and Showdomycin, Carbon Linked Nucleoside Antibiotics. J. Hetero. Chem. **6,** 459 (1969). CRAIN, P. F., J. A. MCCLOSKEY, A. F. LEWIS, K. H. SCHRAM, and L. B. TOWNSEND: Mass Spectra of *C*-Nucleosides II. An Unusual Fragmentation Reaction of the Heterocyclic Moiety of Pyrazomycin and Some Closely Related Compounds. J. Hetero. Chem. **10,** 843 (1973).

*44.* FARKAŠ, J., Z. FLEGELOVÁ, and F. ŠORM: Synthesis of Pyrazomycin. Tetrahedron Letters **1972,** 2279.

*45.* ULBRICHT, T. L. V.: Optical Rotatory Dispersion of Nucleosides and Nucleotides. In: Synthetic Procedures in Nucleic Acid Chemistry, Vol. 2 (W. W. ZORBACH and R. S. TIPSON, eds.), p. 177. New York: J. Wiley. 1973.

*46.* DAVIES, D. B.: Conformations of Nucleosides and Nucleotides. Progr. NMR Spectroscopy **12,** 135 (1978).

*47.* NISHIMURA, H., M. MAYAMA, Y. KOMATSU, H. KATO, N. SHIMAOKA, and Y. TANAKA: Showdomycin, a New Antibiotic from a *Streptomyces Sp.* J. Antibiotics Ser. A **17,** 148 (1964).

*48.* TSUKUDA, Y., Y. NAKAGAWA, H. KANO, T. SATO, M. SHIRO, and H. KOYAMA: The Crystal Structure of Showdomycin and their Derivatives. Chem. Commun. **1967,** 975.

*49.* KALVODA, L., J. FARKAŠ, and F. ŠORM: Synthesis of Showdomycin. Tetrahedron Letters **1970,** 2297.

*50.* KALVODA, L.: Simple Preparative Synthesis of Showdomycin. J. Carbohydrates, Nucleosides, Nucleotides **3,** 47 (1976).

*51.* BOBEK, M., and J. FARKAŠ: Nucleic Acid Components and their Analogues. CXVIII. Synthesis of 8-β-D-Ribofuranosyladenine Starting from 2,5-Anhydro-D-allonic Acid. Collect. Czech. Chem. Comm. **34,** 247 (1969).

*52.* KALVODA, L.: The Synthesis of Pyrazoles. A Simple Preparative Synthesis of C-

Nucleosidic Antibiotics Formycin and Formycin B. Collect. Czech. Chem. Comm. **43,** 1431 (1978).

53. TRUMMLITZ, G., and J. G. MOFFATT: *C*-Glycosyl Nucleosides. III. A Facile Synthesis of the Nucleoside Antibiotic Showdomycin. J. Organ. Chem. (USA) **38,** 1841 (1973).
54. ALBRECHT, H. P., D. B. REPKE, and J. G. MOFFATT: *C*-Glycosyl Nucleosides. II. A Facile Synthesis of Derivatives of 2,5-Anhydro-D-allose. J. Organ. Chem. (USA) **38,** 1836 (1973).
55. BUCHANAN, J. G., A. R. EDGAR, and M. J. POWER: *C*-Nucleoside Studies. Part 1. Synthesis of [2,3,5-Tri-*O*-benzyl-α(and β)-D-ribofuranosyl]ethyne. J. Chem. Soc. Perkin I **1974,** 1943.
56. BUCHANAN, J. G., A. R. EDGAR, M. J. POWER, and C. T. SHANKS: *C*-Nucleoside Studies. Part 7. A New Synthesis of Showdomycin, 2-β-D-Ribofuranosylmaleimide. J. Chem. Soc. Perkin I **1979,** 225.
57. BARKER, R., and H. G. FLETCHER, JR.: 2,3,5-Tri-*O*-benzyl-D-ribosyl and -L-arabinosyl bromides. J. Organ. Chem. (USA) **26,** 4605 (1961).
58. HECK, R. F.: Dicarboalkoxylation of Olefins and Acetylenes. J. Amer. Chem. Soc. **94,** 2712 (1972).
59. INOUE, T., and I. KUWAJIMA: Highly Efficient Method for the Synthesis of Showdomycin. Chem. Commun. **1980,** 251.
60. JUST, G., A. MARTEL, K. GROZINGER, and M. RAMJEESINGH: *C*-Nucleosides and Related Compounds. IV. The Synthesis and Chemistry of D,L-2,5-Anhydroallose Derivatives. Canad. J. Chem. **53,** 131 (1975).
61. JUST, G., T. J. LIAK, M.-I. LIM, P. POTVIN, and Y. S. TSANTRIZOS: *C*-Nucleosides and Related Compounds. XV. The Synthesis of D,L-*epi*-Showdomycin and D,L-Showdomycin. Canad. J. Chem. **58,** 2024 (1980).
62. ITO, Y., T. SHIBATA, M. ARITA, H. SAWAI, and M. OHNO: Chirally Selective Synthesis of Sugar Moiety of Nucleosides by Chemicoenzymatic Approach: L- and D-Riboses, Showdomycin, and Cordycepin. J. Amer. Chem. Soc. **103,** 6739 (1981).
63. JUST, G., and A. MARTEL: *C*-Nucleosides and Related Compounds. Synthesis of D,L-3,4-*O*-Isopropylidene-2,5-anhydroallose: A Novel Periodate Cleavage. Tetrahedron Letters **1973,** 1517.
64. JUST, G., and K. GROZINGER: A Correction to "A Novel Periodate Cleavage". Tetrahedron Letters **1974,** 4165.
65. ALDERSLEY, M. F., A. J. KIRBY, and P. W. LANCASTER: Intramolecular Displacement of Alkoxide Ions by the Ionised Carboxy-group: Hydrolysis of Alkyl Hydrogen Dialkylmaleates. J. Chem. Soc. Perkin II **1974,** 1504.
66. GENSLER, W. J., S. CHAN, and D. B. BALL: Synthesis of a Triazole Homo-*C*-nucleoside. J. Amer. Chem. Soc. **97,** 436 (1975).
67. OHRUI, H., G. H. JONES, J. G. MOFFATT, M. L. MADDOX, A. T. CHRISTENSEN, and S. K. BYRAM: *C*-Glycosyl Nucleosides. V. Some Unexpected Observations on the Relative Stabilities of Compounds Containing Fused Five-Membered Rings with Epimerizable Substituents. J. Amer. Chem. Soc. **97,** 4602 (1975).
68. OHRUI, H., and S. EMOTO: A Rationalization of the Relative Thermodynamic Stabilities of Fused Five-Membered Tetrahydrofurans with Epimerizable Substituents. An Anomeric Effect in Furanoses. J. Organ. Chem. (USA) **42,** 1951 (1977).
69. NOYORI, R., T. SATO, and Y. HAYAKAWA: A Stereocontrolled Synthesis of *C*-Nucleosides. J. Amer. Chem. Soc. **100,** 2561 (1978).
70. SATO, T., R. ITO, Y. HAYAKAWA, and R. NOYORI: Stereocontrolled Synthesis of Showdomycin and 6-Azapseudouridines. Tetrahedron Letters **1978,** 1829.
71. CHU, C. K., I. WEMPEN, K. A. WATANABE, and J. J. FOX: Nucleosides, 100. General Synthesis of Pyrimidine C-5 Nucleosides Related to Pseudouridine. Synthesis of 5-(β-D-Ribofuranosyl)isocytosine (Pseudoisocytidine), 5-(β-D-Ribofuranosyl)-2-thiouracil (2-

Thiopseudouridine) and 5-(β-D-Ribofuranosyl)uracil (Pseudouridine). J. Organ. Chem. (USA) **41,** 2793 (1976).
72. KOZIKOWSKI, A. P., and A. AMES: Total Synthesis of the *C*-Nucleoside *dl*-Showdomycin by a Diels-Alder, Retrograde Dieckmann Strategy. J. Amer. Chem. Soc. **103,** 3923 (1981).
73. MATSUURA, S., O. SHIRATORI, and K. KATAGIRI: Antitumour Activity of Showdomycin. J. Antibiotics Ser. A **17,** 234 (1964).
74. KOMATSU, Y.: Mechanism of Action of Showdomycin. V. Reduced Ability of Showdomycin-resistant Mutants of *Eschericia Coli* K-12 to take up Showdomycin and Nucleosides. J. Antibiotics **24,** 876 (1971).
75. VISSER, D. W., and S. ROY-BURMAN: Showdomycin. In: Antibiotics, Vol. 5, Part 2. Mechanism of Action of Antieukaryotic and Antiviral Compounds (F. E. HAHN, ed.), p. 363. New York: Springer. 1979.
76. OZAKI, M., T. KARIYA, H. KATO, and T. KIMURA: Microbial Transformation of Antibiotics. Part I. Isomerisation of Showdomycin by *Streptomyces* Species. Agr. Biol. Chem. **36,** 451 (1972).
77. NAKAGAWA, Y.: Personal Communication.
78. HORI, M., E. ITO, T. TAKITA, G. KOYAMA, T. TAKEUCHI, and H. UMEZAWA: A New Antibiotic, Formycin. J. Antibiotics Ser. A **17,** 96 (1964).
79. KOYAMA, G., and H. UMEZAWA: Formycin B and Its Relation to Formycin. J. Antibiotics Ser. A **18,** 175 (1965).
80. AIZAWA, A., T. HIDAKA, N. ŌTAKE, H. YONEHARA, K. ISONO, N. IGARISHI, and S. SUZUKI: Studies on a New Antibiotic, Laurusin. Agr. Biol. Chem. **29,** 375 (1965).
81. UMEZAWA, H., T. SAWA, Y. FUKAGAWA, G. KOYAMA, M. MURASE, M. HAMADA, and T. TAKEUCHI: Transformation of Formycin to Formycin B and their Biological Activities. J. Antibiotics Ser. A **18,** 178 (1965).
82. ŌTAKE, N., S. AIZAWA, T. HIDAKA, H. SETO, and H. YONEHARA: Biological and Chemical Transformations of Formycin to Laurusin. Agr. Biol. Chem. **29,** 377 (1965).
83. FUKAGAWA, Y., T. SAWA, T. TAKEUCHI, and H. UMEZAWA: Deamination of Purine Antibiotics by Adenosine Deaminase. J. Antibiotics Ser. A **18,** 191 (1965).
84. SAWA, T., Y. FUKAGAWA, I. HOMMA, T. TAKEUCHI, and H. UMEZAWA: Formycin-deaminating Activity of Microorganisms. J. Antibiotics Ser. A **20,** 317 (1967).
85. ISHIZUKA, M., T. SAWA, G. KOYAMA, T. TAKEUCHI, and H. UMEZAWA: Metabolism of Formycin and Formycin B *In Vivo*. J. Antibiotics **21,** 1 (1968).
86. SAWA, T., Y. KUKAGAWA, I. HOMMA, T. WAKASHIRO, T. TAKEUCHI, M. HORI, and T. KOMAI: Metabolic Conversion of Formycin B to Formycin A and to Oxformycin B in *Nocardia Interforma*. J. Antibiotics **21,** 334 (1968).
87. KAWAMURA, K., S. FUKATSU, M. MURASE, G. KOYAMA, K. MAEDA, and H. UMEZAWA: The Studies on the Degradation Products of Formycin and Formycin B. J. Antibiotics Ser. A **19,** 91 (1966).
88. ROBINS, R. K., F. W. FURCHT, A. D. GRAUER, and J. W. JONES: Potential Purine Antagonists. II. Synthesis of some 7- and 5,7-substituted Pyrazolo[4,3-d]pyrimidines. J. Amer. Chem. Soc. **78,** 2418 (1956).
89. ROBINS, R. K., L. B. HOLUM, and F. W. FURCHT: Potential Purine Antagonists. V. Synthesis of Some 3-Methyl-5,7-substituted Pyrazolo[4,3-d]pyrimidines. J. Organ. Chem. (USA) **21,** 833 (1956).
90. KOYAMA, G., H. UMEZAWA, and Y. IITAKA: Crystal Structure of Formycin Hydrobromide Monohydrate. Acta Crystallogr. Sect. B **30,** 1511 (1974).
91. PRUSINER, P., T. BRENNAN, and M. SUNDARALINGAM: Crystal Structure and Molecular Conformation of Formycin Monohydrates. Possible Origin of the Anomalous Circular Dichroic Spectra in Formycin Mono and Polynucleotides. Biochemistry **12,** 1196 (1973).

92. Koyama, G., H. Nakamura, H. Umezawa, and Y. Iitaka: The Crystal and Molecular Structures of Oxoformycin B and Formycin B. Acta Crystallogr. Sect. B **32,** 813 (1976).
93. Buchanan, J. G., M. R. Hamblin, G. R. Sood, and R. H. Wightman: The Biosynthesis of Pyrazofurin and Formycin. Chem. Commun. **1980,** 917.
94. Farkaš, J., and F. Šorm: Synthesis of 3-(β-D-Ribofuranosyl)-5,7-dihydroxy-1*H*-pyrazolo-[4,3-d]-pyrimidine (Oxoformycin). Collect. Czech. Chem. Comm. **37,** 2798 (1972). Preliminary communication: Bobek, M., J. Farkaš, and F. Šorm, Tetrahedron Letters **1970,** 4611.
95. Acton, E. M., K. J. Ryan, D. W. Henry, and L. Goodman: Synthesis of the Nucleoside Antibiotic Formycin B. Chem. Commun. **1971,** 986.
96. Ogawa, T., Y. Kikuchi, M. Matsui, H. Ohrui, H. Kuzuhara, and S. Emoto: Synthetic Studies on C-Nucleosides. Part I. A Synthesis of Oxoformycin. Agr. Biol. Chem. **35,** 1825 (1971).
97. Lang, R. A., A. F. Lewis, R. K. Robins, and L. B. Townsend: Pyrazolopyrimidine Nucleosides. Part II. 7-Substituted 3-β-D-Ribofuranosylpyrazolo[4,3-d]pyrimidines Related to and Derived from the Nucleoside Antibiotics Formycin and Formycin B. J. Chem. Soc. (C) **1971,** 2443.
98. Kalvoda, L.: The Synthesis of Pyrazoles. A Simple Preparative Synthesis of C-Nucleosidic Antibiotics Formycin and Formycin B. Collect. Czech. Chem. Comm. **43,** 1431 (1978).
99. Buchanan, J. G., A. R. Edgar, R. J. Hutchison, A. Stobie, and R. H. Wightman: *C*-Nucleoside Studies. Part 10. A New Synthesis of 3-(2,3,5-Tri-*O*-benzyl-β-D-ribofuranosyl)pyrazole and its Conversion into 4-Nitro-3(5)-β-D-ribofuranosylpyrazole. J. Chem. Soc. Perkin I **1980,** 2567.
100. Buchanan, J. G., A. Stobie, and R. H. Wightman: *C*-Nucleoside Studies. Part XI. *Cine*-substitution in 1,4-Dinitropyrazoles; Application to the Synthesis of Formycin via Nitropyrazole Derivatives. Canad. J. Chem. **58,** 2624 (1980).
101. Habraken, C. L., and E. K. Poels: Nucleophilic Substitution Reactions on *N*-Nitropyrazoles. J. Organ. Chem. (USA) **42,** 2893 (1977).
102. Buchanan, J. G., A. Stobie, and R. H. Wightman: *C*-Nucleoside Studies. Part 14. A New Synthesis of Pyrazofurin. J. Chem. Soc. Perkin I **1981,** 2374.
103. Ishizuka, M., T. Takeuchi, K. Nitta, G. Koyama, M. Hori, and H. Umezawa: Antitumour Activities of Formycin and Labilomycin. J. Antibiotics Ser. A **17,** 124 1964).
104. Ishizuka, M., T. Sawa, S. Hori, H. Takayama, T. Takeuchi, and H. Umezawa: Biological Studies on Formycin and Formycin B. J. Antibiotics **21,** 5 (1968).
105. Müller, W. E. G., H. J. Rohde, R. Steffen, A. Maidhof, M. Lachmann, R. K. Zahn, and H. Umezawa: Influence of Formycin B on Polyadenosine Diphosphoribose Synthesis *In vitro* and *In vivo*. Cancer Res. **35,** 3673 (1975).
106. Hori, M. T. Wakashiro, E. Ito, T. Sawa, T. Takeuchi, and H. Umezawa: Biochemical Effects of Formycin B on *Xanthomonas Oryzae*. J. Antibiotics **21,** 264 (1968).
107. Takeuchi, T., J. Iwanaga, T. Aoyagi, and H. Umezawa: Antiviral Effect of Formycin and Formycin B. J. Antibiotics Ser. A **19,** 286 (1966).
108. Ishida, N., M. Homma, K. Kumagai, Y. Schimizu, S. Matsumoto, and A. Izawa: Studies on the Antiviral Activity of Formycin. J. Antibiotics Ser. A **20,** 49 (1967).
109. Umezawa, H. T. Sawa, Y. Fukagawa, I. Homma, M. Ishizuka, and T. Takeuchi: Studies on Formycin and Formycin B in Cells of Ehrlich Carcinoma and *E. Coli*. J. Antibiotics Ser. A **20,** 308 (1967).
110. Ward, D. C., A. Cerami, E. Reich, G. Acs, and L. Altwerger: Biochemical Studies of the Nucleoside Analogue, Formycin. J. Biol. Chem. **244,** 3243 (1969).
111. Ikehara, M., K. Murao, F. Harada, and S. Nishimura: Synthesis of Formycin Triphosphate and its Incorporation into Ribopolynucleotide by DNA-Dependent RNA Polymerase. Biochim. Biophys. Acta **155,** 82 (1968).

*112.* MAELICKE, A., M. SPRINZL, F. VON DER HAAR, T. A. KHWAJA, and F. CRAMER: Structural Studies on Phenylalanine Transfer Ribonucleic Acid from Yeast with the Spectroscopic Label Formycin. Eur. J. Biochem. **43,** 617 (1974).

*113.* WARD, D. C., and E. REICH: Conformational Properties of Polyformycin: A Polyribonucleotide with Individual Residues in the *Syn*-Conformation. Proc. Nat. Acad. Sci (USA) **61,** 1494 (1968).

*114.* WARD, D. C., W. FULLER, and E. REICH: Stereochemical Analysis of the Specificity of Pancreatic RNase with Polyformycin as Substrate: Differentiation of the Transphosphorylation and Hydrolysis Reactions. Proc. Nat. Acad. Sci (USA) **62,** 581 (1969).

*115.* WARD, D. C., T. HORN, and E. REICH: Fluorescence Studies of Nucleotides and Polynucleotides. III. Diphosphopyridine Nucleotide Analogues which Contain Fluorescent Purines. J. Biol. Chem. **247,** 4014 (1972).

*116.* VON DER HAAR, F., and E. GAERTNER: Phenylalanyl-tRNA Synthetase from Baker's Yeast: Role of 3′-Terminal Adenosine of $tRNA^{Phe}$ in Enzyme-Substrate Interaction Studied with 3′-Modified $tRNA^{Phe}$ Species. Proc. Nat. Acad. Sci (USA) **72,** 1378 (1975).

*117.* KUMAR, S. A., J. S. KRAKOW, and D. C. WARD: ATP Analogues as Initiation and Elongation Nucleotides for Bacterial DNA-Dependent RNA Polymerase. Biochim. Biophys. Acta **477,** 112 (1977).

*118.* SAWA, T., Y. FUKAGAWA, I. HOMMA, T. TAKEUCHI, and H. UMEZAWA: Mode of Inhibition of Coformycin on Adenosine Deaminase. J. Antibiotics Ser. A **20,** 227 (1967).

*119.* NAKAMURA, H., G. KOYAMA, Y. IITAKA, M. OHNO, N. YAGISAWA, S. KONDO, K. MAEDA, and H. UMEZAWA: Structure of Coformycin, an Unusual Nucleoside of Microbial Origin. J. Amer. Chem. Soc. **96,** 4327 (1974).

*120.* OHNO, M., N. YAGISAWA, S. SHIBAHARA, S. KONDO, K. MAEDA, and H. UMEZAWA: Synthesis of Coformycin. J. Amer. Chem. Soc. **96,** 4326 (1974).

*121.* HENDERSON, J. F., A. R. P. PATERSON, I. C. CALDWELL, and M. HORI: Biochemical Effects of Formycin, an Adenosine Analog. Cancer Res. **27,** 715 (1967).

*122.* GERZON, K., D. C. DELONG, and J. C. CLINE: C-Nucleosides: Aspects of Chemistry and Mode of Action. Pure Appl. Chem. **28,** 489 (1971).

*123.* GERZON, K., R. H. WILLIAMS, M. HOEHN, M. GORMAN, and D. C. DELONG: Pyrazomycin, A C-Nucleoside with Antiviral Activity. 2nd Int. Cong. Hetero. Chem., Montpellier, France, July 1969, Abstr. 30C, p. 131.

*124.* WILLIAMS, R. H., K. GERZON, M. HOEHN, M. GORMAN, and D. C. DELONG: Pyrazomycin — a Novel Carbon-Linked Nucleoside. 158th National Meeting, Amer. Chem. Soc., New York (1969), Abstr. MICR. 38.

*125.* GUTOWSKI, G. E., M. J. SWEENEY, D. C. DELONG, R. L. HAMILL, K. GERZON, and R. W. DYKE: Biochemistry and Biological Effects of the Pyrazofurins (Pyrazomycins): Initial Clinical Trial. Ann. New York Acad. Sci. **255,** 544 (1975).

*126.* WILLIAMS, R. H., and M. M. HOEHN: Pyrazomycin and Process for Production thereof. U.S. Pat. 3,802,999, April 9, 1974 to Eli Lilly and Company.

*127.* JONES, N. D., and M. O. CHANEY: The Crystal Structure of Pyrazomycin, A C-Nucleoside Antiviral Agent. 9th Int. Congr. Crystallogr., Kyoto, Japan, Abstr. S. 48.

*128.* DE BERNARDO, S., and M. WEIGELE: A Synthesis of the Pyrazomycins. J. Organ. Chem. (USA) **41,** 287 (1976).

*129.* OHRUI, H., and J. J. FOX: Nucleosides LXXXI. An Approach to the Synthesis of C-C Linked β-D-Ribofuranosyl Nucleosides from 2,3-*O*-Isopropylidene-5-*O*-trityl-β-D-ribofuranosyl Chloride. Tetrahedron Letters **1973,** 1951.

*130.* KATAGIRI, N., K. TAKASHIMA, and T. KATO: A Simple Synthesis of the Pyrazofurins. Chem. Commun. **1982,** 664.

*131.* DE CLERCQ, E., and P. F. TORRENCE: Nucleoside Analogs with Selective Antiviral Activity. J. Carbohydr. Nucleosides Nucleotides **5,** 187 (1978).

*132.* SWEENEY, M. J., F. A. DAVIS, G. E. GUTOWSKI, R. L. HAMILL, D. H. HOFFMANN, and G.

A. POORE: Experimental Antitumour Activity of Pyrazomycin. Cancer Res. **33,** 2619 (1973).

*133.* CADMAN, E. C., D. E. DIX, and R. E. HANDSCHUMACHER: Clinical, Biological, and Biochemical Effects of Pyrazofurin. Cancer Res. **38,** 682 (1978).

*134.* ROSSI, A.: The Clinical Uses of Nucleoside Analogues in Malignant Disease. In: Nucleoside Analogues. Chemistry, Biology, and Medical Applications (R. T. WALKER, E. DE CLERCQ, and F. ECKSTEIN, eds.), p. 409. NATO Advanced Study Institutes Series. New York and London: Plenum Press. 1979.

*135.* LEVINE, H. L., R. S. BRODY, and F. H. WESTHEIMER: Inhibition of Orotidine-5′-phosphate Decarboxylase by 1-(5′-Phospho-β-D-ribofuranosyl)barbituric Acid, 6-Azauridine 5′-Phosphate and Uridine 5′-Phosphate. Biochemistry **19,** 4993 (1980).

*136.* SHIRATO, S., J. NAGATSU, M. SHIBUYA, and Y. KASUKABE: Antibiotic Minimycin. Ger. Pat. 2,043,946, March 25, 1971 to Kaken Chemical Co. Ltd. Chem. Abs. **74,** 139557h (1971).

*137.* KUSAKABE, Y., J. NAGATSU, M. SHIBUYA, O. KAWAGUCHI, C. HIROSE, and S. SHIRATO: Minimycin, a New Antibiotic. J. Antibiotics **25,** 44 (1972).

*138.* DE BERNARDO, S., and M. WEIGELE: Synthesis of Oxazinomycin (Minimycin). J. Organ. Chem. (USA) **42,** 109 (1977).

*139.* TAKAOKA, K., T. KUWAYAMA, and A. AOKI: Jap. Pat. 615,332 (1971): cited in ref. *140.*

*140.* SAKATA, K., A. SAKURAI, and S. TAMURA: Studies on Ezomycins, Antifungal Antibiotics. Part I. L-Cystathionine as a Component of Ezomycins $A_1$ and $B_1$ from a Streptomyces. Agr. Biol. Chem. **37,** 697 (1973).

*141.* — — — Studies on Ezomycins, Antifungal Antibiotics. Part II. Ezoaminuroic Acid, 3-Amino-3,4-dideoxy-D-*xylo*-hexopyranuroic Acid, as a Constituent of Ezomycins $A_1$ and $A_2$. Tetrahedron Letters **1974,** 1533.

*142.* — — — Studies on Ezomycins, Antifungal Antibiotics. Part III. Isolation of Novel Antifungal Antibiotics, Ezomycins $A_1$, $A_2$, $B_1$ and $B_2$. Agr. Biol. Chem. **38,** 1883 (1974).

*143.* — — — Studies on Ezomycins, Antifungal Antibiotics. Part IV. Structures of Ezomycins $A_1$ and $A_2$. Tetrahedron Letters **1974,** 4327.

*144.* — — — Studies on Ezomycins, Antifungal Antibiotics. Part V. Degradative Studies on Ezomycins $A_1$ and $A_2$.Agr. Biol. Chem. **39,** 885 (1975).

*145.* — — — Studies on Ezomycins, Antifungal Antibiotics. Part VI. Structures of Ezomycins $B_1$, $B_2$, $C_1$, $C_2$, $D_1$, and $D_2$. Tetrahedron Letters **1975,** 3191.

*146.* — — — Studies on Ezomycins, Antifungal Antibiotics. Part VII. Structures of Ezomycins $A_1$ and $A_2$. Agr. Biol. Chem. **40,** 1993 (1976).

*147.* SAKATA, K., and J. UZAWA: Studies on Ezomycins, Antifungal Antibiotics. Part VIII. Application of C-13 NMR Spectrometry to the Structural Investigation of the Novel Bicyclic Anhydrooctose Uronic Acid Nucleosides, Constituents of Ezomycins. Agr. Biol. Chem. **41,** 413 (1977).

*148.* SAKATA, K., A. SAKURAI, and S. TAMURA: Studies on Ezomycins, Antifungal Antibiotics. Part IX. Isolation and Antimicrobial Activities of Ezomycins $B_1$, $B_2$, $C_1$, $C_2$, $D_1$ and $D_2$. Agr. Biol. Chem. **41,** 2027 (1977).

*149.* SAKATA, K., J. UZAWA, and A. SAKURAI: Studies on Ezomycins, Antifungal Antibiotics. Part XI. Application of Carbon-13 N.m.r. Spectroscopy to the Structural Investigation of Ezomycins. Organic Magnetic Resonance **10,** 230 (1977).

*150.* OGAWA, T., M. AKATSU, and M. MATSUI: Synthesis of a Sugar Occurring in an Antibiotic: Ezoaminuroic Acid, the First Example of a Naturally Occurring 3-Amino-3-deoxyhexuronic Acid. Carbohydrate Research **44,** C22 (1975).

*151.* ČERNÝ, M., and J. PAĆAK: Desoxyzucker III. Über Reaktionen der 2-O-Tosyl-1,6:3,4-dianhydro-β-D-galactopyranose Darstellung von 4-Desoxy-D-*xylo*hexose (4-Desoxy-D-glucose) und 4-Desoxy-D-*arabino*hexose (4-Desoxy-D-altrose). Collect. Czech. Chem. Comm. **27,** 94 (1962).

*152.* Mieczowski, J., and A. Zamojski: Total Syntheses of Methyl(methyl 3-amino-3,4-dideoxy-α- and β-DL-*xylo*hexopyranoside)uronates. Bull. Acad. Pol. Sci., Ser. Sci. Chim. **23,** 581 (1975). [Chem. Abs. **84,** 31341a (1976)].

*153.* Isono, K., P. F. Crain, and J. A. McCloskey: Isolation and Structure of Octosyl Acids. Anhydrooctose Uronic Acid Nucleosides. J. Amer. Chem. Soc. **97,** 943 (1975).

*154.* Anzai, K., and T. Saita: Synthesis of 3,7-Anhydrooctose Derivatives Related to Octosyl Acids. Chem. Commun. **1976,** 681.

*155.* — — The Synthesis of Several Octose Derivatives Related to Octosyl Acids A and B. Bull. Chem. Soc. Japan **50,** 169 (1977).

*156.* Kim, K. S., and W. A. Szarek: Synthesis of 3′,7′-Anhydrooctose Nucleosides Related to the Ezomycins and the Octosyl Acids. Canad. J. Chem. **59,** 878 (1981).

*157.* — — Syntheses Related to the 3,7-Anhydrooctose in the Ezomycins and the Octosyl Acids. Carbohydrate Research **100,** 169 (1982).

*158.* Buchanan, J. M., and S. C. Hartman: Enzymic Reactions in the Synthesis of the Purines. Adv. Enzymology **21,** 199 (1959).

*159.* Hartman, S. C.: Purines and Pyrimidines. In: Metabolic Pathways, 3rd Edn., Vol. 4 (D. M. Greenberg, Ed.), p. 1. New York: Academic Press. 1970.

*160.* Elstner, E. F., and R. J. Suhadolnik: Nucleoside Antibiotics. Biosynthesis of the Maleimide Nucleoside Antibiotic, Showdomycin, by *Streptomyces showdoensis.* Biochemistry **10,** 3608 (1971).

*161.* — — Nucleoside Antibiotics. Asymmetric Incorporation of Glutamic Acid and Acetate into the Maleimide Ring of Showdomycin by *Streptomyces showdoensis.* Biochemistry **11,** 2578 (1972).

*162.* Elstner, E. F., R. J. Suhadolnik, and A. Allerhand: Effect of Changes in the Pool of Acetate on the Incorporation and Distribution of $^{13}$C- and $^{14}$C-Labeled Acetate into Showdomycin by *Streptomyces showdoensis.* J. Biol. Chem. **248,** 5385 (1973).

*163.* Buchanan, J. G., M. R. Hamblin, and R. H. Wightman: Heriot-Watt University, unpublished results.

*164.* Sawa, T., Y. Fukagawa, Y. Shimauchi, K. Ito, M. Hamada, T. Takeuchi, and H. Umezawa: Studies on Formycin and Formycin B Phosphates. J. Antibiotics Ser. A **18,** 259 (1965).

*165.* Ochi, K., S. Yashima, and Y. Eguchi: Biosynthesis of Formycin. Formation of Formycin from Formycin B. J. Antibiotics **28,** 965 (1975).

*166.* Kunimoto, T., T. Sawa, T. Wakashiro, M. Hori, and H. Umezawa: Biosynthesis of the Formycin Family. J. Antibiotics **24,** 253 (1971).

*167.* Ochi, K., S. Kikuchi, S. Yashima, and Y. Eguchi: Biosynthesis of Formycin. Incorporation and Distribution of Labeled Compounds into Formycin. J. Antibiotics **29,** 638 (1976).

*168.* Ochi, K., S. Yashima, Y. Eguchi, and K. Matsushita: Biosynthesis of Formycin. Incorporation and Distribution of $^{13}$C-, $^{14}$C-, and $^{15}$N-Labeled Compounds into Formycin. J. Biol. Chem. **254,** 8819 (1979).

*169.* Ochi, K., S. Iwamoto, E. Hayase, S. Yashima, and Y. Okami: Biosynthesis of Formycin. Role of Certain Amino Acids in Formycin Biosynthesis. J. Antibiotics **27,** 909 (1974).

*170.* Krugh, T. R.: Tautomerism of the Nucleoside Antibiotic Formycin, as studied by Carbon-13 Nuclear Magnetic Resonance. J. Amer. Chem. Soc. **95,** 4761 (1973).

*171.* Suhadolnik, R. J., and N. L. Reichenbach: Glutamate as the Common Precursor for the Aglycon of the Naturally Occurring *C*-Nucleoside Antibiotics. Biochemistry **20,** 7042 (1981).

*172.* Isono, K., and R. J. Suhadolnik: The Biosynthesis of the Nucleoside Antibiotics: Minimycin Formation by *Streptomyces hygroscopicus.* Ann. New York Acad. Sci. **255,** 390 (1975).

*173.* — — Biosynthesis of the C-Nucleoside, Minimycin: Asymmetric Incorporation of Glutamate and Acetate into the Oxazine Ring. J. Antibiotics **30,** 272 (1977).
*174.* Isono, K., T. Sato, K. Hirasawa, S. Funayama, and S. Suzuki: Biosynthesis of the Nucleoside Skeleton of Polyoxins. J. Amer. Chem. Soc. **100,** 3937 (1978).
*175.* Sato, T., K. Hirasawa, J. Uzawa, T. Inaba, and K. Isono: Biosynthesis of Octosyl Acid A: Incorporation of C-13 Labeled Glucose. Tetrahedron Letters **1979,** 3441.
*176.* Uematsu, T., and R. J. Suhadolnik: Pseudouridine: Biosynthesis by *Streptoverticillicum Ladakanus*. Biochim. Biophys. Acta **319,** 348 (1973).

*(Received November 3, 1982)*

# Author Index

Page numbers printed in *italics* refer to References


Abbott, I. A. *239*
Abo, K. A. 89, *92, 93, 96*
Acs, C. *295*
Acton, E. M. 265, *295*
Adams, R. M. *237*
Ademchuk. M. R. *183*
Adeoye, S. A. *185*
Adesida, G. A. *185*
Adesogan, E. K. *186*
Adler, M. *290*
Adolf, W. 6, 27, 84, *90, 91, 92, 94, 95, 97*
Ahmed, F. R. 167, *185*
Airy-Shaw, H. K. *91*
Aizawa, A. *294*
Akatsu, M. *297*
Akisanya, A. *187*
Albers, F. *181*
Albrecht, H. P. *293*
Aldersley, M. F. *293*
Allen, F. W. 243, 248, *290*
Allerhand, A. *298*
Altwerger, L. *295*
Ames, A. *294*
Anderson, C. H. *181*
Andrews, A. G. *237*
Anet, F. A. L. *240*
Ansarin, M. *95, 96*
Anton, R. *93*
Anzai, K. 283, *298*
Aoki, A. *297*
Aoyagi, T. *295*
Arai, M. *292*
Argoudelis, A. D. *290*
Arigoni, D. *183, 185*
Arita, M. *293*
Arnott, S. *185*
Arora, S. K. *90*
Arroyo, E. R. *93*
Arteaga, J. M. *241*
Ayafor, J. F. *184*

Baird, A, W. *99*
Baird, W. *98*
Baird, W. M. *97*
Bakhtavar, F. *96*
Baldeo, W. *89*
Balkenhol, W. G. *96*
Ball, D. B. *293*
Balmain, A. *90*
Banerjee, S. *98*
Banthorpe, D. V. *238*
Barker, R. *293*
Baron, F. *291*
Barrett, M. *91*
Barrick, J. C. *97*
Barton, D. H. R. 166, *183, 185*
Bartsch, H. *93, 94*
Bates, P. *240*
Bates, R. B. *90, 186, 187*
Battersby, A. R. *236*
Baxter, R. L. *97*
Beaver, W. A. *90*
Beck, J. P. *93*
Bekoe, D. A. *185*
Ben, T. *98*
Bennington, F. *99*
Benyon, J. H. *240*
Berenblum, I. 3, *90*
Bevan, C. W. L. *185, 187*
Bhattacharyya, P. *182*
Bhattacharyya, S. P. *182*
Biemann, K. *96*

Bijvoet, J. M. *183*
Birnbaum, G. I. *93*
Bloch, A. *290*
Blomster, R. N. *91*
Blount, J. F. *90*
Blumberg, P. M. *92, 93, 98*
Blunt, J. W. *240, 241*
Bobek, M. *292*
Bockel, M. *187*
Bohlmann, F. *90*
Bohm, R. *93*
Boissier, E. *91*
Bordener, J. *182*
Boutwell, R. K. 89, *90, 97, 98*
Bowden, G. T. *98*
Bowers, W. S. *236*
Brandl, F. *92, 93, 95*
Branfman, A. R. *90*
Braslau, R. *238*
Bratley, K. W. *236*
Breen, G. J. W. *186*
Brennan, T. *294*
Bresch, H. *93*
Brody, R. S. *297*
Brown, D. M. *291*
Brown, F. S. *238*
Brown, W. A. C. *182*
Bryan, R. F. *90, 92, 97*
Buchanan, J. G. 243, *291, 293, 295*
Buchanan, J. G. St. C. *186*
Buchanan, J. M. *298*
Burbott, A. J. *238*
Burdon, M. G. *291*
Burke, B. A. *90, 184*
Burreson, B. J. *241*
Busch, H. *93*
Byram, S. K. *293*

Cadman, E. C. *297*
Caglioti, L. *185*
Cairn, B. F. *94*
Caldwell, I. C. *296*
Callsen, H. 104, 106, 112, 119, *181, 182, 183*
Cameron, A. F. *90*
Campbell, L. *239*
Carle, J. C. *237*
Carter, D. *89*
Cashmore, A. R. *94*
Cassady, J. M. *187*
Cassidy, F. *291*
Castaneda, J. P. *181*
Cavill, G. W. K. *236*
Cerami, A. *295*
Černý, M. 280, *297*
Ceroni, M. *186*
Chakraborty, D. P. *182*
Chambers, R. W. *290*
Chan, S. *293*
Chan, W. R. *90, 184, 187*
Chandrasekharen, S. *90*
Chaney, M. O. *291, 296*
Chang, C. *98*
Chapman, O. L. *240*
Chapya, A. *185*
Charlwood, B. V. *238*
Chen, Y.-P. *96*
Cheng, C. C. *291*
Chenon, M.-T. *292*
Chester, R. *237*
Chou, C. H. *237*
Choudhury, A. K. *94*
Christensen, A. T. *293*
Christophersen, C. *237*
Chu, C. K.. *293*
Chu, E. H. Y. *98*
Cimino, G. *185*
Clardy, J. *90, 237, 238, 240*
Cline, J. C. *296*
Coertzer, J. *97*
Coggon, P. *187*
Cohn, W. E. 243, *290*
Cole, J. R. *90, 186, 187*
Connolly, J. D. *90*, 155, 164, 173, *182, 183, 184, 185, 187*
Cook, J. C. *238*
Cooper, E. *89*
Cordell, G. A. *95, 96*
Corey, E. J. *184, 185*
Cotterrell, G. P. *186*
Crain, P. F. *292, 298*
Cramer, F. *296*
Cramer, R. *182, 183*
Crea, A. E. G. *94*
Crews, P. 189, 190, 215, 217, 218, 219, 221, *237, 238, 239, 240, 241*
Crombie, L. *92, 93*
Cronquist, A. 102, *181*
Croteau, R. *238*
Cunliffe, A. V. *240*
Cuno, G.. *181*
Cutler, R. S. *90*

Daebel, W. 104
Dailey, R. G. *90*

Darias, J. *238*
Darnall, K. R. *291*
Daves, G. D. Jr. *291*
Davie, A. W. *185*
Davies, D. B. *292*
Davies, F. A. *291*
Davis, F. A. *296*
Davis, F. F. 248, *290*
Davis, R. E. *241*
Davis, W. *94*
Dean, F. M. *185*
DeBernardis, J. F. *237*
DeBernardo, S. 272, *296, 297*
De Candolle, A. P. *91*
De Clercq, E. *296, 297*
De Clerq, J. P. *182*
De Clerq, P. J. *183*
Declos, B. *99*
Dejongh, D. C. *292*
Delong, D. C. *296*
De Mayo, P. *236, 291*
Deslauriers, R. *291*
De Stefano, S. *185*
Dev, S. *185*
Devon, T. K. 192, *236*
Diamond, L. *97*
Dierks, H. 109, *183*
Dix, D. E. *297*
Dixon, P. S. 201, *239*
Dizaji, F. N. *96*
Dodson, C. H. *237*
Donaldson, C. *97*
Dressler, R. L. *237*
Dreyer, D. L. 166, *182, 183*
Driedger, P. E. *92, 98*
Dublyanskaya, N. F. 13, *92*
Dudek, G. O. *291*
Duncan, A.. *89*
Dunlop, R. W. *241*
Dyer, R. A.. *91*
Dyke, R. W. *296*

Eberwein, H. *238*
Eckstein, F. *297*
Edelman, G. *97*
Edgar, A. R. *293, 295*
Edwards, M. E. *91*
Eguchi, A. *298*
Ehrlich *250*
Ekong, D. E. U. *186, 187*
Elad, D. *183*
El-Missiry, M. M. 89, *96*
Elstner, E. F. *298*
Emerson, O. H. 166, *185*
Emoto, S. *293, 295*
Engel, D. W. *92*
Engler, A. *91*
Enomoto, S. *237, 238*
Epe, B. 101, 104, 112, *162, 181, 182, 183, 184, 186*
Epstein, W. W. *236*
Esumi, S. *291*
Evans, F. J. 1, *90, 91, 92, 93, 94, 95, 96, 97, 98. 99*
Ewards, M. C. *98*

Fairbairn, J. W. *2*
Fakunle, Ch. O. *187*
Fallis, A. G. *185*
Falsone, G. *94*
Farkaš, J. 253, 265, 271, *292, 295*
Farnsworth, N. R. *91, 95, 96*
Fasina, A. K. *187*
Faulkner, D. J. 190, 193, 198, 200, 202, 210, 219, *237, 238, 239, 240, 241*
Fayos, J. *182, 237, 241*
Fenical, W. 192, 201, 208, 218, 221, *237, 238, 239, 240*
Fenzl, W. *187*
Ferguson, G. *93, 183*
Ferrini, P. G. *185*
Finer, J. *238, 240*
Flaschenträger, B. *93*
Flegelová, Z. *292*
Fletcher, H. G. Jr. *293*
Florkin, M. *238*
Fox, J. J. *290, 293, 296*
Fraga, B. M. *181, 182*
Francis, M. J. O. *236*
Freeman, P. W. *96*
Frei, J. V. *97*
Fürstenberger, G.. *93, 94*
Fujita, E. *183*
Fujita, T. *183*
Fukatsu, S. *294*
Fukugawa, Y. *294, 295, 296, 298*
Fukuyama, Y. *187*
Fulke, J. W. B. *183*
Fuller, W. *296*
Funuyama, S. *299*
Furcht, F. W. *294*

Gaertner, E. *296*
Gafur, M. A. *94*

Games, M. L. *93*
Gassmann, I. *93*
Gaudioso, L. A. *236*
Geissman, T. A. *185*
Gensler, W. J. *293*
Germain, G. *182*
Gerster, J. F. *291*
Gerzon, K. *296*
Ghaisarzedeh, M. *96*
Ghisalberti, E. L. *90, 92*
Gibbs, R. D. *181*
Gilmore, C. J. *92*
Glazier, E. R. *185*
Goldberg, E. D. *240*
Gonzales, A. G. 104, *181, 182, 241*
Goodchild, J. *291*
Goodman, L. *295*
Goodwin, T. W. *236*
Gordon, T. *98*
Gorman, M. *296*
Goto, T. *96, 182*
Grant, D. M. *292*
Grauer, A. D. *294*
Greenberg, D. M. *298*
Greenebaum, E. *99*
Grey, A. A. *292*
Grimminger, W. *185, 187*
Grozinger, K. *293*
Gschwendt, M. *93, 94, 95*
Guerriero, A. *185*
Gunasekera, S. P. *96*
Gunn, P. A. *183*
Gunstone, F. D. *237*
Gutman, A. B. *290*
Gutowski, G. E. *291, 292, 296*

Habraken, C. L. *291*
Hackney, J. *98*
Härle, E. *93*
Hafez, A. *94, 97*
Hagaman, E. W. *292*
Hager, L. P. *238, 240*
Halligan, J. P. *236*
Hallsall, T. G. 169, *186, 187*
Haltiwanger, R. C. *90*
Hamada, M. *294, 298*
Hamblin, M. R. *295, 298*
Hamill, R. L. *291, 296*
Hammouda, F. M. *96*
Handschuhmacher, R. E. *297*
Haneishi, T. *291*
Hanessian, S. *291*
Hanke, F. J. 189, *238*
Hanover, J. W. *237*
Harada, F. *295*
Harborne, J. B. *237*
Harris, R. K. *240*
Hartman, S. C. *298*
Hartman, P. *292*
Hartmann, P. *181*
Hartshorn, M. P. *240, 241*
Hashimoto, K. *238*
Hata, T. *291*
Hayakawa, Y. *293*
Hayase, E. *298*
Haynes, H. R. *97*
Hechtfischer, S. *92*
Heck, R. F. *293*
Hecker, E. 2, 6, 27, 28, 58, 59, 84, *90, 91, 92, 93, 94, 95, 96, 97*
Hegnauer, R. *181*
Heidenreich, H. 116, *183*
Henderson, J. F. 270, *296*
Henderson, M. S. *183*
Henkel, G. 109, *183*
Hennessee, G. L. A. *182*
Hennings, H. *98*
Henry, D. W. *295*
Hergenhahn, M. 84, *90, 92*
Hernandez, M. G. *182*
Heron, E. *239*
Herz, W. 143, *184*
Hickernell, G. L. *96*
Hidaka, T. *294*
Higgs, M. D. *239, 240*
Hikino, H. *182*
Hills, H. G. *237*
Hirasawa, K. *299*
Hirata, Y. *90, 92, 96*
Hirose, C. *297*
Hirose, Y. *238, 239*
Hirota, M. 59, *95*
Hirotsu, K. *240*
Hoehn, M. *296*
Hoehn, M. M. *296*
Hoffman, H. 5, *91*
Hoffmann, D. H. *296*
Hoffmann, J. J. *186, 187*
Hogberg, H. E. *239*
Hogue, E. R. *238*
Holcomb, J. *93*
Hollenberg, G. J. *239*
Holum, L. B. *294*
Homma, I. *294, 295, 296*

Homma, M. *295*
Hope, H. *237*
Hoppe, W. *92, 93, 95*
Hori, M. *294, 295, 296, 298*
Hori, S. *295*
Horn, T. *296*
Horner, L. 143, *184*
Horowitz, A. D. *99*
Howard, B. M. 201, *239*
Howden, M. E. H. *97*
Hruska, F. E. *292*
Hsu, H.-Y. *96*
Huffman, J. C. *98*
Hui, W. H. *92*
Hutchinson, R. J. *295*
Hutzinger, D. *237*

Ichikawa, N. 221, *237, 238, 239*
Igarishi, N. *294*
Iitaka, Y. *291, 294, 295, 296*
Ikehara, M. *295*
Imperato, F. 209, *240*
Inaba, T. *299*
Inoue, T. *293*
Ireland, C. *240*
Ishida, N. *295*
Ishizuka, M. *294, 295*
Islam, K. M. S. *93*
Isono, K. 288, 289, *292, 294, 298, 299*
Ito, E. *294, 295*
Ito, K. *298*
Ito, R. *293*
Itô, S. *182*
Ito, Y. *293*
Iwamoto, S. *298*
Iwanaga, J. *295*
Izac, R. R. *238*
Izawa, A. *295*

Jacobi, P. *93*
Jacobs, R. *238*
Jacquemin, H. *186*
Jain, M. K. *183, 184, 187*
Jakupovic, J. *90*
Janoff, A. *99*
Jefferies, P. R. *90, 92*
Jeger, O. *183, 185*
Jenkins, C. L. *237*
Joel, L. I. *99*
Johnson, R. A. *184*
Johnson, R. D. *239*
Jolad, S. D. *186, 187*
Jones, A. J. *240, 292*
Jones, G. H. *293*
Jones, J. W. *294*
Jones, N. D. *291, 296*
Jurgeleit, W. 143, *184*
Just, G. 256, *293*

Kaji, K. *238*
Kakkar, V. V. *99*
Kalvoda, L. 253, 256, *292, 295*
Kamga, C. S. *184*
Kamikawa, T. *185*
Kano, H. *291, 292*
Kariya, T. *294*
Kasukabe, Y. *291*
Katagiri, K. *294*
Katagiri, N. *296*
Katayama, C. *92*
Katayama, T. 190, *237*
Kato, H. *292, 294*
Kato, T. *296*
Katsumata, H. *96*
Kawaguchi, O. *297*
Kawamura, K. *294*
Kawazu, K. *96*
Kazlauskas, R. *241*
Kelly, W. R. *96*
Kho, E. 190, 221, *237, 238, 240*
Kho-Wiseman, E. 201, *239, 240, 241*
Khwaja, T. A. *296*
Kiamuddin, M. *94*
Kidd, J. G. *90*
Kikuchi, S. *298*
Kikuchi, Y. *295*
Kim, K. S. *298*
Kimbu, S. F. *184*
Kimelberg, H. K. *98*
Kimura, T. *294*
King, J. C. *91*
King, T. J. *239*
Kinghorn, A. D. 89, *91, 92, 94, 95*
Kirby, A. J. *293*
Kirson, I. *183*
Kishi, Y. *96*
Klassen, A. *99*
Knaen, G. *92*
Kogiso, S. *97*
Koike, K. *95*
Koike, S. *95*
Komai, T. *294*
Komatsu, Y. *292, 294*
Kondo, K. *184*

Kondo, S. *296*
Koshimizu, K. *95, 96*
Koyama, G. *291, 294, 295, 296*
Koyama, H. *291, 292*
Kozikowski, A. P. 260, *294*
Krakow, J. S. *296*
Kraus, W. 166, 176, *182, 183, 184, 185, 187*
Kreibich, G. *93*
Krejcarek, G. E. *239*
Krugh, T. R. *298*
Kubinski, H. *98*
Kubinyi, E. *93*
Kubinyi, H. *92, 93*
Kubo, J. *185*
Kubota, T. *177, 185, 187*
Kukagawa, Y. *294*
Kumagai, K. *295*
Kumar, S. A. *296*
Kunimoto, T. 287, *298*
Kupchan, S. M. 13, 77, 85, *90, 92, 97*
Kusakabe, Y. *291, 297*
Kusumoto, S. *92*
Kusyynski, C. *98*
Kuwajima, I. *293*
Kuwayama, T. *297*
Kuzuhara, H. *295*
Kypke, K. *184, 187*

Labbé, C. *90, 184, 187*
Lachmann, M. *295*
La Cour, T. *97*
Lallemand, J.-Y. 171, *186*
Lancaster, P. W. *293*
Lang, R. A. *295*
Langenback, R. *98*
Lavie, D. 169, *183, 184, 186, 187*
Le Cocq, C. 171, *186*
Lee, L. S. *97*
Lee, Y. W. *185*
Leighton, A. P. *97*
Lemieux, R. U. *291*
Lendle, L. *93*
Lerch, U. *291*
Levine, H. L. *297*
Levine, L. *99*
Levy, E. C.. *186*
Levy, G. C. *241*
Lewis, A. F. *291, 292, 295*
Lewis, R. A. *237*
Lhomme, M. F. *90*
Liak, T. J. *293*
Lichter, R. L. *241*
Lieberman, M. W. *98*
Lievschitz, Y. *182*
Likhacher, A. Y. *90*
Lim, M.-I. *293*
Liu, Ch.-Sh. *187*
Lobreau-Callen, D. 102, *181*
Loomis, W. D. *237, 238*
Lotter, H. *92*
Lyons, C. W. 169, *186*

Mabry, T. J. 209, *240*
Mack, H. *94*
Maddox, M. L. *293*
Maeda, K. *291, 294, 296*
Maelicke, A. *296*
Magnus, K. E. *184*
Mahindran, M. *241*
Maidhof, A. *295*
Maltz, A. *90*
Manchand, P. S. *90*
Manes, L. V. *189*
Marcelle, G. B. 167, *185*
Marsh, N. *93*
Marsh, W. C. *183*
Marshall, G. T. *91*
Martel, A. *293*
Martin, J. D. *238, 241*
Martines-Ripolls, M. *241*
Masaki, N. *96*
Masaki, Y. 199, *238*
Mason, H. S. *238*
Matsui, M. *295, 297*
Matsumoto, M. *184*
Matsumoto, S. *295*
Matsumoto, T. *185*
Matsushita, K. *298*
Matsuura, S. *294*
Matsuura, T. *185*
Matyukhina, L. G. *90*
Matz, M. J. *90, 92*
Maurer, R. *97*
Mayama, M. *292*
Mayhew, E. *98*
McCloskey, J. A. *292, 298*
McClure, T. H. *96*
McConnell, O. J. *237, 239*
McCormick, I. R. N. *94*
McCrindle, R. *183, 185*
McMillan, J. A. *240*
McPhail, A. T. *187*
Melera, A. *183, 185*
Melera, P. W. *97*

Merrien, A. *186*
Meshulam, M. *184*
Messmer, W. M. *91*
Meunier, B. *186*
Miana, G. A. *94*
Michelson, A. M. *290, 291*
Mieczowski, J. *298*
Miller, R. D. *291*
Miller, R. L. *291*
Minale, L. *185, 240*
Mitsui, T. *91, 96*
Miura, I. *185*
Mizsak, S. A. *290*
Moffatt, J. G. 253, 260, *291, 293*
Mohaddes, G. *96*
Mohseni, M. *96*
Mondon, A. 101, *181, 182, 183, 184, 186*
Moniot, J. L. *94*
Montana, P. *239*
Moore, R. E. *237, 240, 241*
Mootoo, B. S. 167, *185*
Moretti, Ch. *182*
Morin, R. D. *99*
Morris, D. R. *238*
Müller, W. E. G. *295*
Muller, C. H. *237*
Munakata, K. *97*
Munro, M. H. G. *240, 241*
Murae, T. *97*
Murao, K. *295*
Murase, M. *294*
Murphy, P. T. *241*
Mynderse, J. S. 202, *238, 239, 240*

Nadig, H. *186*
Nagatsu, J. *297*
Naito, A. *291*
Nakagawa, Y. 251, *291, 292, 294*
Nakamura, H. *295, 296*
Nakanishi, K. 167, *182, 185*
Nakatani, Y. *90*
Nakayama, Y. *91*
Narasimhan, N. S. *183*
Narayanan, C. R. *183*
Narayanan, P. *92*
Natori, S. *182*
Naya, Y. *237, 238, 239*
Naylor, S. 189, *238*
Neeman, M. 33, *95*
Nelson, G. L. *241*
Nevling, L. I. *91*
Newman, A. A. *182, 237*
Ng, A. S. *185*
Ng, K. K. *92*
Ng, P. *239*
Nidy, E. G. *184*
Nilsson, S. *181*
Nishida, T. *238*
Nishimura, H. 250, *292*
Nishimura, S. *295*
Nitta, K. *295*
Nixon, P. E. *94*
Nobuhara, K. *91*
Nomura, M. *291*
Nonomura, A. M. *239*
Norris, J. N. 201, *239*
Norton, S. *240*
Nouri, A. M. E. *98*
Noyori, R. *293*
Nozoe, S. *182*
Nyborg, J. *97*

O'Brien, S. *97*
Ochi, K. 286, 287, *298*
Ocken, P. R. *91*
Oelbermann, U. 104, 105, 162, *181, 182, 184, 186*
Ogawa, T. *295, 297*
Ogden, R. C. *291*
Ogura, M. *95*
Ohigashi, H. *91, 95, 96*
Ohno, M. *293, 296*
Ohochuku, N. S. *185*
Ohrui, H. 272, *293, 295, 296*
Ohta, K. *239*
Ohta, T. *182*
Ohuchi, K. *99*
Ohwaki, H. *96*
Okami, Y. *298*
Okazaki, T. *291*
Oki, Y. *95*
Okogun, J. I. *184, 187*
Okorie, D. A. *183*
Okuda, T. *95*
Opferkuch, H. J. 59, *92, 95*
Otake, N. *294*
Ott, H. H. *96*
Ourisson, G. *90*
Overton, K. H. *182, 185*
Ozaki, M. *294*

Paćak, J. *297*
Pace, C. *239*
Pachapurkar, R. V. *183*

Pack, G. R. *93*
Pailer, M.. *187*
Panzicka, R. P. *292*
Parker, K. A. *183*
Pascard, Cl. *182, 186*
Pascoe, K. O. *90*
Paterson, A. R. P. *296*
Pattenden, G. *92*
Paul, I. C. *240*
Paul, J. V. *237*
Pax, F. 5, *91*
Payne, T. G. *90, 92*
Pearce, J. B. *99*
Pedersen, M. *238*
Pernet, A. G. *291*
Persinos, G. J. *91*
Pettersen, R. C. *93*
Pferdeman, A. F. *183*
Pieterse, M. J. *97*
Pino, O. *182*
Plouvier, C. *181*
Poels, E. K. *295*
Pointer, D. J. *93*
Poirier, M. C. *98*
Poling, M. *96*
Polonsky, J. 169, *182, 186*
Ponsinet, G. *90*
Poore, G. A. *297*
Porter, J. W. *238*
Potvin, P. *293*
Poulter, C. D. *236*
Powell, J. W. *185, 187*
Power, M. J. *293*
Pradhan, S. K. *183, 185*
Prangé, Th. *182*
Prantl, K. *91*
Pridham, J. B. *237*
Prince, E. C. *90*
Pruisiner, P. *294*
Pryce, R. J. *90*
Pugmire, R. J. *292*
Pukoshataman, K. K. *90*

Quinn, R. J. *241*

Rabanal, R. M. *186*
Raick, A. N. *97*
Ramjeesingh, M. *293*
Rao, M. M. *184*
Rastogi, R. P. *90*
Ravelo, A. G. *182*
Reich, E. *295, 296*
Reichenbach, N. L. *298*
Reinhold, L. *182*
Remberg, G. 169, *186*
Renauld, J. A. S. *90*
Repke, D. B. *293*
Restivo, R. *183*
Riccio, R. *240*
Rice, J. M. *291*
Riley, J. P. *237*
Rilling, H. C. *236*
Rinehart, K. Jr. 208, *239*
Riszk, A. *96*
Ritchie, A.-C. *97*
Ritchie, E. *96, 186*
Roberts, H. B. *96*
Robertson, J. M. *185*
Robins, R. K. 250, 251, 264, *291, 292, 294, 295*
Robinson, D. R. *91*
Robinson, W. T. *240*
Rodriguez, M. L. *241*
Röhrl, M. *92, 93*
Rohde, H. J. *295*
Rohmer, M. *93*
Rohrschneider, L. R. 89, *98*
Ronlán, A. 76, 77, *96*
Rose, A. F. *238*
Rossana, D. M. *183*
Rossi, A. *297*
Rossow, P. W. *98*
Rothschild, M. 25, *93*
Rous, P. *90*
Roy-Burman, S. *294*
Ruzicka, L. *236*
Ryan, K. J. *295*
Rycroft, D. S. *90, 184, 187*

Sahai, R. *90*
Saita, T. *298*
Sakata, K. 74, *96*, 279, 283, *292, 297*
Sakuma, K. *238*
Sakurai, A. *292, 297*
Saleh, N. *96*
Saltikova, I. A. *90*
Samiyeh, R. *96*
Sammes, P. G. *291*
Sandermann, W. *238*
Santhanakrishnan, T. S. *90*
Sasaki, K. *92, 291*
Sater, A. M. *96*
Sato, T. *292, 293, 299*
Sawa, T. *294, 295, 296, 298*

Sawai, H. *293*
Sawitzki, G. *185, 187*
Sayed, M. D. *96*
Schaden, G. *187*
Schaffner, K. *183, 185*
Schairer, H. U. *93, 94*
Scheuer, P. 189
Scheuer, P. J. *238, 240*
Schildknecht, H. *97*
Schimizu, Y. *295*
Schmidt, R. *94*
Schmidt, R. J. *89, 90, 91, 92, 95, 97, 98*
Schram, K. H. *292*
Schramm, K. H. *186*
Schroeder, G. *93*
Schulte, G. R. *240*
Schwarzmaier, U. *181*
Schwinger, G. *187*
Scopes, P. *89*
Scott, A. I. 183, 192, *236*
Scribner, J. D. *98, 99*
Scutt, A. *93*
Seawright, A. A. *96*
Sedgwick, J. A. *97*
Seelye, R. N. *94*
Seki, I. *291*
Selover, S. J. *238*
Séquin, U. *186*
Seto, H. *294*
Sevenet, Th. *182*
Shah, V. R. *183*
Shakui, P. *96*
Shamma, M. *94*
Shanks, C. T. *293*
Shaw, P. D. *239*
Shen, M. S. *97*
Shepherd, J. *237*
Sheridan, R. S. *240*
Shibahara, S. *296*
Shibata, T. *293*
Shibuja, M. *297*
Shimaoka, N. *292*
Shimauchi, Y. *298*
Shimizu, Y. *97*
Shinozuka, H. *97*
Shirato, S. *297*
Shiratori, O. *294*
Shiro, M. *292*
Shizuri, Y. *91, 97*
Shpan-Gabrielith, S. R. *187*
Shubik, P. *90*
Sickles, B. R. *97*
Sidwell, W. T. L. *186*
Sigel, C. W. *90, 92*
Sim, G. A. *185*
Sime, J. G. *93*
Simm, G. A. *182*
Simmons, O. D. 33, *95*
Sims, J. J. 221, *238, 239, 240*
Sindemann, C. J. *238*
Sinnott, M. L. *276*
Sinnwell, V. 169, *186*
Sivak, A. *90, 97, 98, 98*
Sivapalan, A. *241*
Siuda, J. F. *237, 239*
Siuder, J. S. *238*
Slaga, T. G. *99*
Slaga, T. J. *90, 98*
Slatcher, R. P. *291*
Sloane, B. L. *91*
Smith, G, M. *239*
Smith, I. C. P. *291, 292*
Smith, J. C. *292*
Smythies, J. R. 89, *99*
Sondengam, B. L. *184*
Sood, G. R. *295*
Soper, C. J. *90, 98*
Sorg, B. *84*
Šorm, F. 253, 265, 271, *292, 295*
Spiteller, G. *187*
Springer, J. P. *90*
Sprinzl, M. *296*
Spurgeon, S. L. *238*
Squillacote, M. E. *240*
Stallard, M. O. 193, *237, 238, 240, 241*
Starr, M. P. *237*
Steffen, R.. *295*
Sternbach, D. D. *183*
Sternhell, S. *185*
Stevens, P. *97*
Stierle, D. B. *239, 240*
Stobie, A. *295*
Storer, R. *187*
Stout, G. H. *96*
Straka, H 102, *181*
Strangstalien, M. A. *98*
Strell, I. *93*
Stroke, T. A. 169, *186*
Stuart, K. L. *91*
Suggs, J. W. *184*
Suhadolnik, R. J. 284, 288, 289, *290, 298, 299*
Sumner, W. C. *97*
Sundaralingam, M. *294*
Suzuki, S. *294, 299*

Sweeney, J. G. *97*
Sweeney, M. J. *296*
Szarek, W. A. 283, *298*
Szczepanski, Ch. V. *93, 94*

Tafazuli, A. *96*
Takagi, M. *239*
Takahashi, S. *96*
Takaoka, K. *297*
Takashima, K. *296*
Takayama, H. *295*
Takemoto, T. *182*
Takeuchi, T. *294, 295, 296, 298*
Takhtajan, A. 102, *167, 181*
Takita, T. *294*
Tamura, C. *291*
Tamura, S. *292, 297*
Tanaka, Y. *292*
Tanis, S. P. *185*
Taylor, D. A. H. 164, *183, 184, 185, 186, 187*
Taylor, D. R. 169, *184, 186, 187*
Taylor, S. E. 1, *91, 94, 98*
Taylor, W. C. *96, 186*
Taylor, W. I. *236*
Tedstone, A. *93*
Tempesta, M. S. *186, 187*
Templeton, J. F. *185*
Theiler, R. F. *238*
Thielmann, H. W. *93*
Thompson, J. E. *240*
Thompson, R. H. *239*
Thoms, H. *187*
Thornton, I. M. S. 164, *184*
Tilabi, J. *96*
Tipson, R. S. *291, 292*
Tobinga, S. *185*
Toft, P. *186*
Toh, N. *95*
Toia, R. F. *92*
Tokoroyama, T. *185, 187*
Torrance, S. J. *90*
Torrence, P. F. *296*
Torres, R. *181*
Townsend, L. B. *291, 292, 295*
Trashjian, A. H. *99*
Trautmann, D. 104, 124, 131, 136, 147, 150, 162, 166, *181, 182, 184*
Troll, W. *99*
Tronchet, J. M. J. *291*
Trosko, J. E. *98*
Trummlitz, G. *293*
Tsantrizos, Y. S. *293*
Tsukuda, Y. *291, 292*
Turner, J. A. 143, *184*
Tursch, B. *239*
Tyler, M. I. *97*

Uchida, I. *90, 183*
Uematsu, T. 289, *290, 299*
Uemura, D. *90, 91, 92, 96*
Ulbricht, T. L. V. *292*
Umezawa, H. *291, 294, 295, 296, 298*
Uno, E. *92*
Upadhyay, R. R. *93, 95, 96*
Uzawa, J. 288, *292, 297, 299*

Van Bommel, A. J. *183*
Vanderah, D. J. *239*
Van der Helm, D. *97*
Van Duuren, B. L. *97, 98*
Van Engen, D. *240*
Van Royen, L. A. *183*
Van Tri, M. *182*
Varon, Z. *182, 186*
Visser, D. W. *294*
Volkin, E. *290*
Von der Haar, F. *296*

Wada, K. *97*
Wakashiro, T. *294, 295, 298*
Walker, R. P. *240*
Walker, R. T. *297*
Ward, D. C. *295, 296*
Warren, R. G. *240*
Watanabe, K. A. *290, 293*
Waters, T. N. *94*
Watson, D. G. *185*
Weatherston, J. *237*
Weber, J. *94*
Webster, G. L. *91*
Wehrli, F. W. *238*
Weigele, M. 272, *296, 297*
Weinstein, B. *182*
Weinstein, I. B. *97, 99*
Wells, R. J. *240, 241*
Wempen, I. *293*
Wenkert, E. *292*
Wenner, C. E. *98*
West, C. A. *91*
Westheimer, F. H. *297*
Westwick, J. *98, 99*
Whittaker, D. *237*
White, A. *91*
Whybrow, D. *92*

Wickberg, B. 76, 77, *96*
Wiedhopf, R. M. *90*
Wightman, R. H. *291, 295, 298*
Wilcott, M. R. *241*
Wilkes, J. D. *91*
Williams, N. H. *237*
Williams, R. H. *296*
Williams, T. J. *99*
Willamson, E. M. *91, 96, 98, 99*
Wilson, S. R. *98*
Wing, R. M. *239, 240*
Winkley, M. W. *292*
Witz, G. *98*
Wolff, Ch. *184*
Woolard, F. X. *240, 241*
Worth, G. K. *90, 92*
Wratten, S. J. *240*
Wriglesworth, M. J. *186*
Wu, K. K. *97*

Yagisawa, N. *296*
Yamagiwa, K. *90*
Yashima, S. *298*
Yee, T. *187*
Yefei, B. *90*
Yokotani, K. *187*
Yonehara, H. *294*
Yorke, J. C. *240, 241*
Yoshida, T. *95*
Yoshihara, K. 198, *238*
Yotti, L. P. *98*
Young, D. W. *187*
Yu, C.-T. *290*
Yuspa, S. H. *98*

Zahn, R. K. *295*
Zamojski, A. 280, *298*
Zarintan, M. H. *95, 96*
Zayed, S. *94, 97*
Zechmeister, K. *92, 93, 95*
Zehnder, M. *186*
Zelnik, R. *184*
Zimmerly, V. A. *97*
Zorbach, W. W. *291, 292*

# Subject Index

By

A. Siegel, Wien

*Acalypha* 4
Acetic acid 167, 173, 200, 253, 254, 267, 268, 275
Acetic anhydride 38, 151, 251, 252, 253, 257, 258, 259, 260, 266, 268, 271, 272
Acetone 9, 10, 11, 12, 68, 125, 130, 131, 138, 145, 150, 151, 157, 160, 222, 259, 275
Acetonedimethylketal 257, 258, 262, 266
Acetonitrile 259, 274
7-α-Acetoxydihydronomolin 167, 179
2-Acetoxy-6-methyl-2,5-heptadiene 198
Acetylation 122
Acetyl chloride 255
13-Acetyl-16-hydroxyphorbol 57
8-Acetylingol-12-tiglate 25
12-O-Acetylingol-3,7,8-tribenzoate 22
Adenosine 243, 249, 264, 269
Adenosine deaminase 264, 265, 269
Adenylosuccinate synthetase 286
Adenylosuccinic acid 284
*Aleurites fordii* 56, 57
Algae 191
n-Alkanes 104
n-Alkanols 104
1-Alkyl-daphnanes 8, 9, 84
Alumina E 11
Aluminium chloride 262
Ammonia 22, 252, 255, 256, 266, 272, 273, 274, 275, 277
Ammonium chloride 267, 281
Ammonium hydroxide 137
Ammonium nitrate 112, 114, 125
*Amphiroa zonata* 208
Analgesic activity 15
*Anaspidea* 191
2,5-Anhydro-D-allonic acid 253, 267
Anhydrocneorins 120
*Anisophyllum* 62
*Anthacantha* 63
*Anthonomus grandis* 192
Antibacterial activity 263, 270, 277
Antibiotic MSD 125A 250
Antifeedant activity 211
Antifungal activity 13, 211, 277, 279, 283
Antiinsect activity 211
Antileukemic activity 13, 77, 86
Antimicrobial activity 211
Antimony pentachloride 280, 281
Antitumour activity 263, 264, 269, 270, 275, 277
Antitumour diterpenes 1f, 6, 13, 14, 25, 87
Antiviral activity 269, 270, 275
Antiwrithing activity 15
Aquilarioideae 5
*Aplysia californica* 190, 191, 193, 203, 209, 210, 211
*Aplysia limacina* 191, 203, 209, 210, 211
*Aplysia sp.* 191, 210, 211
Aplysiidae 191
Apotirucallane 105
Apotirucallane skeleton 112
Apotirucallan(20S)-triterpenoids 172
Apotirucallol 156
Arachidonic acid 89
L-Aspartic acid 284
Azadiradione 163, 174
Azadirone 174

**B**aeyer-Villiger oxidation 155, 164, 166, 260
Baliospermin 54, 56
*Baliospermum montanum* 47, 48, 53, 54, 55, 75, 76
Barbituric acid nucleotide 276
Benzaldehyde 248
Benzene 11, 104, 252, 273
Benzoyl bromide 266
Benzoyl chloride 254, 280, 281
N-Benzoyl-hydrazino-acetic acid 172
*Bertya cuppressoidea* 17, 18
— *sp.* 18
Bertyadionol 7, 8, 17, 18, 19
Bicyclononanolides 164
Biological activity 209, 244, 263, 269, 275, 279, 283
Biological methylation 244
Bis-dimethylamino-t-butoxymethane 261, 277, 278
Bitter principles in Cneoraceae 10f
— in Meliaceae 104
— in Rutaceae 104
—, R-Series 105
—, S-Series 105
Borane 258, 262
Borohydride reduction 260
Boron trichloride 253, 255, 268
Boron trifluoride etherate 125, 132
*Botrytis sp.* 283
Bourjotinolan-A 171
Bromine 109, 110, 111
Bromine in seaweed terpenes 196
N-Bromoacetamide 198
5-Bromo-4,4-dimethyl-2-vinyl-1-cyclohexane 199
Bromohydroquinones, prenylated 209
3-Bromomyrcene 198
Bromoperoxidases 196
Bryozoan 191, 208
t-Butyl-dimethyl silyl chloride 257, 261
t-Butyl-diphenyl silyl chloride 262

**C**almodulin inhibitors 88
Calodendrolite 176, 177
*Calodendron capense* 176
Canarian stoneberry 133
Candletoxin A 53
Candletoxin B 53, 54
*Carapa procera* 156, 161
Carbon dioxide 284, 288
Carbon monoxide 255
Carbonyldiimidazole 277
Carboxylic acids 104
Carcinogenesis 3
Cartilagineal 190, 210, 219, 225
Casbane 4, 7, 10, 13, 14
Casbene 6, 7, 8, 13, 14
*Cedrela mexicana* 167
Cembrene 6, 8
Centrifugal liquid chromatography 12
Ceramiaceae 191, 207
Ceramiales 191, 208
Charcoal, activated 104
*Chisochetum paniculatus* 173, 179
Chitin biosynthesis 244
Chlorine in marine terpenes 196
Chloroform 11, 12, 28, 31, 37, 40, 43, 52, 54, 75, 78, 109, 115, 125, 126, 131, 132, 133, 134, 139, 141, 145, 146, 147, 150, 151, 278
7-Chloromyrcene 193
m-Chloroperbenzoic acid 129, 130, 156
Chloroperoxidase 196
Chlorophyll 104
Chlorophyta 191
*Chondrococcus hornemanni* 191, 199, 203, 206, 207
— *japonicus* 191, 221
— *sp.* 192, 196, 206
Chondrocolactone 217, 228
Chondrocole A 206, 213, 216, 217, 228
Chondrocole B 216, 217, 228
Chondrocole C 206, 228, 338
Chromatography of cneorans 153
Chromatography of diterpenes 9, 10, 12, 75
Chromic acid 264
Chromium perchlorate 218
Chromones 104
*cis*-Chrystanthemic acid 13
1,8-Cineol 191
Circular dichroism
Daphnetoxin 74
5-Deoxyingenol diacetate 61
5-Deoxyphorbol diacetate 48
12-Deoxyphorbol diacetate 48
4α-Deoxyphorbol esters 44
4-Deoxyphorbol esters 44
4-Deoxyphorbol triacetate 37
16-Hydroxy-ingenol 68
Ingenol triacetate 59, 60
C-Nucleosides 250, 280
Phorbol esters 33, 34
4α-Phorbol esters 33, 34
Proresiniferatoxin 83

Circular dichroism
Resiniferonol esters 82
Tiglane esters 34
cis-Citral 191
trans-Citral 191
Citronellol 191
$^{13}C$-labelled compounds
Acetate 284
Glutamate 287
Lysine 287
$^{14}C$-labelled compounds
Acetate 288
Glutamate 284, 288
α-Ketoglutarate 284
Ribose 284, 287, 288
Clausenolide 105
Cneoraceae 102, 128, 133, 141, 144, 155, 166, 167, 168, 169, 173, 174
—, bitter principles in 101f
—, tetranortriterpenoids in 155, 166
Cneoran 112
Cneoran skeleton 151, 153
Cneorin A 105, 106, 108, 112, 122, 128, 134
Cneorin B 105, 106, 108, 112, 114, 115, 116, 118, 120, 121, 122, 124, 125, 126, 134, 136, 141, 142, 143, 180
Cneorin $B_I$ 106, 114, 115, 116, 118, 120, 124, 126, 134, 135, 136, 141, 143, 145, 180
Cneorin $B_{II}$ 119, 120, 121, 122, 125, 134
Cneorin $B_{III}$ 106, 114, 115, 118, 119, 120, 121, 125, 128, 129, 132, 134, 136, 145, 147
Cneorin C 105, 106, 107, 108, 109, 111, 112, 113, 114, 115, 116, 120, 121, 122, 124, 125, 126, 134, 136, 141, 142
Cneorin $C_I$ 106, 112, 113, 114, 115, 120, 124, 126, 136, 141, 179
Cneorin $C_{II}$ 112, 119, 120, 121, 122
Cneorin $C_{III}$ 106, 107, 109, 111, 116, 118, 119, 120, 121, 125, 128, 136
Cneorin D 105, 106, 108, 112, 122, 126, 127, 128, 134
Cneorin E 112
Cneorin F 128, 129, 130, 131, 135, 145
— epoxide 130, 132
— ethylene acetal 129
Cneorin G 167
Cneorin H 126, 127, 128, 136
— methyl ether 128
Cneorin I 167
Cneorin K 133, 135, 143
Cneorin $K_I$ 133, 134, 135
— epoxide 134
Cneorin M 134, 180
Cneorin N 126, 127, 128, 136, 145
Cneorin $NP_{27}$—$NP_{38}$ 104
Cneorin $NP_{27}$ 128, 166
Cneorin $NP_{29}$ 141, 142, 143
Cneorin $NP_{32}$ 169, 170, 171
Cneorin $NP_{34}$ 169, 170, 171, 172
Cneorin $NP_{35}$ 172, 173
Cneorin $NP_{36}$ 169, 171
Cneorin $NP_{38}$ 169
Cneorin O 126, 127, 128, 136
Cneorin P 114
Cneorin Q 141, 142, 143
Cneorin R 168
Cneorin $R_9$ 131
Cneorin S 166
Cneorins 105, 178, 179
—, biosynthesis 173
Cneoroids 105, 112, 118, 144
—, precursors 144
Cneorubins 104
*Cneorum pulverulentum* 102
— *tricoccon* 102, 104, 105, 114, 130, 131, 134, 135, 150, 155, 160, 166, 167, 173, 179
— *trimerum* 102
Co-carcinogenesis 3
*Codium fragile* 191
Coformycin 270
Column chromatography of bitter principles 104, 131, 153
Compound A 173, 179
Copper 259
Copper nitrate 268
Copper sulfate 251
*Coralline pilulifera* 208
Corticosteroids 88
Costatol 225
Costatolide 222, 232
Costatone 222, 225
Cotton effect
5-Deoxyingenol diacetate 68
Deoxyphorbol ester 44
Ingenol triacetate 60, 68
Phorbol esters 33
Coumarins 104
Craig liquid-liquid-distribution 12
Crotofolanes 4, 25f
Crotofolin 26
Croton factor $F_1$ 57
*Croton corylifolius* 26
— *flavens* 36, 56, 57
— *nitens* 13

— *rhamnifolius* 26
— *sp.* 4, 5, 25
— *sparciflorus* 29
— *tiglium* 2, 27, 29, 34
Crotoneae 5
Crotonitenone 13, 14
Crotonoideae 5
Croton oil 28, 87
Crotophorbolone 52
18-Crown-6 272, 273
Cryptic irritants 28
*Cunuria spruceana* 75, 76
Curtius degradation 267
Curtius rearrangement 265
Cyclic peroxides 141
Cycloacetals 147
Cyclohexane 11
Cyclosporin-A 89
p-Cymene 191
*Cymopolia barbata* 209
Cymopols 209, 232
L-Cystathionine 279, 282
Cytidine 275
Cytosine 279
Cytotoxic activity 13, 25

**D**aphnane 3, 4, 5, 6, 8, 9, 10, 87
—, derivatives 73f
*Daphne gnidium* 73, 76
— *laureola* 73, 76
— *mezereum* 73, 76, 79
— *odora* 78, 79
Daphnetoxin 73, 74, 76
— bromoacetate 74
— -12-O-cinnamoyl-dienacetyl ester 76
Daphnetoxins 73f
Daphnopsis factor $R_1$ 86, 87
*Daphnopsis sp.* 86, 87
Dasycladaceae 200
7-Deacetyl proceranone 156
Decarbocylation 124, 175, 177, 258, 264, 265, 275
12-O-[n-Deca-2,4,6-trienoyl]-4-deoxyphorbol-13-acetate 34
Decussatae 63
1,2-Dehydro-4,15-dihydroxybertyadionol 17, 18
2′-Deoxycytidine 279
2-Deoxy-D-erythropentose 243
12-Deoxy-5-hydroxy-6,7-epoxyphorbol 55
4-Deoxy-5-hydroxyphorbol 29, 36
— ester 39
4-Deoxy-16-hydroxyphorbol ester 36
12-Deoxy-5-hydroxyphorbol 54, 55
12-Deoxy-16-hydroxyphorbol 62, 63
— esters 52f
5-Deoxyingenol 63, 67f
— -3,20-diacetate 61, 67
20-Deoxyingenol 72f
— monoesters 72, 73
— 3-hexanoate 73
4-Deoxyphorbaldehyde 37
— diester 37
4-α-Deoxyphorbaldehyde 37
4-Deoxyphorbols 10, 33, 36, 37, 41
— acetates 11
— esters 34f, 37f
— triacetate 37, 38
4-α-Deoxyphorbol 36, 37, 38, 41
— esters 35, 41
— triacetate 11, 37, 38, 40
12-Deoxyphorbol 47, 48, 62, 63, 64, 83
— diacetate 11, 48
— diester 47, 49, 50, 52
— esters 47, 48, 51
— monoesters 49, 50
— phenylacetate 51, 89
2-Deoxyribose 243
Dephosphorylation 243
20-Desacetylresiniferatoxin 84
Desmethylimipramine 88
Deuterated compounds
  Acetone 130, 131, 138, 145, 150, 151, 160
  Chloroform 31; 32, 40, 43, 52, 54, 75, 78, 82, 86, 106, 115, 125, 126, 132, 141, 145, 146, 147, 150, 151
  Dimethyl sulfoxide 108, 130, 141, 145
  Deuterium oxide 54, 130, 145, 149, 160, 165, 168
  Ribose 288
*Diacanthium* 62
7α,11β-Diacetoxydihydronomilin 167
3,7-Diacetylingol-12-tiglate 25
Diazoacetonitrile 267
Diazomethane 106, 107, 116, 117, 118, 270, 271, 272
Diborane 260
1,3-Dibromoacetone 259
1,3-Dicarbomethoxyallene 260, 262
Dicarbomethoxylation 256
*Dictamus albus* 176
4,20-Dideoxy-5-hydroxyphorbol 36
— diester 39
4,20-Dideoxyphorbol 39

13,15-seco-4,12-Dideoxy-4α-phorbol-20-acetate 26, 27
12,20-Dideoxyphorbol esters 54, 55
Diels-Alder reaction 121, 256, 258, 260
Diethyl acetonedicarboxylate, potassium salt 272
Diethylene glycol 12
*Digenea simplex* 191
13, 19-Dihydroxyingenol 71f
—, esters 72
—, triesters 71
Dimethylacetylenedicarboxylate 258, 265, 266
1,5-Dimethyl-1-ethylcyclohexane 194
Dimethylformamide 218, 252, 255, 261, 278
2,6-Dimethyloctadienes 224, 225
2,6-Dimethyloctatrienes 225, 226
2,6-Dimethyloctenes 223
2,2-Dimethylpropionyl chloride 261
Dimethyl sulfate 251
Dimethyl sulfide 252, 257, 261, 262
Dimethyl sulfoxide 108, 130, 141, 145, 254, 257
Dinitrofluorobenzene 268
Dinitrogen tetroxide 266
Dioxane 108, 113, 116, 120
N,N′-Diphenylethylenediamine 253
Diterpenes 1f, 104, 194
—, biosynthetic relationships 6
—, isolation 9f
—, monocyclic 13f
—, polyhydroxylated 3
—, structural types 4
—, toxic 5, 6
DNA-synthesis 88
Dowex resin 279, 280
Dreiding models 138

Ehrlich ascites tumour 250
Ehrlich carcinoma 270
*Elaeophorbia drupifera* 58
— *grandifolia* 58
— *sp.* 5, 64
*Enteromorpha sp.* 191
Epoxydation 135
6,17-Epoxylathyrol 7, 16, 17, 18
—, -3-phenylacetate-5, 10-diacetate 17
14β,15β-Epoxymeliacolide 175
7α,8α-Epoxymelianone 169
7α,8α-Epoxytirucallane 169
*Escherichia coli* 269
Esculae 63
*Etandrophragma cylindricum* 171
Ethanol 28, 273
Ether 11, 12
Ethyl acetate 11, 72, 258
Ethyl diazoacetate 267
1-Ethyl-1,3-dimethylcyclohexanes 200, 231
1-Ethyl-2,4-dimethylcyclohexanes 231, 232
1-Ethyl-3,3-dimethylcyclohexanes 228, 229, 230
Ethynyl magnesium bromide 255, 256
*Euphorbia balsamifera* 47, 48
— *biglandulosa* 34, 35, 67
— *coerulescens* 29, 30, 46, 48
— *cooperii* 52, 53
— *cyparissias* 69, 70, 71, 72
— *deightonii* 58
— *desmondi* 58
— *fortissima* 45, 48
— *frankiana* 29, 30
— *helioscopia* 46, 48
— *ingens* 20, 21, 58, 68
— *jolkini* 17, 19
— *kamerunica* 21, 22, 25, 58
— *kansui* 14, 15, 69, 70, 72
— *lactea* 21, 22, 68
— *lathyris* 16, 17, 58
— *maddeni* 14, 15
— *marginata* 25
— *millii* 66
— *myrsinites* 67
— *paralias* 72
— *peplus* 68
— *poissonii* 46, 48, 52, 53, 79, 81, 84
— *polyacantha* 46, 48
— *resinifera* 22, 45, 48, 54, 55, 79, 81, 84
— *seguieriana* 58
— *sibthorpii* 68
— *sp.* 4, 5, 47, 62f, 79, 83
— *tirucalli* 29, 30, 34, 35, 36
— *triangularis* 45, 47, 53
— *unispina* 25, 47, 48, 53, 79, 84
Euphorbiaceae 1f, 4f, 6, 75
Euphorbia steroid 13, 16
Euphorbieae 5
Euphorbiinae 5
Euphorbium 22, 62
Euphornin 14, 15, 16
Evodulone 161
Exchange chromatography of pseudouridines 246
*Excoecaria agallocha* 75, 76
— *sp.* 5

Excoecariatoxin 75, 76
Ezoaminuroic acid 279, 280, 282
—, -TMS ether 279
Ezomycin $A_1$ 279, 280, 281, 283, 289
Ezomycin $A_2$ 279, 281, 283
Ezomycin $B_1$ 245, 247, 249, 279, 282, 283, 289
Ezomycin $B_2$ 282, 283
Ezomycin $C_1$ 282, 283
Ezomycin $C_2$ 282, 283
Ezomycin $D_1$ 283
Ezomycin $D_2$ 283
Ezomycins 279f
—, biological activity 283
—, synthesis 283

Fischer glycosidation 247
Fish antifeedant activity 193, 211
Fish toxicity 193
Flavonoids 104
*Flindersia bourjotiana* 171
Florosil 12
*Flustra foliacea* 191
Formamide 265, 266
Formic acid 267
Formycins 245, 249, 264, 265, 267, 269, 270, 274, 284
—, biological activity 269
—, biosynthesis 284, 288
—, isolation 264
—, -phosphates 269, 284
—, structure determination 264
—, synthesis 253, 265
Fraxinellone 176, 177
Friedel-Crafts-reaction 253
Furfuraldehyde 260, 261
Furan 256, 258, 260, 262
3-(3-Furylmethyl)-acetophenone 124

Galarrhaei 63
Gas chromatography of terpenoids 203, 208, 209
Gas-liquid chromatography 12, 75
Gedunin 174
Geraniol 191, 192
Geranyl-geranylpyrophosphate 6, 8
Geranyl pyrophosphate 192, 196
Gigartinales 191, 208
Gilgiodaphoideae 5
Glabretal 112
*Gleosporium sp.* 283
*Glochidon* 4
D-Glucose 265
L-Glutamate 287, 288
L-Glutamine 284, 285
Glycine 285
Glycol dimethyl ether 112, 114, 120, 121, 129, 140
1-Glycosylcytosin 279
5-Glycosyluracil 279
*Gnidia glaucus* 79
— *lamprantha* 79
— *latifolia* 79
— *sp.* 6, 77
— *subcordata* 85, 87
Gnidicin 77, 79
Gnididin 77, 78, 79
Gnidiglaucin 78, 79
Gnidilatidin 78, 79
—, -20-palmitate 78, 79
Gnidilatin 78, 79
— -20-palmitate 78, 79
Gnidimacrin 85, 86, 87
— -20-p-iodobenzoate 85
— -20-O-palmitate analogue 86, 87
Gniditrin 77, 79
Gonystyloideae 5
Grandifoliae 62
*Guarea glabra* 112

Haloperoxidase 196
*Harrisonia abyssinica* 167
Hedamycin 171
Hexane 9, 10, 153, 222
High pressure liquid chromatography 12
*Hippomanae* 5
Hippomane factor $M_2$ 75, 76
Hippomane factor $M_x$' 75, 76
*Hippomane mancinella* 54, 55, 75, 76
Histamine induced acid secretion 89
*Homolanthus* 5
*Hura crepitans* 74, 76
Huratoxin 74, 75, 76
Hydrazine 248, 266, 267, 268
Hydrazinolysis 251
Hydrocarbons 9
Hydrogenation 72, 266, 268, 272, 277, 278, 280, 281
Hydrogen bromide 110, 111, 273
Hydrogen chloride 113, 120, 121, 125, 129, 246, 252, 253, 258, 267, 268, 280
Hydrogen cyanide 253, 267
Hydrogenolysis 118
Hydrogen peroxide 254, 257, 259, 262
Hydrogen sulfide 108, 113

7α-Hydroxy-$\Delta^{14}$-apotirucallans 169
8-Hydroxycneorins 126
12-Hydroxydaphnetoxin 76f, 86
—, -bromoacetate 77
15-Hydroxy-4-hydro-bertyadional 17, 18
13-Hydroxyingenol 69
—, diesters 69, 70
—, monoesters 69, 70
16-Hydroxyingenol 68
—, -3(2,4,6-decatrienoate)-16-angelate 68
—, -3,5,16-20-tetraacetate 59, 69, 86
Hydroxyingenol tetraacetate 59
8-Hydroxyingol 25
Hydroxylamine 277
7-Hydroxylathyrol 7, 17
—, diacetate-nicotinate 17
—, dibenzoate-diacetate 17, 18
—, tetraesters 16
4-Hydroxyphenylacetic acid 82
—, methyl ester 82
Hydroxyphorbol 10
16-Hydroxyphorbol 56
—, esters 56
3-Hydroxytirucalla-7,24-diene 173
8-Hydroxytricoccins 126
*Hyles Euphorbiae* 25
Hypericifoliae 62

Ichthotoxicity 211
Imipramine 88
Influenza A1 virus 269
Infrared spectra
  13-Acetyl-16-hydroxyphorbol 57
  Baliospermin 56
  Bertyadionol 19
  Cneorins 107, 114, 128, 131, 143, 171
  Crotofolin 26
  Croton factor $F_1$ 57
  20-Deoxyingenol 73
  5-Deoxyingenol diacetate 67
  4-Deoxyphorbol triacetate 37
  13,19-Dihydroxyingenol ester 71
  12-Dioxyphorbol diacetate 48
  Euphornin 16
  Gnidimacrin 85
  13-Hydroxyingenol esters 70
  16-Hydroxyingenol tetraacetate 69
  Ingol tetraacetate 21
  Ingenol triacetate 59, 60
  Jatropholone-B 26
  Jatrophone 16
  Jolkinols 19, 20
  Lathyrane diterpenes 18
  Mancinellin 56
  Marine monoterpenes 213
  seco-Meliacanes 156, 158
  Mezerein 77
  4α-Phorbol 32
  Phorbol esters 33
  Proresiniferatoxin 83
  Pyrazofurin 270
  Resiniferatoxin 80
  Rhamnofolane 27
  Tinyatoxin 81
  Tricoccins 149, 151, 152, 153, 156, 158, 162, 164
Ingenane 3, 4, 6, 8, 10, 87
—, derivatives 58f
Ingenol 58f, 62, 63, 64
—, diesters 10
—, distribution in plants 66
—, esters 65
—, monoesters 10
—, triacetate 11, 58, 59, 60, 61, 67, 68, 72
Ingol 8, 20, 21, 24
—, acetate 25
—, derivatives 25
—, esters 21, 22
—, tetraacetate 21, 23, 24
—, triacetate-nicotinate tetraester 20
Inosinic acid 284
Iodine in marine terpenes 196
Iodoacetonitrile 260
IRC-50 72
Iron carbonyl 259
Iron(II) sulfate 143
IR-spectra s. infrared spectra
Isobutane 22
Isobutene 258
Isophthalic acid 124
Isopropanol 11
2,3-O-Isopropylidenepyrazofurin 272
2,3-O-Isopropylidene-D-ribose 260, 272
2,3-O-Isopropylidene-5-O-trityl-D-ribofuranose 274
5-(2,3-O-Isopropylidene-5-O-trityl-D-ribofuranosyl)-barbituric acid 272
Isoshowdomycin 263
—, triacetate 263

*Jatropha gossypiifolia* 13, 14, 25
— *macrorhiza* 14
— *sp.* 5, 25
Jatrophane A 4

Jatrophane B 4
Jatrophatrione 14, 15
Jatropholane A 25, 26
Jatropholane B 25, 26
Jatropholanes 4, 25f
Jatrophone 6, 8, 13, 14, 15, 16
Jolkinol A 7, 17, 19
Jolkinol B 7, 17, 19, 20
Jolkinol C 7, 17, 19, 20
Jolkinol D 7, 17, 19, 20
Jolkinols 7, 8, 19

Kansuinine A 14, 15
Kansuinine B 14, 15
Karplus rule 212
Ketoglutarate 284
Kieselgur 12

γ-Lactols 144
*Laminaria sp.* 191
*Lasiosiphon burchellii* 77
— *sp.* 6
Lathyrane 4, 7, 8, 10, 16f
—, benzoate-diacetate 18
—, biosynthesis 7
Lathyrol 7, 16, 17
—, triester 17
*Laurencia pacifica* 201
— *sp.* 208
Laurusin 264
L-Lysine 287
Lead tetraacetate 171
d-Limonene 191
Limonin 166, 174
Limonin group 105, 166
Limonoids 105, 112, 128, 155, 173
Linalool 191
Linifolin A 86, 87
Linifolin B 86, 87
Lipids 10
Lithium aluminium hydride 251, 259, 265, 266
Lithium carbonate 218
Lithium hexamethyldisilazide 256
Lithium hydride 259
Lithium tri-t-butylaluminiumhydride 257
Lupeol 104
Lupeone 104
Lysine 287

*Macaranga tanarius* 16, 17
— *sp.* 5
Macrocyclic diterpenes 13f
Maleimide 251
Mancinellin 54, 56
*Manihot* 5
*Marginosporum aberanns* 208
Marine terpenes 189f
—, artifacts 221
—, biogenesis 195
—, biological activity 209
—, chemical properties 212
—, diterpenes 194
—, halogen content 215
—, metabolite transfer 209
—, regiochemistry 215
—, sesquiterpenes 194
—, spectroscopic properties 212
—, stereochemistry 219
Markovnikov addition 196
Mass spectra
  13-Acetyl-16-hydroxyphorbol 57
  Bertyadionol 19
  Candletoxins 53
  Cneorins 106, 116, 129, 133, 141, 143, 169, 171, 173
  Croton factor $F_1$ 57
  4-Deoxy-5-hydroxyphorbol ester 39
  12-Deoxy-16-hydroxyphorbol esters 52
  20-Deoxyingenol 73
  5-Deoxyingenol diacetate 68
  12-Deoxyphorbol diacetate 48
  12-Deoxyphorbol esters 49, 50
  4-Deoxyphorbol triacetate 37
  4-α-Deoxyphorbol triacetate 37, 38
  Euphornin 13
  Ezoaminuroic acid 279
  Ezomycin 282
  Formycin 265
  Gnidimacrin 85
  8-Hydroxycneorins 126, 127
  16-Hydroxyingenol 68
  13-Hydroxyingenol ester 70
  16-Hydroxyingenol tetraacetate 69
  8-Hydroxytricoccins 126, 127
  Ingenol triacetate 59, 60, 68
  Ingol tetraesters 21, 22
  Jatrophone 13
  Jolkinols 19, 20
  Lathyrane diterpenes 18
  Mancinellin 56
  *seco*-Meliacanes 156, 159

Mass spectra
  Mezerein 77
  Nucleosides 248, 249
  Phorbol esters 30, 31
  Proresiniferatoxin 83
  Resiniferatoxin 80, 82
  Terpenes 203, 207, 209, 212, 213, 214, 215
  Tinyatoxin 81, 82
  Tricoccins 125, 131, 150, 152, 153, 156, 159, 160, 164, 165, 166
*Maytenus piperita* 193
*Melia azadirachta* 163
— *azedarach* 169, 176
Meliacane 105, 112
3,4-*seco*-Meliacanes 155, 174
7,8-*seco*-Meliacanes 162, 174
3,4-16,17-*seco*-Meliacanes 166
3,4-7,8-16,17-*seco*-Meliacanes 168
Meliaceae 104, 105, 112, 155, 156, 161, 163, 167, 169, 171, 173, 176
Meliacins 112, 128, 155, 173
Melianodiol 171
Melianone 105, 169, 179
— epoxide 169
Mercuric chloride 255
Mercuric cyanide 253
Mertensene 217, 218, 221, 231
Methane 22
Methanol 9, 10, 24, 32, 38, 69, 104, 109, 118, 130, 131, 133, 135, 222, 247, 252, 254, 255, 258, 259, 261, 263, 267, 268, 272, 273, 275, 277, 280, 281
12-O-[2-Methylaminobenzoyl]-phorbol-13-acetate 30, 31
12-O-[2-Methylaminobenzoyl]-4-deoxyphorbol-13-acetate 38, 40
12-O-[2-Methylaminobenzoyl]-4-deoxyphorbol 13,20-diacetate 43, 44
12-O-[2-Methylaminobenzoyl]-4-deoxy-5-hydroxyphorbol 39
12-O-[2-Methylaminobenzoyl]-4,20-dideoxy-5-hydroxyphorbol 39
Methylene chloride 12, 130, 139, 256, 273
Methyl ivorensate 168
Methyl nitroacrylate 256
Mexicanolide 164
Mezerein 76, 77, 79
Michael addition 9, 88, 251
*Microcladia borealis* 191, 207, 208
— *californica* 191, 207, 208
— *coulteria* 191, 207, 208
— *sp.* 207, 208
Minimycin 245, 248, 276, 288
Monoterpenes 190f
—, degraded 209, 232, 233
—, halogenated 192, 201, 208, 209
Montanin 75, 76
Morris rat hepatoma cells 25
Mouse leukaemia L 1210 269
Mouse leukaemia L 5178 Y 269
Mouse P-388 lymphocytic leukaemia 13, 77
Mouse TLX/5 lymphoma cells 25
Myrcenes 192, 193, 194, 196, 227, 228
—, brominated 198
—, cyclization 199
—, dihalogenated 198
—, halogenated 197
Myrsiniteae 63

Nematicidal properties 78
*Neochamaelea pulverulenta* 102, 104, 105, 112, 114, 128, 133, 134, 166, 167, 168, 173, 178
Neophorbol-13,20-diacetate-3p-bromobenzoate 28
Nerol 191
Nickel acetate 275
Nimbinene 105
Nitrogen gas 12
Nitromethane 253
Nitrosyl chloride 256
Nitrous acid 264, 275
Nitrourea 265, 266
NMR-spectra s. nuclear magnetic resonance spectra
*Nocardia interforma* 264, 270, 284, 287
Nuclear magnetic resonance spectra
  13-Acetyl-16-hydroxyphorbol 57
  Adenosine 264
  1-Alkyldaphnane derivatives 86
  Azadiradione 163
  Baliospermin 56
  Bertyadionol 19
  Candletoxin B 54
  Cneorins 109, 112, 114, 115, 118, 120, 121, 122, 126, 128, 130, 131, 133, 134, 141, 142, 143, 145, 168, 169, 171, 172, 173
  Crotofolin 26
  Croton factor $F_1$ 57
  Daphnetoxin 74, 75, 76
  12-Deoxy-16-hydroxyphorbol esters 52
  20-Deoxyingenol 73
  5-Deoxyingenol diacetate 67, 68
  20-Deoxyingenol esters 72, 73

Deoxyphorbol derivatives 40, 41
4-Deoxyphorbol diesters 41
Deoxyphorbol esters 50, 51, 52
12-Deoxyphorbol diacetate 48
4-Deoxyphorbol triacetate 37
13,19-Dihydroxyingenol ester 71
Euphornin 16
Ezomycins 280, 282
Formycins 264
Gnidimacrin 85
Hydroxycneorins 126
12-Hydroxydaphnetoxin derivatives 78
18-Hydroxyingenol 68
13-Hydroxyingenol esters 70
16-Hydroxyingenol triacetate 69
4-Hydroxyphenylacetic acid 82
Hydroxytricoccins 126
Ingenol triacetate 58, 59, 60, 68
Ingol esters 25
Ingol tetraacetate 21, 23, 24
Jatropholone B 26
Jatrophone 16
Jolkinols 19, 20
Lathyrane diterpenes 18
Mancinellin 56
*seco*-Meliacanes 156, 158, 159
Mertensene 221
Mezerein 76, 77
Monoterpenes 193, 212, 213, 214, 215, 217, 218, 219, 222
Monoterpenes, halogenated 208
Nucleosides 248, 239, 250
Oxacinomycin 277
4α-Phorbol 33
Phorbol esters 31, 32, 33, 58
Proresiniferatoxin 83
Pyrazofurin 270
Resiniferatoxin 80, 81
Rhamnofolane 27
Sapatoxins 41
4α-Sapinene acetate 42
Sapinaldehyde 41
Sapintoxin-A acetate 42
Showdomycin 251, 284
Tinyatoxin 81, 82
Tricoccins 125, 130, 132, 136, 137, 139, 147, 148, 149, 150, 151, 152, 153, 154, 155, 156, 158, 159, 160, 162, 163, 164, 166
Tricoccin acetates 146
Nucleic acids 253
—, metabolism 244

C-Nucleoside antibiotics 143f
—, behaviour towards chemical reagents 246
— biological activity 263
—, biosynthesis 284
—, spectroscopic properties 248
— synthesis 251
Nucleotide coenzymes 243
Nucleotide metabolism 244, 270
Nucleotide reductase 279

α-**O**bacunol 105, 167, 180
—, acetate 105, 167, 179, 180
Obacunone 105, 150, 166, 167, 174, 175, 179, 180
Ochtodane derivatives 192, 199, 228, 229, 230
Ochtodene 192, 218, 229
*Ochtodes crockeri* 191, 206
— *secundirames* 191, 192, 206
— *sp.* 206
Ochtodiol 229
Ocimene 196, 197, 198
—, halogenated derivatives 196, 197
Odoracin 78, 79
O'Keefe liquid-liquid-distributions 12
OMP decarboxylase 275, 276
ORD-spectra of C-Nucleosides 250
Oregonene A 213, 224
Orotodine-5′-phosphate 275
Orthoester diterpenes 8
9,13,14-Orthophenylacetylresiniferol 82, 83
Osmium tetroxide 108, 113, 116, 120, 129, 257, 258, 259, 262
Oxazinomycin 245, 248, 249, 276f
—, biological activities 279
—, biosynthesis 288, 289
—, isolation 276
—, structure determination 276
—, synthesis 277
Oxoformycin B 264, 265, 284
Ozone 252, 253, 256, 257, 258, 259, 260, 261, 262
Ozonolysis of cneorins 106, 116

**P**achycladae 63
Palladium 72, 118, 123, 124, 268, 272, 281
Palladium dichloride 255, 266
*Pedilanthus* 5
Pentachlormonoterpene 209
Pentane 132
Pentanotriterpenes 105, 166
Pentitols 247
Peptide from *Euphorbia millii* 66

Peracids 134, 135
Perchloric acid 32
Perhydrol oxidation 106
Periodate oxidation 116, 264
Phaeophyta 191
Pharmacological activity 193
*Phasoelus vulgaris* 279
Phenols 104
Phenylhydrazine 282
Phenylselenyl chloride 262
Phorbol 3, 13, 27f, 33, 62
—, bromofuroate 28
—, diesters 2, 5, 10, 27, 88
—, esters 27, 29, 30, 50, 87f
—, tetraacetates 11
—, triacetates 11, 26, 31, 32, 33, 58
—, triesters 10
4-α-Phorbol 28, 32, 33
Phosphoenolpyruvate 289
Phospholipase $A_2$ 88
Phosphorane 252, 253, 267, 274
5-Phospho-α-D-ribofuranosyl pyrophosphate 264, 270, 284
Phosphorus oxychloride 266
Phosphorus pentoxide 252
Phosphoryl chloride 106, 120, 126, 139, 158
Photooxidation 143
*Phyllanthus sp.* 4
Phyllantoideae 5
Phytoalexins 13
Phytosterols 104
Picrasin A 105
Pig liver esterase 258
*Pimelea ligustrina* 86, 87
— *linifolia* 86, 87
— *prostrata* 47, 55, 56, 76, 86, 87
— *simplex* 76, 86, 87
— *sp.* 6, 75, 86
Pimelea factor $P_1$ 75, 76
Pimelea factor $P_2$ 86, 87
Pimelea factor $P_4$ 75, 76
Pimelea factor $P_5$ 56
Pimelea factor $P_6$ 86, 87
Pimelea factor $S_7$ 86, 87
Pimelodendrinae 5
α-Pinene 191
Platinum 278
Plocamene 192
Plocamene A 232
Plocamene B 231
Plocamene C 231
Plocamene D 200, 203, 231
Plocamene D′ 231
Plocamene E 203, 232
epi-Plocamene B 231
epi-Plocamene C 231
epi-Plocamene D 200, 231
epi-Plocamene D′ 231
epi-Plocamene E 203
Plocamiaceae 191, 196, 198, 201, 203, 208
*Plocamium cartilagineum* 190, 191, 201, 202, 203, 206, 207, 208, 210
— *angustum* 191
— *costatum* 191
— *cruciferum* 191, 209
— *mertensii* 191
— *oregonum* 191, 202, 207, 208
— *sandvicense* 191, 202
— *sp.* 194, 201, 202, 207, 209
— *violaceum* 191, 201, 202, 203, 206, 207, 208
*Poinsettia* 62
Polygonae 62
Polyhydroxy diterpenes 27
Polynucleotide phosphorylase 269
Poranteroideae 5
*Porphyra tenera* 191
Potassium bromide 114, 131, 162
Potassium carbonate 251
Potassium cyanide 268
Potassium hydroxide 24, 32, 38, 130, 266
Potassium iodide 142
Potassium permanganate 123, 124, 270, 271
Preplocamene A 224
Preplocamene B 198, 224
Preplocamene C 198, 224
Preplocamenes 199
Proceranone 156
Pro-inflammatory activity 87
Pro-inflammatory diterpenes 1f
Propylene glycol 12
Proresiniferatoxin 82, 83, 84
Prostaglandin 89
Protease inhibitors 88
Protein biosyntheses 244, 269
Protein synthesis 88
Protolimonoids 168, 169
Pseudouridine 244, 246, 247, 248, 249, 251, 260, 263, 270, 272, 277, 279, 282, 283
—, 1-methyl derivative 244
Pseudouridine B 246
Pseudouridine C 246
Purines 243, 246
—, biosynthesis 287
—, N-ribonucleotides 284

Pyranose isomers $A_S$ and $A_F$ 247
Pyrazofurin 245, 247, 249, 267, 270f
—, biological activity 275
—, biosynthesis 288
—, dimethyl derivative 270
—, isolation 270
—, nucleotide 276
—, -5′-phosphate 275
—, structure determination 270
—, synthesis 271
Pyrazofurin B 270
Pyrazolo[4,3-d]-pyrimidines 264
Pyrazomycin 245, 270
Pyridine 38, 106, 108, 113, 120, 126, 139, 151, 158, 251, 252, 253, 254, 255, 256, 258, 259, 260, 266, 268, 273, 280, 281
Pyridinium chlorochromate 145, 149
Pyrimidines 243, 246
—, biosynthesis 275, 284
—, nucleotides 275, 284

Quassinoids 105

Raney nickel 254
Resiniferatoxin 63, 64, 79, 80, 81, 82, 83, 84
Resiniferonols 79f
Retro-Dieckmann cleavage 260
Retro-Diels-Alder reaction 124
Rhamnofolane 4, 26
—, acetate 26
Rhizophyllidaceae 191, 196, 199, 206, 208
Rhodomelaceae 200
Rhodophyta 191
Ribofuranosylacetonitriles 277
Ribofuranosylamine 284
2-α-D-Ribofuranosylmaleimide 250
5-β-D-Ribofuranosyluracil 243
Ribonucleosides 243, 246
D-Ribose 243, 246, 247, 248, 251, 253, 256, 277, 284, 288
Ricinocarpoideae 5
*Ricinus communis* 6, 13
RNA polymerase 269
RNA synthesis 88
Rutaceae 104, 105, 155, 156, 166, 167, 169, 171, 173, 176
Rutales 102, 105

Sapelin A 171
Sapinaldehyde 42
Sapindales 102, 105
4α-Sapinine acetate 43
Sapintoxin A 34, 44
—, acetate 43
Sapintoxin B 41
Sapintoxin C 42
*Sapium indicum* 29, 34, 35, 36, 37
— *japonicum* 29
— *sp.* 5
*Sargassum sp.* 91
*Sclerotinia sclerotiorum* 279
— *sp.* 283
*Sebastiana* 5
Selenenyl-selenoxide dehydrogenation 260
Sesquiterpenes 194
—, halogenated 210
Sesterpenes 105
Sex pheromones 192
Showdomycin 245, 248, 249, 250f, 277
—, acetonide 251
—, biological activity 263
—, biosynthesis 281, 284, 288, 289
—, isolation 250
—, spiro-derivatives 251
—, structure determination 250
—, synthesis 251, 271
Silica gel 12, 104
Silica gel G 11
Silica gel H 11
Silver acetate 200
Silver nitrate 109
Simarolide 105
*Simarouba amara* 169
Simaroubaceae 105, 167, 169, 173
Simplexin 75, 76
Sodium 266
Sodium acetate 271, 272
Sodium azide 281
Sodium bromide 114
Sodium carbonate 9, 10, 262, 275
Sodium chloride 12
Sodium cyanide 254
Sodium diethyl cyanomethylphosphonate 277
Sodium dihydrogen phosphite 254
Sodium ethoxide 273
Sodium hydride 135, 140, 168, 254, 258, 266, 273, 274, 278
Sodium hydrogen carbonate 262
Sodium hydroxide 75, 252, 265, 281
Sodium methoxide 38, 69, 72, 75, 252, 253, 254, 261, 266, 267, 268, 272
Sodium nitrite 266
Sodium periodate 117, 262, 265, 281, 282
Sodium phenylhydrazine 282

Splendentes 62
Steroids 9
Sterols 10
St. Georg's disease 75
*Stillingia sp.* 5
— *sylvativa* 54, 55, 78, 79
Stylingia factors 78, 79
*Streptococcus hemolyticus* 250
*Streptomyces cacaoi* 281
— *candidus* 270, 277, 288
— *hygroscopicus* 276, 288
— *lavendulae* 264
— *platensis* 244
— *showdoensis* 250
— *sp.* 279, 283
— sp. MA 406-A-1 286, 287
— S. No. 383 263
— *tanesashinensis* 276
*Streptoverticillium ladakanus* 244, 289
Sulfuric acid 71, 104, 253
Surenolactone 176
Surenone 161
*Synadenium* 5
*Synaptolepsis sp.* 86, 87
Synaptolepsis factor $K_1$ 86, 87
Syncarcinogenesis 3

*Teclea grandifolia* 156
Terpenes 104
Terpenes, halogenated 201, 210, 211
2-Terpineol 191
3,12,13,20-Tetraacetyl-3-deoxo-4-deoxy-3-hydroxyphorbol 38
Tetracarbonyl nickel 13
Tetrachlorocyclopropene 260
12-O-Tetradecanoyl-phorbol-13-acetate 27
Tetrahydrofuran 143, 266
Tetranortriterpenes 105, 168, 173, 176
Tetranortriterpenoids 155, 156, 162, 176
Tetrodotoxin 74
Thin layer chromatography
  bitter principles 104
  diterpene acetates 11
  ingol esters 24
  phorbol esters 24
  tricoccins 132, 137, 138
Thionyl chloride 267
Thymelaeaceae 1f
Thymelaeoideae 5, 6
Tigliane 3, 4, 6, 8, 9, 10, 27, 87f
—, esters 27, 34
8-Tigloylingol-12-acetate 25
7-O-Tigloylingol-3,12-diacetate 25
12-O-Tigloylingol-7,8-diacetate 25
Tin tetrachloride 256
Tinyatoxin 64, 81, 82, 84
Tirucallanes 105
Tirucallan-(20S)-triterpenoids 168
*Tirucalli* 62
Tirucallol 156
Tirucallol-7-ene 169
Tirucallon-7-ene 169
*Tithymalus* 63
Tocotrienols 104
Toluene 11
Toluenesulfonyl azide 272, 273, 274
Toluenesulfonyl chloride 255, 256
*Toona ciliata* 166
— *sureni* 161, 176
Toonacilin 166
Toonafolin 166
*Tragia* 5
3,7,8-Triacetylingol-12-tiglate 25
Tri-O-benzyl-D-ribose 256, 267
*Trichilia hispida* 171
Trichloroethanol 266
Tricoccin B 141
— $B_I$ 141
— $R_0$ 124
— $R_1$ 134, 135, 145, 179
— $R_2$ 124, 125, 126
— $R_3$ 124
— $R_5$ 124, 125, 141, 143
— $R_6$ 126, 127, 128, 145
— $R_8$ 126, 127, 128
— $R_9$ 131, 132, 140
— $R_{10}$ 134, 135
— $R_{12}$ 131, 132, 140, 145
— epi-$R_2$ 125
— epi-$R_5$ 125, 141
— epi-$R_9$ 131, 140, 143,
— epi-$R_{12}$ 131, 132, 140
— $S_1$ 136, 137, 138, 139, 140
— $S_2$ 136, 137, 138, 139, 140
— $S_3$ 166
— $S_4$ 147, 148
— $S_4$-cis-diol 150
— $S_5$ 136, 137, 138, 139, 140
— $S_6$ 145
— $S_7$ 162, 163, 164, 166, 174
— $S_8$ 128, 162, 164, 165, 166
— $S_8$ acetate 164, 165
— $S_{10}$ 137, 139
— $S_{11}$ 137, 139

— $S_{12}$ 138, 139
— $S_{13}$ 155, 156, 159, 160, 174
— $S_{13}$ epoxide 156
— $S_{13}$ monoacetate 159
— $S_{14}$ 128, 130, 132, 135
— $S_{15}$ 137, 138, 139, 140
— $S_{16}$ 150, 151, 152, 153
— $S_{16}$ acetal 153
— $S_{16}$ monoacetate 151
— $S_{17}$ 145
— $S_{18}$ 145, 147
— $S_{19}$ 128, 162, 164, 165, 166
— $S_{20}$ 137, 138, 139, 140
— $S_{22}$ 160, 161
— $S_{22}$ diacetate 160, 162
— $S_{22}$ monoacetate 160, 162
— $S_{23}$ 167
— $S_{24}$ 145
— $S_{24}$ acetate 146
— $S_{26}$ 138, 139
— $S_{27}$ 150, 152, 153
— $S_{27}$ cycloacetal 153
— $S_{27}$ diacetate 151, 152, 153
— $S_{28}$ 245
— $S_{28}$ acetate 146
— $S_{30}$ 153
— $S_{30}$ monoacetate 153
— $S_{31}$ 153
— $S_{31}$ monoacetate 143
— $S_{32}$ 160, 162
— $S_{32}$ triacetate 162
— $S_{33}$ 147, 149, 150
— $S_{35}$ 145, 147
— $S_{36}$ 145
— $S_{36}$ acetate 146
— $S_{38}$ 156, 158, 174
— $S_{38}$ monoacetate 158
— $S_{39}$ 147, 150
— $S_{39}$ acetate 147
— $S_{40}$ 156, 159, 160, 174
— $S_{40}$ diacetate 159
— $S_{40}$ monoacetate 159
— $S_{41}$ 167
— $S_{42}$ 147, 148
— $S_{43}$ 162, 166, 174
— epi-$S_{26}$ 138, 139
Tricoccins 105, 124, 144, 151, 179, 180, 181
—, biosynthesis 173
Triethylamine 263, 267, 268, 274
Trifluoroacetic acid 131, 261
Trifluoroacetic anhydride 252, 256
Trifluoroperacetic acid 259
Trigonae 62
1,3,5-Trimethoxybenzene 252, 253
Triphenylphosphin 142
Triterpenes 104, 173
$C_{30}$-Triterpenes 105
Triterpenoids 9, 10
Tritium labelled compounds
    Formycins 284
    Glutamate 284, 288
Tumour promoting diterpenes 1f, 27, 87
Turraeanthin 172
— acetate 173
*Turraeanthus africanus* 169

Ultraviolet spectra
    13-Acetyl-16-hydroxyphorbol 57
    Baliospermin 56
    Bertyadionol 19
    Cneorins 109, 122, 128, 131, 143
    Crotofolin 26
    Croton factor $F_1$ 57
    Daphnetoxin 74
    5-Deoxyingenol diacetate 67
    12-Deoxyphorbol diacetate 48
    4-Deoxyphorbol triacetate 27
    13,19-Dihydroxyingenol ester 71
    Diterpenes 12
    Euphornin 16
    Ezomycin 282
    Formycins 264, 269
    13-Hydroxyingenol ester 70
    16-Hydroxyingenol tetraacetate 69
    Ingenol triacetate 59
    Ingol tetraacetate 21
    Jatropholone B 26
    Jatrophone 16
    Lathyrane diterpenes 18
    Mancinellin 56
    Marine monoterpenes 213
    *seco*-Meliacanes 156, 158
    Mezerein 77
    C-Nucleosides 250
    4α-Phorbol 32
    Phorbol esters 32
    Pyrazofurin 270
    Resiniferatoxin 80
    Rhamnofolane 27
    Sapintoxin A 34
    Showdomycin 251
    Tricoccins 151, 154, 156, 160, 162, 163, 164

*Ulva pertusa* 190, 191
Uracil 243
Uridine 243, 248, 249, 263, 275
—, -5′-phosphonate 275
Urticaceae 5
*Urtica dioica* 5
UV-spectra s. Ultraviolet spectra

Vanillin 71
Violacene 195, 208, 213, 216, 218, 231
Violacene-2 231

Wax esters 104
Wittig reaction 260

*Xanthomonas oryzae* 264, 269, 270
X-ray crystallography
Cneorins 108f, 120, 133, 134
Cryptofolin 26
Crotonitenone 13
Daphnetoxin 74
Gnidimacrin 85
12-Hydroxydaphnetoxin 77
Huratoxin 74
Ingenol triacetate 58
Ingol tetraacetate 21
Jatrophatrione 14
Jatropholane 25
Kansuinines 15
*seco*-Meliacanes 156
Mezerein 76
C-Nucleosides 250, 264, 270, 277
Phorbol 28, 32
Terpenes 195, 212, 215, 217, 219, 223, 224
Violacene 195

Zinc 259, 267
Zinc oxide 252

---

Satz: Austro-Filmsatz Richard Gerin, A-1020 Wien
Druck: Paul Gerin, A-1021 Wien

*Fortschritte der Chemie organischer Naturstoffe*

# *Progress in the Chemistry of Organic Natural Products*

## Volume 43:

1983. VIII, 383 pages.
Cloth DM 208,—. ISBN 3-211-81741-7

*Contents:* J. L. INGHAM: Naturally Occurring Isoflavonoids (1855—1981). — A. KOSKINEN and M. LOUNASMAA: The Sarpagine-Ajmaline Group of Indole Alkaloids.

## Volume 42:

1982. VII, 323 pages.
Cloth DM 164,—. ISBN 3-211-81706-9

*Contents:* Y. ASAKAWA: Chemical Constituents of the Hepaticae. — M. HEIDELBERGER: Cross-Reactions of Plant Polysaccharides in Antipneumococcal and Other Antisera, an Update.

## Volume 41:

1982. 37 figures. VIII, 373 pages.
Cloth DM 196,—. ISBN 3-211-81690-9

*Contents:* E. HASLAM: The Metabolism of Gallic Acid and Hexahydroxydiphenic Acid in Higher Plants. — D. G. ROUX and D. FERREIRA: The Direct Biomimetic Synthesis, Structure and Absolute Configuration of Angular and Linear Condensed Tannins. — ST. J. GOULD and ST. M. WEINREB: Streptonigrin. — D. J. ROBINS: The Pyrrolizidine Alkaloids. — J. W. DALY: Alkaloids of Neotropical Poison Frogs (Dendrobatidae).

All Volumes and Cumulative Index 1—20 available
*Price reduction for subscribers: 10%.*

**Special reduced price (20% reduction) for the complete Series Vols. 1—44 incl. the Cumulative Index to Vols. 1—20**

## Volume 40:

1981. 21 figures. IX, 295 pages.
Cloth DM 158,—. ISBN 3-211-81624-0

*Contents:* P. LEFRANCIER and E. LEDERER: Chemistry of Synthetic Immunomodulant Muramyl Peptides. — SUKH DEV: The Chemistry of Longifolene and Its Derivatives. — W. HELLER and CH. TAMM: Homoisoflavanones and Biogenetically Related Compounds. — R. G. COOKE and J. M. EDWARDS: Naturally Occurring Phenalenones and Related Compounds. — C. W. JEFFORD and P. A. CADBY: Molecular Mechanisms of Enzyme-Catalyzed Dioxygenation (An Interdisciplinary Review).

## Volume 39:

1980. 5 figures. XI, 316 pages.
Cloth DM 158,—. ISBN 3-211-81530-9

*Contents:* B. FRASER-REID and R. C. ANDERSON: Carbohydrate Derivatives in the Asymmetric Synthesis of Natural Products. — H. JONES and G. H. RASMUSSON: Recent Advances in the Biology and Chemistry of Vitamin D. — S. LIAAEN-JENSEN: Stereochemistry of Naturally Occurring Carotenoids. — T. KASAI and P. O. LARSEN: Chemistry and Biochemistry of $\gamma$-Glutamyl Derivatives from Plants Including Mushrooms (Basidiomycetes).

## Volume 38:

1979. 5 figures. VII, 430 pages.
Cloth DM 195,—. ISBN 3-211-81529-5

*Contents:* R. W. FRANCK: The Mitomycin Antibiotics. — N. H. FISCHER, E. J. OLIVIER, and H. D. FISCHER: The Biogenesis and Chemistry of Sesquiterpene Lactones.

***Springer-Verlag Wien New York***